Modern Methods of Plant Analysis

New Series Volume 6

Editors
H.F. Linskens, Erlangen/Nijmegen
J.F. Jackson, Adelaide

Volumes Already Published in this Series:

Volume 1: Cell Components
1985, ISBN 3-540-15822-7

Volume 2: Nuclear Magnetic Resonance
1986, ISBN 3-540-15910-X

Volume 3: Gas Chromatography/ Mass Spectrometry
1986, ISBN 3-540-15911-8

Volume 4: Immunology in Plant Sciences
1986, ISBN 3-540-16842-7

Volume 5: High Performance Liquid Chromatography in Plant Sciences
1987, ISBN 3-540-17243-2

Volume 6: Wine Analysis
1988, ISBN 3-540-18819-3

Volume 7: Beer Analysis
1988, ISBN 3-540-18308-6

Volume 8: Analysis of Nonalcoholic Beverages
1988, ISBN 3-540-18820-7

Forthcoming:

Volume 9: Gases in Plant and Microbial Cells
ISBN 3-540-18821-5

Volume 10: Plant Fiber Analysis
ISBN 3-540-18822-3

Wine Analysis

Edited by
H. F. Linskens and J. F. Jackson

Contributors

L. S. Conte H. Eschnauer Th. Henick-Kling J. F. Jackson H. F. Linskens
G. J. Martin M. L. Martin A. Minguzzi R. Neeb A. C. Noble C. S. Ough
A. Rapp W. Simpkins V. L. Singleton T. C. Somers W. R. Sponholz
E. Vérette

With 86 Figures

Springer-Verlag Berlin Heidelberg GmbH

Professor Dr. HANS-FERDINAND LINSKENS
Goldberglein 7
D-8520 Erlangen

Professor Dr. JOHN F. JACKSON
Department of Biochemistry
Waite Agricultural Research Institute
University of Adelaide
Glen Osmond, S.A. 5064
Australia

ISBN 978-3-642-83342-7 ISBN 978-3-642-83340-3 (eBook)
DOI 10.1007 / 978-3-642-83340-3

Softcover reprint of the hardcover 1st edition 1988

2131/3130-543210 – Printed on acid-free paper

Introduction

Modern Methods of Plant Analysis

When the handbook *Modern Methods of Plant Analysis* was first introduced in 1954 the considerations were:

1. the dependence of scientific progress in biology on the improvement of existing and the introduction of new methods;
2. the difficulty in finding many new analytical methods in specialized journals which are normally not accessible to experimental plant biologists;
3. the fact that in the methods sections of papers the description of methods is frequently so compact, or even sometimes so incomplete that it is difficult to reproduce experiments.

These considerations still stand today.

The series was highly successful, seven volumes appearing between 1956 and 1964. Since there is still today a demand for the old series, the publisher has decided to resume publication of *Modern Methods of Plant Analysis*. It is hoped that the New Series will be just as acceptable to those working in plant sciences and related fields as the early volumes undoubtedly were. It is difficult to single out the major reasons for success of any publication, but we believe that the methods published in the first series were up-to-date at the time and presented in a way that made description, as applied to plant material, complete in itself with little need to consult other publications.

Contributing authors have attempted to follow these guidelines in this New Series of volumes.

Editorial

The earlier series *Modern Methods of Plant Analysis* was initiated by Michel V. Tracey, at that time in Rothamsted, later in Sydney, and by the late Karl Paech (1910–1955), at that time at Tübingen. The New Series will be edited by Paech's successor H.F. Linskens (Nijmegen, The Netherlands) and John F. Jackson (Adelaide, South Australia). As were the earlier editors, we are convinced "that there is a real need for a collection of reliable up-to-date methods for plant analysis in large areas of applied biology ranging from agriculture and horticultural experiment stations to pharmaceutical and technical institutes concerned with raw material of plant origin". The recent developments in the fields of plant biotechnology and genetic engineering make it even more important for workers in the plant sciences to become acquainted with the more sophisticated methods, which sometimes come from biochemistry and biophysics, but which also have been developed in commercial firms, space science laboratories, non-university research institutes, and medical establishments.

Concept of the New Series

Many methods described in the biochemical, biophysical, and medical literature cannot be applied directly to plant material because of the special cell structure, surrounded by a tough cell wall, and the general lack of knowledge of the specific behavior of plant raw material during extraction procedures. Therefore all authors of this New Series have been chosen because of their special experience with handling plant material, resulting in the adaptation of methods to problems of plant metabolism. Nevertheless, each particular material from a plant species may require some modification of described methods and usual techniques. The methods are described critically, with hints as to their limitations. In general it will be possible to adapt the methods described to the specific needs of the users of this series, but nevertheless references have been made to the original papers and authors. While the editors have worked to plan in this New Series and made efforts to ensure that the aims and general layout of the contributions are within the general guidelines indicated above, we have tried not to interfere too much with the personal style of each author.

There are several ways of classifying the methods used in modern plant analysis. The first is according to the technological and instrumental progress made over recent years. These aspects were used for the first five volumes in this series describing methods in a systematic way according to the basic principles of the methods.

A second classification is according to the plant material that has to undergo analysis. The specific application of the analytical method is determined by the special anatomical, physiological, and biochemical properties of the raw material and the technology used in processing. This classification will be used in Volumes 6 to 8, and for some later volumes in the series. A third way of arranging a description of methods is according to the classes of substances present in the plant material and the subject of analytic methods. The latter will be used for later volumes of the series, which will describe modern analytical methods for alkaloids, drugs, hormones, etc.

Naturally, these three approaches to developments in analytical techniques for plant materials cannot exclude some small overlap and repetition; but careful selection of the authors of individual chapters, according to their expertise and experience with the specific methodological technique, the group of substances to be analyzed, or the plant material which is the subject of chemical and physical analysis, guarantees that recent developments in analytical methodology are described in an optimal way.

Volume Six – Wine Analysis

Chemical analysis of plant products is vitally important to the field of food regulation and for the protection of public health and safety generally. Over several centuries it has been the advances in chemical analysis that have dictated and led to the formulation of laws and regulations governing food and beverages, and not the other way round. Adulteration of food has always occurred, by accident or design, but regulations covering this are not enforceable unless analysis can discriminate and detect such adulteration. It is up to the analytical chemist to develop and test accurate methods for analysis of food and beverages where it affects public health and safety, or the economic protection of the consumer. The food analyst should also carry out research in the analytical sciences where it impinges on agriculture, public health and regulatory controls of raw materials and products.

The editors have planned the present volume in such a way as to illustrate the sophistication and diversity that exists in the present-day application of chemical and microbiological analysis to wines and spirits. It will be of interest to students (both undergraduate and graduate) in the fields of agriculture and food technology, as well as to analytical scientists involved in particular areas of the wine industry and wanting a handy reference to methods and applications in others. As can be seen from Volume 7 in the Series, which deals with analysis of beer, and from Volume 8, which covers analysis of non-alcoholic beverages, analysis plays an important part in regulating the products being offered to the public and in our understanding of the factors involved in the attractiveness of the product to the public. The latter may in time lead to a more economic production through replacement of expensive natural products with simpler or cheaper, and also safer, materials. Medically safer products may also be developed for those members of the general public who are suffering from particular disabilities. An example of this is the replacement of natural sugars, which cannot be taken by the diabetic population, with artificial sweeteners. These products were not available a few decades ago, which illustrates the importance of research in chemical analysis to our food and beverage industries.

We have gathered together in this volume chapters by some world authorities on various aspects of analysis of wines and spirits, all of them dealing with analytical methods for regulating control, quality-assurance or research. These eminent scientists have been chosen from a number of countries and they deal with a very wide spectrum of wines and spirits. The editors hope that, together with Volumes 7 and 8 in this series, this volume will provide an up-to-date account of analysis in various countries, the variety of methods available today for control, and the direction that research in the area is taking us. These three volumes should prove invaluable to scientists working in these and allied industries, as well as to students who are looking for some guidance for a career in chemical analysis of food and drinks. We would point out that although regulations and laws governing food, wines, spirits and other drinks vary from country to country, and also with time, it is not the aim of this book to dwell on these factors, but rather to illustrate the large number and the diversity of methods available today which can be of service to the general public.

Acknowledgements. The editors express their thanks to all contributors for their efforts in keeping to production schedules, and to Dr. Dieter Czeschlik, Ms. K. Gödel, Ms. J. v. d. Bussche, and Ms. E. Göhringer of Springer-Verlag for their cooperation with this and other volumes in Modern Methods of Plant Analysis. The constant help of José Broekmans is gratefully acknowledged.

Nijmegen/Siena and Adelaide, Summer 1988

H. F. Linskens
J. F. Jackson

Contents

Micro-Element Analysis in Wine and Grapes
H. ESCHNAUER and R. NEEB (With 1 Figure)

Acids and Amino Acids in Grapes and Wines
C. S. OUGH (With 5 Figures)

List of Contributors

CONTE, LANFRANCO S., Ispettorato Centrale Prevenzione e Repressione Delle Frodi, Agro-Alimentari – Ufficio di Bologna, Via S. Giacomo, 7, 40126 Bologna, Italy

ESCHNAUER, HEINZ, Gelnhäuser Str. 15, 6463 Freigericht, FRG

HENICK-KLING, THOMAS, Cornell University, Dept. of Food Science & Technology, Geneva, NY 14456-0462, USA

JACKSON, JOHN F., Department of Biochemistry, Waite Agriculture Research Institute, University of Adelaide, Glen Osmond, S.A. 5064, Australia

LINSKENS, HANS-FERDINAND, Goldberglein 7, 8530 Erlangen, FRG

MARTIN, GÉRARD J., Université de Nantes-CNRS, Laboratoire de RMN et de Réactivité Chimique, 2 rue de la Houssiniere, 44072 Nantes, France

MARTIN, MARYVONNE L., Université de Nantes-CNRS, Laboratoire de RMN et de Réactivité Chimique, 2 rue de la Houssiniere, 44072 Nantes, France

MINGUZZI, ATTILE, Ispettorato Centrale Prevenzione e Repressione Delle Frodi, Agro-Alimentari – Ufficio di Bologna, Via S. Giacomo, 7, 40126 Bologna, Italy

NEEB, ROLF, Carl-Orff-Str. 22, 6500 Mainz 33, FRG

NOBLE, ANN CURTIS, Department of Viticulture and Enology, University of California, Davis, CA 95616, USA

OUGH, CORNELIUS S., Department of Viticulture and Enology, University of California, Davis, CA 95616, USA

RAPP, ADOLF, Bundesforschungsanstalt für Rebenzüchtung, Geilweilerhof, 6741 Siebeldingen, FRG

SIMPKINS, WAYNE ANTHONY, Australian Government Analytical Laboratories, 1 Suakin Street, Pymple NSW 2073, Australia

SINGLETON, VERNON LEROY, Department of Viticulture and Enology, University of California, Davis, CA 95616, USA

SOMERS, THOMAS CHRISTOPHER, The Australian Wine Research Institute, Waite Road, Urrbrae, Glen Osmond, S.A. 5064, Australia

SPONHOLZ, WOLF RÜDIGER, Forschungsanstalt Geisenheim, Fachgebiet Mikrobiologie und Biochemie, 6222 Geisenheim/Rhein, FRG

VÉRETTE, ERIC, Laboratoire de Chimie Analytique et Toxicologie, Faculte de Pharmacie, Avenue Charles Flahault, 34060 Montpellier Cedex, France

Wine Analysis

H. F. LINSKENS and J. F. JACKSON

Wine is an alcoholic beverage which results from the fermentation of sugar-containing plant fruit. More strictly speaking, wine is the product of enzymic transformation of the juice of the berries of grape into a beverage containing alcohol. Originally, the fermentation was brought about by the natural yeasts attached to the surface of the ripe harvested berries. The activity of yeast cells on the sugars of the grape juice results in the production of alcohols, carbon dioxide, heat and many other secondary plant products in small quantities. A secondary or malolactic fermentation often takes place, particularly in red wines, whereby lactic acid bacteria convert malic acid into lactic acid. Natural yeast found on the skin of the berries comes in many different strains and ecotypes, each of which will produce a different assortment of by-products during the fermentation reactions, characteristic for the particular types of wine. *Saccharomyces cerevisiae* is the main species of yeast important in wine-making, although many other yeasts may be present at the start of fermentation. As fermentation proceeds, these other yeast species are strongly inhibited. Modern wine-makers prefer to introduce specially cultivated yeast strains, which help to guarantee the product they are aiming for. These yeast strains may have been selected for certain desirable characteristics such as resistance to high sugar concentrations or to sulfur dioxide. The sulfur dioxide is often added to prevent undesirable bacterial fermentation of the grape juice.

The purity of wine is in many countries guaranteed by law. But the purity of wine as a product of biological processes is a special problem for the controlling authorities. The complexity of the starting product, the vast number of varieties of the botanical species *Vitis vinifera* L. with its cultivars and races, the use of specific races of the yeast *Saccharomyces* for induced controlled and directed fermentation, the many different possible treatments, and the introduction of new technology make analysis of the final product essential. This in turn helps to advance our understanding of vinification and further improve the process. For example, the various aliphatic alcohol esters and certain aromatic compounds in wines can now be routinely estimated by glass capillary gas chromatography.

1 Analytical Methods and Wine Laws

Wine analysis is an old branch of analytical chemistry. One hundred years ago a well-known laboratory under the directorship of Charles Remigius Fresenius (1818–1897) was devoted to the specific problems of wine analysis. The first compilation of customary methods used at that time was written by Eugen Borgmann,

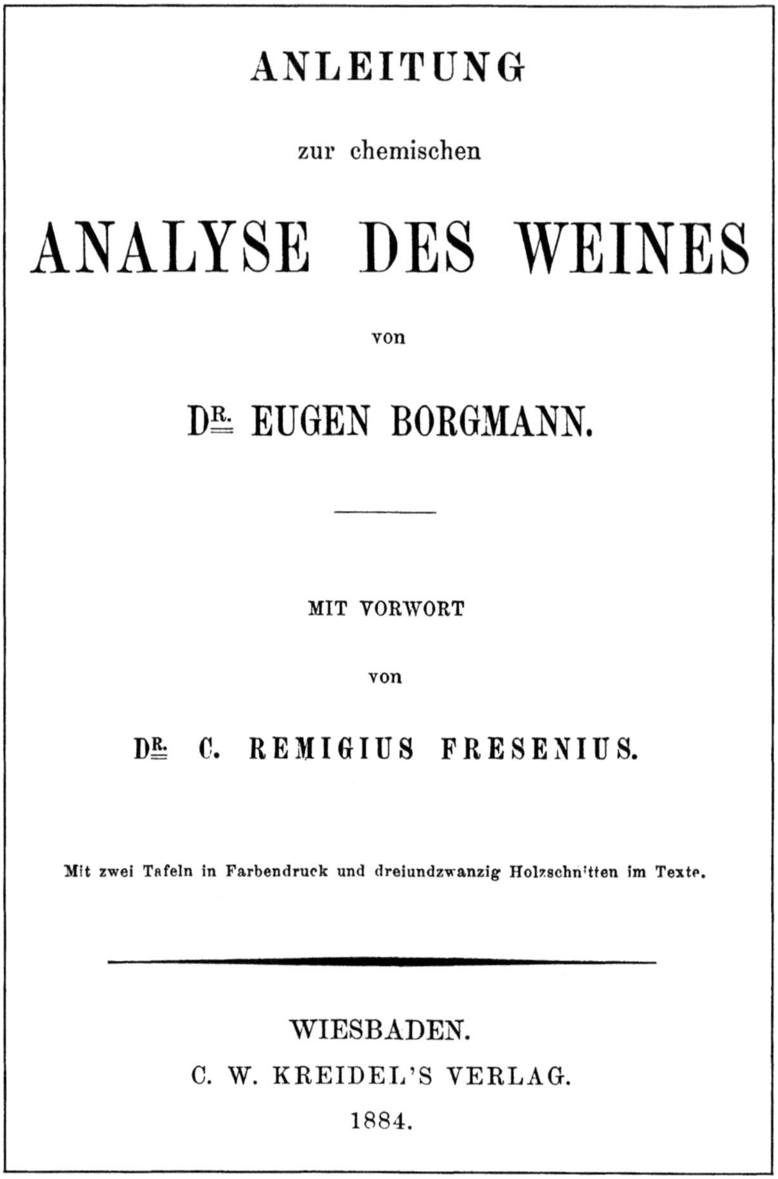

ANLEITUNG

zur chemischen

ANALYSE DES WEINES

von

DR. EUGEN BORGMANN.

MIT VORWORT

von

DR. C. REMIGIUS FRESENIUS.

Mit zwei Tafeln in Farbendruck und dreiundzwanzig Holzschnitten im Texte.

WIESBADEN.
C. W. KREIDEL'S VERLAG.
1884.

Fig. 1. Front page of the classical instruction for wine analysis, published more than 100 years ago with two color tables and 23 wood cuttings. The manual is preceded by an Introduction by Charles Remigius Fresenius (1818–1897); in the 19th century he was considered one of the stars of analytical chemistry, on which subject he wrote a famous textbook and in 1882 founded the first journal of analytical chemistry. In his commercial laboratory he introduced the chemical control of spas, and the author, Eugen Borgmann, was his younger collaborator, who specialized in wine analysis together with his colleague C. Neubauer, who wrote the first monograph on the chemistry of wine. This book, published in 1884 (Fig. 1), described in detail the analytical methods for the determination of alcohol extract, free volatile and bound acids, sugars, glycol, total minerals, sulfuric, phosphoric and boric acids, chloride, nitrogen, heavy metals, lime and arsenic iron. Also included were methods for organic acids such as citric, succinic and salicylic as well as proofs of evidence for rubber, dextrine, tannins and various natural and synthetic pigments. These analytical methods (Figs. 2–4) used in the wine industry became the basic prerequisite for the control of legislation made for food and semi-luxurious items intended for human consumption over the last century. The earliest laws and regulations were more concerned with protection of the prorogatives of the rulers. It is only in the last 150 years or so that the regulations began to have more to do with the quality of wine, and this was only possible as analytical techniques became available to detect differences. In France, laws were proclaimed in 1919 to bring into correspondance the naming of certain wines and their originating district and grapes used. These are known under the term „Appellation d'Origine Controlée".

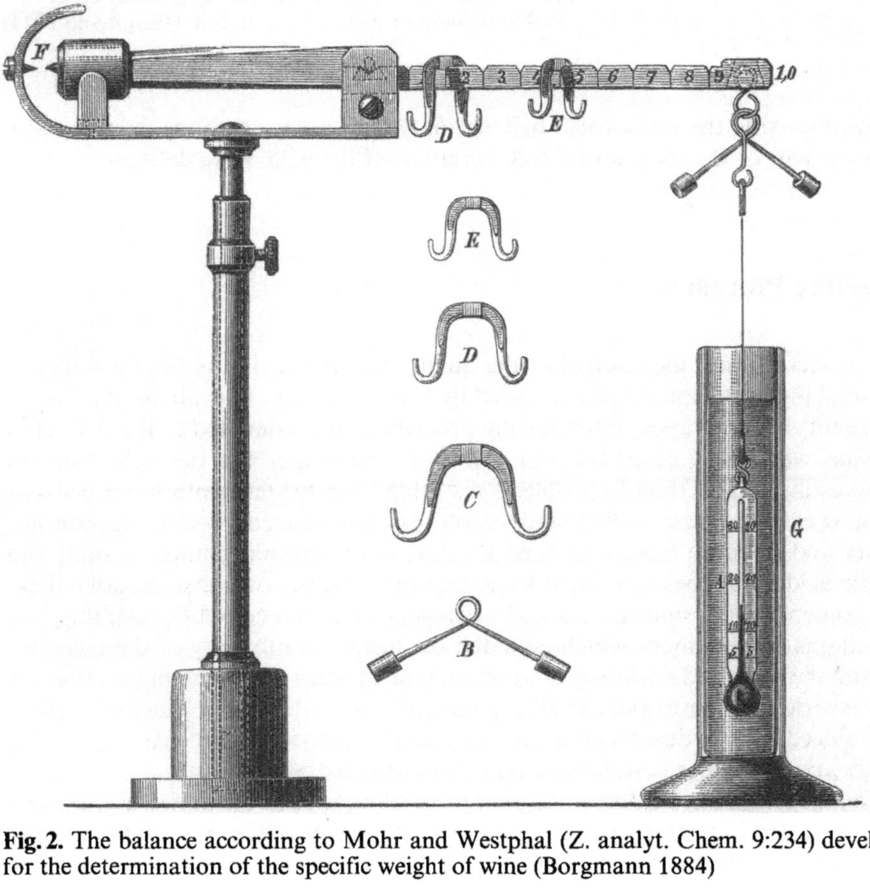

Fig. 2. The balance according to Mohr and Westphal (Z. analyt. Chem. 9:234) developed for the determination of the specific weight of wine (Borgmann 1884)

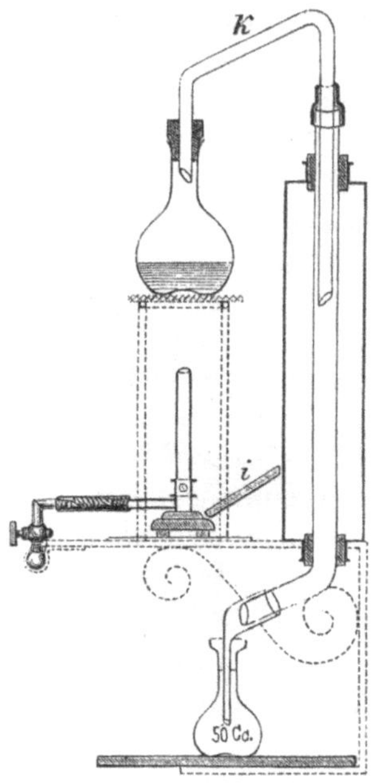

Fig. 3. Experimental device for the determination of the alcohol content of wine by distillation (Borgmann 1884)

Additionally, these laws also had an analytical side to them, as the minimum sugar content of the must and alcohol content of the wine were defined.

2 Tasting Procedure

Since ancient times, the control of the quality was made on the final product of wine-making itself, sometimes preceded by tasting in order to evaluate the variety and quality of the grapes. Wine-testing procedures were devoted to the defects of the wine, which originated from the choice of the grapes and the right moment of harvesting, rather than to problems resulting from the fermentation process or the aging events. The so-called organoleptic examination checked for appearance (clarity and freedom from sediments), color, odor (aroma, bouquet), total and volatile acidity, dryness (residual sugar), body (residues of nonsugar solids) extract content, taste, smoothness and astringency (tannin content), resulting in a general quality judgment which included the proper conformity of the constituents and the balanced harmony of a certain type of wine. These tasting procedures need experienced controllers as all organoleptic tests are more or less subjective. Experienced tasters developed a high degree of conformity, so that even today classification of wines is still made on a physiological basis.

Chemical and physical analysis over the last decade has been used as an essential supplement to tasting procedures, especially in view of the increasing manipulation of the fermentation procedure.

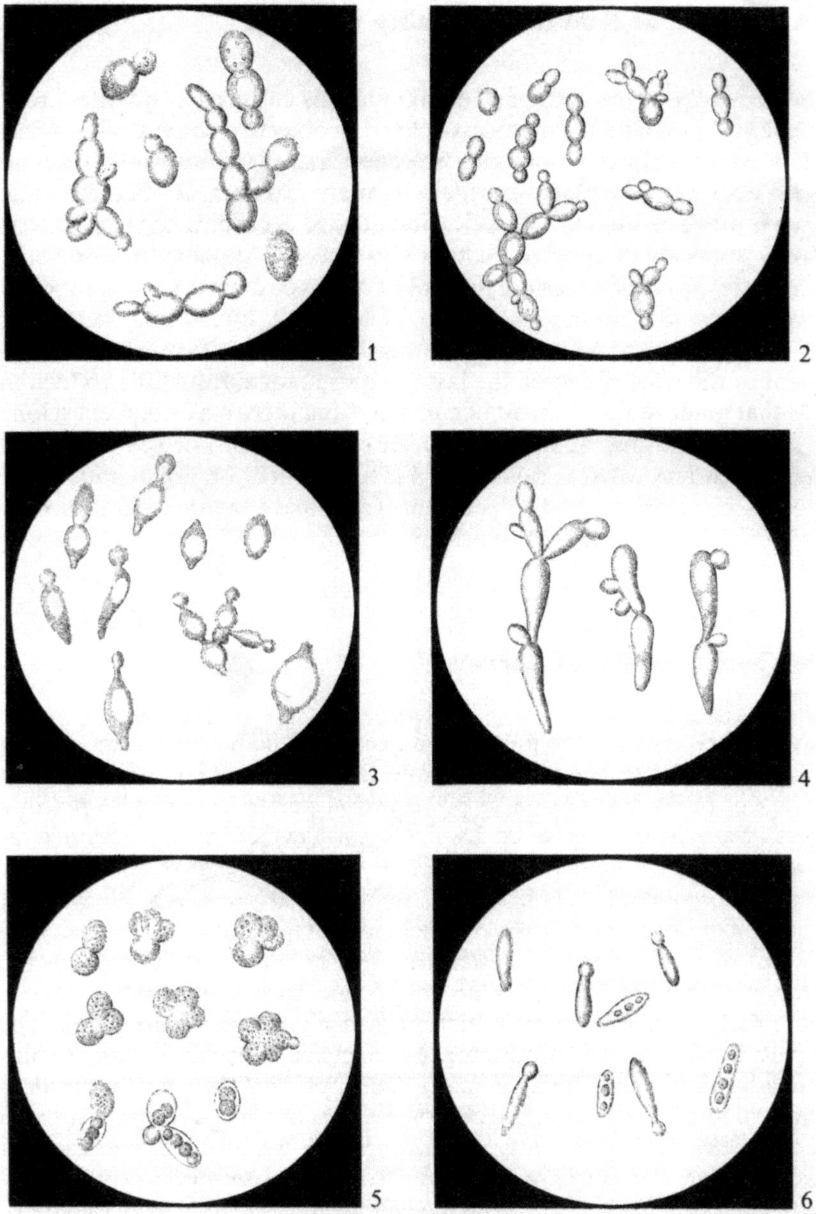

Fig. 4. Color table from Borgmann's book, demonstrating the various types of yeast found in the sediment of "turbid wine": the explanation says that Fig. 1 is *Saccharomyces cerevisiae*, Fig. 2 *S. ellipsoideus*, Fig. 3 *S. apiculatus*, Fig. 4 *S. Pastorianus*, Fig. 5 *S. conglometratus*, Fig. 6 *S. Reessii Blankenhorn*. (Ann. Oenolog. 3:11) (Borgmann 1884)

3 Wine Analysis as an Additional Quality Criterion

Drinking wine is part of the culture of mankind. This culture has no direct relation with wine analysis, which is a process delivering objective criteria, while wine-drinking is a partly subjective process. Selection and classification of various kinds of wine does not take place according to analytical data. Rather, the judgment of wine is made by tasting. Classification of wine according to the usual criteria, whereby wines are designated Premier Cru Classes, Appellation Controlée, Vins Délimités de Qualité Superieure, Grand Cru, Cuvée dans la Cave and so on, give information on the origin and the way of handling, but not on the quality as given by chemical analysis except for the sugar content of the must and the alcohol content of the wine. Between the last analysis made during the production process and that made at the moment of consumption, there may intervene a long period of storage, bottling, aging, transportation, formation of sediments and even inexperienced and careless treatment. Final judgment on the quality of the final product is, however, an organoleptic one. Chemical analytical data can only help to discriminate real wine from synthetic.

4 Has the Quality of Wines Improved?

Wine analysis is the only way to answer the question whether or not the quality of wine has improved over the years. This question can be asked about a certain kind of wine during its maturation or preservation period or it can be put in a more general way. It was answered by Professor Peynaud in an interview at Chateau Margaux in December, 1984, when he concluded that the high quality of the wines of today could not have been achieved 40 years ago. He was not speaking of the quality of wine in the pre-technical civilizations and antiquity, so one should not decry the wines of yesteryear. The quality of wines today is the result of the way they are treated. One has to take into consideration that in former times there were no general rules and instructions on wine-making. The date of harvest was left to each proprietor of a vineyard to decide for himself, unlike the officially agreed date today in many countries. In former times, the alcoholic strength of wines was lower, because the harvest products, the grapes, came from other varieties, were frequently undernourished, with smaller berries, with lower sugar content, and perhaps even infected. Clonal selection of wine grapevines began perhaps 100 years ago and in general has resulted in increased berry yield, by as much as fivefold. There is probably room for further improvement through clonal selection. The fermentation process was not well understood. For instance, it should be an anaerobic process, and it was common to use a plaster cap or even a piece of rag on the top of the vat as a lid. Both are not good hermetic seals, so that an excess aeration took place, and the result was that the top few inches of the wine contained much volatile acidity. Under aerobic conditions bacteria from the grape skins (lactic acid and acetic acid bacteria) produce large amounts of acetic acid, giving rise to a vinegar taste and wine spoilage. Nowadays, volatile acidity of claret is about 0.4; one generation ago the average was 0.75, and this was thought to be quite acceptable. Only two generations ago there was no way of cooling the wine during the fermentation process, except by placing the wooden vats in a dark, generally contaminated cellar or in grottos. As a result, fermentation temperature was uncontrolled, perhaps not even measured. The period in which the grapes have to remain in contact with the skins (cuvaison) was not varied according to the needs of the particular vat and a certain lot of the harvest.

Progress in the general quality of wines and the reduction of faults during wine production is closely linked with progress in analytical techniques and the introduction of these in the control of the fermentation processes. Professor Peynaud came to the conclusion that "wine is the reflection of certain circumstances, and the technical methods associated with it".

5 Wine as an Object of Culture

Special demands are put on wine as soon as it is used in religious ceremonies. Especially for the services in the Jewish Communities and in many Christian denominations, wine is used as an offering. Of course the highest quality is expected for the product used as a basic symbol for devotion. Special rules are followed. In the Catholic Church, originally preferences were given to wine made by the clergy, but over the last centuries and coincident with changes in human society, the responsibility for purity has been delegated to local bishops (Can 336 2 Corpus Juris Canonicus). This seems sufficient for wine-producing countries. The Roman dioecesan synod of 1960 declares in art. 403, that the local priest (rector ecclesiae) is responsible for the reliable and confidential source of the wine. Objective criteria and analytical data for wine to be used in cultus (vinum de vite) are not given. Locally, statutory declarations on purity and origin have been asked from wine-dressers and wine-merchants.

6 Analysis in the Detection of Wine Forgery

Before speaking about the falsification or forgery of wine, one needs a clear answer to the question "what is wine"? Deviation from the qualities described by definition should be disclosed. The objective way to trace mixtures, illegal additions and falsification of origin is by chemical analysis.

By definition, wine production does not allow

- addition of water
- addition of chemicals, other that those necessary during the fermentation process,
- addition of sugars in certain cases, depending on the legal regulations,

- blending of wine with juice, only under certain legal regulations,
- blending of wine of various origins or various vintages only under certain conditions and when a declaration is given on the label.

In recent years, forgery in wine-making has become more evident. For example, criminal addition of polyethyleneglycol to wine of lower quality has been carried out in order to make it appear to be of better quality and to realize better prices. These additives, sometimes even dangerous to the health of the consumer, can only be detected by chemical analysis. Recent progress in analytical methods has made it possible to detect these chemicals. The greatest problem is to know what foreign chemical has to be sought. As a result, the whole spectrum of analytical methods, with high accuracy and very low detection limits, has to be applied in wine analysis.

Modern methods in wine analysis are therefore more important than ever for maintaining the high quality of wine, to protect consumers from falsified and minor quality products, and also to safeguard bona fide wine-makers from dishonest competition.

Analysis of Wine Sensory Properties

A. C. Noble

1 Introduction

Traditionally, sensory evaluation of wine is perceived as being that done by a wine expert, who evaluates the "quality" of wine. However, this is only one isolated and, in fact, poor example of the activity. Sensory evaluation is a scientific discipline used to evoke, quantitate, analyze and interpret reactions to the characteristics of wines as perceived by the senses of sight, smell, taste, and touch.

Because of the extraordinary number of interferences that can bias sensory evaluation, it is conducted under controlled conditions using trained subjects. The type of test and judge depends on the purpose of the test. Most analytical tests answer one of the following questions: "is there a difference?", "what is the difference?" and "how much is it?". This is done in sequential fashion and can provide very detailed information about the sensory properties of wines.

In contrast, hedonic or quality evaluation is done with a different purpose. Rather than defining the sensory characteristics of a wine, the consumer is asked "which wine do you prefer?" or an expert wine judge is asked "which wine has higher quality?".

The purpose of this chapter is to introduce the basic principles of sensory evaluation of wine and to give specific examples of its application. For more detailed information, consult the following reference texts: Amerine et al. 1965; Amerine and Roessler 1983; O'Mahony 1986; Stone and Sidel 1985.

2 Types of Sensory Tests

In analytical sensory evaluation, the first step is to determine if a difference exists, then to determine the nature and magnitude of the difference. In consumer evaluation, if a sufficiently large difference exists, the preference of the consumer can be determined in marketing surveys. In wine judging at competitions or in retail and export tastings, the assessment is more subjective and generally utilizes a "quality" scorecard for overall evaluation.

2.1 Bench Tests

For any analysis, the wines should be evaluated informally before doing any testing. By bench testing, considerable time can be saved. Wines which have a defect

Table 1. Suggested guidelines for maximum number of samples presented in one session and minimum number of judges (n)

Test	n	White	Red
Duo-trio	10–20	4– 5 sets	3– 5 sets
Triangle	10–20	3– 5 sets	3– 4 sets
Pair	10–20	4– 6 sets	3– 5 sets
Intensity scaling	10–15	9–12 wines	7–10 wines
Time-intensity	8–12	10–12 wines	8–10 wines
Descriptive	8–12	6– 7 wines	3– 5 wines

or spoilage irrelevant to the treatment being studied should be eliminated. If color differences are observed in this preliminary evaluation, and the object of the experiment is to determine aroma or taste differences, subsequent testing should be conducted with the color of the wine masked by the use of black glasses or red lights. If the wines seem to have no obvious difference among them, than duo-trio difference tests should be performed before planning any further evaluation. If the differences are obvious, it may be possible to skip the difference testing step and determine a specific aroma or taste characteristic which can be rated for intensity, such as the astringency of wines made with increasing skin contact. Also, with a bench testing evaluation, it is possible to estimate how many wines can be evaluated without fatigue. A general guideline is presented in Table 1.

2.2 Discrimination Tests

2.2.1 Unspecified Difference Tests

Initially, a difference test must be conducted, whether it is done informally in the bench testing or with formal tests. The two most common tests are the duo-trio test and the triangle test. In both cases, three samples are presented; two are the same and one is different. In the duo-trio test, one wine is identified as the reference and the other two are coded. The judge is asked to select the wine which is different from the reference. The probability of a correct response due to chance alone is 50% ($p = \frac{1}{2}$). For a triangle test, all three samples are coded and the judge is required to select the odd sample. The probability of a correct response here is $33^1/_3\%$. The results can be interpreted by consulting tables constructed from a normal probability curve or binomial distribution to determine the number of correct responses needed for a statistically significant difference. Generally the differences are considered to be significant at $p < 0.05$. Tables for determining significance levels and exact probability levels have been provided (Amerine and Roessler 1983; Roessler et al. 1978).

If differences are significant at $p < 0.05$, but not at $p < 0.01$ or higher levels of significance, then generally one can only conclude that there is a significant difference between the wines, but that it is too small to be characterized further.

2.2.2 Directional Differences Tests

When the difference can be defined, pair tests are performed to ask which sample is higher in a specified attribute. The probability of a correct response is 50% (one-tailed, since there is only one correct answer). If two products are evaluated for preference, then the pair test is used and the target consumer asked which wine is preferred. Since either answer is correct, the significance of the paired-preference test is interpreted using the two-tailed test, where $p = ½$. It is crucial to recognize that asking for a preference will not determine if the wines are different. People have different preferences. For a group of judges one sample of wine may not be preferred over the other; however, the samples may be completely different.

2.2.3 Threshold Tests

Although the results of threshold tests are of little value, they are routinely performed to determine at what concentration a specific compound can be perceived. The methods for threshold tests vary considerably; however, the most sensitive method is to present sample sets in order of ascending concentration. The detection threshold is the level at which the compound produces a detectable difference in aroma or taste. The recognition threshold, which is higher than the detection threshold, is that level at which the specific odor or taste of the compound can be recognized. To determine a detection threshold, duo-trio or triangle difference tests can be utilized. To determine a recognition threshold, pair tests are used and the judge asked to identify which of the coded samples contain the test compound. The threshold level is determined in a variety of ways (Amerine et al. 1965). Depending on the response, which is a function of the concentration range tested, a linear relationship or semi-log relationship is most often utilized, where the threshold concentration is defined at that at which 75% correct responses occur. This corresponds to 50% correct response above chance.

To determine the odor threshold for dimethyl sulfide, five duo-trio sets were presented in order of increasing concentration. Each set contained one labeled reference wine and two coded samples, one of which was spiked with a known amount of DMS. In Fig. 1, the log concentration of dimethyl sulfide is plotted against the percentage of correct responses observed. By this method, an odor threshold value of 25 $\mu g\ l^{-1}$ for dimethyl sulfide in white wine was found.

However, the threshold results are valid for the test wine only under specific testing conditions. Clearly, at another temperature in another wine of different ethanol concentration or acidity, the threshold concentration will vary. It should be noted that determination of the threshold value for the compound does not allow one to predict its intensity or odor quality at higher concentrations. Guadagni et al. (1968) proposed the concept of odor units, whereby the concentration of a component was expressed as the concentration divided by the threshold value. However, different compounds increase in intensity at different rates and can demonstrate marked differences in quality at different concentrations. Therefore, inferring the intensity of an odor by this means should be avoided.

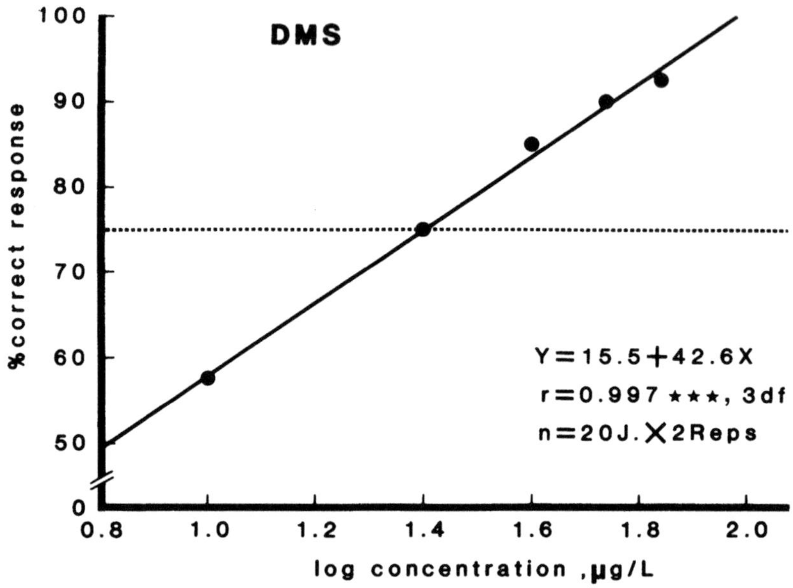

Fig. 1. Determination of dimethyl sulfide (*DMS*) odor threshold in white wine. Percentage of judges selecting with wine DMS versus log concentration of DMS. (Goniak and Noble 1987)

2.3 Intensity Scaling

Rating the intensity of a specific attribute can be done using several different procedures such as category scaling or magnitude estimation. However, the results are similar and the former method is easier to use and to analyze (Giovanni and Pangborn 1983).

2.3.1 Scalar Rating

Rating the intensity of specified attributes requires the use of trained judges familiar with the definition of the characteristic being rated and with the use of the scale. Two types of category scales are commonly used and yield similar results. Intensity can be rated on numerical or graphic scales. For example, use of a nine-point scale provides an adequate range of values. Alternatively, a line scale, anchored at the ends with the terms "low" and "high", can be used. The main advantage of the line scale is absence of numerical or word bias. The ratings are then analyzed by analysis of variance (AOV) to see if a significant difference exists among the samples and if the judges are reproducible. Additionally, the Judge × Wine interaction should be examined to ascertain that the judges are using the term consistently (Stone et al. 1974). Using tests such as Fischer's least significant difference (LSD) or Duncan's multiple range test, wines which differ significantly can be identified. Further details about performing analyses of variance and interpreting results of AOV's are available in Amerine and Roessler (1983) and O'Mahony (1986).

Table 2. Mean sourness ratings[a] in wine (n = 14 judges × 3 reps)

Code	pH	TA[b]	Sourness intensity rating[c]
A	4.00	4.99	2.22[a]
B	3.75	5.93	3.26[a, b]
Ref[d]	3.50	6.60	(5.0)
C	3.25	7.48	5.10[c, d]
D	3.00	7.94	6.10[e]
E	3.50	7.84	4.27[b, c]
F	3.50	8.12	5,01[d, e]
G	3.50	9.32	5.77[d, e]

[a] Where 1 = low sourness and 9 = high sourness.
[b] TA = Titratable acidity as tartaric acid $g\,l^{-1}$.
[c] Means with the same superscript are not significantly different, $p < 0.001$.
[d] Samples rated in sourness relative the reference which was assigned a sourness intensity of 5.

To examine the effect of pH and titratable acidity (undissociated protons) on perceived sourness, the pH and titratable acidity (TA) of white wine were adjusted with citric acid or sodium hydroxide. Trained judges rated the sourness intensity of the samples in triplicate. Ratings were made relative to a reference wine which was assigned a sourness value of 5, midpoint on the 10-cm sourness line scale. As shown in Table 2, as the pH decreased and the TA increased in samples A to D, the sourness increased. For samples E, F and G, which have the same pH, as the titratable acidity increased the samples were perceived as increasing in sourness. The most sour sample, D, had the lowest pH, but a TA close to those in E and F (Noble and Schmidt 1981).

A scalar rating test was used to characterize the taste effect of two phenolic fractions in wine (Leach 1985). Bitterness and astringency were rated in white wines to which grape seed tannin and catechin, a naturally occurring grape flavonoid, had been added. As illustrated in Fig. 2, as the concentration of catechin is increased, bitterness and astringency increase. Perhaps more important is the result that bitterness increases more rapidly than astringency ($p < 0.05$). In all samples above a concentration of 100 $mg\,l^{-1}$ catechin is more bitter than astringent. In contrast, as shown in Fig. 3, astringency increased more rapidly than bitterness when a commercial extract of grape seed tannin was added to the same white wine (Leach and Noble, in preparation).

2.3.2 Time-Intensity Procedures

It is often important to evaluate the temporal properties of a sensation. For example, attributes such as astringency and bitterness are characterized by persistent aftertastes. Various time-intensity (T-I) systems have been devised which record the change in intensity over time, yielding curves similar to the representative one shown in Fig. 4 (Larson-Powers and Pangborn 1978; Guinard et al. 1986; Lee 1985). In addition to determining the intensity of a stimulus, parameters including

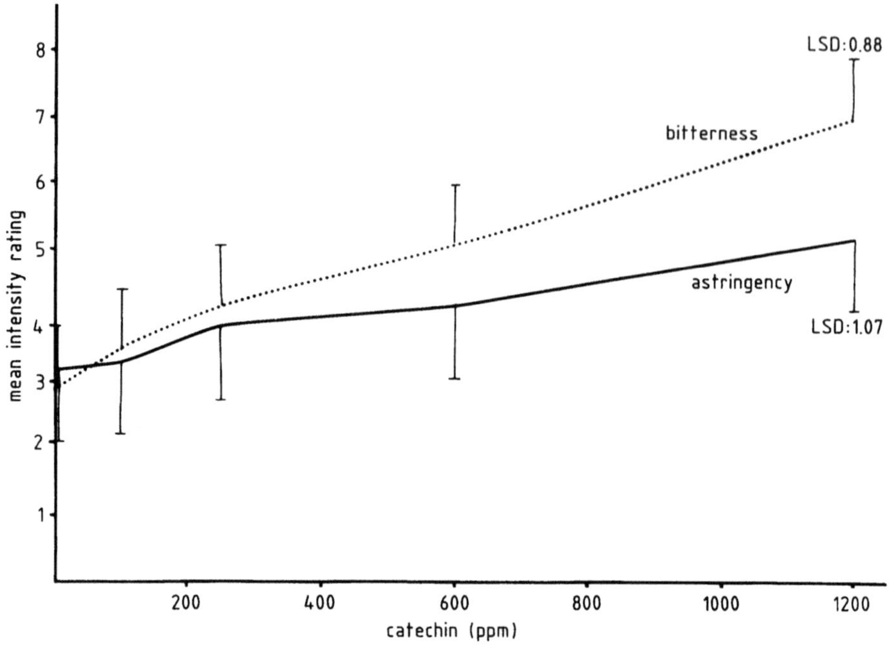

Fig. 2. Mean intensity ratings for bitterness and astringency of increasing concentrations of catechin in white wine (n = 14 judges × 3 reps). (Leach 1985)

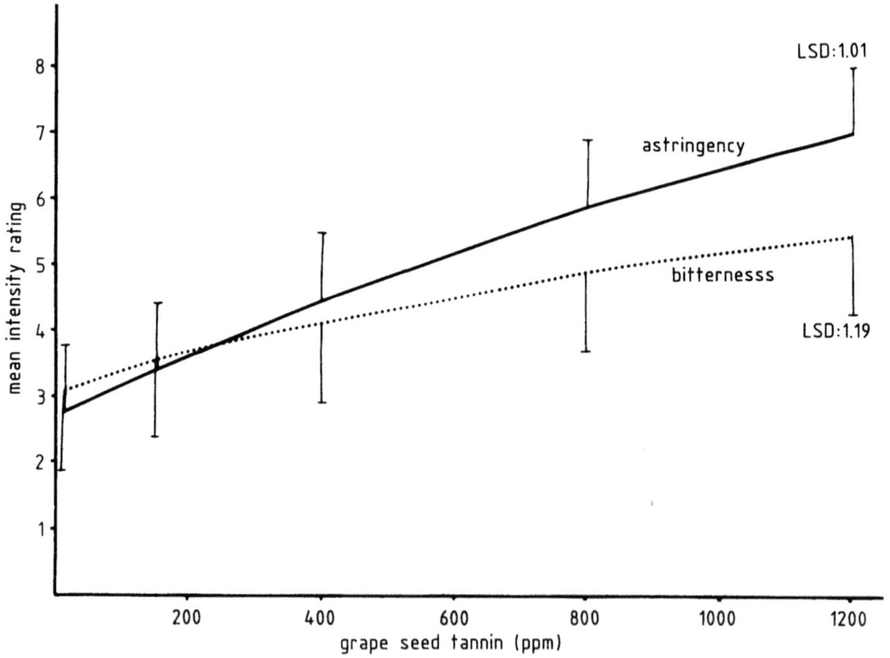

Fig. 3. Mean intensity ratings for bitterness and astringency of increasing concentrations of grape seed tannin in white wine (n = 14 judges × 3 reps). (Leach 1985)

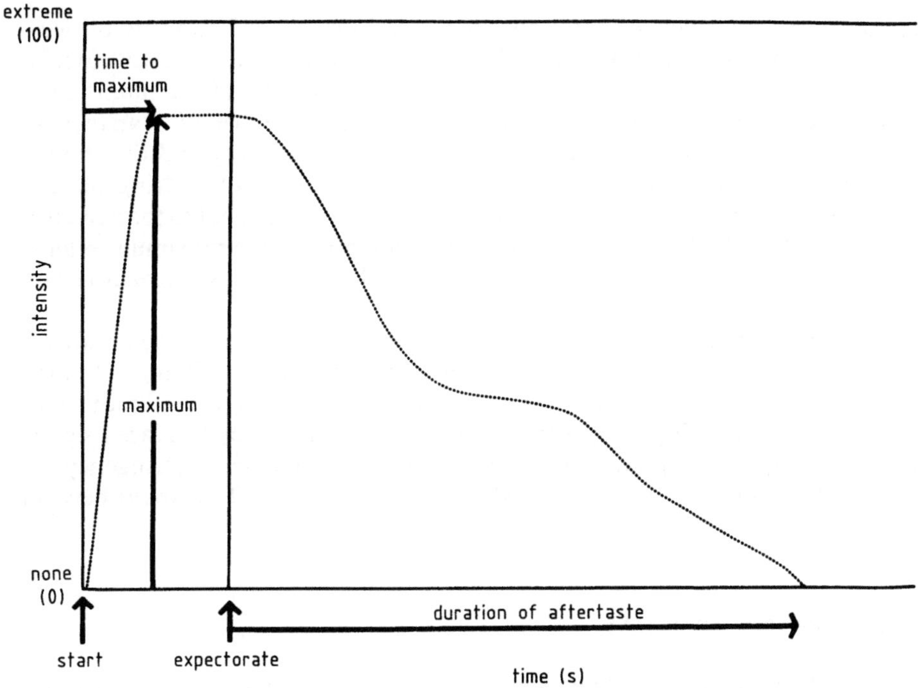

Fig. 4. Representative time-intensity curve illustrating one point measures: maximum intensity, time to maximum, and duration of aftertaste. (Leach 1985)

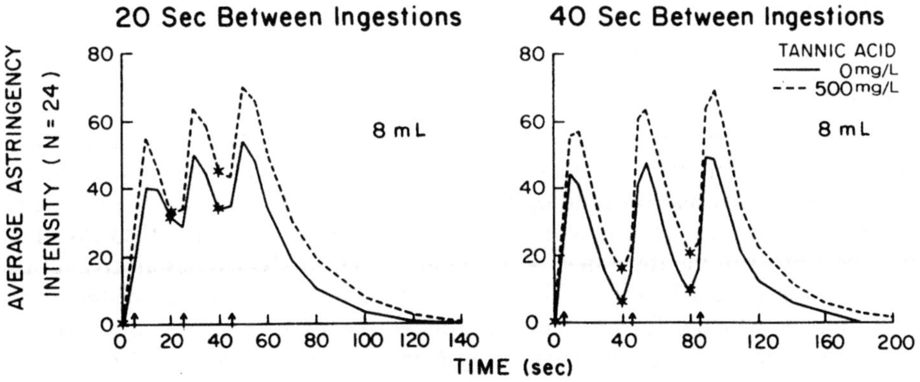

Fig. 5. Average time-intensity curves for astringency in wine upon three succcessive ingestions: left 20 s between ingestions; right 40 s between ingestions. Sample uptake and swallowing are indicated by a *star* and an *arrow*, respectively. (Guinard et al. 1986)

time to maximum intensity, total duration, and rate of onset and rate of decay can be obtained from the T-I curves. When bitterness and astringency of catechin and grape seed tannin in white wine were evaluated by T-I, the maximum intensities showed the same pattern as the scalar data reported above. With an increase in concentration of either component, the maximum intensity and duration of aftertaste of both bitterness and astringency increased. For both astringency and bitterness, the duration of the aftertaste was directly proportional to the perceived maximum intensity. However, for samples eliciting the same maximum intensities, the duration of bitter aftertaste was 10–15 s shorter than astringency (Leach 1985).

For the first time, the cumulative effect of astringency was demonstrated in wine by T-I evaluation (Guinard et al. 1986). Wines with 0 and 500 $mg\,l^{-1}$ tannic acid were ingested at 20- or 40-s intervals, and the astringency continuously recorded. As shown in Fig. 5, maximum intensity of astringency increased significantly upon repeated ingestion of the same wine. The increase in astringency was greater (although not significantly) when 20 s rather than 40 s elapsed between samples.

2.4 Descriptive Analysis (DA)

Probably the most important new sensory technique for evaluation of wine flavor is that of descriptive analysis, the quantitative characterization of wine sensory properties. By this method, profiles of wine flavors are developed which permit analytical evaluation of the differences among wines. For good descriptive analysis, the wines must have significant differences and the judges must be well trained.

2.4.1 Descriptive Terminology

Initially descriptive terms or attributes are identified which are necessary to describe and differentiate the wines. In Fig. 6, a modification of the proposed Standardized Wine Aroma Terminology, better known as the wine aroma wheel, is shown, with the most general words being listed in the first tier and the most specific descriptors in the third tier (Noble et al. 1987). Analytical descriptive terms such as these should be utilized in DA. For each attribute, reference standards are developed which define the specific aroma or taste characteristic. Suggestions for preparation of standards listed in Fig. 6 are given by Noble et al. 1987.

2.4.2 Use of Descriptive Analysis

Judges are trained in the use of these terms in discussion sessions and in formal scoring tests, with reference standards available to clearly define each attribute. The intensity of each term is then rated in each of the wines. The data are analyzed by analysis of variance as described above (Sect. 2.3.1).

In descriptive analysis of white wines containing ethanethiol or dimethyl sulfide, a clear difference between aromas produced by the two compounds was

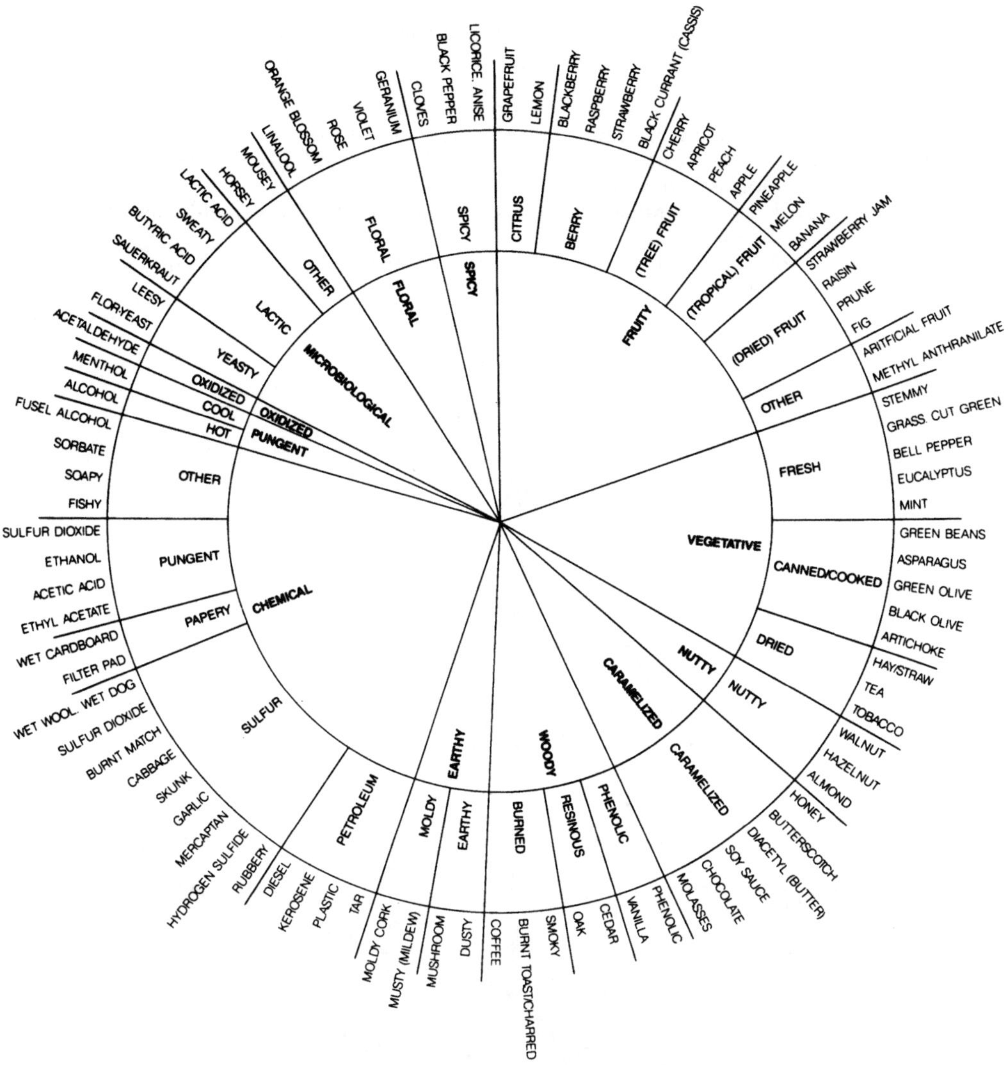

Fig. 6. Wine aroma wheel. Standardized terminology showing first, second, and third tier terms. (Noble et al. 1987)

found (Goniak and Noble 1987). Judges were trained to rate five aroma attributes using the reference standards, described in Table 3. The results are illustrated in the polar coordinate graph provided in Fig. 7. The center of the figure represents low intensity with the distance from the center to the intensity rating corresponding to the relative intensity of each wine for the attribute. By connecting the mean ratings, the profiles of wines containing two levels of dimethyl sulfide and ethanethiol are shown. Dimethyl sulfide produces an aroma characterized by "canned corn", "canned asparagus" and molasses notes. In contrast, wine containing ethanethiol has distinct rubber and onion aromas.

Table 3. Composition of descriptive analysis reference standards

Attribute	Composition
Asparagus	0.5 ml brine from canned asparagus in 50 ml wine
Corn	2.0 ml brine from canned corn in 20 ml wine and 30 ml water
Molasses	1 g molasses in 50 ml wine
Onion	Soak 1 g macerated fresh onion for 12 h in 50 ml wine; add 2 ml of this to 50 ml wine
Rubber	Soak 10 g of small pieces of black rubber laboratory hose in 50 ml wine; decant wine from rubber after 30 min

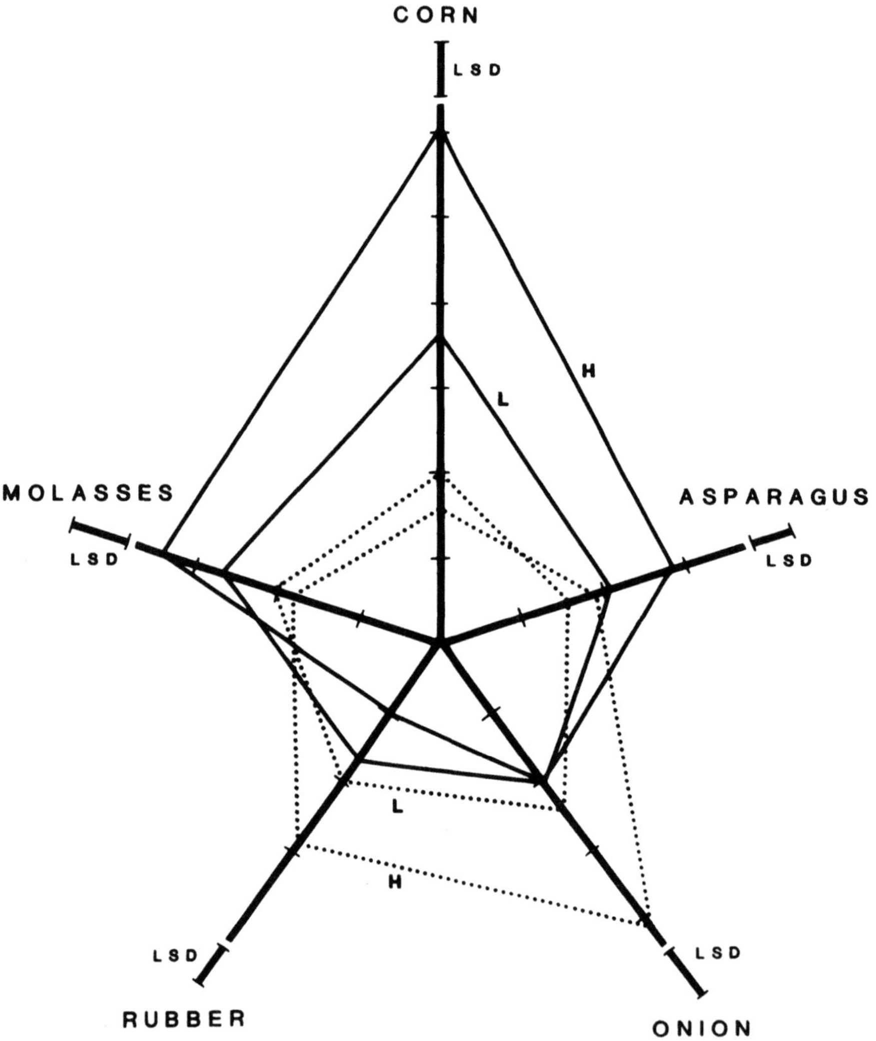

Fig. 7. Comparison of aroma profiles of white wine and least significant differences (*LSD*) for wines with dimethyl sulfide (µ); *L*,*H* = 500 and 650 mg l^{-1}, respectively and ethanethiol (...); *L*,*H* = 5.0 and 6.5 µg l^{-1}, respectively (n = 12 judges × 3 reps). (Goniak and Noble 1987)

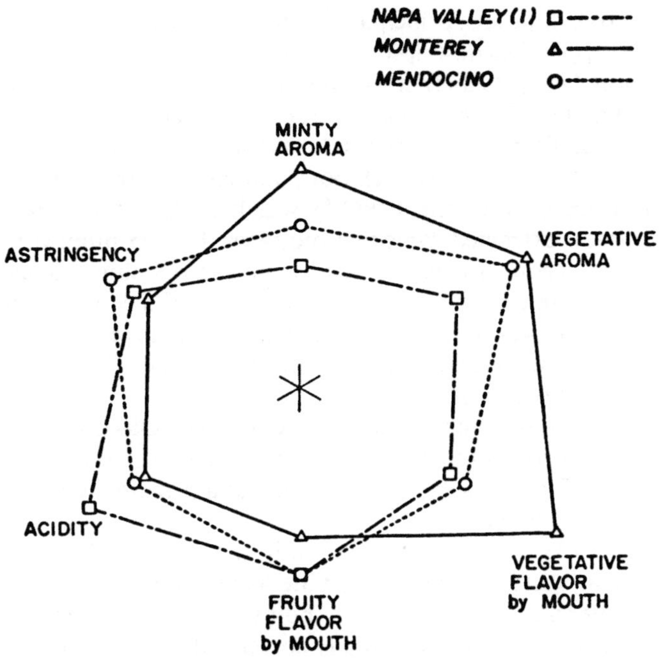

Fig. 8. Comparison of descriptive profiles of Cabernet Sauvignon wines from Napa Valley, Monterey, and Mendocino (n = 11 judges × 4 reps). (Schmidt 1981)

The information provided by descriptive analysis of Californian Cabernet Sauvignon wines is shown in a polar representation in Fig. 8 (Schmidt 1981). The Monterey wine, which is made in a very cool region, is characterized by high intensity of minty and vegetative aromas, and by a vegetative flavor by mouth, while it has the lowest fruitiness by mouth. In contrast, the Napa Valley wine is low in the minty and vegetative attributes, and highest in acidity. The Mendocino wine, also made from a cool region, differs from the Monterey wine primarily by having a more intense fruity flavor and less intense vegetative flavor by mouth.

In a larger study, in which 21 Cabernet Sauvignon wines were evaluated by DA, (Heymann and Noble 1987), a significant correlation was found between the temperature zone of the grape-growing area and the eucalyptus aroma and vegetative flavor by mouth. Wines from cooler areas generally had higher intensities of both attributes. In addition, the effect of vine age was demonstrated. Wines made from older vines had higher intensities of berry aroma and lower intensity of green bean aroma and vegetative flavor by mouth. In contrast, in a DA of Bordeaux wines, wines from the coolest area, St. Estephe, were not significantly more vegetative (canned green bean/green olive) than those from three other communes (Noble et al. 1984).

2.4.3 Principal Component Analysis of Descriptive Data

When large numbers of wines are examined by descriptive analysis, it is not feasible to evaluate the aromas by the simple cobweb plots of Fig. 7 and 8. To simplify the interpretation of large data sets, principal component analysis (Chatfield and Collins 1980) has been applied to descriptive analysis of wine flavor for Zinfandel (Noble and Shannon 1987), Cabernet Sauvignon (Heymann and Noble 1987), Pinot noir (Guinard and Cliff 1987), and red Bordeaux wines (Noble et al. 1984). Principal component analysis is a multivariate statistical technique which reduces the dimensionality of the data, shows relations among the attributes and among the wines. A first principal component (PC 1), which is a linear combination of the sensory attributes, is extracted, which accounts for the maximum variation in the original data. A second PC is extracted from the remaining unexplained variance in the same manner, with the final number of PC's equal to the number of original variables, However, generally the first two or three principal components will explain most of the variation in the data, so that the information can be plotted in two or three dimensions.

Descriptive analysis of 58 Chardonnay wines was conducted (Ohkubo et al. 1987). Seven attributes were scored in the wines which came from three vintages and four locations within California. In Fig. 9, the results of a principal component analysis of these data are shown. For the first two principal components, which together account for 51% of the total variation in the data, the loadings for the attributes are shown as vectors, together with the scores for means of each region and vintage. In the interest of clarity, the data points for each wine are not plotted. Peach, floral, and citrus aromas were highly correlated with each other, as suggested by the small angle between their vectors. They are highly correlated with the first principal component to which they are closely aligned. As indicated by the short length of its vector, the green pepper attribute, while negatively correlated with the peach, floral and citrus aromas, is relatively unimportant in explaining differences in these first two dimensions. The second PC is primarily a function of sweetness and vanilla aroma contrasted with bitterness. As shown by the nearly 180° alignment of their vectors, sweetness and bitterness are inversely correlated.

From the location of the means shown in Fig. 9, we can make inferences about the properties of these wines. The first PC classifies the wines by vintage. The 1983 wines, located on the right, are high in peach, floral and citrus aromas, while the 1981 wines on the left are lowest in these three terms, while being slightly higher in the green pepper note. When the data is examined by region, the wines from the cool region were highest in fruity-floral character, although a significant difference among regions was not found (Ohkubo et al. 1987).

The Chardonnay descriptive analysis given above was part of a larger study in which an additional 103 wines from three other varieties were examined: White Riesling, Sauvignon blanc and Chenin blanc (Noble and Ohkubo, in preparation 1987). For each variety, wines could not be grouped by location of origin; however, a separation by vintage was observed. For each variety, a decrease of the floral, citrus and peach aromas occurred as the wines aged.

In Fig. 10, the principal component analysis of the entire data set is displayed in three dimensions. For clarity, only the mean values for each variety are indicated. Although there is considerable overlap of the varieties when individual wines are plotted in this space, clusters emerged around the mean point for all varieties except the Chenin blanc. The Sauvignon blanc wines were shown to be

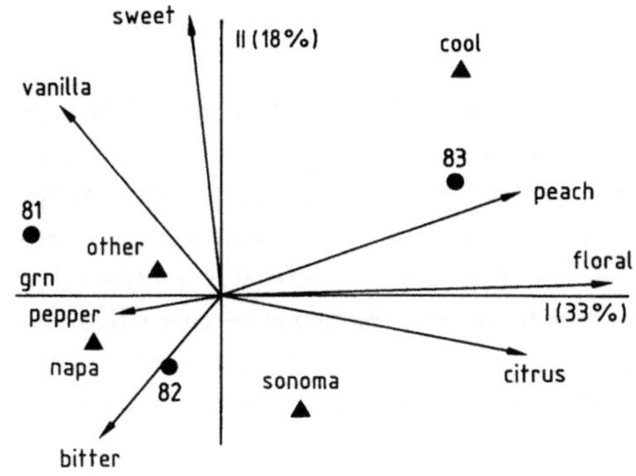

Fig. 9. Principal component analysis of Chardonnay wines. Projection of sensory data on principal components I and II. Attribute loadings (vectors) and mean factor scores for wines from 1981, 1982, and 1983 vintages (●) and from Napa Valley, Sonoma Valley, cool regions and "other" locations (▲). (Ohkubo and Noble unpubl.)

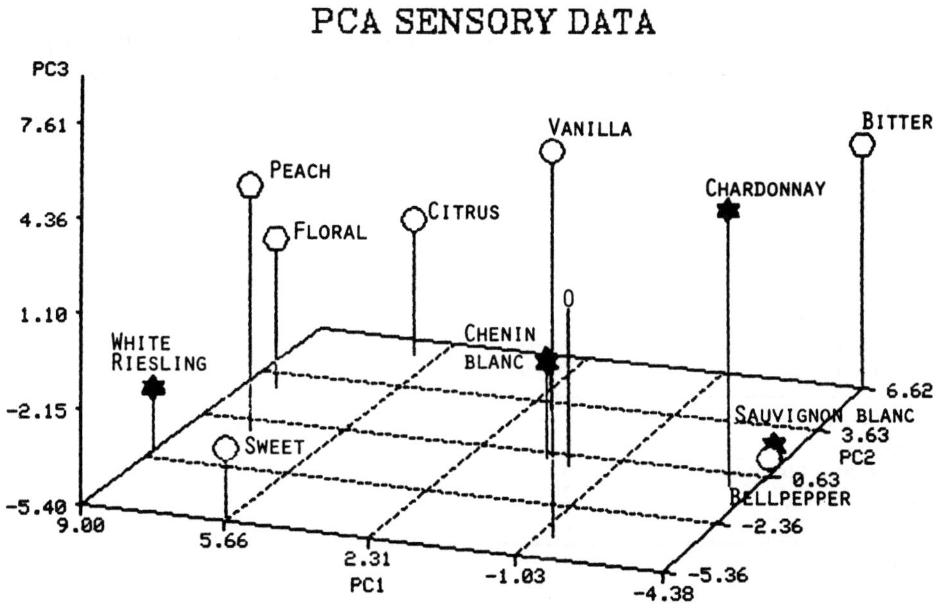

Fig. 10. Principal component analysis of four white wine varietals. Attribute loadings are shown as *spheres,* means for each wine variety and *stars,* and the origin of the three-dimensional space is indicated by the letter *"O".* (Noble and Ohkubo 1987)

highest in bell pepper aroma, while the White Riesling wines were characterized as being sweet, floral, citrus and low in vanilla. Chardonnay wines stand out in this space as being high in vanilla, produced as a result of being aged in oak barrels. In contrast, White Rieslings are rarely aged in oak and have very low vanilla aroma.

2.5 Quality Evaluations Using Scorecards

Unfortunately, quality evaluations of wine are a "Catch 22" situation. Whereas descriptive analysis will provide precise information, it does not yield an overall preference or quality rating. Quality is a composite response to the sensory properties of wine based on one's expectations for a given wine type, which are a function of one's previous experience with wines (Noble 1981). Quality is a subjective and individual response, varying from person to person because of our different experiences, expectations, and preferences. Therefore, the most experienced, skilled and sensitive of wine judges will have differences of opinion about wine quality.

One of the most widely used scorecards is the Davis 20-point scorecard, the use of which is explained in detail by Amerine and Roessler (1983). It serves as a training device by forcing judges to examine the same factors, but seldom will experienced judges rate each category using the weighting assignments listed in Table 4. Instead, judges tend to evaluate the wine and come up with a final number using their own criteria. Recently, a modification of the Davis 20-point scorecard was developed by the Oenology faculty at Roseworthy Agriculture College (Rankine 1986). As shown in Table 5, it provides more specific point assignments for rating of positive or negative characteristics, but is subject to the same misuse as the Davis 20-point scorecard.

Despite the difficulty in obtaining quality evaluations, it is a necessary evil. To optimize the value of quality scoring one should test the judges to be certain that they can reproduce their own results. Judges should have experience with the

Table 4. Point assignment of the Davis 20-point scorecard[a]

Characteristic	Weight
Appearance	2
Color	2
Aroma and bouquet	4
Volatile acidity	2
Total acid	2
Sugar	1
Body	1
Flavor	2
Astringency	2
General quality	2
Potential total	20

[a] 17 to 20, superior wines, must have some outstanding characteristic and no marked defect; 13 to 16, standard wines with neither an outstanding character or defect; 9 to 12, wines of commercial acceptability but with a noticeable defect; 5 to 8, wines of below commercial acceptability; 1 to 4, completely spoiled wines.

Table 5. Point assignment of the Roseworthy 20-point scorecard

Characteristics	Potential score
Appearance	Maximum 3
Color	(Max. 2)
Meets specifications	2
Some incorrect tints	0.5–1.5
Fails specifications	0
Clarity	(Max. 1)
Brilliant	1
Clear	0.5
Dull	0
Aroma	Maximum 7
Intensity	(Max. 4)
Grape aroma/oak/bottle complexity	2 –4
Vinous	0.5–1.5
Neutral	0 –0.5
Faults	(Max. 3)
None	3
Detectable	1 –2.5
Serious (Acetic acid, ethyl acetate SO_2, moldy, bacterial, oxidized, yeasty, corky, etc.)	0 –0.5
Flavor by mouth	Maximum 10
Intensity	(Max. 3)
Full flavor/persistence	2 –3
Medium	1 –1.5
Neutral/thin	0 –0.5
Acid balance	(Max. 2)
Deduct for excess or low acidity	
Faults	(Max. 3)
Absence	3
Detectable	1 –2.5
Serious (bitterness, astringency, etc.)	0 –0.5
Overall quality and balance	Maximum 2
Superior wine	2
Adequate	1
Neutral or flawed	0

scorecard being used, and discussion among the judges should never be permitted, to avoid the biasing influence of dominant judges and the power of suggestion.

As an example of the use of scorecards, the Chardonnay wines which were evaluated by descriptive analysis of Ohkubo et al. (1987) were also rated for quality, using the 20-point scorecard. Across the 58 wines, there was no significance difference in quality ratings, because of the large variability among the judges. Similarly, mean scores for wines from each location of origin did not differ significantly. However, when the data was examined by vintage, a significant difference emerged: wines for the 1981 vintage were scored significantly higher, 14.8, than those of the 1982 or 1983 vintages which had mean scores of 14.22 and 14.08, respectively (Noble and Ohkubo 1987). Examining Fig. 9, it appears that the judges rated the less fruity and less floral 1981 wines higher than the fruitier younger wines.

3 Correlation of Sensory and Instrumental Data

Attempts have been made for several years to relate the composition of wine to its sensory properties, from simple examples such as relating pH and titratable acidity to sourness, to examining more complex relationships between volatile composition and descriptive ratings. To avoid developing spurious relationships, careful selection of the chemical variables should be made where possible. When volatiles are analyzed by gas chromatography, important, "aroma significant" volatiles can be selected by sniffing the split effluent at the exit port of the gas chromatograph. Multiple regression analysis has been used to derive equations in which "aroma significant" volatile components were used to predict the intensity of specific aroma attributes in Cabernet Sauvignon wines (Noble 1978) and for Bordeaux wines (Williams et al. 1984). The limitations of such approaches are numerous. The equations describe predictive relationships, but not causal ones. Further, because most of the volatiles are highly correlated, several solutions to each equation are possible. For any multivariate technique to yield meaningful results in examining problems such as these, a very large number of samples must be evaluated, more than the 10 Cabernets and 24 Bordeaux wines being used in these studies.

Visual comparison of principal component analyses of wine chemical data with PCA's of descriptive analysis data has been made for Chardonnay (Ohkubo et al. 1987) and Zinfandel (Noble and Shannon 1987). In both cases, the chemical variables included no volatile data, therefore relationship between the sensory and chemical spaces was observed.

Two techniques have been used to statistically compare wine configurations developed from two or more data sets. Procrustes analyses is a technique by which the spaces derived by principal component analyses are matched (Gower 1975). Spaces are rotated, reflected and elongated or compressed mathematically and the best fit between two or more figurations determined. Where the two or more configurations can be fitted, it suggests that both (all) data sets are measuring related variables. Examination of the chemical and sensory loadings for each of the principal axes can indicate which volatiles are related to each sensory attribute. It has been used to compare the configurations of 24 Bordeaux wines derived from PCA's of volatile data and descriptive analysis (Williams et al. 1984). Williams and Langron (1984) have also used Procrustes analysis to compare port wine configurations from descriptive analyses by individual tasters.

The method of partial least squares (PLS) analysis of latent variables (Wold 1982; Martens and Martens 1986) has also been used to attempt to uncover relationships between sensory and instrumental analyses of Italian sparkling wines (Bertuccioli et al. 1987) and Chianti (Bertuccioli 1987). This technique indicates how well each of the variables in one data set predicts the variation among the variables in the other data set. From PLS of sparkling wines, the first latent variable explained 63% of the variation in the total data set. The fruity aroma was correlated with isoamyl acetate; the cheesy aftertaste was related to ethyl caprate and ethyl laurate.

The application of multivariate techniques to wine data to find patterns in the sensory and instrumental variables has only been used in an exploratory way. It has great potential for understanding the relationships between wine aroma and wine volatile compounds and their precursors in grapes.

4 Conducting Sensory Tests

4.1 Facilities

The most important criteria for a tasting room is that a quiet, odor-free, controlled-temperature location be provided in which the judge can evaluate wines under controlled lighting and free of distraction. Ideally, booths with partitions between the judges should be provided to eliminate interaction between judges and facilitate concentration. The booth area should be separated from the preparation area so that the judges are not inadvertently given clues about the experiment.

In the booths, controlled lighting permits evaluation of wine color under a constant light source, or permits the color to be masked under red light. A system, such as a light switch, should be provided in each booth for the judge to signal the experimentor. A small hatch door in each booth permits a the administrator to communicate with a judge and to change samples without distracting other judges. Spitoons can be provided for expectorating samples if the cost of sinks is prohibitive.

The preparation area should contain adequate counter space for efficient operation and ready access to the hatch doors to the booths. Sufficient space for storage of glassware, trays and score sheets should be provided, as well as facilities for dishwashing and data analysis. Further details about sensory facilities are discussed by Larmond (1973).

4.2 Protocol

To remove all extraneous clues, samples should always be presented coded with random numbers in identical containers. Standard tulip-shaped clear wine glasses are necessary for any evaluation involving aroma, while 50-ml beakers are excellent if only taste is being evaluated. To enhance aroma evaluation, providing watch glasses or Petri dishes as lids on the wine glasses is very effective.

For several reasons, it is imperative to randomize the order of presentation to each judge. For example, serving a high acid wine followed by a low acid wine will cause the second wine to be rated disproportionately low in sourness (Norris 1982). By randomizing the order over all judges, this bias is eliminated. In the case of pair or duo-trio presentations, the order of presentation of the samples must be randomized between right and left to avoid a judge recognizing that the correct sample is always on the left. Randomizing also prevents a sequence effect, as in the evaluation of astringency and bitterness, where the second sample is almost always perceived as higher in intensity.

For difference tests, if 15 to 20 judges are available, no replications need to be made. If there are only ten judges, having each person perform the evaluations in duplicate will provide an "n" (number of judgements) of 20. For rating tests or descriptive analysis, it is imperative to obtain two to three replications to be able to demonstrate the reliability of the data.

4.3 Judge Selection and Training

Initially, judges should be selected on the basis of availability and motivation. Recruit more judges than are needed, using the guidelines in Table 1. At the beginning of any new testing, introduce each judge to the scorecard and the procedures. Define any descriptive terms verbally and present physical reference standards if possible. In the first session of any test, have samples which differ significantly in the test variable. This will assist in training the judge, and helps morale. Keeping records on judge performance will permit selection of judges who are consistent, sensitive, and reproducible.

The requirements for training judges vary with the type of test. For simple difference tests, once judges are familiar with the testing procedure, only one training session, if any, is required. For rating a single attribute, "high" and "low" reference standards should be presented to define the term. One to three sessions should be held in which the judges rate wines, similar to those that will be evaluated formally, using the same procedures as those that will be used in formal testing.

For descriptive analysis, one to six weeks of training are required. The first sessions are discussions, in which reference standards are presented and their appropriateness for rating the flavor of selected wines is reviewed. Once suitable references have been developed and final terms selected, training sessions resemble formal testing. Prior to each session, the judge should smell and/or taste each of the references, then rate the intensity in coded wines. Further guidelines are provided by Zook and Wessman (1977).

For any type of test, giving judges feedback improves their performance and motivates them. If a judge does not rate a term properly, he/she should be presented with wines or reference standards which illustrate "low" and "high" intensity of the term. After familiarizing himself/herself with these, the judge should rate coded samples, and then his/her results reviewed.

Whether for long-term projects or for routine short tests, the regular attendance and serious concentration of each judge are essential for obtaining sound data. This is achieved by positively motivating judges through feedback about their performance, providing rewards after each session and conducting the tests in a scientific manner.

5 Conclusion

The sensory properties of wine can be evaluated analytically, just as wine chemical and physical properties are evaluated analytically by chemical tests and instrumental analyses. It is necessary to "calibrate" judges through training, a procedure analogous to standardizing of instruments. To avoid any sequence or time-order effects, samples must be randomized. To avoid any bias, all samples should be coded. The results must be analyzed statistically to determine if significant differences among wines exist and to test the reproducibility and consistency of the judges. Despite the sophistication of new instrumentation, only by the use of humans for sensory analysis can the sensory properties of wine be evaluated.

References

Amerine MA, Roessler EB (1983) Wines, their sensory evaluation. Freeman, San Francisco

Amerine MA, Pangborn RM, Roessler EB (1965) Principles of sensory evaluation of food. Academic Press, New York

Bertolucci M (1987) Sensory evaluation applied to enological industry and research. In: Scienza A (ed) Symposium: The aroma components in grapes and wines. 25–27 June 1987. San Michele all'Adige-Trento, Italy

Bertuccioli M, Clementi S, Cruciani G (1987) Impiego dell' analisi multivariate per lo studio di relazioni quantitative tra serie dati sperimentali. In: Chirotti G (ed) Proc 3rd Int Symp of Wine, 15–16 May 1987, Pavia, Italy

Chatfield C, Collins AJ (1980) Introduction to multivariate analysis. Chapman and Hall, London

Giovanni ME, Pangborn RM (1983) Measurement of taste intensity and degree of liking of beverages by graphic scales and magnitude estimation. J Food Sci 48:1175–1182

Goniak OJ, Noble AC (1987) Sensory study of selected volatile sulfur compounds in wine. Am J Enol Vitic 38:223–227

Gower JC (1975) Generalized procrustes analysis. Psychometrica 40:35–51

Guadagni DB, Miers JC, Venstrom D (1968) Methyl sulfide concentration, odor intensity and aroma quality in canned tomato juice. Food Technol 22:1003–1006

Guinard JX, Cliff MJ (1987) Descriptive analysis of Pinot noir wines from Carneros, Napa and Sonoma. Am J Enol Vitic 38:211–215

Guinard JX, Pangborn RM, Lewis MJ (1986) The time-course of astringency in wine upon repeated ingestion. Am J Enol Vitic 37:184–189

Heymann H, Noble AC (1987) Descriptive analysis of commercial Cabernet Sauvignon wines from California. Am J Enol Vitic 38:41–44

Larmond E (1973) Physical requirements for sensory testing. Food Technol 27(11):28–32

Larson-Powers N, Pangborn RM (1978) Paired comparison and time-intensity measurements of the sensory properties of beverages and gelatins containing sucrose or synthetic sweeteners. J Food Sci 43:41–46

Leach EJ (1985) Evaluation of astringency and bitterness by scalar and time-intensity procedures. Thesis, University of California-Davis, California

Lee WE (1985) Evaluation of time-intensity sensory response using a personal computer. J Food Sci 50:1750–1751

Martens M, Martens H (1986) Partial least squares regression. In: Piggott JR (ed) Statistical procedures in food research. Elsevier Applied Science, London, pp 292–359

Noble AC (1978) Sensory and instrumental evaluation of wine aroma. In: Charalambous G (ed) Analysis of foods and beverages, Academic Press, New York, pp 203–228

Noble AC (1981) Defining wine "quality" versus wine "sensory properties". In: Beech FW, Redmond WJ (eds) Fifth wine subject day on bulk wine transport and storage, 27–28 Oct. 1980. Long Ashton Research Station, Bristol

Noble AC, Ohkubo T (1987) Evaluation of flavor of California Chardonnay wines. In: Scienza A (ed) Symposium: The aroma components in grapes and wines, 25–27 June 1987. San Michele all'Adige-Trento, Italy

Noble AC, Schmidt JO (1981) Comparison of structured and unstructured category scales in the evaluation of sourness intensity. In: Schreier P (ed) Flavour '81. de Gruyter, New York, pp 63–68

Noble AC, Shannon M (1987) Profiling Zinfandel wines by sensory and chemical analyses. Am J Enol Vitic 38:1–5

Noble AC, Williams AA, Langron SP (1984) Descriptive analysis and quality ratings of 1976 wines from four Bordeaux communes. J Sci Food Agric 35:88–98

Noble AC, Arnold RA, Buechsenstein J, Leach EJ, Schmidt JO, Stern PM (1987) Modification of a standardized system of wine aroma terminology. Am J Enol Vitic 38:143–146

Norris MB (1982) Salivary and gustatory responses in aqueous solutions as a function of anionic species, titratable acidity, and pH. Thesis, University of California-Davis, California

Ohkubo T, Noble AC, Ough CS (1987) Evaluation of California Chardonnay wines by sensory and chemical analyses. Sci Aliment 7:573–587

O'Mahony M (1986) Sensory evaluation of food, statistical methods and procedures. Marcel Dekker, New York

Rankine B (1986) Roseworthy develops new wine scorecard. Australian Grapegrower & Winemaker, Feb, p 16

Roessler EB, Pangborn RM, Sidel JL, Stone H (1978) Expanded tables for estimating significance in paired-preference, paired-difference, duotrio and triangle tests. J Food Sci 43:940–943, 947

Schmidt JO (1981) Comparison of methods and rating scales used for sensory evaluation of wine. Thesis, University of California-Davis, California

Stone H, Sidel JL (1985) Sensory evaluation practices. Academic Press, New York

Stone H, Sidel JL, Oliver S, Woolsey A, Singleton RC (1974) Sensory evaluation by quantitative descriptive analysis. Food Technol 28(11):24, 26, 28, 29, 32, 34

Williams AA, Langron SP (1984) The use of free-choice profiling for the evaluation of commercial Ports. J Sci Food Agric 35:558–568

Williams AA, Rogers C, Noble AC (1984) Characterization of flavour in alcoholic beverages. In: Nykanen L, Lehtonen P (eds) Flavour research of alcoholic beverages. Instrumental and sensory analysis. Foundation for Biotechnical and Industrial Research, Helsinki 3:235–254

Wold H (1982) Soft modelling: The basic design and some extensions. In: Joreskog KG, Wold H (eds) Systems under indirect observation. North-Holland, Amsterdam

Zook K, Wessman C (1977) Selection and use of judges for descriptive panels. Food Technol 31(11):56–60

Wine Aroma Substances from Gas Chromatographic Analysis

A. RAPP

Aroma compounds, as a result of their pronounced effect on our sensory organs, play a definitive role in the quality of our food and luxury products.

As is the case with most food products, the aroma or "bouquet" of a wine is influenced by the action of several hundred different compounds on the sensory organs. The total content of aroma compounds in wine amounts to approximately 0.8–1.2 g l^{-1}, which is equivalent to about 1% of the ethanol concentration. The fusel oils, formed during fermentation, are responsible for about 50% of this content. The concentration of the remaining aroma compounds range from 10^{-4}–10^{-9} g l^{-1} (Rapp et al. 1973; 1978).

The human sensory organs display somewhat sensitive and variable reactions to these amounts of aroma compounds. Thus threshold values, i.e., concentrations just barely sufficient to achieve sensory recognition of substances, differ considerably and could vary between 10^{-4} and 10^{-12} g l^{-1} (Boeckh 1972; Guadagni et al. 1963). The organoleptic threshold value for butyric acid is thus quoted as 10^{-5} g kg^{-1} (Rothe 1974), whereas 1-p-menthen-8-thiol is given as 10^{-13} g kg^{-1} (Demole et al. 1982).

Early investigations into volatile constituents of wine date back to the year 1942, referring to the research of Hennig and Villforth. Bayer et al. (1958, 1961) instituted gas chromatography for the first time towards the end of the 1950's for the determination of wine aroma compounds. In the following years, several research institutions used this method. Constantly improving physicochemical methods of analysis (gas chromatography), combined with the various possibilities of spectroscopic structure identification (gas chromatography-mass spectrometry), especially the introduction of highly efficient capillary columns and detectors (FID, NPD, FDP, SIM), led to an ever-improving insight into the complex constitution of wine aroma.

In various synoptic reviews (Drawert and Rapp 1968; Rapp 1972, 1981; Rapp and Mandery 1986; Webb and Muller 1972; Muller et al. 1973; Schreier 1979; Nykänen and Suomaleinen 1983; Van Straten and Maarse 1983), the aroma compounds identified by many authors up to the present in grape must and wine have been listed (Table 1). These authors include amongst others; Chaudhary et al. 1964; Stevens et al. 1965; Stern et al. 1967; Bayonove et al. 1971 a; Terrier et al. 1972; Ribereau-Gayon et al. 1975; Schreier et al. 1976 a, b; Williams et al. 1980 a, b; Webb and Kepner 1957; Peynaud 1965; Lemperle and Mecke 1965; Drawert and Rapp 1966; Bertrand et al. 1967; Boidron and Ribereau-Gayon 1967; Van Wyk et al. 1967 a, b; Drawert and Rapp 1968; Webb et al. 1969; Stevens et al. 1969; Hardy and Ramshaw 1970; Muller et al. 1972; Bayonove and Cordonnier 1971 b; Webb and Muller 1972; Dubois et al. 1971; Bertuccioli and Montedoro 1974; Schreier and Drawert 1974 a; Schreier et al. 1975 a; Schreier and Drawert 1974 b; Schreier et al. 1974, 1975 b; Drawert et al. 1974; Dubois et al. 1976; Meunier and Boot 1979; Schreier et al. 1976 a, b, c; Bertuccioli and Viani 1976; Simpson 1978, 1979 a, 1980; Noble et al. 1980; Drawert et al. 1976 a, b; Rapp and Knipser 1979; Rapp et al. 1980 a, 1984 b, c, 1986.

Table 1. Volatile compounds in grape, must and wine

Hydrocarbons

ethylene
1-butene
hexane
1-hexene
2-hexene
3-hexene
2-methylhexa-2,4-diene
3-methylhepta-1,4-diene
decane
1-decene
undecane
dodecane
1-dodecene
tridecane
tetradecane
1-tetradecene
pentadecane
hexadecane
1-hexadecene
heptadecane
1-heptadecene
octadecane
1-octadecene
nonadecane
1-nonadecene
eicosene
1-eicosene
heneicosane
1-heneicosene
docosane
1-docosene
tricosane
1-tricosene
tetracosane
1-tetracosene
pentacosane
1-pentacosene
hexacosane
1-hexacosene
heptacosane
1-heptacosene
octacosane
1-octacosene
nonacosane
1-nonacosene
triacontane
1-triacontene
hentriacontane
1-hentriacontene
dotriacontane
1-dotriacontene
myrcene
α-farnesene
cyclohexane
α-terpinene
limonene
germacrene D
α-humulene
3-carene
α-cadinene
γ-cadinene
α-muurolene
γ-muurolene
β-selinene
α-guaiene
β-caryophyllene
copaene
β-ylangene
β-bourbonene
toluene
ethylbenzene
vinylbenzene
propylbenzene
isopropylbenzene
isobutylbenzene
tert. butylbenzene
1,2-dimethylbenzene
1,3-dimethylbenzene
1,4-dimethylbenzene
1-ethyl-2-methylbenzene
1-ethyl-3-methylbenzene
1-ethyl-4-methylbenzene
1-methyl-3-propylbenzene
4-isopropyl-1-methylbenzene
1,2,3-trimethylbenzene
1,2,4-trimethylbenzene
1,3,5-trimethylbenzene
1,4-dimethyl-2-ethylbenzene
biphenyl
3-methylbiphenyl
calamenene
1-methylnaphthalene
2-methylnaphthalene
2-methylbut-2-ene
1,3-dimethylcyclopentane
isopropylcyclohexane
1,1,3,5-tetramethylcyclohexane
cyclooctatetraene
γ-terpinene
terpinolene
benzene
dimethylbenzene (unkn. str.)
trimethylbenzene (unkn. str.)
1,1,6-trimethyl-1,2,3,4-tetrahydronaphthalene
1,1,6-trimethyltetrahydronaphthalene
1,1,6-trimethyl-1,2-dihydronaphthalene
indene
naphthalene
dimethylnaphthalene (unkn. str.)
2,3-dimethylbuta-1,3-diene
5-methylhex-1-ene
3,5,5-trimethylhex-1-ene
isopropylcyclopropane
isopropylcyclohexane

Alcohols

methanol
ethanol
1-propanol
2-propanol
2-methylpropan-1-ol
2-methylpropan-2-ol
1-butanol
2-butanol
2-methylbutan-1-ol
3-methylbutan-1-ol
2-methylbutan-2-ol
3-methylbutan-2-ol
2-methylbut-2-en-1-ol
3-methylbut-2-en-1-ol
2-methylbut-3-en-2-ol
1-pentanol
2-pentanol
3-pentanol
trans-2-penten-1-ol
3-hexen-1-ol
1-penten-3-ol
4-methylpentan-1-ol
1-hexanol
cis-2-hexen-1-ol
trans-2-hexen-1-ol
cis-3-hexen-1-ol
trans-3-hexen-1-ol
5-methylhexan-1-ol
2-ethylhexan-1-ol
1-heptanol
2-heptanol
1-octanol
2-octanol
3-octanol
trans-2-octen-1-ol
1-octen-3-ol

Table 1 (continued)

trans-3,7-dimethylocta-1,5,7-trien-3-ol
3,7-dimethyloct-1-en-3,7-diol
3,7-dimethylocta-1,7-dien-3,6-diol
3,7-dimethylocta-1,5-dien-3,7-diol
3,7-dimethyloct-1-en-3,6,7-triol
1-nonanol
1-decanol
2-decanol
1-undecanol
citronellol
geraniol
nerol
linalool
cis-ocimenol
trans-ocimenol
myrcenol
benzyl alcohol
1-phenylethanol
2-phenylethanol
2-hydroxy-2-methoxy-1-phenylethane
α-terpineol
terpineol-4
cis-terpine
trans-terpine
ionol
borneol
fenchyl alcohol
sabinene hydrate
α-cadinol
aminoethanol
3-ethoxypropan-1-ol
1,2,3-propanetriol
2-ethylbutan-1-ol
2,3-butanediol
3-methylpentan-1-ol
4-methylpentan-1-ol
cis-2-hexen-1-ol
3-methylhexan-1-ol
5-methylhexan-2-ol
2,4-dimethyl-3-ethoxy-hex-4-en-1-ol
2-hepten-1-ol
6-methylhept-5-en-2-ol
2,6-dimethylheptan-4-ol
3,7-dimethylocta-2,5,7-trien-1-ol
cis-3,7-dimethylocta-1,5,7-trien-3-ol
3,7-dimethyloctane-1,7-diol
2-nonanol
2-nonen-1-ol
9-methylnonen-1-ol
1-dodecanol
2-(4-hydroxyphenyl)ethanol
3-phenylpropan-1-ol
p-menthen-9-ol (unkn. str.)
δ-cadinol
1-penten-3-ol
2-(4-hydroxy-methoxyphenyl)-ethanol (unkn. str.)
1-ethoxypropan-2-ol
1-methoxybutan-2-ol
3-methoxybutan-2-ol
7-methylheptan-1-ol
cis-farnesol
trans-farnesol
nerolidol

Carbonyls, aldehydes

acetaldehyde
propanal
2-propenal
2-methylpropanal
butanal
2-butenal
2-methylbutanal
3-methylbutanal
2-methylbut-2-enal
pentanal
trans-2-pentenal
cis-2-pentenal
2-methylpent-2-enal
hexanal
trans-2-hexenal
cis-2-hexenal
cis-3-hexenal
trans-2,cis-4-hexadienal
trans-2,trans-4-hexadienal
heptanal
trans-2-heptenal
trans-2,cis-4-heptadienal
trans-2,trans-4-heptadienal
octanal
trans-2-octanal
trans-2,trans-4-octadienal
trans-2,cis-4-octadienal
nonanal
trans-2,trans-4-nonadienal
decanal
trans-2,trans-4-decadienal
dodecanal
citronellal
geranial
neral
citral
benzaldehyde
2-hydroxybenzaldehyde
2-phenylacetaldehyde
4-hydroxybenzaldehyde
3,4-dihydroxybenzaldehyde
4-hydroxy-3-methoxy-benzaldehyde
3,5-dimethoxy-4-hydroxy-benzaldehyde
cinnamaldehyde

Carbonyls, ketones

2-propanone
2-butanone
3-buten-2-one
3-methylbutan-2-one
3-methylbut-3-en-2-one
3-hydroxybutan-2-one
4-ethoxybutan-2-one
2,3-butanedione
2-pentanone
3-pentanone
1-penten-3-one
3-methylpentan-2-one
4-methylpentan-2-one
2-methylpentan-3-one
2,4-pentanedione
2-hexanone
2-heptanone
3-heptanone
6-methylhept-5-en-2-one
6-methylhepta-3,5-dien-2-one
2-octanone
2-nonanone
2-decanone
6,10,14-trimethylpentadecan-2-one
9-heptadecanone
acetophenone
4-methoxyacetophenone
acetovanillone
1-(2,6,6-trimethylcyclohexa-1,3-dienyl-1)but-2-en-1-one
benzophenone

carvenone
1,1-diethoxypropan-2-one

Table 1 (continued)

4-hydroxy-4-methylpentan-2-one
2,3-pentanedione
2-methylheptan-3-one
3-octen-2-one
2-nonen-4-one
trimethylcyclopentenone
2,2,6-trimethylcyclohexan-1-one
3,5,5-trimethylcyclohex-2-en-1-one
2,4-dimethylacetophenone
2-hydroxy-methyl-acetophenone
α-ionone
β-ionone
3-hydroxy-4-phenylbutan-2-one
3-ethoxybutan-2-one
3-ethoxypentan-2-one
acetosyringone
propiovanillone
2-undecanone

Acids

formic
acetic
propanoic
2-methylpropanoic
2-hydroxypropanoic
butanoic
2-methylbutanoic
3-methylbutanoic
pentanoic
4-methylpentanoic
hexanoic
trans-2-hexenoic
cis-3-hexenoic
2-ethylhexanoic
heptanoic
octanoic
nonanoic
3-nonenoic
decanoic
undecanoic
dodecanoic
tridecanoic
tedradecanoic
pentadecanoic
hexadecanoic
9-hexadecanoic
heptadecanoic
octadecanoic
9-octadecenoic
9,12-octadecadienoic
octadecadienoic (unkn. str.)
9,12,15-octadecatrienoic
eicosanoic
eicosenoic (unkn. str.)
docosanoic
trans-geranic
benzoic
4-methylbenzoic
2-hydroxybenzoic
4-hydroxybenzoic
2-phenylacetic
3-phenylpropanoic
trans-cinnamic
2,2-dimethyl-3-(2-methylpropyl)-cyclopropane-carboxylic acid
2,2-dimethylpropanoic
pentadecenoic (unkn. str.)
2-oxopropanoic
2-(N-acetylamino)-propanoic
2-hydroxy-2-methyl-propanoic
2-hydroxy-2-methyl-butanoic
2-hydroxy-3-methyl-butanoic
3-oxobutanoic
2-hydroxy-3-methyl-pentanoic
2-hydroxy-4-methyl-pentanoic
4-methyl-2-oxopentanoic
cis-3-hexenoic
2-hydroxyhexanoic
cis-2-pentenoic
trans-2-heptenoic
cis-4-octenoic
nonenoic (unkn. str.)
9-decenoic
dodecenoic (unkn. str.)
2,3-dihydroxybenzoic
2,4-dihydroxybenzoic
2,5-dihydroxybenzoic
2,6-dihydroxybenzoic
3,4-dihydroxybenzoic
3,5-dihydroxybenzoic
4-hydroxy-3-methoxy-benzoic
3,4-dimethoxybenzoic
3,4,5-trihydroxybenzoic
3,4,5-trihydroxycyclohex-1-ene-carboxylic
3,5-dimethoxy-4-hydroxy-benzoic
2-phenylacetic
2-(4-hydroxyphenyl)acetic
1-(4-hydroxy-3-methoxy-phenyl)-acetic
2-oxo-3-phenylpropanoic
3-(4-hydroxyphenyl)-2-oxopropanoic
2-hydroxy-3-phenyl-propanoic
3-(4-hydroxyphenyl)-propanoic
2-hydroxycinnamic
3-hydroxycinnamic
4-hydroxycinnamic
3,4-dihydroxycinnamic
4-hydroxy-3-methoxy-cinnamic
3,5-dimethoxy-4-hydroxy-cinnamic
phthalic

Esters

diethyl carbonate
methyl formate
ethyl formate
isopropyl formate
hexyl formate
benzyl formate
methyl acetate
ethyl acetate
propyl acetate
isopropyl acetate
butyl acetate
isobutyl acetate
2-butyl acetate
butenyl acetate (unkn. str.)
2-methylbutyl acetate
3-methylbutyl-2 acetate
amyl acetate
isoamyl acetate
hexyl acetate
trans-2-hexenyl acetate
cis-3-hexenyl acetate
heptyl acetate
octyl acetate
nonyl acetate
geranyl acetate
neryl acetate

Table 1 (continued)

benzyl acetate
methylbenzyl acetate (unkn. str.)
phenethyl acetate
ethyl propanoate
propyl propanoate
isobutyl propanoate
amyl propanoate
methyl 2-methylpropanoate
ethyl 2-methylpropanoate
isobutyl 2-methylpropanoate
isoamyl 2-methylpropanoate
hexyl 2-methylpropanoate
ethyl 2-hydroxypropanoate
methyl butanoate
ethyl butanoate
butyl butanoate
isoamyl butanoate
hexyl butanoate
cis-3-hexenyl butanoate
phenethyl butanoate
methyl 2-butenoate
ethyl 2-butenoate
isopropyl 2-butenoate
isobutyl 2-butenoate
ethyl 2-methylbutanoate
hexyl 2-methylbutanoate
methyl 3-hydroxybutanoate
ethyl 3-hydroxybutanoate
ethyl 3-acetyloxybutanoate
ethyl 2-hydroxy-2-methylbutanoate
ethyl pentanoate
hexyl pentanoate
heptyl pentanoate
methyl hexanoate
ethyl hexanoate
propyl hexanoate
isopropyl hexanoate
butyl hexanoate
isobutyl hexanoate
amyl hexanoate
isoamyl hexanoate
hexyl hexanoate
2-hexenyl hexanoate
ethyl trans-2-hexenoate
ethyl 3-hydroxy-hexanoate
ethyl heptanoate
hexyl heptanoate
methyl octanoate
ethyl octanoate
isoamyl octanoate
hexyl octanoate
ethyl trans-2-octenoate
ethyl octadienoate (unkn. str.)
methyl nonanoate
ethyl nonanoate
methyl decanoate
ethyl decanoate
hexyl decanoate
ethyl cis-4-decenoate
ethyl trans-2,cis-4-decadienoate
ethyl trans-2,trans-4-decadienoate
ethyl cis-2,trans-4-decadienoate
ethyl trans-2,cis-6-decadienoate
ethyl 2,4,7-decatrienoate
methyl dodecanoate
ethyldodecanoate
isobutyl dodecanoate
isoamyl dodecanoate
hexyl dodecanoate
ethyl 2,6-dodecadienoate
methyl tetradecanoate
ethyl tetradecanoate
isobutyl tetradecanoate
isoamyl tetradecanoate
hexyl tetradecanoate
ethyl tetradecadienoate (unkn. str.)
methyl hexadecanoate
ethyl hexadecanoate
methyl octadecanoate
ethyl octadecanoate
methyl 9-octadecenoate
methyl 9,12-octadecadienoate
ethyl 9,12,-octadecadienoate
methyl 9,12,15-octadecatrienoate
methyl trans-geranate
diethyl malonate
ethyl acid succinate
diethyl succinate
diethyl glutarate
methyl benzoate
ethyl benzoate
methyl salicylate
ethyl salicylate
isoamyl salicylate
methyl 2-methoxybenzoate
ethyl 2-methoxybenzoate
methyl anthranilate
ethyl anthranilate
methyl 4-hydroxy-3-methoxybenzoate
ethyl 2-phenylacetate
hexyl 2-phenylacetate
dimethyl phthalate
diethyl phthalate
dibutyl phthalate
geranyl formate
vinyl acetate
1,3-propanediol monoacetate
2,3-butanediol monoacetate
trans-3-hexenyl acetate
citronellyl acetate
linalyl acetate
isoamyl propanoate
butyl 2-methylpropanoate
methyl 2-methylpropanoate
methyl 2-hydroxypropanoate
propyl 2-hydroxypropanoate
isobutyl 2-hydroxypropanoate
isoamyl 2-hydroxypropanoate
ethyl 2-oxopropanoate
isopropyl butanoate
2-hydroxypropyl butanoate
2-methylbutyl butanoate
decyl butanoate
phenethyl butanoate
ethyl trans-2-butenoate
ethyl 3-methylbutanoate
butyl 3-methylbutanoate
isobutyl 3-methylbutanoate
hexyl 3-methylbutanoate
methyl 3-oxobutanoate
ethyl 2-hydroxy-3-methylbutanoate
methyl pentanoate
butyl pentanoate
isobutyl pentanoate
isoamyl pentanoate
ethyl 4-hydroxypentanoate
ethyl 4-oxopentanoate
ethyl 2-hydroxy-3-methylpentanoate
ethyl 2-hydroxy-4-methylpentanoate
heptyl hexanoate

Table 1 (continued)

phenethyl hexanoate
ethyl 2,4-hexadienoate
ethyl 1,6-hexadienoate
ethyl methylhexanoate (unkn. str.)
methyl heptanoate
ethyl trans-2-heptenoate
propyl octanoate
isobutyl octanoate
2-methylbutyl octanoate
phenethyl octanoate
propyl decanoate
isobutyl decanoate
isoamyl decanoate
isoamyl decenoate
hexyl decenoate
ethyl 9-decenoate
isobutyl decenoate (unkn. str.)
ethyl undecanoate
2-methylbutyl dodecanoate
ethyl 9-hexadecenoate
ethyl 9-octadecenoate
diethyl oxalate
ethyl methyl succinate
propyl acid succinate
ethyl propyl succinate
ethyl isopropyl succinate
isobutyl acid succinate
butyl ethyl succinate
ethyl isobutyl succinate
isoamyl acid succinate
ethyl amyl succinate
ethyl isoamyl succinate
ethyl hexyl succinate
diisoamyl succinate
ethyl phenethyl succinate
ethyl acid glutarate
diethyl glutarate
diethyl 2-hydroxy-pentanedioate
diethyl adipate
diethyl heptanedioate
ethyl acid nonanedioate
diethyl nonanedioate
ethyl acid malate
ethyl methyl malate
diethyl malate
ethyl acid 2-hydroxy-2-methylbutanedioate
ethyl acid 2-hydroxy-2-isopropylbutanedioate
methyl 4-hydroxybenzoate
ethyl cinnamate
ethyl 2-hydroxy-3-phenyl-propanoate
ethyl 3,4-dihydroxy-cinnamate
ethyl-4-hydroxy-3-methoxy-cinnamate
ethyl 4-phenylbut-3-enoate
ethyl acid phthalate
isobutyl acetate
ethyl 2-acetoxyacetate
isobutyl 2-hydroxy-propanoate
hexyl 2-hydroxy-propanoate
ethyl 2-acetoxy-propanoate
amyl butanoate
isoamyl 2-methylbutanoate
ethyl 4-hydroxybutanoate
ethyl 3-acetoxybutanoate
ethyl 4-acetoxybutanoate
isoamyl 4-acetoxybutanoate
ethyl 2-acetoxy-3-methylbutanoate
ethyl 2-acetoxy-4-methyl-pentanoate
ethyl heptanoate
ethyl 2-methylbutyl succinate
diethyl 2-acetoxy-succinate
diethyl maleate
diethyl 2-methyl-butanedioate
ethyl methyl glutarate
dimethyl malate
diethyl 2-hydroxy-2-methylbutanedioate
ethyl isoamyl 2-hydroxy-2-methylbutanedioate
diisoamyl 2-hydroxy-2-methylbutanedioate
diethyl tartrate
ethyl 4-hydroxy-3-methoxybenzoate
ethyl 3,5-dimethoxy-4-hydroxybenzoate
2-methylbutyl formate
hexenyl acetate (unkn. str.)
2-methylbutyl propanoate
2-methylbutyl 2-hydroxy-propanoate
amyl pentanoate
2-methylbutyl pentanoate
ethyl 4-oxopentanoate
2-methylbutyl hexanoate
isoamyl octanoate
2-methylbutyl decanoate
2-methylbutyl tetradecanoate
isoamyl tetradecanoate
2-methylbutyl pentadecanoate
2-methylbutyl hexadecanoate
isoamyl hexadecanoate
2-methylbutyl octadecanoate
isoamyl octadecanoate
dibutyl succinate
diisobutyl succinate
acid 2-methylbutyl succinate
ethyl 2-methylbutyl succinate
butyl 2-methylbutyl succinate
di-2-methylbutyl succinate
butyl isoamyl succinate
ethyl isoamyl glutarate
ethyl propyl malate
butyl propyl malate (unkn. str.)
ethyl isobutyl malate
ethyl 2-methylbutyl malate
di-2-methylbutyl malate
ethyl isoamyl malate
diisoamyl malate
methyl salicylate
ethyl 2-phenylacetate
2-methylbutyl cinnamate
isoamyl cinnamate
diisobutyl phthalate
phenethyl formate
methyl pentanoate
methyl hexadecanoate
methyl anthranilate
amyl octanoate
butyl decanoate
ethyl dodecenoate

Esters, lactones

4-hydroxyhexanoic acid lactone
4-hydroxy-4-methyl-hex-5-enoic acid lactone
dihydroactinidiolide
5-hydroxypentanoic acid lactone

Table 1 (continued)

4-hydroxy-4-methylpentanoic acid lactone
5-hydroxy-6-ethylpentanoic acid lactone
4-hydroxy-5-oxohexanoic acid lactone (5 ring)
4-hydroxyheptanoic acid lactone
5-hydroxy-octanoic acid lactone
cis-4-hydroxy-3-methyloctanoic acid lactone
4-hydroxynonanoic acid lactone
5-hydroxynonanoic acid lactone
5-hydroxydecanoic acid lactone
4-hydroxy-2-methylbutanoic acid lactone
4-ethoxy-4-hydroxybutanoic acid lactone
4-carboethoxy-4-hydroxybutanoic acid lactone
4-hydroxypentanoic acid lactone
4-hydroxyhexanoic acid lactone
4,5-dihydroxyhexanoic acid lactone (5 ring)
4-hydroxy-3-methyloctanoic acid lactone
trans-4-hydroxy-3-methyloctanoic acid lactone
4-hydroxynonanoic acid lactone
4-hydroxydecanoic acid lactone

Bases

ammonia
methylamine
ethylamine
propylamine
isobutylamine
2-methylbutylamine
isoamylamine
dimethylamine
diethylamine
phenethylamine
1-methylpyrrole
indole
pyridine
2-isobutyl-3-methoxypyrazine
1,4-diaminobutane
1,5-diaminopentane
histamine
pyrrolidine
2-formyl-1-methylpyrrole
2-formyl-5-methylpyrrole
1-ethyl-2-formylpyrrole
quinoline
1,2-diaminoethane
octopamine
3-(4,5-dimethyl-1,3-dioxolane-2)propanamine
serotonine
ethyl nicotinate
ethyl 2-pyrrolidone-5-carboxylate

S-Compounds

ethyl 3-mercaptopropanoate
methylsulfonylmethane
benzothiazole
methylthiomethane
carbondisulfide
methyldithiomethane
3-(methylthio)propan-1-ol
3-(ethylthio)propan-1-ol
3-(methylthio)propyl acetate
3-(methylthio)propanoic acid
methyl thiolacetate
ethyl thiolacetate
methyl 3-(methylthio) propanoate
ethyl 3-(methylthio) propanoate
N-(3-(methylthio)propyl)-acetamide
ethyl methanedisulfonate
2,4-ditert. butylthiophene
2,5-di-(2-methylpropyl)-thiophene
2-methyltetrahydro-thiophen-3-one
2-(1-butyl)-5-(2-methylpropyl)-thiophene

Acetals

1-ethoxy-1-methoxyethane
1,1-diethoxyethane
1-ethoxy-1-(2-hexenoxy)-ethane
1,1-diethoxymethane
1,1-dimethoxyethane
1-ethoxy-1-propoxyethane
1-ethoxy-1-(2-methylbutoxy)-ethane
1-ethoxy-1-(3-methylbutoxy)-ethane
1-ethoxy-1-hexoxyethane
1,1-di-(2-methylbutoxy)-ethane
1,1-di-(3-methylbutoxy)-ethane
1,1-dipentoxyethane
1,1-diethoxypropane
1,1-di-(2-phenylethoxy)-ethane
1,1-diethoxy-2-methylpropane
1,1-diethoxybutane
1,1-diethoxy-3-methylbutane
1,1-diethoxypentane
1,1-dipropoxyethane
1-ethoxy-1-pentoxyethane

Ethers

1,2-dimethoxy-1-phenylethane
diphenyl ether
1,4-cineole
1,8-cineole
benzyl methyl ether
3-ethoxy-1-(2,3,6-trimethylphenyl)but-1-ene
ethyl tetrahydropyranyl ether (unkn. str.)
2-butyl ethyl ether
tert. butyl ethyl ether

Halogenes

trichloromethane
1-chloroethane

Nitriles

benzylcyanide
N,N-dimethylacetamide
N,N-dimethylformamide
N-ethylacetamide
N-butylacetamide
N-isobutylacetamide
N-(2-methylbutyl)-acetamide

Table 1 (continued)

N-isoamylacetamide
N-(2-phenethyl)acetamide
N-butyl-N-methyl-acetamide
nitrobenzene

Phenols

phenol
3-methylphenol
4-methylphenol
2-(1,1-dimethylethyl)-6-methylphenol
3,5-ditert. butyl-4-hydroxytoluene
1-allyl-4-methoxybenzene
2-methoxy-4-vinylphenol
tert. butyl-2-methoxyphenol (unkn. str.)
1,2-dimethoxybenzone
2-methylphenol
4-methylphenol
2-ethylphenol
4-ethylphenol
4-vinylphenol
3,4-dimethylphenol
2-isopropyl-5-methyl-phenol
2,6-ditert. butyl-4-ethylphenol-naphtol
1,2-dihydroxybenzene
1,3-dihydroxybenzene
1,3-dihydroxy-5-methyl-benzene
2-methoxyphenol
4-ethyl-2-methoxyphenol
2-methoxy-5-vinylphenol
4-allyl-2-methoxyphenol
2-butoxy-tert. butylphenol
1,2,3-trihydroxybenzene
1,3,5-trihydroxybenzene
4-allyl-2,6-dimethoxy-phenol
2,6-dimethoxyphenol

Furans

2,2-dimethyl-5-(1-methyl-prop-1-enyl)tetra-hydrofuran
5-isopropenyl-2-methyl-2-vinyl-tetra-hydrofuran
trans-5-isopropenyl-2-methyl-2-vinyl-tetrahydrofuran
cis-5-isopropenyl-2-methyl-2-vinyltetrahydrofuran
dehydrofuranlinalool oxide (unkn. str.)
cis-5-(2-hydroxyisopropyl)-2-methyl-2-vinyl-tetrahydrofuran
trans-5-(2-hydroxyiso-propyl)-2-methyl-2-vinyltetrahydrofuran
2-amylfuran
furfural
2,5-dimethyl-4-hydroxy-2H-furan-3-one
2,5-dimethyl-4-methoxy-2H-furan-3-one
2-acetylfuran
2-butyltetrahydrofuran
2,3,4,5-tetramethyl-2,3-dihydrofuran
2-(ethoxymethyl)furan
di-(furyl-2)methane
5-methylfurfural
5-hydroxymethylfurfural
5-(ethoxymethyl)furfural
5-methyl-5-vinyltetra-hydrofuran-2-one
2-acetylfuran
furfuryl alcohol
2-furancarboxylic acid
ethyl 2-furancarboxylate
ethyl furfuryl ether

(Ep)oxides

cis-rose oxide
trans-rose oxide
nerol oxide
2,6,6-trimethyl-2-vinyl-tetrahydropyran
4-hydroxy-2,6,6-trimethyl-2-vinyltetrahydropyran
cis-5-hydroxy-2,6,6-tri-methyl-2-vinyltetra-hydropyran
trans-5-hydroxy-2,6,6-tri-methyl-2-vinyltetra-hydropyran
1-oxa-2,6,10-tetramethyl-spiro(4.5)dec-6-ene
6-methylene-1-oxa-2,10,10-trimethylspiro(4,5)dec-7-ene
ethylmethyl-1,3-dioxolane (unkn. str.)
2,4,5-trimethyl-1,3-dioxo-lane
limonene oxide
5-hydroxy-2,6,6-trimethyl-2-vinyltetrahydropyran
4-acetoxy-2,6,6-trimethyl-2-vinyltetrahydropyran
cis-linalool oxide
trans-linalool oxide
7-hydroxycoumarin
6,7-dihydroxycoumarin
7,8-dihydroxycoumarin
7-hydroxy-6-methoxy-coumarin
cis-6-methylene-1-oxa-2,10,10-trimethylspiro (4,5) dec-7-ene
trans-6-methylene-1-oxa-2,10,10-trimethylspiro (4,5)-dec-7-ene
cis-5-hydroxy-2-methyl-1,3-dioxane
2,4-dimethyl-5-ethyl-1,3-dioxolane
2-methyl-1,3-dioxolane
2-ethyl-4-methyl-1,3-dioxolane
4-methyl-2-propyl-1,3-dioxolane
2-butyl-4-methyl-1,3-dioxolane
2-isobutyl-4-methyl-1,3-dioxolane
4-methyl-2-phenyl-1,3-dioxolane
4,5-dimethyl-2-propyl-1,3-dioxolane
4,5-dimethyl-2-isobutyl-1,3-dioxolane
4-hydroxymethyl-2-methyl-1,3-dioxolane
cis-4-hydroxymethyl-2-methyl-1,3-dioxolane
cis-5-hydroxy-2-methyl-1,3-dioxane
ethyl methyl 1,4-dioxane

Anhydrides

phthalide
5-methylphthalide

In classifying the aroma compounds of wine, one distinguishes between the following:

- *primary or grape aroma:* compounds as they are to be found in the undamaged plant cells of the grape;
- *secondary grape aroma:* aroma compounds formed during the processing of the grapes (destalking, crushing, pressing), and by chemical, enzymatic-chemical and thermal reactions in grape must;
- *fermentation bouquet:* aroma compounds formed during the alcoholic fermentation;
- *maturation bouquet:* caused by chemical reactions during maturation of the wine in the bottle.

With the aid of suitable extraction methods, wine aroma compounds can be sufficiently concentrated without the formation of artefacts (Rapp et al. 1976).

These aroma concentrations can be separated into 600–800 single components by means of gas chromatography, using capillary columns. This creates a thorough foundation for the analytical evaluation of varietal character and wine quality.

Significant differences exist amongst the individual grape varieties pertaining to the composition of the aroma compounds. In the aromagrammes ("finger print patterns") it is clear that the components ("key substances") vary greatly from one another in quantity (Fig. 1). Between the varieties Morio-Muskat (Fig. 1 below) and Riesling (Fig. 1 above), obvious differences in the quantity of several components can be recognized (Fig. 1, marked with arrows). The varietal char-

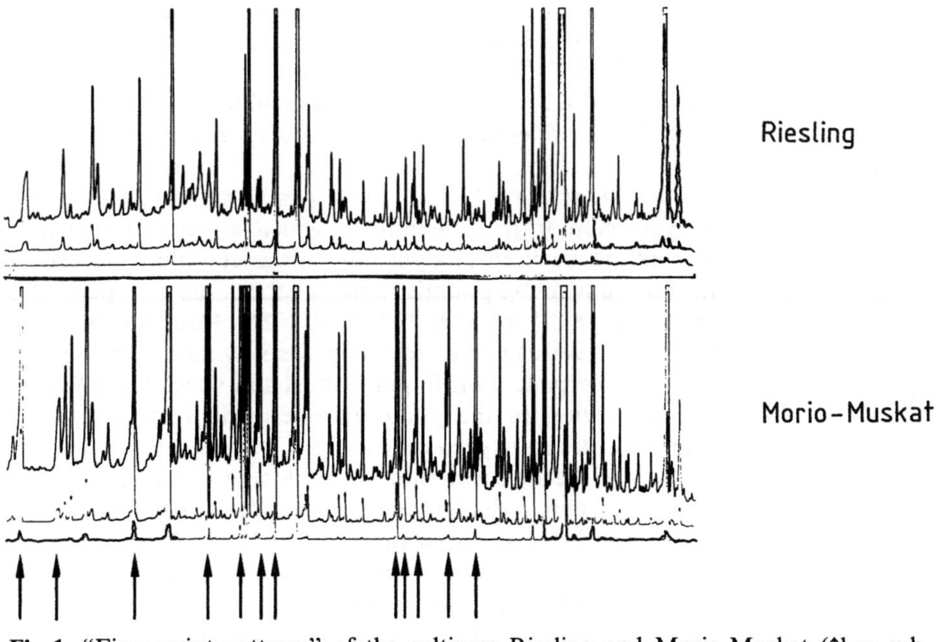

Fig. 1. "Fingerprint patterns" of the cultivars Riesling and Morio-Muskat (↑key substances)

Table 2. Determination of characteristic components (multiple regression). (Noble 1978)

Aroma attribute	GC Peak	Compound	Multiple correlation coefficient (R)	Significance
Fruit	22	i-Amylacetate	0.765	11.3*
	48	Not identified	0.965	47.9***
Varietal character	26	Not identified	0.736	9.4**
	49	Not identified	0.898	14.6**
	42	Not identified	0.950	18.4**

acter can be analytically ascertained by the content of certain of these "key substances" (Rapp et al. 1985c, Rapp and Güntert 1985). Noble (1978) discovered correlations between certain analytically determined components and sensory wine evaluation. In the case of Cabernet Sauvignon wines from the vintages 1970–1973, a significant correlation was ascertained between the components 26, 49, 42 (not identified), and sensory evaluation of "varietal character" (Table 2). The component 48 was found to be closely related to the fruity character of Cabernet wines. Marais et al. (1979) also ascertained a relationship between gas chromatographic results and wine quality. They found in the case of the variety Pinotage that it is possible to predict the quality of the wine from the content of n-hexyl acetate and ethyl octanoate.

Esters (mainly acetates) are to be found only in limited quantities in grapes. However, in the case of the American varieties *V. labrusca* and *V. rotundifolia,* they are characteristic of the variety. The importance of methyl anthranillate in the aroma of *V. labrusca* (Delaware, Niagara) has been reported by several authors (Neudoerffer et al. 1965; Stern et al. 1967; Stevens et al. 1965).

Welch et al. (1982) found that hexanol, iso-amyl alcohol, benzaldehyde and 2-phenyl ethanol, as well as their derivates, are typical components for Muscadine grapes.

It has emerged from numerous studies that the terpenoid compounds form the axis for the sensory expression of the wine bouquet which is typical of its variety and that they can therefore be used analytically for varietal characterization. Apart from the hitherto known compounds (terpene ethers, monoterpene alcohols) Rapp and Knipser (1978); Rapp et al. (1984c, d, 1986); Rapp and Mandery (1986) and Williams et al. (1980a) identified numerous new monoterpene compounds, in particular monoterpene diols in grape must and wine. The monoterpene diols can form according to the photohydroperoxide synthesis scheme (Schenk 1957) by photo-sensitized oxygen transfer onto the trisubstituted double bond – in the 6-position – of the acyclic monoterpene alcohols (linalool, geraniol, nerol, citronellol) and subsequent reduction of the corresponding hydroperoxides to diols. Another formation involving an acid-catalysed H_2O addition onto the double bond – in the 6-position – of the monoterpene alcohols – yields the following monoterpene diols: hydroxylinalool, hydroxynerol, hydroxygeraniol and hydroxycitronellol. At present about 50 monoterpene compounds are known of which the most important are presented in Fig. 2.

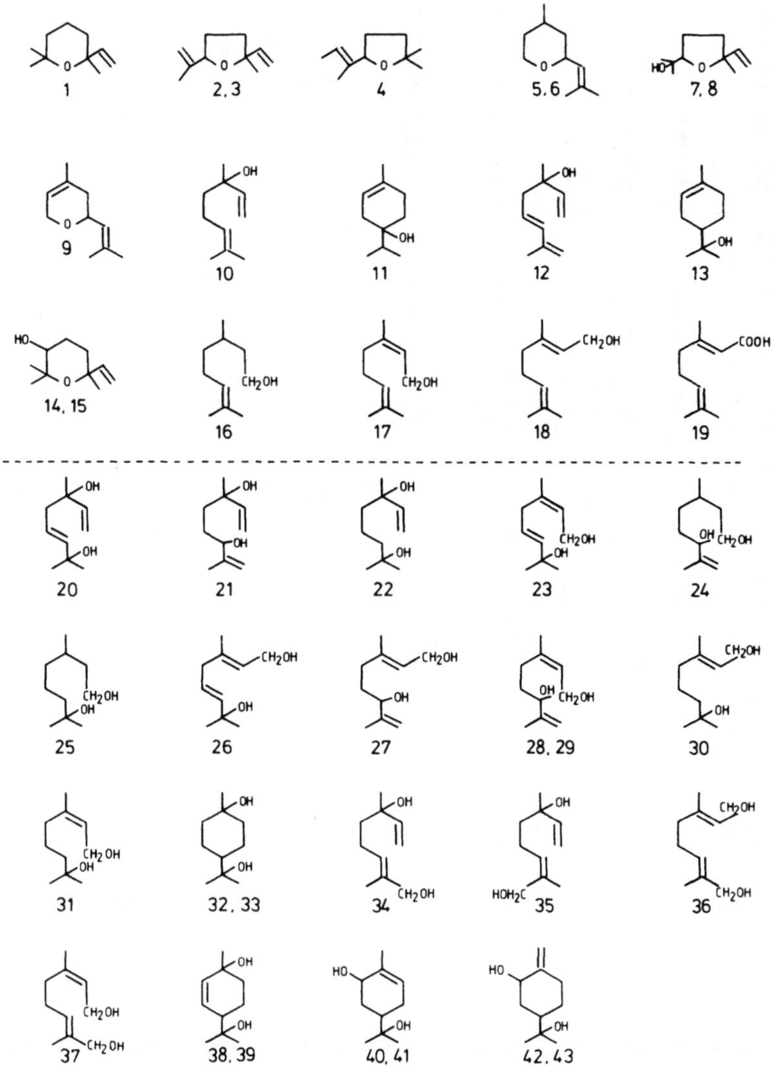

Fig. 2. Volatile monoterpene compounds in must and wine

The dominating monoterpene alcohols, particularly for muscat varieties, are linalool (10), geraniol (18), nerol (17), citronellol (16), α-terpineol (13). In the case of Koshu, an indigenous Japanese variety, terpinene-4-ol was identified as the dominating monoterpene compound (Shimizu and Watanabe 1981). The flavour threshold of nerol and alpha-terpineol is three to four times higher than that of linalool (100 µg l^{-1}). The linalool oxides (5, 6, 7, 8) have flavor thresholds of 3000 to 5000 µg l^{-1}.

The amounts of wine aroma components can be influenced by various factors, amongst others, the environment (climate, soil), grape variety, the degree of ripeness, fermentation conditions (pH, temperature, yeast flora), wine production

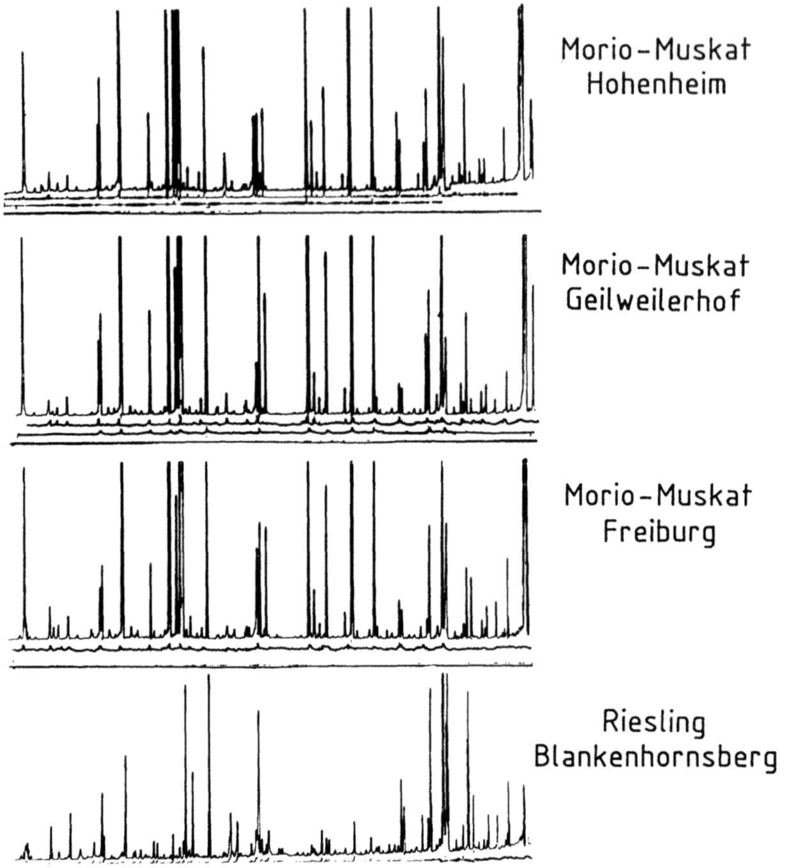

Fig. 3. "Fingerprint patterns" of the varieties Morio-Muskat (different growing areas in Germany) and Riesling

(enological methods, treatment substances) and aging (bottle maturation) of the wine.

The characteristic varietal composition of the monoterpene compounds in the various grape varieties (monoterpene patterns) is only marginally influenced by the *growing area* (Fig. 3). The "fingerprint patterns" of Morio-Muskat samples of different growing areas in Germany show a strong similarity in prominent characteristic components (mainly monoterpene compounds) (Rapp and Hastrich 1978; Rapp 1984; Rapp et al. 1985c).

The fungus *Botrytis cinerea* is responsible for the rotting of grapes. Under special climatic conditions it can cause noble rot, which is a prerequisite for the production of botrytized wines having a distinct aroma. Only a few papers deal with the volatiles produced by this fungus. Boidron (1978) observed decreases in monoterpenes of *Botrytis cinerea*-infected Muscat varieties. Shimizu et al. (1982) report that *B. cinerea* does not produce terpenoids in grapes without terpenes, but transforms linalool, which has been added to grape must into some other monoterpenes. Apart from quantitative differences in the aroma composition of botrytized wines compared to normal wines, Masuda et al. (1984) found two compounds to be responsible for the flavour of botrytized wines. They are ethyl-9-hydroxynonanoate and 4,5-dimethyl-3-hydroxy-2(5H)-furanone (sotolone). The latter compound has a sweet, sugar- and caramel-like aroma and its threshold value is 2.5 ppb. The sotolone content in botrytized wine was about 5–20 ppb, in normal wines below 1 ppb (Masuda 1984). This compound is also reported to contribute to the flavour of sugar molasses, aged rice wine and flor sherry (Masuda 1984). Rapp and Mandery (1986) observed a significantly higher amount of 1-octen-3-ol in *Botrytis*-infected grapes and wines. This alcohol is called "mushroom alcohol" on account of its typical odor.

Rapp et al. (1986) found that the monoterpene alcohols are changed by *Botrytis cinerea* (Fig. 3 a) by:

- addition of a hydroxy-group in Δ^1-double-bond of α-terpineol, six new diastereomeric compounds are produced
- addition of a hydroxy-group in Δ^6-double-bond of the acyclic monoterpene alcohols (linalool, citronellol, geraniol, nerol), two monoterpenediols are produced of each monoterpene alcohol
- hydroxylation in the C-8-position of the acyclic monoterpene alcohols, a new group of monoterpene alcohols are produced, e.g. from linalool resulted (E)-8-hydroxylinalool and (Z)-8-hydroxylinalool [in a *Botrytis* culture 95 : 5 (E/Z)]. In must and wine Rapp et al. (1986) found (E)- and (Z)-(8)-hydroxylinalool in the ratio of about 3 – 1 (E/Z).

In the case of increased *skin contact,* the content of caproic, caprylic and capric acid, their ethyl esters, as well as the acetic acid esters of the higher alcohols (acetates) decreased. A contrasting behavior is shown by the detectable N-acetamides in wine [N-(2-methylbutyl)-acetamide, N-(2-phenyl ethyl) acetamide, N-i-butyl acetamide, N-(3-methyl butyl)-acetamide, N-methionyl-acetamide]. The concentrations of these compounds increase with augmented skin contact time (Rapp 1985 b; Güntert 1984; Rieth 1984).

On the other hand, skin contact time did not significantly change the contents of monoterpene compounds, which are partly glycosidically bound in the grape berry (Bayonove et al. 1984; Günata 1984; Williams et al. 1982 a, b). The release of the glycosidically bound terpenes by means of natural β-glucosidases in the berry, as can be observed in the case of a pH change in the must (to pH 5.0), (Bayonove et al. 1984; Günata 1984; Williams et al. 1982 a, b), does not occur in the mash.

The essential part of the wine flavor is formed during the *alcoholic fermentation.* Figure 4 shows the changes caused by the yeast during fermentation and the change caused in the flavor composition between a grape must (Fig. 4 above) and the resultant wine (Fig. 4 below) (Rapp et al. 1984 a; Mandery 1983; Rapp and Mandery 1986). Apart from ethanol and glycerin, as well as diols and higher alcohols (2-methyl-1-propanol, 3-methyl-1-butanol and 2-methyl-1-butanol), numerous other wine constituents are formed by yeast metabolism (especially acids, esters, aldehydes, ketones, and S-compounds). These components can contribute

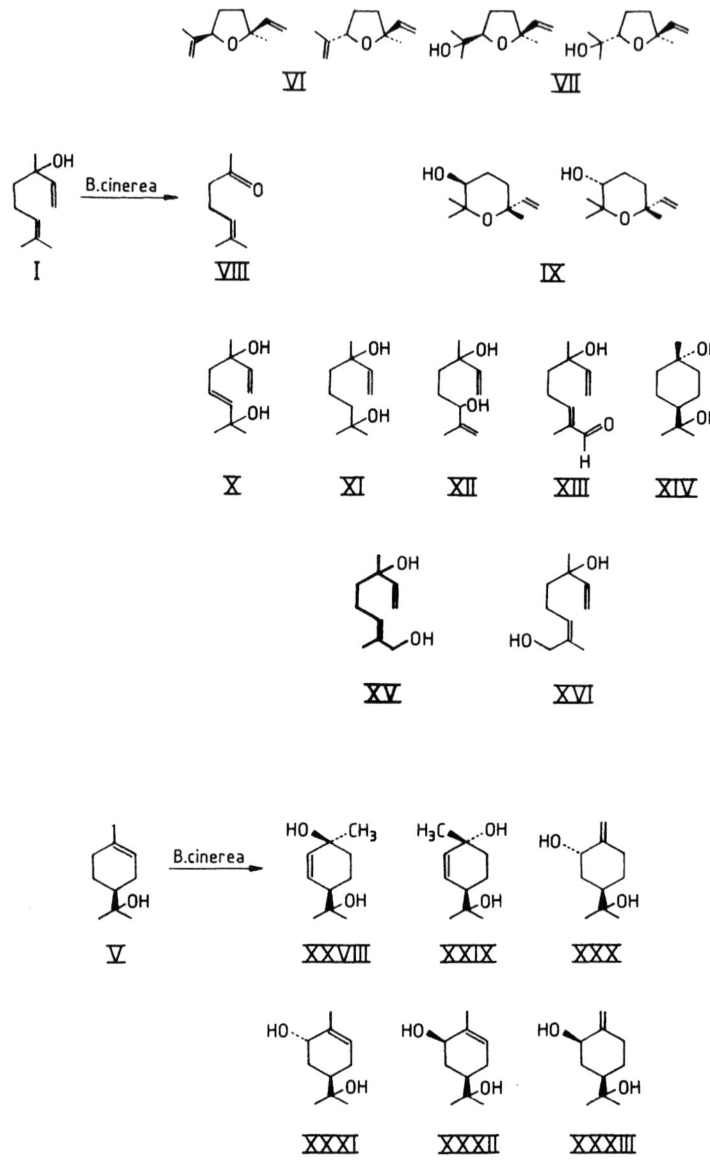

Fig. 3. Monoterpene compounds produced by *Botrytis cinerea* from linalool (*I*) and α-terpineol (*V*)

to the evaluation of optimal wine technology, but are, however, not suitable for a varietal characterization.

The *higher alcohols*, fatty acids and esters are the most important groups of the yeast-synthesized aroma substances of the fermentation bouquet, whereby the fusel alcohols quantitatively predominate and the esters qualitatively. The formation of 1-propanol, 2-methyl-1-propanol, 3-methyl-1-butanol and 2-methyl-1-bu-

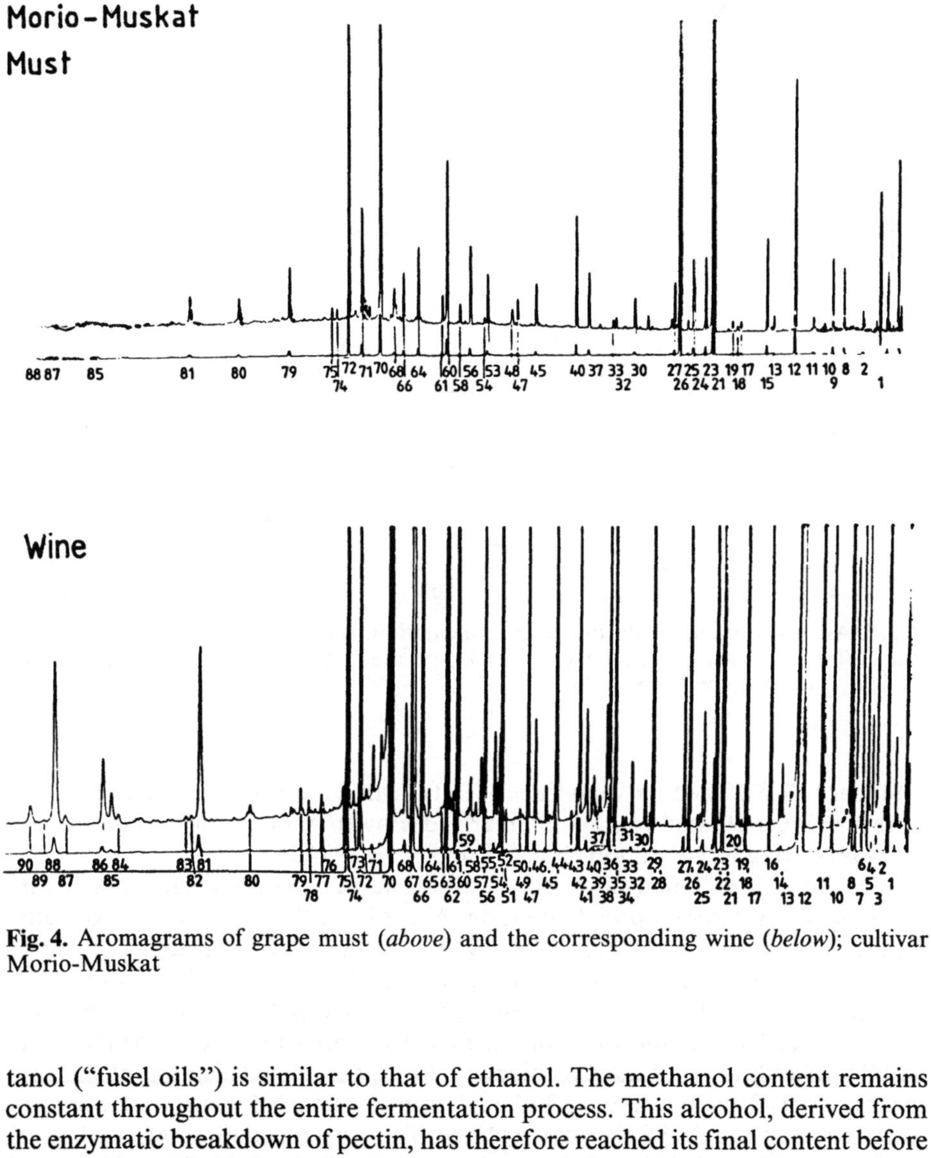

Fig. 4. Aromagrams of grape must (*above*) and the corresponding wine (*below*); cultivar Morio-Muskat

tanol ("fusel oils") is similar to that of ethanol. The methanol content remains constant throughout the entire fermentation process. This alcohol, derived from the enzymatic breakdown of pectin, has therefore reached its final content before fermentation. The behaviour of the unsaturated C_5 and C_6-alcohols is similar to methanol (Fig. 5; Rapp et al. 1984 a, Rapp and Mandery 1986).

Ethyl esters of straight-chain fatty acids and *acetates* of higher alcohols are the dominating esters in wine and they are formed during the alcoholic fermentation. The yeast synthesizes the acetates in the same way as the fatty acids: the fatty acids originate from acetyl-coA-derivate following hydrolysis with water and the splitting of coA-SH through alcohols leads to the formation of esters. The fatty acids are formed in large quantities during fermentation at an earlier stage than the respective ethyl esters. Only when substantial amounts of ethanol have been formed do the fatty acid ethyl esters increase (Fig. 6, Rapp et al. 1984 a;

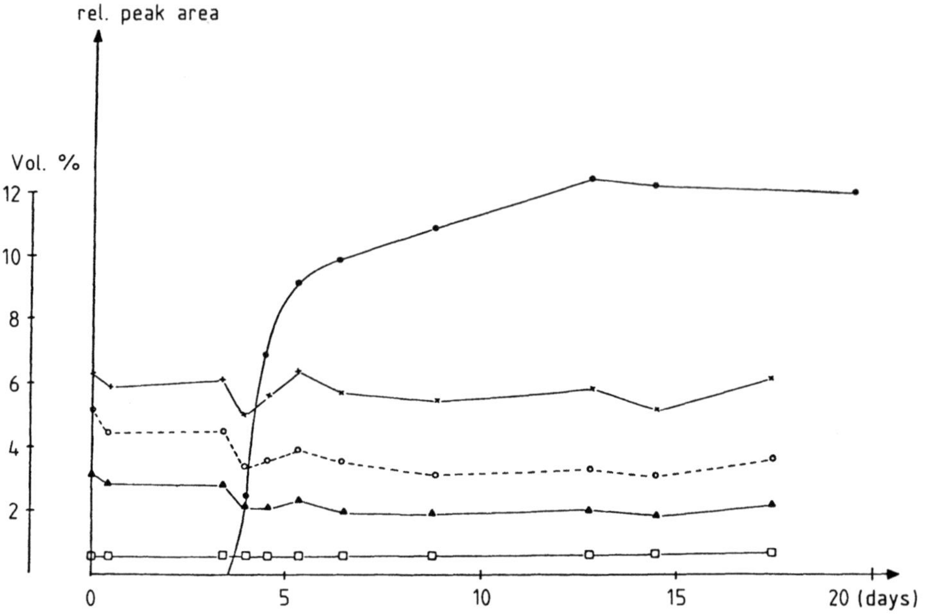

Fig. 5. Aroma compounds during yeast fermentation. ● ethanol (vol.%); + trans 3-hexenol-1 (rel. peak area); ○ cis 3-hexenol-1 (rel. peak area); ▲ cis 2-hexenol-1 (rel. peak area); □ cis 2-pentenol-1 (rel. peak area)

Mandery 1983). The final quantities of fatty acids have already been achieved before maximum ethanol content. Similarly, during ethanol production the acetates increase during fermentation (Fig. 6, Rapp et al. 1984a).

Due to their extremely low odor threshold values, natural volatile organic *sulfur compounds* deserve special attention. Apart from thioethers such as dimethyl sulfide and diethyl sulfide (Schreier 1979), a group of less volatile sulfur compounds was discovered in certain wines also (Schreier et al. 1974; Muller et al. 1971). The highest content of these compounds was the thioether-alcohol 3-(methylthio)-1-propanol (methionol), first discovered in wine by Muller et al. (1971). In addition to 2-methyl-thiophen-3-one, the related alcohols of this thiophenone (cis-2-methyl-thiolan-3-ol and trans-2-methyl-thiolan-3-ol) were also detected (Rapp et al. 1984b; Güntert 1984). The yeast metabolism plays a key role in the formation of these substances in wine. The formation of several other S-compounds during the fermentation is similar to that of ethanol. These are, together with methionol, N-[3-(methylthio)-propyl]-acetamide and 2-methyl-thiolan-3-ol.

Terpene compounds as a group form an important part of the grape bouquet; they belong to the secondary plant constituents, of which the biosynthesis begins with acetyl-coA. Microorganisms are also able to synthesize terpene compounds (Hock et al. 1984), but the formation of terpenes by *Saccharomyces cerevisiae* has not yet been observed. Figure 7 shows the development of several terpene compounds (terpene oxides) during fermentation. These compounds are not changed

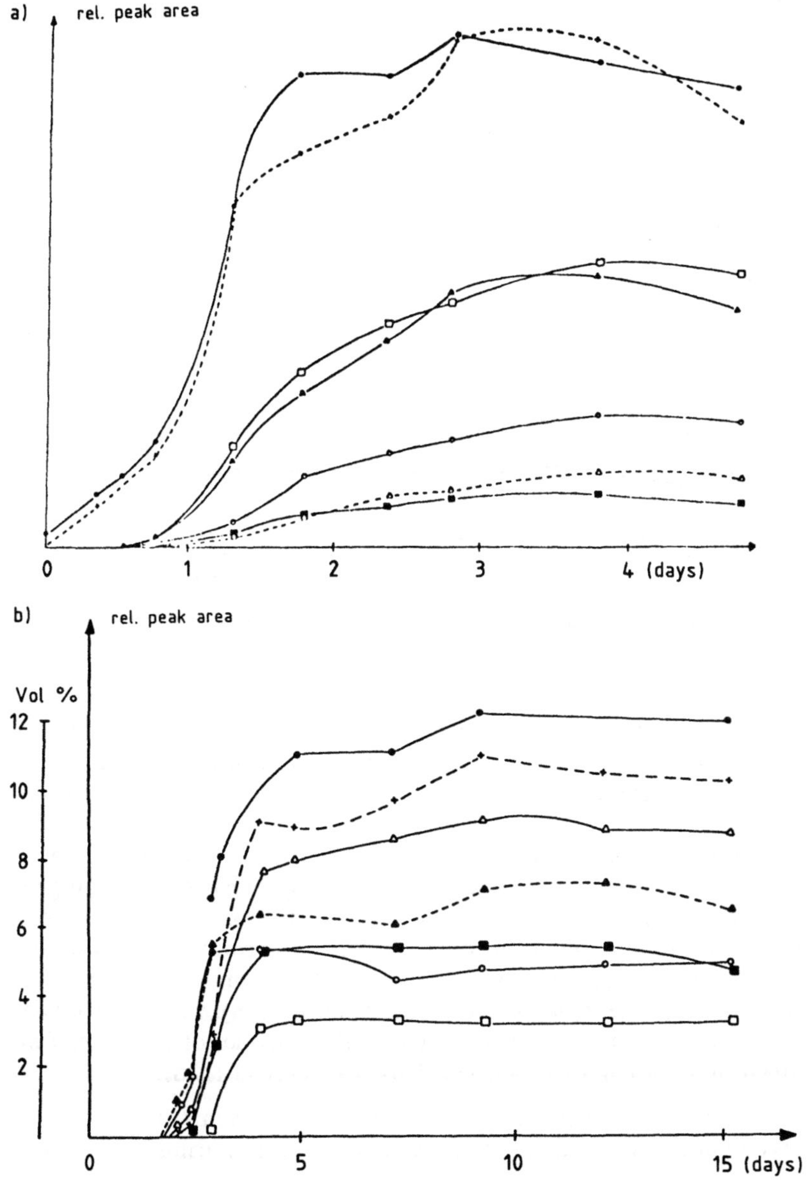

Fig. 6 a, b. Formation of aroma compounds during yeast fermentation. **a** △ butanoic acid; + hexanoic acid; ● octanoic acid; ○ ethyl butyrate; □ ethyl hexanoate; ▲ ethyl octanoate; ■ ethyl decanoate; **b** ● ethanol (vol.%); + isoamyl acetate (rel. peak area); ○ isobutyl acetate (rel. peak area); ▲ hexyl acetate (rel. peak area), ○ 1,3-propandiol monoacetate (rel. peak area); ■ 2-phenylethyl acetate (rel. peak area); □ propyl acetate (rel. peak area)

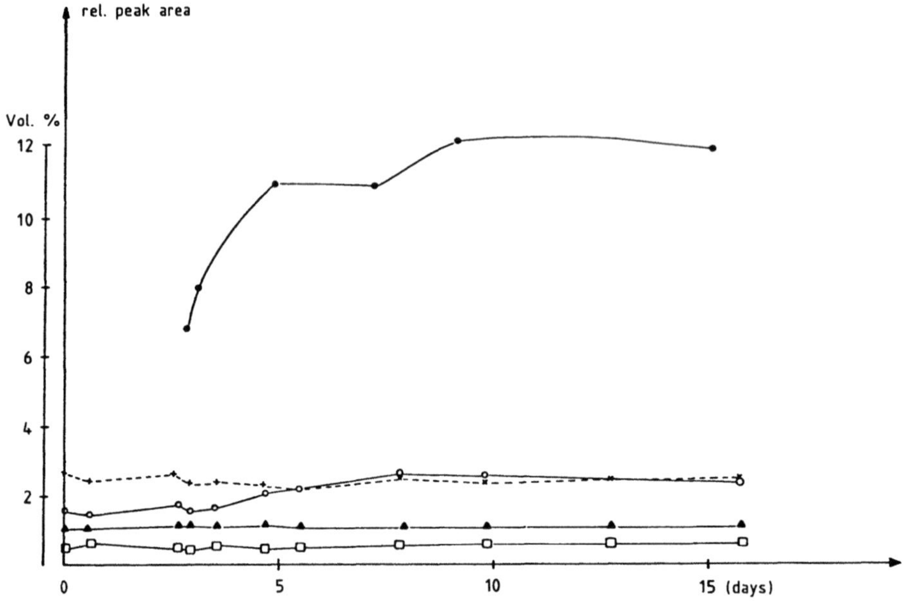

Fig. 7. Monoterpene compounds during yeast fermentation. ● ethanol (vol.%); + cis rose oxide; ○ trans linalool oxide (f); ▲ trans linalool oxide (p); □ cis linalool oxide (f)

by the yeast metabolism during fermentation (Rapp et al. 1984 a; Mandery 1983). The monoterpene compounds as typical primary aroma components are therefore suitable for the varietal characterization of wines made from different grape varieties.

Most of the volatile *phenol compounds* discovered in wine are not present in grape berries. They are formed by yeasts, bacteria or by the hydrolysis of higher phenols. 4-vinyl guaiacol and 4-vinyl phenol can be formed from cinnamic acid, p-cumaric acid or ferulic acid by means of enzymatic or thermal decarboxylation (Albagnac 1975). Volatile phenols have a definite influence on wine flavor.

During the *maturation or aging* of wine (bottle maturation), various processes influence the volatile substances and thereby also the bouquet of the wine. Two types of maturation bouquet are described (Ribereau-Gayon 1978):

- the oxidative bouquet, determined by acetaldehyde and acetals, and
- the reductive bouquet, formed under reductive conditions during bottle maturation.

In the case of wines, the bouquet is further influenced by the maturation in wooden barrels and by the components derived from the wood (e.g., vanillin, eugenol).

Marais (1979) found a highly significant relationship in South African wines between the dimethyl sulfide concentration and sensory evaluation of bottle bouquet, underlining the importance of dimethyl sulfide for the character of the typical maturation bouquet. The formation is temperature-dependant and takes place in different white wines at a varying tempo (Table 3).

Table 3. The influence of maturation time and temperature on the maturation bouquet of wines made from various grape varieties. (Marais 1979)

Storage temperature	Dimethyl sulfide ($\mu g\, l^{-1}$)				
	Control	After 1 week	3 weeks	7 weeks	16 weeks
Chenin blanc					
10°	0	0	0	0	0
20°	0	0	0	1.7	6.9
30°	0	1.9	4.6	12.3	25.7
Riesling					
10°	0	0	0	0	0
20°	0	0	2.7	9.9	29.0
30°	0	5.6	11.3	47.9	85.8

Table 4. Changes in the acetate content during bottle maturation (grape variety: Riesling). Relative peak heights (mm)

Vintage	1982	1981	1980	1979	1978	1977	1975	1974	1973	1964
i-Amyl acetate	107	211	120	33.6	58.4	13.5	8.1	14.3	5.9	10.9
2-Phenylethyl acetate	38.7	77.5	42.3	17.2	25.1	7.0	3.6	5.3	1.9	5.7

Rapp et al. (1985a), Rapp and Güntert (1986) noted that numerous changes take place in the content of single aroma substances during bottle maturation of wine and that these can be essentially divided into four aspects:

- a change in ester content
  decrease in acetates
  increase in mono-and dicarboxyllic acid ethyl esters
- formation of substances from carotene breakdown
- formation of substances from carbohydrate breakdown
- acid-catalyzed reactions of monoterpene compounds

In the first few years after the wine has been made, the content of the *acetates* (acetic acid esters) decreases continuously (to maintain chemical equilibrium: law of mass action) until after ca. 4–6 years, constant values are achieved (Rapp et al. 1985a, Rapp and Güntert 1986).

This clear decrease in the acetate content (Table 4) is correlated to the decrease in the freshness and fruitiness of matured wines. Analytical data also unequivocally support the organoleptically detectable decrease in the fruitiness of the wines.

The *fatty acid ethyl esters* and other mono- and dicarboxyllic acids display a totally different behavior. The amount of these compounds increases significantly during aging (Table 5; Rapp et al. 1985a, Güntert 1984).

Table 5. Changes in diethylsuccinate during bottle maturation (grape variety: Riesling). Relative peak heights (mm)

Vintage	1982	1981	1980	1979	1978	1977	1975	1974	1973	1964
Diethyl succinate	41	66	104	266	384	256	476	400	656	738

The results which have been quoted regarding the behavior of acetates, fatty acid esters, as well as dicarboxylic acid esters as a part of the bottle maturation process in wine, can be attributed to chemical equilibrium adjustment (law of mass action) in an aqueous acid ethanol medium. The acetates are enzymatically formed during fermentation from acetyl co-A and higher alcohols present in the reaction medium. After fermentation, acetates are hydrolyzed to reach an equilibrium – by means of the equilibrium concentrations of acetic acid, acetates and alcohols later found in wine – after approximately 6 years.

The increase in the concentrations of the ethyl esters, on the contrary, can be attributed to a chemical esterification. This begins after completion of fermentation, during which the said compounds are only formed in limited concentrations from the enzyme complex of the yeasts and in the case of certain compounds is not yet complete even after 10 years (as the example of diethyl succinate clearly shows; Table 5).

The biogenesis of the isomers of vitispirane, 1,1,6-trimethyl-1,2-dihydronaphtalene (TDN), damascenone and dihydroactinidiolide, is generally due to *carotene metabolism* during the aging of wine (Simpson 1979b; Ohloff 1978). Williams et al. (1982a, b) suggest other precursors and have reported on glycosides as possible precursors for vitispirane and damascenone in wine grapes.

The formation of TDN is clearly correlated to the organoleptic change in wine during aging and is responsible for a pronounced kerosene or petrol note (Table 6).

Of all the compounds formed during the aging of wine as a result of *carbohydrate degradation,* 2-furfural and furan-2-carboxylic acid ethylester could already be identified in young wines. With aging, the amounts of these components increase significantly. In addition to this, certain other compounds can be detected after further years of storage, such as 2-acetyl furan, 2-hydroxymethyl-furan (furfuryl alcohol), 2-ethoxymethyl-5-furfural and 2-hydroxymethyl-5-furfural. The development of these components, which cause a caramel-like taste reminiscent of burnt sugar, can be explained by dehydration reactions from sugar in an acid medium. These can take place during maturation over a period of several years

Table 6. Correlation between vintage and petrol note. Relative peak heights (mm)

Vintage	1982	1981	1980	1979	1978	1977	1975	1974	1973	1964
1,1,6-TDN	–	–	0.4	0.8	0.5	1.3	1.7	0.5	3.0	3.4

Table 7. Correlation between vintage and furfural content. Relative peak heights (mm)

Vintage	1982	1981	1980	1979	1978	1977	1975	1974	1973	1964
2-Furfural	4	6	6	9	14	34	33	26	39	44

as is indicated by an 11-fold increase of 2-furfural in a 1964 Riesling compared to a 1982 Riesling (Rapp et al. 1985a, Rapp and Güntert 1986 (Table 7).

The content of various *monoterpene* components changes during bottle storage or rather during the maturation of wine by means of acid-catalyzed reactions (Rapp et al. 1985a; Güntert 1984; Rapp and Güntert 1986). In the case of several components (e.g., linalool, geraniol, hotrienol and isomers of linalool oxide), an obvious decrease in concentration can be ascertained during the course of aging. In addition to this, compounds are formed which are not present in young wines: amongst others the formation of cis- and trans 1,8-terpine. During the maturation of wine, monoterpene alcohols can be formed from linalool, as indicated in the reaction diagram (Fig. 8). According to this scheme, linalool is transformed in an aqueous acid medium to α-terpineol by cyclization, to hydroxy linalool through hydration in the C-7 position, and to geraniol and nerol by a nucleophilic 1,3-transition (allyl).

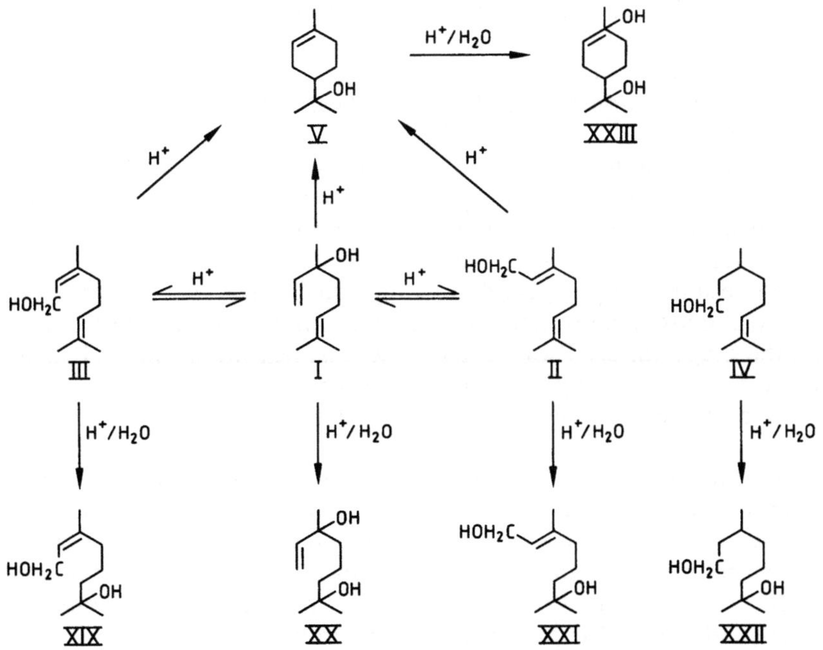

Fig. 8. Proposed reaction scheme of monoterpene alcohols during bottle aging of white wine. *III* nerol; *I* linalool; *II* geraniol; *IV* citronellol; *V* α-terpineol; *XXIII* cis-1,8-terpine; *XIX* hydroxynerol; *XX* hydroxylinalool; *XXI* hydroxygeraniol; *XXII* hydroxycitronellol

Table 8. Correlation between vintage and terpine content. Relative peak heights (mm)

Vintage	1982	1981	1980	1979	1978	1977	1975	1974	1973	1964
Cis-1,8-terpine	–	–	0,4	0,8	0,5	1,3	1,7	0,5	3,0	3,4

1,8-terpine is a typical component of matured wines (Table 8). Up till now this compound was only detected in musts of the grape variety Muscat of Alexandria, which had been heat-treated or to which acid additions had been made (Williams et al. 1980a). The isomers of 1,8-terpine were also identified in rectified grape juice concentrate (Rapp et al. 1983a; Mandery 1983).

These variations in the content of certain terpene compounds, as well as the formation of new components (e.g. 1,8-terpine), provide an explanation for the detectable organoleptic change which can be observed during bottle maturation and the subsequent loss of the wine bouquet typical of grape variety. After all, linalool displays a totally different aroma note and taste intensity compared with 1,8-terpine.

These results indicate that numerous monoterpene components are suitable for the varietal characterisation of wines (at a maximum age of 4–5 years).

Utilizing only 12 monoterpene compounds (e.g., linalool, trans-linalooloxide (f), cis-linalooloxide (p), nerol, geraniol, 3,7-dimethyl-octa-1,5-dien-3,7-diol, Rapp and Mandery (1986), Rapp et al. (1982, 1983b, 1984d), Rapp (1984a, b) succeeded in classifying the grape varieties into various aroma types (Fig. 9). In the terpene patterns ("terpene profiles"), clear differences exist between the grape varieties with a muscat-related aroma, ("Muscat type"; Fig. 9 above: e.g., Muskateller, Morio-Muskat, Schönburger, Würzer), and varieties with a fruity Riesling-related aroma, ("Riesling type"; Fig. 9 below: e.g., Riesling, Müller-Thurgau, Kerner, Scheurebe), and then grape varieties with a neutral bouquet ("Sylvaner" or "Pinot blanc type"; not shown in figure). An accurate classification of wines into distinct sensory detectable wine types can therefore be effected analytically, only by the content of 12 components (monoterpene compounds).

From these results, Rapp et al. (1985c, Rapp and Güntert 1986) in further research were able to attain a pronounced analytical differentiation between wines made from the grape variety Riesling and wines just bearing the name Riesling and not made from this variety. Also on taste it is possible to differentiate analytically between varieties by using only a few, but important, aroma components.

All analyzed wines from the grape variety Riesling (Weisser Riesling) possess a similar terpene pattern ("terpene profile") which is displayed by the monoterpene content with normal degrees of variation (caused by varying maturation and aging), but in a composition which is typical of this variety. With the aid of these terpene profiles, a significant analytical differentiation could be ascertained between Riesling and Welsch Riesling from different growing regions (Austria, Italy and Yugoslavia) (Fig. 10; Rapp et al. 1985c).

In all the Riesling wines, the selected monoterpene compounds [e.g., linalool, trans-linalooloxide (f), α-terpineol, terpineol I] were present in a ten-to 50-fold

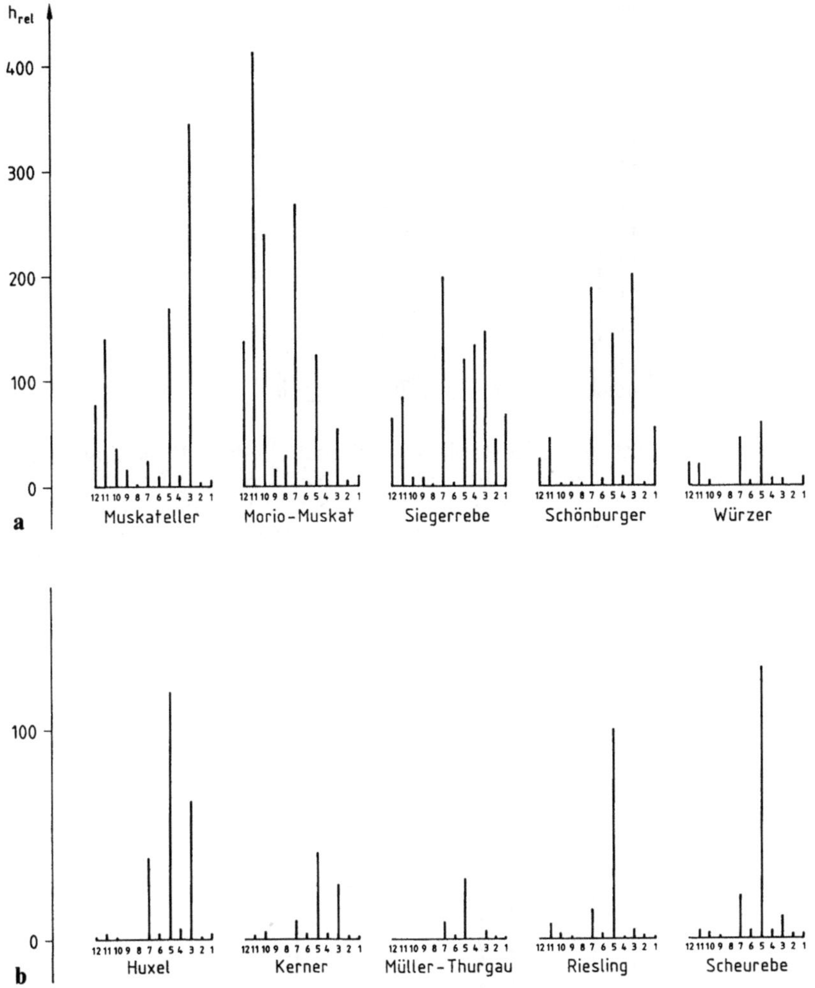

Fig. 9a, b. "Terpene profiles" of Muscat type (**a**) and Riesling type (**b**) varieties. *1* trans linalool oxide (f); *2* cis linalool oxide (f); *3* linalool; *4* hotrienol; *5* trans linalool oxide (p); *6* cis linalool oxide (p); *7* 3,7-dimethylocta-1,5-diene-3,7-diol; *8* nerol oxide; *9* citronellol; *10* nerol; *11* geraniol; *12* (E)-geranoic acid

higher concentration than in the wines made from Welsch Riesling. In comparison with Riesling (White Riesling), significant differences also exist in the "terpene profiles" of the Hunter Valley Riesling (Australia) and the Emerald Riesling (California), which makes a definite analytical differentiation also in this case possible (Fig. 11; Rapp et al. 1965c).

Using an example of South African wines, it was also demonstrated that a distinct differentiation between various Riesling wines is possible using the terpene profiles typical of the variety: e.g., using Weisser Riesling (WR) and Cape Riesling [also sometimes classified as Riesling or Paarl Riesling (R)] (Fig. 12). The ter-

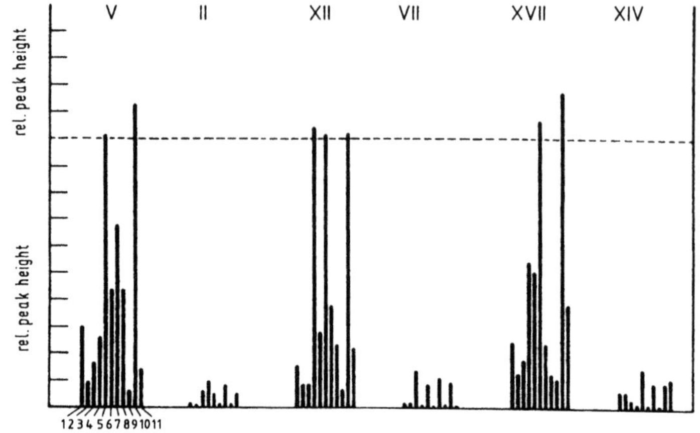

Fig. 10. "Terpene profiles" of the varieties Riesling and Welsch Riesling. *V* Riesling/Österreich; *XII* Riesling/Südtirol; *XVII* Riesling/Jugoslawien; *II* Welsch Riesling/Österreich; *VII* Welsch Riesling/Südtirol (=Riesling Italico); *XIV* Welsch Riesling/Jugosl. (=Laski Riesling); *1* trans linalool oxide (f); *2* cis linalool oxide (f); *3* nerol oxide; *4* linalool; *5* hotrienol; *6* α-terpineol; *8* trans linalool oxide (p); *9* cis linalool oxide (p); *10* terpendiol I; *11* hydroxylinalool

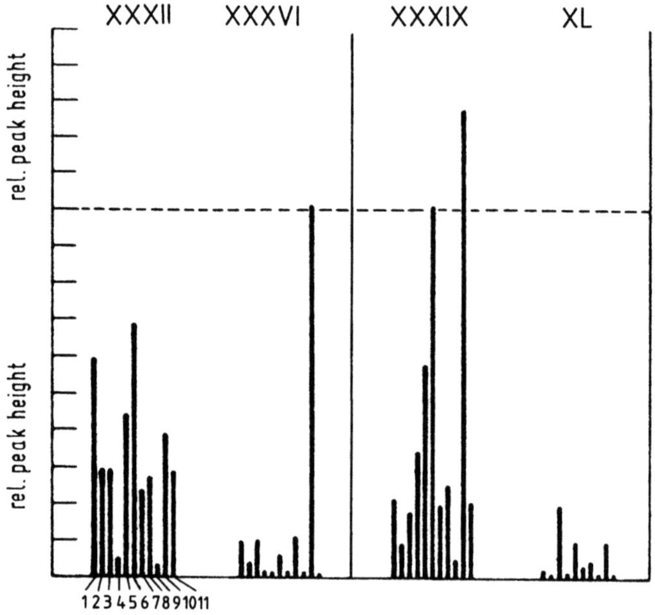

Fig. 11. "Terpene profiles" of the varieties Riesling, "Hunter Valley Riesling", and Emerald Riesling. *XXXII* Rhein Riesling Barossa/Australia; *XXXVI* Hunter Valley Riesling/Australia; *XXXIX* Johannisberg Riesling: Sonoma/California/USA; *XL* Emerald Riesling: Fresno/Calif./USA. *1* trans-linalooloxide (f); *2* cis-linalooloxide (f); *3* neroloxide; *4* linalool; *5* hotrienol; *6* α-terpineol; *7* unknown; *8* trans-linalooloxide (p); *9* cis-linalooloxide (p); *10* terpendiol I; *11* hydroxy-linalool

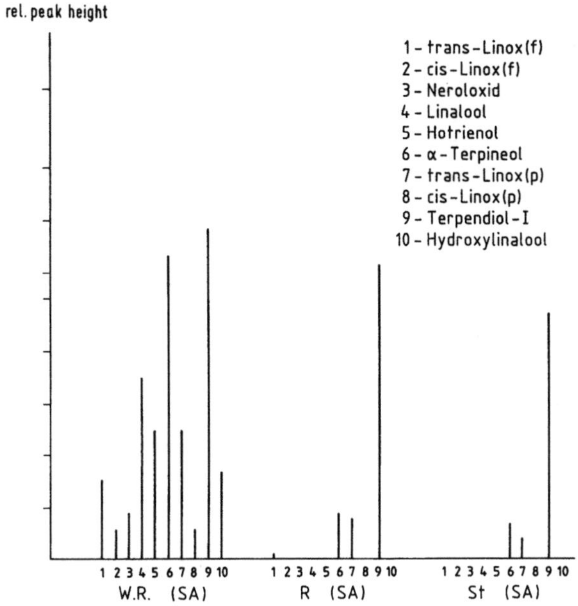

Fig. 12. "Terpene profiles" of the cultivars Riesling (*W. R.*), Paarl Riesling (*R*), and Chenin blanc (*St*)

pene profile of Cape Riesling wine (R), made from the variety with the synonym Riesling vert or Cruchen blanc, is very similar to that of wines made from the grape variety Chenin Blanc (St) (Rapp et al. 1985c; Güntert 1984).

Schreier et al. (1976d), using the quantitative distribution of certain monoterpenes (rose oxide, nerol oxide, geraniol, trans-linalool oxide, linalool, hotrienol, α-terpineol, cis-linalool oxide and nerol), were able to significantly separate the wines made from Morio-Muskat, Gewürztraminer, Scheurebe, and Ruländer from each other by using multiple discriminant analysis. The genetically closely related varieties Riesling, Scheurebe and Müller-Thurgau could not be distinguished by the use of these selected components.

Selecting further suitable monoterpene compounds (trans-linalool oxide (furanoid), cis-linalool oxide (furanoid), nerol oxide, linalool, hotrienol, α-terpineol, peak × (an ethoxy monoterpene ether), trans-linalool oxide (pyranoid), cis-linalool oxide (pyranoid), 3,7-dimethyl-1,5-octadien-3,7-diol and 3,7-dimethyl-1-octen-3,7-diol) and aided by a statistical computer program (linear discriminant analysis), Rapp and Güntert (1985) were able to achieve a significant analytical differentiation between the dominating white wine varieties in the Federal Republic of Germany (Fig. 13) viz. Riesling (1), Müller-Thurgau (2) and Sylvaner (3) (Rapp and Güntert 1985; Güntert 1984). The group centers of authentically pure varietal wines are clearly separated.

Gas chromatographic analysis followed by discriminant analysis of Riesling, Müller-Thurgau and Sylvaner trade wines resulted in approximately 60% of the samples tested being recognized for their grape variety. A complete classification of all wines is not yet possible as a result of the existing wine laws, whereby a varietal wine can contain up to 25% of other varieties through blending and the addition of sweet reserve.

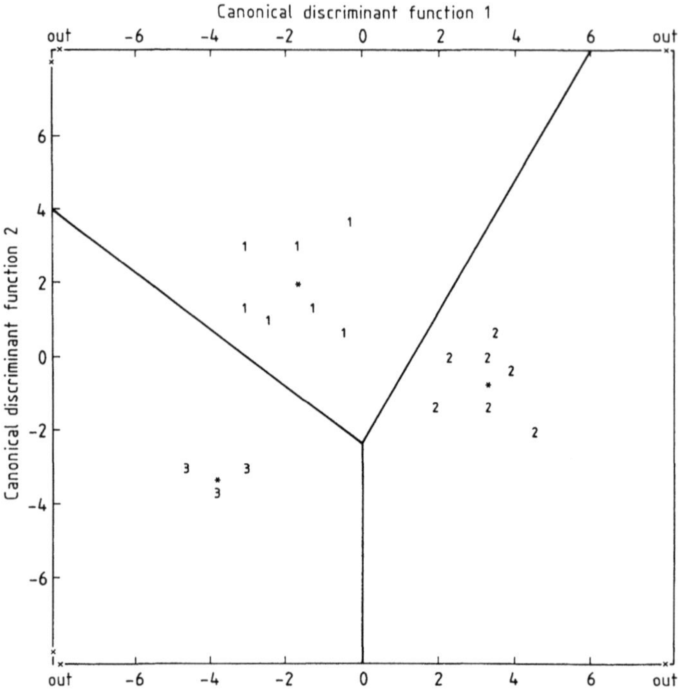

Fig. 13. Differentiation of the varieties Riesling (*1*), Müller-Thurgau (*2*), and Silvaner (*3*) by discriminant analysis

With more recent computer programs, it was possible to ascertain the importance of the individual components in the analytical differentiation of grape varieties. Rapp (1987) discovered that hydroxylinalool displays a high significance in differentiating between Riesling and Müller-Thurgau (Fig. 14). In fact, using only seven such significant components it is possible to achieve a clear differentiation between Riesling and Müller-Thurgau. In Fig. 15 the analytical differentiation between Riesling and Müller-Thurgau wines is outlined according to the Chernoff-Gesichter-Model (Rapp 1987). Both head types differ significantly from each other and differences can even be recognized within each wine type.

For wine and similar products of high quality, an important prerequisite for related marketing and commercial considerations is that the product be free of imperfections. Faults in wine detract from the flavor and taste enjoyment thereof. The causes are mostly of a chemical nature as opposed to wine sickness, which has a biological origin. In many cases a fault in wine is of a complex nature, which makes the diagnosis as well as the treatment exceptionally difficult. Should several faults exist simultaneously, it is then extremely difficult for the taster to make on accurate judgement.

In Fig. 16, several undesirable aroma and taste characteristics of wine *(wine faults, "off flavors")* are listed. The off-flavors listed in part A are varietal-specific aromatic flavors (e.g., strawberry flavor). Those in part B can be formed during the production of wine (e.g., "böckser", corkiness, mousiness).

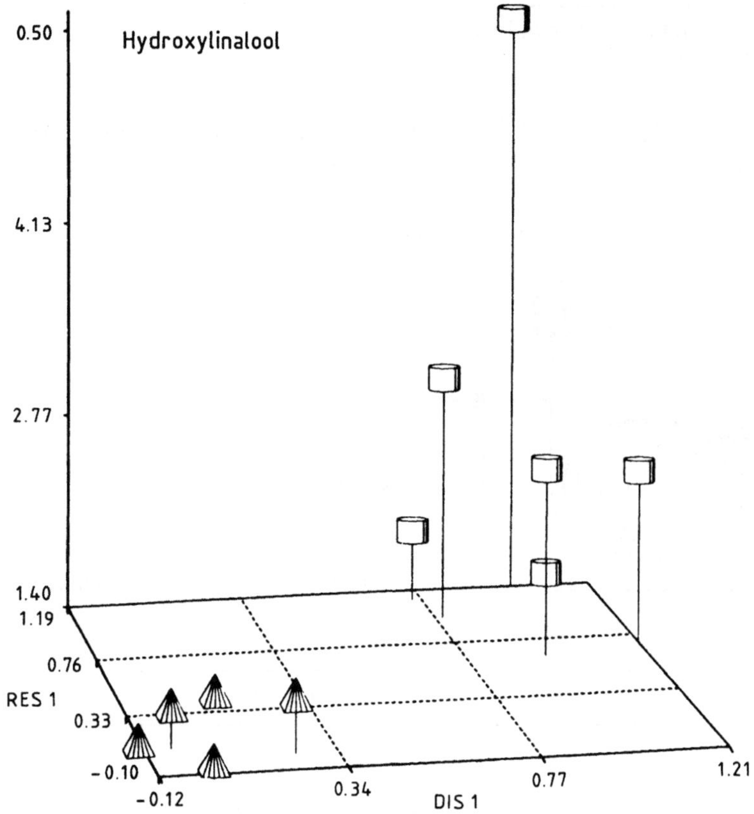

Fig. 14. Differentiation of the varieties Riesling and Müller-Thurgau by hydroxylinalool

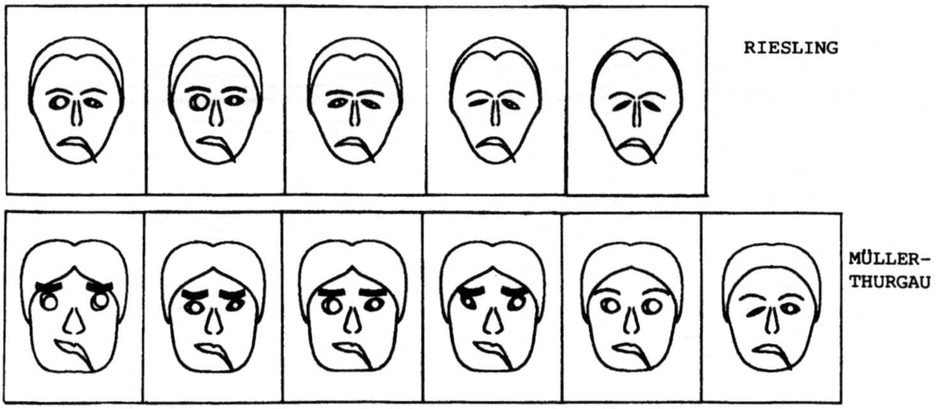

Fig. 15. Differentiation of the varieties Riesling and Müller-Thurgau by using only seven monoterpene compounds (Chernoff faces)

A	Strawberry, phenolic (medicine), black currant, green pepper green, grassy, bitter
B	Böckser, sauerkraut, corky, woody rotten eggs, geranium-like, mousy

Fig. 16. Foreign and undesirable odors in wine ("off-flavor")

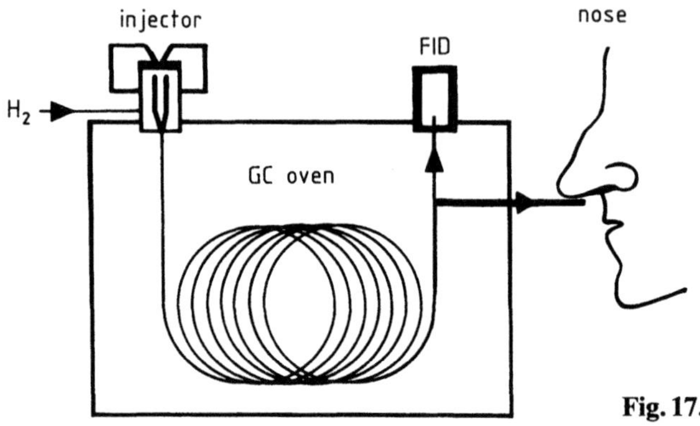

Fig. 17. Scheme of a "sniff-detector"

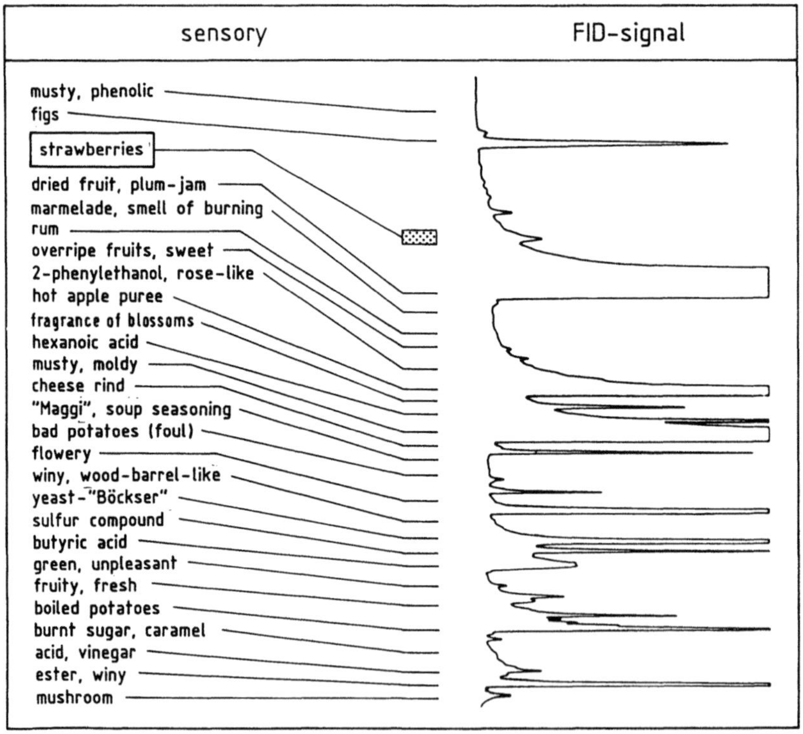

Fig. 18. Section of a "sniff"-chromatogram of an aroma extract of a grape variety with a typical strawberry-like flavour (cultivar: Pollux)

With the aid of modern methods of analysis, it is becoming increasingly possible to recognize the causes of the various wine faults. It is an important requirement to successfully eliminate or prevent wine faults. Thus, with the aid of the "sniffing technique", components causing undesired flavors can now be found. The separated components eluting from the effluent end of the gas chromatographic column are simultaneously detected by the detector (FID) and the nose ("sniff detector"), and the odor impression recorded on the chromatogram (Fig. 17; Rapp 1981; Rapp et al. 1980 b).

By means of this method, the components causing the undesired aromas can be specifically traced out of the complex aroma mixture and subsequently be identified by using gas chromatography-mass spectrometry. This method creates a clear relationship between instrumental analysis and organoleptic assessment of wines.

Rapp et al. (1980 b) were therefore able to isolate the undesired *strawberry-like flavor* (Fig. 18) from extracts of specific wine varieties where after it was identified by means of gas chromatography-mass spectrometry. The component concerned is 2,5-dimethyl-4-hydroxy-2,3-dihydro-3-furanone ("furaneol"), identified up till now only in wines made from certain interspecific hybrids. By means of gas chromatographic determination of this component, wines can be tested analytically for the presence of this undesired aroma. The gas chromatographic detection point for 2,5-dimethyl-4-hydroxy-2,3-dihydro-3-furanone (ca. 2 ppb) is lower than the sensory perception (50–100 ppb) (Fig. 19), thereby making an analytical evaluation of wine quality substantially more accurate and sensitive than the organoleptic assessment. As the amount of "furaneol" present in the grape variety Pollux is most vintages beneath the taste threshold (Fig. 19), the strawberry aroma can therefore not be detected. However, in the case of the grape variety Castor, this aroma can be detected organoleptically in most vintages.

Since the strawberry-like flavor is unknown in European varieties (*Vitis vinifera*), investigations have been carried out in order to identify the crossing partner with the component (2,5-dimethyl-4-hydroxy-2,3-dihydro-3-furanone) causing this aroma note in fungus resistant cultivars as Castor (B-7-2) or Pollux (B-6-18). Figure 20 shows the genetic descent of a Castor – a fungus-resistant variety by which the strawberry-like flavor can be detected sensorially in most vintages – together with sections of chromatogram containing 2,5-dimethyl-4-hydroxy-2,3-dihydro-3-furanone (marked peak). This compound is not present in the cultivars *Vitis riparia* (an American wild variety), Gamay noir (*V. vinifera*), Oberlin 595, or Forster's White Seedling (*V. vinifera*), the parents of the cultivar Castor (B-7-2) (Fig. 20). However, Rapp et al. (1980 b, 1983 b) could identify "furaneol" in the cv. Vi 5861 (Fig. 20), as well as in the American wild variety *Vitis labrusca* and its descendants (e.g., Niagara, Buffalo, Seneca). Based on these results, it can be suggested that the origin of the cv. Vi 5861 could not have come from self-fertilization of Oberlin 595 but must have originated from uncontrolled pollination of Oberlin 595 with pollen of *V. labrusca* varieties.

In the case of new grape crossings, especially those resistant to fungal disease, the gas chromatographic determination of "furaneol" is utilized for the early detection of new varieties free from undesired flavors. In this way, the tediously long course for the crossing to be released as a new variety is substantially shortened.

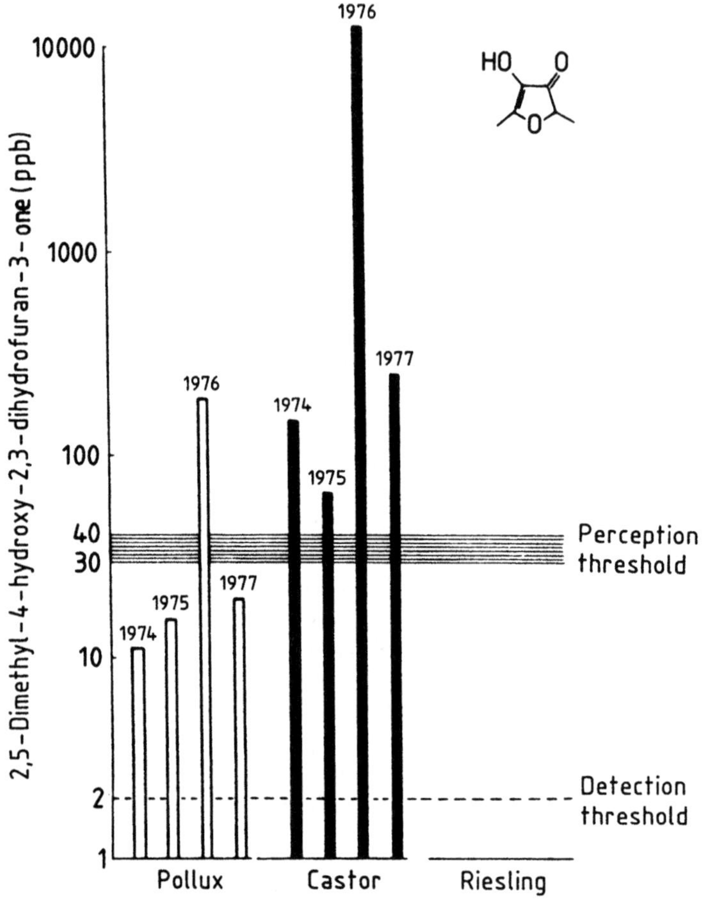

Fig. 19. Furaneol in wines of cultivars Pollux, Castor, and Riesling

The *foxy* or *hybrid* character of earlier hybrid varieties (crossings of American indigenous varieties with European *Vitis* varieties), and *Vitis labrusca* types is caused by anthranilic acid methyl ester (Power and Chesnut 1921; Fuleki 1972). This component was not yet detected in European varieties *(V. vinifera)*. More recent results show, however, that this component alone is not responsible for the foxy character (Nelson et al. 1977).

The cause of the *potato-related* or *herbaceous-grassy* character, also reminiscent of green peppers, which, apart from Cabernet wines, can also occur in white wines, was identified by means of gas chromatography-mass spectrometry as methoxy-pyrazines (3-ethyl-2-methoxypyrazine, 3-i-propyl-2-methoxypyrazine, 3-i-butyl-2-methoxypyrazine) (Bayonove et al. 1975; Augustyn et al. 1982). These compounds have very low threshold values. The 3-i-butyl-2-methoxypyrazine is a component which is very characteristic of wines made from Cabernet Sauvignon, but is undesirable in white wines.

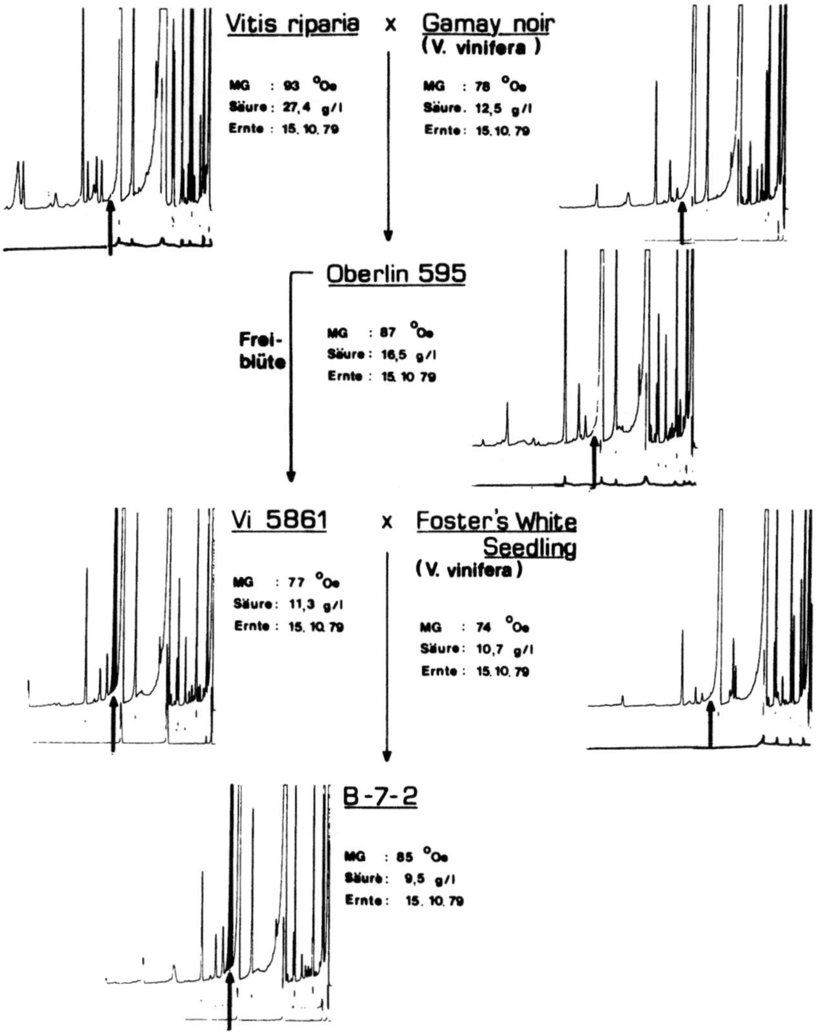

Fig. 20. Origin of the interspecific cv. B-7-2 (Castor). The retention time for "furaneol" is marked with an ↑ in each chromatogram

Despite great progress made in the field of wine technology, the unpleasant *musty-moldy corky* taste or musty character remains a feared and aggravating problem. Although the corky taste occurs in only 2% of all bottled wines sealed with cork, it results in heavy economic losses. Heimann et al. (1981, 1983) were able to isolate the fungus *Penicillium roquefortii* from cork stoppers extracted from bottles in which the wine had displayed a typical corky character. If this fungus is cultivated in a synthetic growth medium together with cork oak wood, it develops a typical corky taste. With the aid of the "sniffing technique" and mass spectrometry, several sesquiterpenes were recognized to be the cause of this musty-moldy aroma note (Rapp and Mandery 1986) (Fig. 21, 22).

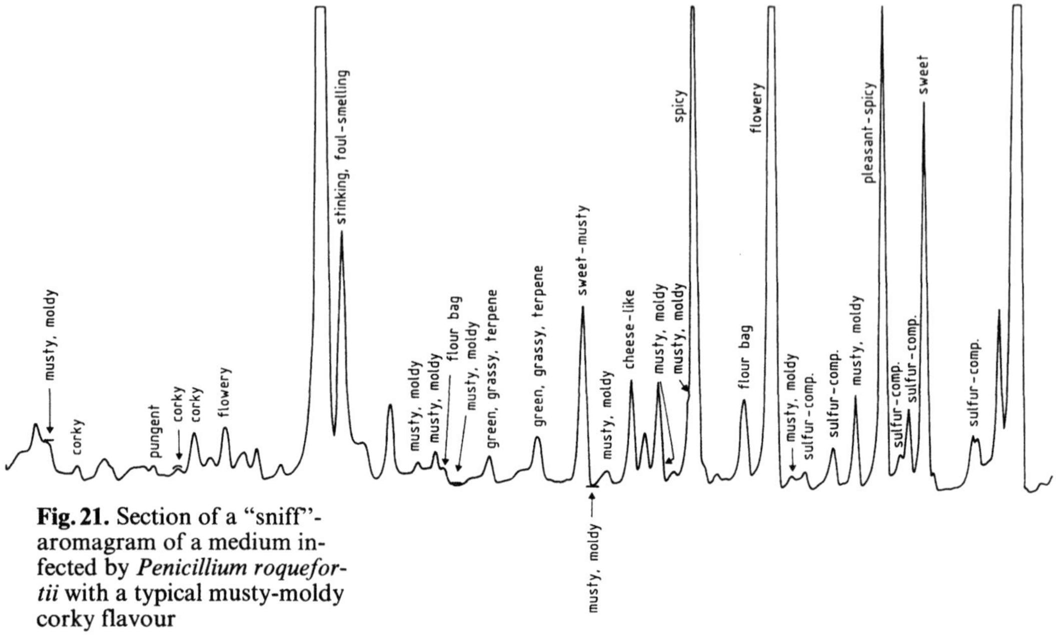

Fig. 21. Section of a "sniff"-aromagram of a medium infected by *Penicillium roquefortii* with a typical musty-moldy corky flavour

Rel. ret. time	Conc.	Sensorial Analysis "Sniff-detection"	Instrumental Analysis GC-MS-results
34.8	++++	Fresh mushroom	Octan-3-one
47.9	++++	Mushroom	Octan-3-ol
52.5	++++	Champignon	Oct-1-en-3-ol
63.9	+++	Campher-like, musty-moldy	Sesquiterpene (204)
65.0	++	Campher-like, musty, corky	Sesquiterpene (204)
67.9	++	Musty	Sesquiterpene (204)
71.9	++	Musty-moldy, corky	Sesquiterpene (204)
81.1	++	Musty-moldy, corky	Sesquiterpene (202)
100.0	++++	Rose-like	2-phenylethanol

Fig. 22. Volatile compounds with a typical musty-moldy corky flavour produced by *Penicillium roquefortii*

Tanner et al. (1981) discovered in cork and wine substances displaying a typical musty character e.g., chloroanisole, especially 2,4,6-trichloroanisole, which can be formed from phenols during the regular treatment of the raw cork material with chlorine and the subsequent microbiological methylation. The flavor threshold of 2,4,6-trichloroanisole is found to be in the ppt range.

Another off-flavor is the *geranium character,* which is caused by 2-ethoxyhexa-3,4-diene (Crowell and Guymon 1975). This component is formed by bacterial action from sorbic acid.

Table 9. Foreign and undesirable odors in wine ("off-flavour")

Green pepper, grassy	3-Isobutyl-2-methoxypyrazine 3-Isopropyl-2-methoxypyrazine 3-Ethyl-2-methoxypyrazine	Bayonove et al. (1975) Augustyn et al. (1982)
Green	Hexanal, hexenals	Drawert (1974)
Strawberry	2,5-Dimethyl-4-methoxy-2,3-dihydro-3-furanone 2,5-Dimethyl-4-hydroxy-2,3-dihydro-3-furanone	Rapp et al. (1980b)
Geranium-like	2-Ethoxy-hexa-3,5-diene	Crowell and Guymon (1975)
Mousy	2-Ethyl-3,4,5,6-tetrahydropyridine 2-Acetyl-3,4,5,6-tetrahydropyridine 2-Acetyl-1,4,5,6-tetrahydropyridine	Tucknott (1977) Rapp et al., unpubl.
Rotten eggs	Hydrogen sulfide	
Onion-like/ garlic-like	Mercaptans	
Vinegary	Ethyl acetate, acetic acid	
Sauerkraut	Lactic acid, diacetyl + other components	
Corky	Methyl tetrahydronaphthalene 2,4,6-Trichloroanisole Sesquiterpenes produced by *P. roquefortii* 2,3,4,6-Tetrachloroanisole and pentachloroanisole	Dubois and Rigaud (1981) Tanner and Zannier (1981) Heimann (1981, 1983) Tanner et al. (1981)
Woody	3-Methyl-γ-octalactone	Kepner et al. (1972)

Mousiness is another unpleasant aroma character which can be formed during the wine-making process and is more pronounced in certain years. The wines display an odor reminiscent of mouse urine and a repulsive aftertaste. This taste characteristic can be caused by the infection of the wine with *Lactobacillus* bacteria or *Brettanomyces* yeasts in oxidative circumstances. The responsible components were identified as 2-acetyl tetrahydropyridine and 2-ethyl tetrahydropyridine (Tucknott 1977; Strauss and Heresztyn 1984; Rapp and Mandery 1986). Up till now these components could not be identified in normal, healthy wines.

In Table 9, several undesirable aromas in wine as well as the responsible components have been summarized.

References

Albagnac G (1975) La décarboxylation des acides cinnamiques substitués par les levures. Ann Technol Agric 24:133–141

Augustyn OPH, Rapp A, van Wyk CJ (1982) Some volatile aroma components of *Vitis vinifera* L. Cv. Sauvignon blanc. S. Afr J Enol Vitic 3:53–59

Bayer E (1958) Anwendung chromatographischer Methoden zur Qualitätsbeurteilung von Weinen und Mosten. Vitis 1:298–312

Bayer E, Bässler E (1961) Systematische Identifizierung von Estern im Weinaroma. II. Mitteilung zur systematischen Identifizierung verdampfbarer organischer Substanzen. Z Anal Chem 181:418–424
Bayonove C, Cordonnier R (1971 a) Recherches sur l'arôme du Muscat. III. Etude de la fraction terpénique. Ann Technol Agric 20(4):347–355
Bayonove C, Cordonnier R (1971 b) Le linalool, constituant important mais non spécifique de l'arôme des muscats. CR Acad Agric Fr 57:1374–1377
Bayonove C, Cordonnier R, Dubois P (1975) Etude d'une fraction caractéristique de l'arôme du raisin de la variété Cabernet Sauvignon: mise en évidence de la 2-methoxy-3-i-butylpyrazine. CR Acad Sci Ser D 281:75–78
Bayonove C, Günata YZ, Cordonnier R (1984) Mise en évidence de l'intervention des enzymes dans le développment de l'arôme du jus de muscat avant fermentation: la production des terpénols. Bull OIV 57:741–758
Bertrand A, Boidron JN, Ribéreau-Gayon P (1967) No 560. – Méthode d'extraction des consituants volatils des fruits et de leur dérivés en vue d'une étude par chromatographie en phase gazeuse. Bull Soc France 9:3149–3151
Bertuccioli M, Montedoro G (1974) Concentration of the headspace volatiles above wine for direct chromatographic analysis. J Sci Food Agric 25:675–687
Bertuccioli M, Viani R (1976) Red wine aroma: identification of headspace constituents. J Sci Food Agric 27:1035–1038
Boeckh J (1972) Die chemischen Sinne – Geruch und Geschmack. In: Gauer OH, Cramer K, June (eds) Handbuch Physiologie des Menschen. Urban und Schwarzenberg, München
Boidron JN (1978) Relation entre les substances terpéniques et la qualité du raisin (Role du *Botrytis cinerea*). Ann Technol Agric 27:141–145
Boidron JN, Ribéreau-Gayon P (1967) Les techniques de laboratoire appliquées à l'identification des arômes des vins. Ind Aliment Agric 84:883–893
Chaudhary SS, Kepner RE, Webb AD (1964) Identification of some volatile compounds in an extract of the grape *Vitis vinifera*. Var. Sauvignon blanc. Am J Enol Vitic 15:190–198
Crowell EA, Guymon JF (1975) Wine constituents arising from sorbic acid addition, and identification of 2-ethoxyhexa-3,5-diene as a source of geranium-like off-odour. Am J Enol Vitic 26:97–102
Demole E, Engist P, Ohloff G (1982) 1-p-Menthen-8-thiol: powerful flavour impact constituent of grape fruit juice *(Sitrus paradisi)*. Helv Chim Acta 65:1785–1794
Drawert F (1974) Winemaking as a biotechnological sequence. Adv Chem Ser 137:1–10
Drawert F, Rapp A (1966): Über Inhaltsstoffe von Mosten und Weinen. VII. Gaschromatographische Untersuchung der Aromastoffe des Weines und ihrer Biogenese. Vitis 5:351–376
Drawert F, Rapp A (1968) Gaschromatographische Untersuchung pflanzlicher Aromen. I. Anreicherung, Trennung und Identifizierung von flüchtigen Aromastoffen in Traubenmosten und Weinen. Chromatographia 1:446–457
Drawert F, Schreier P, Scherer W (1974) Gaschromatographisch-massenspektrometrische Untersuchung flüchtiger Inhaltsstoffe des Weines. III. Säuren des Weinaromas. Z. Lebensm Unters Forsch 155:342–347
Drawert F, Schreier P, Leupold G, Kerényi Z, Lessing V, Junker A (1976 a) Gaschromatographisch-massenspektrometrische Untersuchung flüchtiger Inhaltsstoffe des Weines. VII. Aromastoffe in Tokajer Weinen. b) organische Säuren. Z Lebens Unters Forsch 162:11–20
Drawert F, Leupold G, Lessing V (1976 b) Gaschromatographische Bestimmung der Inhaltsstoffe von Gärungsgetränken. VI. Quantitative gaschromatographische Bestimmung von Neutralstoffen (Kohlenhydraten) und phenolischen Verbindungen in Tokajer Weinen. Z Lebensm Unters Forsch 162:407–414
Dubois P, Rigaud JA (1981) A propos de goût de bouchon. Vignes Vins 301:48–49
Dubois P, Brulé G, Ilic M (1971) Étude des phénols volatils de deux vins rouges. Ann Technol Agric 20:131–139
Dubois P, Rigaud J, de Kimpe J (1976) Identification de la dimethyl-4,5-tetra-hydro-furandione-2,3 dans le vin jaune des Jura. Lebensm Wiss Technol 9:366–368

Fuleki T (1972) Changes in the chemical composition of Concord grapes grown in Ontario during ripening in the 1970 season. Can J Plant Sci 52:863–868
Guadagni DG, Buttery RG, Okano S (1963) Odour threshold of some organic compounds associated with food flavours. J Sci Food Agric 14:761–765
Günata YZ (1984) Recherches sur la fraction liée de nature glycosidique de l'arôme du raisin. Thése, Univ. des Sciences et Techniques du Languedoc/France
Güntert M (1984) Gaschromatographisch-massenspektrometrische Untersuchungen flüchtiger Inhaltsstoffe des Weinaromas: Beitrag zur Sortencharakterisierung der Rebsorte Riesling. Dissertation, Univ. Karlsruhe
Hardy PJ, Ramshaw EH (1970) Analysis of minor volatile constituents of wine. J Sci Food Agric 21:39–41
Heimann W, Rapp A, Völter I, Knipser W (1981) Beitrag zum derzeitigen Stand der Entstehung des Korktons. 6. Internat önolog Symp, Mainz, pp 215–228
Heimann W, Rapp A, Völter I, Knipser W (1983) Beitrag zur Entstehung des Korktons im Wein. Dtsch Lebensm Rundsch 79:103–107
Hennig K, Villforth F (1942) Die Aromastoffe der Weine. Vorratspflege und Lebensmittelforschung, pp 181–200, pp 312–333
Hock R, Benda J, Schreier P (1984) Formation of terpenes by yeasts during alcoholic fermentation. Z Lebensm Unters Forsch 179:450–452
Kepner RE, Webb AD, Muller CJ (1972) Identification of 4-hydroxy-3-methyloctanoic acid γ-lactone (5-butyl-4-methyl-dihydro-2-(3B)-furanone) as a volatile component of oak-wood aged wines of *Vitis vinifera* var. Cabernet sauvignon. Am J Enol Vitic 23:103–105
Lemperle E, Mecke R (1965) Gaschromatographische Analyse der flüchtigen Inhaltsstoffe von Weinen, Mosten und Spirituosen. Z Anal Chem 212:18–30
Mandery H (1983) Gaschromatographisch-massenspektrometrische Untersuchungen flüchtiger Inhaltsstoffe des Traubenmost- und Weinaromas: Auswirkung der Süßung auf die Aromazusammensetzung. Dissertation, Univ. Karlsruhe
Marais J (1979) Effect of storage time and temperature on the volatile composition and quality of dry white table wines. Vitis 18:254–260
Marais J, van Rooyen PC, Plessis CS (1979) Objective quality rating of pinotage wine. Vitis 18:31–39
Masuda M, Okawa E, Nishimura K, Yunome H (1984) Identification of 4,5-dimethyl-3-hydroxy-2(5H)-furanone (Sotolone) and ethyl-9-hydroxynonanoate in botrytised wine and evaluation of the role of compounds characteristic of it. Agric Biol Chem 48:2707–2710
Meunier JM, Boot EW (1979) Das Verhalten verschiedener Aromastoffe in Burgunderweinen im Verlauf des biologischen Säureabbaus. Chem Mikrobiol Technol Lebensm 6:92–95
Muller CJ, Kepner RE, Webb AD (1971) Identification of 3-(methylthio)-propanol as an aroma constituent in Cabernet-Sauvignon and Ruby Cabernet wines. Am J Enol Vitic 22:156–160
Muller CJ, Kepner RE, Webb AD (1972) Identification of 4-Ethoxy-4-hydroxybutyric acid-γ-lactone [5-Ethoxydihydro-2(3H)-furanone] as an aroma component of wine from *Vitis vinifera* variety Ruby Cabernet. J Agric Food Chem 20:193–195
Muller CJ, Kepner RE, Webb AD (1973) Lactones in wines. A Review Am J Enol Vitic 24:5–9
Nelson RR, Acree TE, Lee CY, Butts RM (1977) Methylanthranilate as an aroma constituent of American wine. J Food Sci 42:57–59
Neudoerffer TS, Sandler S, Zubechis E, Smith MD (1965) Detection of an undesirable abnormally in concord grape by gaschromatography. J Agric Food Chem 13:584–588
Noble AC (1978) Sensory and instrumental evaluation of wine aroma. In: Charalambous G (ed) Analysis of food and beverages headspace technique. Academic Press, New York, pp 203–228
Noble AC, Robert AF, Forrey RR (1980) Wine headspace analysis. Reproducibility and application to varietal classification. J Agric Food Chem 28:346–353
Nykänen L, Suomaleinen H (1983) Aroma of beer, wine and distilled alcoholic beverages. D Reidel, Dordrecht, Holland, p 413

Ohloff G (1978) Recent developments in the field of naturally occurring aroma components. Fortschr Chem Org Naturst 35:431–527
Peynaud E (1965) Le goût et l'odeur du vin. Bull Soc Sci Hyg Aliment 53:249–260
Power FB, Chesnut VK (1921) The occurrence of methyl anthranilate in grape juice. J Am Chem Soc 43:1741–1742
Rapp A (1972) Les arômes des vins et des eaux-de-vie. Leur formation et leur évolution. Bull OIV 45(1):151–166
Rapp A (1981) Analysis of Grapes, Wines and Brandies. In: Jennings WG (ed) Applications of glass capillary gas-chromatography. Academic Press, New York, pp 597–621
Rapp A (1984a) Composants aromatique du vin. Possibilités de classification et appréciation qualitative. L'Enotecnico 20:25–33
Rapp A (1984b) Composants aromatiques du vin. Possibilités de classification et d'appreciation qualitative. Rev Franc Oenologie 93:65–70
Rapp A (1987) Beitrag zur Sortencharakterisierung verschiedener Rebsorten. Vitis (in preparation)
Rapp A, Güntert M (1985) Beitrag zur Sortencharakterisierung der Rebsorte Weißer Riesling. II. Untersuchung der Aromastoffzusammensetzung in inländischen Weißweinen der Rebsorten Weißer Riesling, Müller-Thurgau und Silvaner. Vitis 24:139–150
Rapp A, Güntert M (1986) Changes in aroma substances during the storage of white wines in bottles. In: Charalambous G (ed) The shelf life of foods and beverages, Proceedings of Flavour Conference, Rhodos. Elsevier Science, BV, Amsterdam, pp 141–167
Rapp A, Hastrich H (1978) Gaschromatographische Untersuchungen über die Aromastoffzusammensetzung der Rebsorte Riesling. Vitis 17:288–298
Rapp A, Knipser W (1979) 3,7-Dimethyl-okta-1,5-dien-3,7-diol. Eine neue terpenoide Verbindung des Trauben- und Weinaromas. Vitis 18:229–233
Rapp A, Mandery H (1986) New progress in vine and wine research: wine aroma. Experentia 42:857–966
Rapp A, Hoevermann W, Jecht U, Franck H (1973) Gaschromatographische Untersuchungen an Aromastoffen von Traubenmosten, Weinen und Branntweinen. Chem Z 97:29–36
Rapp A, Hastrich H, Engel L (1976) Gaschromatographische Untersuchungen über die Aromastoffe von Weinbeeren. I. Anreicherung und kapillarchromatographische Auftrennung. Vitis 15:29–36
Rapp A, Hastrich H, Engel L, Knipser W (1978) Possibilities of characterizing wine quality and wine varieties by means of capillary chromatography. Charalambous G (ed) Flavor of foods and beverages. Academic Press, New York, pp 391–417
Rapp A, Knipser W, Engel L (1980a) Identifizierung von 3,7-Dimethyl-okta-1,7-dien-3,6-diol in Trauben- und Weinaroma von Muskatsorten. Vitis 19:226–229
Rapp A, Knipser W, Engel L, Ullemeyer H, Heinmann W (1980b) Fremdkomponenten im Aroma von Trauben und Weinen interspezifischer Rebsorten. I. Die Erdbeernote. Vitis 19:13–23
Rapp A, Knipser W, Hastrich H, Engel L (1982) Possibilities of characterizing wine quality and wine varieties by means of capillary chromatography. In: Webb Ad (ed) Symposium Proceedings University of California, Grape and Wine Centennial (1980) Davis, pp 304–316
Rapp A, Mandery H, Heimann W (1983a) Untersuchungen über die flüchtigen Inhaltsstoffe von rektifiziertem Traubenmostkonzentrat. Dtsch Lebensm Rundsch 79:361–365
Rapp A, Knipser W, Engel L, Hastrich H (1983b) Capillarychromatographic Investigations on various grape varieties. In: Charalambous G, Inglett G (eds) Instrumental analysis of foods. Academic Press, New York, pp 435–454
Rapp A, Mandery H, Ullemeyer H (1984a) Neuere Ergebnisse über die Aromastoffe des Weines. Oenolog. Symposium, Rom, pp 157–196
Rapp A, Güntert M, Almy J (1984b) Neue schwefelhaltige Aromastoffe im Wein -cis- und trans-2-Methyl-thiolan-3-ol. Vitis 23:66–72
Rapp A, Mandery H, Ullemeyer H (1984c) Neue Monoterpendiole in Traubenmosten und Weinen und ihre Bedeutung für die Genese cyclischer Monoterpenäther. Vitis 23:84–92

Rapp A, Mandery H, Güntert M (1984d) Terpene compounds in Wine. In: Nykänen L, Lehtonen P (eds) Flavour research of alcoholic beverages. Instrumental and sensory analysis. Kauppakirjapino Oy Helsinki, pp 255–274

Rapp A, Güntert M, Ullemeyer H (1985a) Über Veränderungen der Aromastoffe während der Flaschenlagerung von Weißweinen der Rebsorte Riesling. Z Lebensm Unters Forsch 180:109–116

Rapp A, Güntert M, Rieth W (1985b) Einfluß der Maischestandzeit auf die Aromastoffzusammensetzung des Traubenmostes and Weines. Dtsch Lebensm Rundsch 81:69–72

Rapp A, Güntert M, Heimann W (1985c) Beitrag zur Sortencharakterisierung der Rebsorte Weißer Riesling. I. Untersuchung der Aromastoffzusammensetzung von ausländischen Weißweinen, die als Sortenbezeichnung den Begriff Riesling tragen. Z Lebensm Unter Forsch 181:357–361

Rapp A, Mandery H, Niebergall H (1986): Neue Monoterpendiole in Traubenmost und Wein sowie in Kulturen von *Botrytis cinerea*. Vitis 25:79–84

Ribéreau-Gayon P (1978) Wine flavour. In: Charalambous G, Inglett GE (eds) Flavour of foods and beverages. Academic Press, New York

Ribéreau-Gayon P, Boidron JN, Terrier A (1975) Aroma of Muscat grape varieties. J Agric Food Chem 23:1042–1047

Rieth W (1984) Gaschromatographisch-massenspektrometrische Untersuchungen flüchtiger Inhaltsstoffe des Weinaromas: Einfluß oenologischer Verfahren und Behandlungsstoffe auf die Aromastoffzusammensetzung. Dissertation. Univ. Karlsruhe

Rothe M (1974) Handbuch der Aromaforschung. Aroma von Brot, Berlin

Schenck OG (1957) Aufgaben und Möglichkeiten der Strahlenchemie. Angew Chem 69:579–599

Schreier P (1979) Flavor compositions of wines: a review. CRC Crit. Rev Food Sci Nutr 12:59–111

Schreier P, Drawert F (1974a) 3-(Methylthio)-1-propanol, eine flüchtige Komponente des Weinaromas. Z Lebensm Unters Forsch 154:27–28

Schreier P, Drawert F (1974b) Gaschromatographisch-massenspektrometrische Untersuchung flüchtiger Inhaltsstoffe des Weines. I. Unpolare Verbindungen des Weinaromas. Z Lebensm Unters Forsch 154:273–278

Schreier P, Drawert F, Junker A (1974) Gaschromatographisch-massenspektrometrische Untersuchung flüchtiger Inhaltsstoffe des Weines. II. Thioether-Verbindungen des Weinaromas. Z Lebensm Unters Forsch 154:279–284

Schreier P, Drawert F, Junker A (1975a) Gaschromatographisch-massenspektrometrische Untersuchung flüchtiger Inhaltsstoffe des Weines. IV. Nachweis sekundärer Amide im Wein. Z Lebensm Unters Forsch 157:34–37

Schreier P, Drawert F, Junker A (1975b) Über die Biosynthese von Aromastoffen durch Mikroorganismen. I. Bildung von N-Acetylaminen durch *Saccharomyces cerevisiae*. Z Lebensm Unters Forsch 158:351–360

Schreier P, Drawert F, Junker A (1976a) Sesquiterpen-Kohlenwasserstoffe in Trauben. Z Lebensm Unters Forsch 160:271–274

Schreier P, Drawert F, Junker A (1976b) Identification of volatile constituents from grapes. J Agric Food Chem 24:331–336

Schreier P, Drawert F, Kerènyi Z, Junker A (1976c) Gaschromatographisch-massenspektrometrische Untersuchung flüchtiger Inhaltsstoffe des Weines. VI. Aromastoffe in Tokajer Trockenbeerenauslese (Aszu)-Weinen. a) Neutralstoffe. Z Lebensm Unters Forsch 161:249–258

Schreier P, Drawer F, Junker A, Reiner L (1976d) Anwendung der multiplen Diskriminanzanalyse zur Differenzierung von Rebsorten anhand der quantitativen Verteilung flüchtiger Weininhaltsstoffe. Mitt Klosterneuburg 26:225–234

Shimizu J, Watanabe M (1981) Neutral volatile components in wines of Koshu and Zenkoji grapes. Agric Biol Chem 45:2797–2803

Shimizu J, Nokora M, Watanabe M (1982) Transformation of terpenoids in grape and must by *Botrytis cinerea*. Agric Biol Chem 46:1339–1344

Simpson RF (1978) 1,1,6-Trimethyl-1,2-dihydronaphtalene: an important contributor to the bottle aged bouquet of wine. Chem Ind, pp 37

Simpson RF (1979a) Some important aroma components of white wine. Food Technol Aust 31:516–521
Simpson RF (1979b) Aroma composition of bottle aged white wine. Vitis 18:148–154
Simpson RF (1980) Volatile aroma components of Australian Port wines. J Sci Food Agric 31:214–222
Stern DJ, Lee A, Mc Fadden WH, Stevens KL (1967) Volatiles from grapes. Identification of volatiles from Concord essence. J Agric Food Chem 15:1100–1103
Stevens KL, Lee A, Mc Fadden WH, Teranishi R (1965) Volatiles from grapes. I. Volatiles from Concord essence. J Food Sci 30:1006–1007
Stevens KL, Flath A, Lee A, Stern DJ (1969) Volatiles from grapes. Comparison of Grenache juice and Grenache rosé wine. J Agric Food Chem 17:1102-1106
Strauss CR, Heresztyn T (1984) 2-Acetyltetrahydropyridines – a cause of the "mousy" taint in wine. Chem Ind, pp 109–110
Tanner H, Zanier C, Buser HR (1981) 2,4,6-Trichloranisol: Eine dominierende Komponente des Korkgeschmacks. Schweiz Z Obst- Weinbau 117:97–103
Terrier A, Boidron JN, Ribéreau-Gayon P (1972) L'identification des composés terpéniques dans les raisins de Vitis vinifera. CR Acad Sci Ser D 275:405
Tucknott OG (1977) The mousy taint in fermented beverages: its nature and origin. Thesis, Univ. Bristol/England
Van Straten S, Maarse H (1983) Volatile compounds in food qualitative data. Div. for nutrition and food research TNO
Van Wyk CJ, Webb AD, Kepner RE (1967a) Some volatile components of *Vitis vinifera* variety White Riesling. I. Grape Juice. J Food Sci 32:660–664
Van Wyk CJ, Kepner RE, Webb AD (1967b) Some velatile components of *Vitis vinifera* variety White Riesling. II. Organic acids extracted from wine. J Food Sci 32:664–668
Webb AD, Kepner RE (1957) Some volatile aroma constituents of *Vitis vinifera* var. Muscat of Alexandria. Food Res 22:384
Webb AD, Muller CJ (1972) Volatile aroma compounds of wines and other fermented beverages. Adv Appl Microbiol 15:75–146
Webb Ad, Kepner RE, Maggiora L (1969) Some volatile components of wines of *Vitis vinifera* varieties. Cabernet-Sauvignon and Ruby-Cabernet. I. Neutral compounds. Am J Enol Vitic 20:16
Welch RC, Johnston JC, Hunter GLK (1982) Volatile constituents of the Muscadine grape *(V. rotundifolia)*, J Agric Food Chem 30:681–684
Williams PJ, Strauss CR, Wilson B (1980a) Hydroxylated linalool derivatives as precursors of volatile monoterpenes of muscat grapes. J Agric Food Chem 28:766–771
Williams PJ, Strauss CR, Wilson B (1980b) New linalool derivates in muscat of Alexandria grapes and wines. Phytochemistry 19:1137–1139
Williams PJ, Strauss CR, Wilson B, Massy-Westropp PA (1982a) Novel monoterpene disaccharide glycosides of *Vitis vinifera* grapes and wines. Phytochemistry 21:2013–2020
Williams PJ, Strauss Cr, Wilson B, Massy Westropp PA (1982b) Use of C_{18}-reversed-phase liquid chromatography for the isolation of monoterpene glycosides and norisoprenoid precursors from grape and wines. J Chromatogr 235:471–480

Micro-Element Analysis in Wine and Grapes

H. ESCHNAUER and R. NEEB

1 Introduction

The number of inorganic and organic compounds which have been identified and determined in wine so far may well approach 1000. Only about 50 of these are inorganic compounds, and their number is limited by the number of elements in the Periodic Table. "Per definitionem vinum", the concentration of a mineral in wine is at most 1000 mg l^{-1}, that of a trace element lies at around 1 mg l^{-1} or below (ppm range) and that of an ultra-trace element at around 1 µg l^{-1} or below (ppb range). The number of inorganic substances present and the concentrations thus outline the scope of trace element analysis in wine.

As a rule the concentrations of all mineral compounds (ash constituents) do not exceed 5 g (5000 mg) per liter of wine. These substances include four cations and four anions and represent the most important constituents as far as weight is concerned.

The total content of all trace elements is below 50 mg l^{-1} (1% of the mineral amount), that of all ultra-trace elements lies at 50 µg l^{-1} (0.001% of the mineral amount).

So far, about 50 inorganic substances have been identified and determined in wine, viz. eight minerals (ash constituents), 25 trace elements, and about 20 ultra-trace elements (and additionally, radioactive elements and the 14 rare earths).

A summary of the typical concentrations of all inorganic compounds normally present in commercially available wines (normal content) is given in Table 1. The concentrations of the trace and ultra-trace elements vary over a wide range, viz. from a few milligrams (10^{-3} g) to a fraction of a nanogram (10^{-9} g) per liter. Table 1 is, therefore, subdivided into certain concentration ranges for the sake of better presentation.

The total content of a trace or ultra-trace element in commercially available wines (normal content) is made up of the sum of the primary and secondary contents. As a rule, wine is analyzed for total content. The correct primary content can only be determined with difficulty and much effort, the secondary content becomes important if an assessment of the contaminants ("contamination") is to be made (Fig. 1). The primary content of a trace or ultra-trace element in wine is that content which is naturally present by transfer from the soil via the roots to the grapes and from there to the wine. Often it constitutes the characteristic and largest part of the total content and it often fluctuates within a wide margin, depending on the geological formation in which the vineyards are located.

The secondary content of a trace or ultra-trace element is due to contamination by geogenous (natural) or anthropogenic (artificial) sources. Figure 1 gives more information about its causes.

Table 1. Inorganic constituents ($mg\,l^{-1}$) of wine (typical values of normal contents). (Eschnauer 1986)

1000–10	10–1	1–0.1	0.1–0.01	0.01–0.001	≦0.001
K 370–1120	B 5 –2	Al 0.9–0.5	As 0.02–0.003	Co 0.02–0.001	Sb 0.006
Mg 60– 140	Fe 1	F 0.5–0.05	Ba 0.3 –0.04	Mo 0.01–0.001	Be 0.00008
Ca 70– 140	Cu 0.5	I 0.6–0.1	Pb 0.1– 0.03	Ag 0.02–0.005	Cd 0.001
Na 7– 15	Mn 5 –1.5	Rb 4.2–0.2	Br 0.7 –0.01		Cs 0.0027
C 100– 120	Si 6 –1.5	Sr 3.5–0.2	Cr 0.06–0.03		Au 0.00006
P 130– 230	Zn 3.5–0.5	Ti 0.3–0.04	Li 0.2 –0.01		Hf 0.0007
S 5– 10			Ni 0.05–0.03		Nb 0.001
Cl 20– 80			V 0.26–0.06		Hg 0.00005
			Sn 0.7 –0.01		Se 0.0006
					Ta 0.0005
					Tl 0.0001
					Bi 0.00015
					W 0.003
					Rare earths
					Radioactive elements

Natural contamination is of geogenous origin and practically unavoidable. It is connected with the geographic location (natural location) of the vineyard. If the grapes are grown in the vicinity of the sea, of salt lakes, or of active or inactive vulcanoes (exhalation), then wines from these regions will often show an increased secondary content of arsenic, boron, chlorine, iodine, iron, manganese, mercury, sodium or zinc. The increase may be small but it can be determined quite easily. It is caused by an additional intake of these elements via the grapes rather than the roots.

Artificial contamination is always of anthropogenic origin with a variety of causes which can be divided into three groups, namely industrial environment, enologic contamination, and adulteration. It can almost always be avoided or at least greatly reduced. Increased secondary contents of arsenic, boron, cadmium, lead, mercury, and thallium may be found in wines originating from vineyards in the vicinity of factories, metallurgical plants, cement mills, brick works, coal-fired power plants, mining operations, or highways.

Enologic contamination is caused by a combination of all factors connected with wine-growing (fertilizers and pest control) and wine production (vintage, treatment, storage, shipment). These activities will seldom, if ever, be the cause of secondary contents.

A high secondary content occurs if the wine is "improved" by an intentional but strictly forbidden addition of adulteration substances. In the past, numerous treatments of wine have been carried out by adding compounds of aluminum, barium, bismuth, bromine, fluorine, lead, manganese, or strontium.

The identification and determination of trace and ultra-trace elements in wine is also of importance from two more points of view, viz. with reference to a sufficient supply of micronutrients to the grapevine and the contribution of essential elements by wine for human nutritional requirements.

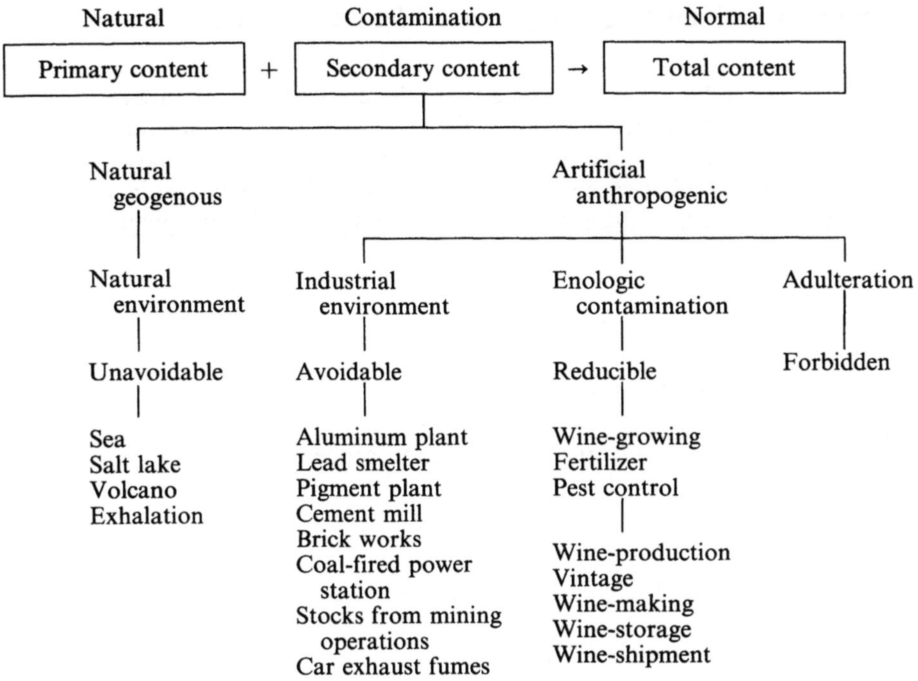

Fig. 1. Secondary content of trace elements in wine (contamination). (Eschnauer 1986)

Not only the main or macronutrients (according to J. von Liebig: nitrogen, phosphorus, potassium, magnesium) but also trace or micronutrients, such as boron, manganese, copper, cobalt, and zinc, are important for the healthy growth of the vine.

Several minerals and trace elements are vital to the human organism and must be ingested with the daily food in sufficient amounts. Wine can contribute not only minerals, containing potassium, calcium and magnesium, but also numerous essential trace elements, such as chromium, cobalt, iron, fluorine, iodine, copper, manganese, molybdenum, nickel, selenium, zinc (tin, silicium, vanadium).

The identification and determination of inorganic compounds, i.e. minerals, trace and ultra-trace elements, can only be carried out by suitable methods of analysis. A total of about 500 papers dealing with the determination and content of trace elements in wine were published until 1965. These described mainly colorimetric and spectrophotometric but also gravimetric and titrimetric methods. Advanced physical methods, however, were hardly used. Since 1965 about 20–30 papers have been published per year, in which only modern methods of trace analysis have been applied, including single and multi-element analysis and even pattern recognition and speciation analysis. In the last few years the number of papers published and methods employed have increased even further.

2 General Problems Regarding Trace Element Analysis

The contents found in wine and related samples are very low and are in the ppm to ppb range, sometimes even lower (cf. Table 1). Therefore, the general problems of trace analysis are also encountered when samples of wine are analyzed (Koch and Koch-Dedic 1974; Tschöpel 1982; Van Loon 1985). They will be discussed at first by describing how the analysis is carried out. The following diagram shows the various steps of a trace analysis:

Outline of the problem – analysis planning
↓
Sampling – – → sample storage
↓
Digestion (decomposition) – separation procedure
↓
Analytical determination
Direct determination
↓ ↓
Evaluation (reproducibility, accuracy).

Depending on the sample matrix, certain steps may be omitted. For example, spectrochemical and spectrophotometric determinations can be carried out directly in wine without digestion. The errors of each single step are cumulative and result from manipulations with solutions of very low concentrations. They can be summarized as follows:

Source of error	Causes of error
Sampling	Inhomogeneous sample
Sample treatment	Interaction with equipment used (crushing, grinding, cutting)
Sample storage	± Interaction with container; instability of sample (aqueous solution)
Dissolution digestion	± Interaction with equipment used; losses by volatilization, adsorption, dissolution of Vessel walls
Determination measurement	Errors regarding instrument readout and recording of results; calibration errors; wrong standard solution, unsuitable standard; change of instrument parameters.

The sampling of wine is without problems, but that of must already becomes more difficult, since a separation of phases (cloudy/clear) might be necessary. Completely homogeneous samples of wood and leaves can be obtained only with

Table 2. Summary of decomposition methods

Principle/procedure	Application	Remarks
Wet digestion methods		
Open system with oxidizing acids (H_2SO_4–HNO_3, $HClO_3$, H_2O_2	Suitable for almost all samples sometimes; however, time-consuming	Simple procedure, big problems with contamination especially by ubiquitous elements (Pb, Zn, Cu)
Catalytic: $H_2O_2+Fe^{2+}$	Not universally applicable	Suitable for wines
Closed systems (pressurized) Containers made of various plastics Oxidizing acids	Versatile; however, also danger of explosions (e.g., lipids with HNO_3)	Little contamination if small amount of acid is used, no losses by volatilization
Closed high pressure system (100 bar, 320° C) Quartz containers	Universally applicable also for difficult samples	Little contamination, no losses by volatilization
UV-photolysis + H_2O_2	Not universally applicable only for samples with little organic content or after dilution of the sample	Simple procedure, little contamination, only for liquid samples
Dry ashing methods		
O_2, air Open system	Universally applicable, larger amounts	Simple procedure, not suitable for some elements
Closed quartz system in O_2 atmosphere Cooling (Trace-O-Mat)	For small amounts of solid samples	Limited amount of sample, not so suitable for liquid samples, little contamination
Low temperature ashing in O_2 plasma	Not suitable for all samples	In some cases time-consuming and not complete

difficulty. It goes without saying that, aside from the actual sampling, a pre-treatment of these latter samples, e.g., washing, must also be carried out. It should also be pointed out that errors result from contamination by the apparatus used for crushing or grinding the sample and that it will often be necessary to use zirconium or titanium tools. A direct determination of trace elements without digestion, i.e., without dissolving the sample, will be possible for only a few samples. The digestion methods (Bock 1979; Gorsuch 1970) and the errors resulting from them hold, therefore, positions of vital importance in trace analysis and are frequently the limiting factors in the determination of ultra-trace elements.

A summary of the most important digestion methods is presented in Table 2.

Especially the digestion methods are, of course, susceptible to the errors mentioned above. Contamination from reagents used for digesting the sample, dissolution of vessel materials, and losses, particularly by volatilization, are worth mentioning. The use of clean rooms to prevent contamination (air, aerosols) is a conditio sine qua non for modern ultra-trace analysis (Mitchell 1982). The choice of suitable vessel material is especially important with respect not only to the digestion but also to all other analytical manipulation. Glass can only rarely be used (contamination by glass components but also adsorption onto glass walls result in errors). Quartz and certain plastics, e.g., polycarbonates or materials based on Teflon, are better suited. It will often be necessary to clean the equipment and condition it with the same solution as is used during the analysis. This can be carried out by using apparatus especially designed for this purpose (Tschöpel et al. 1980).

The usual digestion of wine is carried out in a Kjeldahl flask with oxidizing acids. Foam formation can be avoided by adding a few drops of a higher alcohol, e.g., octylalcohol. Commonly used pressurized digestions in plastic vessels allow working temperatures up to 165–170° C (Jackwert and Gomicek 1984). Somewhat higher temperatures can be reached if glassy carbon vessels in safety enclosures are used. Safety valves should always be installed. The advantage of these pressurized digestions is that smaller amounts of reagents are required and losses by volatilization are avoided. This is especially true for digestions under high pressure where the sample – up to 500 mg – can be heated in a high pressure vessel (100 bar) with nitric acid to 320° C (Knapp 1984; Knapp 1985; Schramel et al. 1987). This digestion method is very efficient, has low blank values, and can also be used successfully for samples which are difficult to digest otherwise. A disadvantage is only its high price. Low temperature ashing in an oxygen plasma is more convenient and low blank values are still achieved with it. This method, however, does not mineralize all samples completely, even with long ashing times. Dry ashing at high temperatures in a stream of oxygen or air can be applied in general to solid samples and to liquid samples after evaporation to dryness. Losses by volatilization and reactions with the vessel walls are prevented by adding ashing aids to the sample. It should be evident that this dry ashing method cannot be used for elements such as Hg, Tl, and Se, or to a certain extent for Cd, Pb, and Zn.

Volatile elements can be recovered quantitatively in a closed system by condensation on cold fingers, cooled by liquid nitrogen, after the sample has been burned (Knapp et al. 1981).

Wet digestions are universally applicable. The accuracy of the method can be easily checked by determining recovery rates. The digestion after Fenton, which is carried out with H_2O_2 and Fe^{2+} as catalyst (Sansoni and Kracken 1968), proved to be a suitable method for the determination of heavy metals in wine when compared to other methods (Cela et al. 1983).

UV photolysis in the presence of oxidants (e.g., H_2O_2) is another convenient method to remove organic material from solution, provided the amount of organics is not too large. Digestion times can be considerably shortened by using strong UV lamps placed at a short distance from the sample. The voltammetric determination of heavy metals in wine has been carried out successfully after digestion according to this method (Dorten et al. 1984).

A critical review of the analytical results obtained is always necessary, also in the case of trace analysis. The following errors can occur:

Random errors: scattering – reproducibility (precision) statistics; standard deviation.

Systematic errors: deviations from the true value (accuracy) caused by errors related to the method itself (often difficult to detect – detection of error by applying different methods, and using calibrated samples or standards).

Systematic errors are difficult to find and the determination of detection limits is difficult. The latter can be determined by statistical methods (Ebel and Kamm 1983). Systematic errors, however, cannot be detected by these methods.

3 Methods of Trace Element Analysis

3.1 Introduction

The necessity of a reliable and sensitive detection poses a great challenge to any method employed in trace analysis. A variety of methods is available today for determining small amounts and low concentrations of elements. Table 3 gives an overview of these methods.

The classification into single (S)-, oligo (O)- and multi (M)-element methods, which gives an indication of how many elements can be determined simultaneously, is somewhat arbitrary, although it is helpful for deciding whether a method is suitable for a particular task or not. Estimates on the universal applicability, sample throughput and costs of a method can be made from Table 3, although they are always somewhat subjected to personal judgement. Aspects related to the reliability of a method and ease with which results can be obtained must also be examined and not only the price of an instrument but also the operating costs (materials and personnel) must be taken into account to evaluate a particular method. It is thus difficult to present all the information necessary for deciding which would be the "best possible" method. Often compromises must be made, which are dictated by each specific case. Although one or several of the methods cited might be preferred today by the majority of analysts, "less used" methods still have a rightful place as reference methods. In contrast to the "usual" methods, they must provide results which can be calibrated. Chromatographic and – apart from their application as routine methods of analysis – electrochemical methods can be cited in this context. Often a powerful method is not fully taken advantage of if it is employed merely for routine determinations of an element. It should be emphasized here that a combination of methods can be used, whereby the actual determination is preceded by a chemical separation or preconcentration procedure carried out either directly with the sample or with a solution obtained by decomposition of the sample. A suitable combination of an instrumental method with a chemical processing step can often decrease the detection limit and increase the reliability of the analysis. Such chemical steps are, for instance, preconcentration on ion exchangers or by chelate extraction, co-precipitation, and selective volatilization of elements or compounds from the sample, e.g., by the hydride method as used in conjunction with AAS.

Table 3. Most important methods of trace element analysis

Method	Range/detection limit[a]	Price (DM)	Especially suited for	Remarks
Atomic emission spectroscopy (AES)	1 –10 ppm[b] (M)	200–400 000	Almost all metals	For solid samples (DC arc)
AES with plasma burner (ICP-AES)	1 –50 ppb (M)	200–800 000	Almost all metals	For liquid samples
Atomic absorption spectroscopy (AAS)	1 –50 ppb (S)	80–100 000	Almost all metals	Single element method
AAS with electrothermal atomization (ETA-AAS)	0.005– 0.5 ppb (S)	120–200 000	Almost all metals	Single element method
Zeemann-AAS	0.005– 0.5 ppb (S)	150–300 000	Almost all metals	Single element method
X-ray fluorescence analysis (XRA)	0.1 –10 ppm (M)	300–400 000	Fe, Zn, Cu, Mn. Ni, Co etc. Also nonmetals	Sensitivity increased by pre-concentration
Particle-induced X-ray emission (PIXE)	0.01 – 1 ppm (M)	>1 000 000	Max. sensitivity for $25 < Z < 35$	Complicated instrumentation
Spectral photometry	0.1 –10 ppm (S)	5– 10 000	Pb, Cu, Cr, Fe, Ti, Mo etc.	
Differential pulse polarography (DPP)	1 –10 ppb (O)	20– 30 000	Almost all heavy metals	
Stripping voltametry (e.g. DPASV)	0.01 – 1 ppb (O)	30– 50 000	Cu, Sb, Cd, Sn, Pb, Tl, Ti, Ni, Co etc.	
Neutron activation analysis (NAA)	Variable (ppm-ppb) (M)	See Remarks	(Variable sensitivity)	High sensitivity only with nuclear reactors
Gaschromatography (GC)	0.01 – 1 ppm (O)	30– 50 000	Some heavy metals (Ni, Co, Cu, Cr)	Only after derivatization
High performance liquid chromatography (HPLC)	0.01 – 1 ppm (O)	30– 50 000	Many heavy metals	Mostly only after derivatization

[a] With reference to the final sample solution.
[b] With reference of solid samples.

3.2 Spectroscopic Methods/Emission and Atomic Absorption Spectrometry

During the last decades, the use of optical emission methods with classical sources of excitation (arc, spark, flame) has declined (Addink 1971; Cresser et al. 1986; Galan 1986; Mankopf and Friede 1975). Only flame excitation is still used today to some extent for determining the alkali and alkali earth metals. The determination of trace elements has largely been carried out by atomic absorption spectroscopy (AAS) (Welz 1986). This method is based on the absorption of light by gaseous atoms (Walsh 1955). After "atomization" of the sample, light of sharply bundled wavelengths – preferably monochromatic light – is passed through it. In the beginning a flame, e.g., an air-acetylene flame, was used for vaporizing the sample and also today this method is still often used. Mercury can be vaporized at normal temperatures (cold vaporization). Insufficient atomization and chemical interferences, which can only partially be compensated for instrumentally, limit the sensitivity of classical AAS. The introduction of "electrothermal atomization" (ETA-AAS) (L'vov, Massmann) has brought a considerable increase in sensitivity. The sample is now vaporized in a graphite furnace which is heated electrically by a termperature program. Atomization is almost complete and a high concentration of atoms is obtained due to the geometry of the design. The application of AAS can be extended to the direct analysis of various (liquid) samples by optimizing the drying, thermal decomposition and atomization of the sample in the graphite furnace (or similar configurations, e.g., L'vov platform).

Unspecific absorption caused by matrix components can be corrected electronically by comparing the background absorption of a deuterium or hydrogen lamp. More recently, it has become possible to solve this problem by Zeemann-AAS. Here, line-splitting by a strong magnetic field (Zeemann effect) not only provides the line necessary for the measurement, but also additional lines which are used to make a background correction with the same lamp (Stephen 1980). Even solid samples can be analyzed by Zeemann-AAS.

A serious disadvantage is the fact that AAS is a pronounced single rather than multi-element method, no matter which atomization is used. This is connected with the light source used most frequently, viz. the hollow-cathode lamp, which produces constant radiation only for that element from which it is made. Apart from the rather unsuccessful attempts to design a constant hollow-cathode for multi-element AAS analysis, the application of a strong continuum source is worth mentioning in this context. For instance, a simultaneous multi-element atomic absorption continuum source (SIMAAC) spectrometer has been developed which is based on the use of a "synergistic combination of a high resolution echelle polychromator, wave length modulation and computerized high speed data acquisition" (Harnly 1986). Up to 20 elements can be determined simultaneously with good sensitivity. A background correction can also be made. Information on how widely it is used in practice is still lacking.

Solid samples must be decomposed prior to the determination of trace elements (Blake 1980). No results are known about the direct analysis of solid samples so far. Trace elements in wine can be determined directly. With flame AAS, where the sample is fed through nebulizers, the signal decreases with increasing sugar content due to an increased viscosity. With increasing alcohol content it decreases again (Meranger and Somers 1968). These matrix effects do not occur with ETA-AAS. If a flame is used, these effects can be avoided by digesting the sample or by first extracting the elements from the sample. For instance, after acidification of the wine sample, Pb and Cd can be directly extracted as their substituted thiocarbamates into methyl-isobutyl-ketone and determined by flame AAS (Anders and Hailer 1976). The sensitivity of the determination thus also increases. With ETA-AAS, the acidified wine sample can be brought directly into the graphite furnace (Medina and Sudraud 1980). However, the conditions for drying, decomposition and atomization must be very carefully checked by round robins (Haller 1971; Mack 1975; Ristow and Bernau 1982). Other authors, however, have strongly recommended mineralization of wine samples also in the case of ETA-AAS analysis (Gallego 1983).

The addition of modifiers can prevent losses caused by directly placing the sample into the graphite furnace. If very small amounts of cadmium are determined in wine, it is advantageous, for instance, to add $Al(NO_3)_3$ which reacts to form γ-Al_2O_3 on heating. Thus, in an atmosphere containing N_2O, cadmium losses are avoided while the organic matrix is completely destroyed (Severin et al. 1982). Elements which form volatile hydrogen compounds can be determined by the hydride method. By reacting the element with a strong reducing agent (e.g., sodium borohydride), a volatile hydride is formed in solution. A stream of argon sweeps it from solution to the nebulizer where it is decomposed thermally. In this way, As, Se (Sb, Bi, Sn, Te) can be determined with increased sensitivity. The direct hydride method without prior decomposition, however, produced arsenic values that were too low (Simer et al. 1977). Only dry ashing with $Mg(NO_3)_2$ as ashing aid prior to generating the arsine yielded satisfactory results.

Sources of excitation for atomic emission spectroscopy (AES), such as arc, spark and flame, which have been known for a long time (see above), produce only rather low temperatures or are not suited for larger amounts of sample, especially not of solutions. Considerable progress was, however, made by the introduction of the "plasma burner". After nebulization, the sample solution is injected into the hot center (5000–10000 K) of a plasma. The energy needed for generating the plasma is provided to the system by an inductive or capacitive transmittance of HF-energy or, less frequently, by DC-arc. The various plasmas show different sensitivities and performance capabilities (Zander 1986). Most frequently, an argon plasma is used with inductive energy transmittance (inductively coupled plasma, ICP). The high temperature of the plasma produces an almost complete atomization and spectra with many lines. The lines are measured simultaneously with high resolution spectrometers by a series of fixed photocells. Sequential spectrometers which measure the intensity of the spectral line in a certain wave length range are especially suited for carrying out precise background corrections.

Combinations of both types of instruments are also available. The sensitivity of ICP-AES is nearly equal for all elements. Increased sensitivities are obtained if ultrasonic nebulizers are used instead of pneumatic ones. Concentrations below 1 ng ml^{-1} can then be determined (Fassel 1986). With ICP-AES, liquid samples, e.g., wine, can be injected without pretreatment (Interesse et al. 1984; Schramel et al. 1982). Decompositions with concentrating effects and other enrichment procedures can be used to increase the sensitivity of the determination. Concentrations are determined by adding standard solutions or – in the case of wine – from calibration curves prepared from alcoholic solutions of the metals (Interessse et al. 1984). The possibility of overlapping lines by interfering elements, e.g., from matrix constituents, should always be checked by consulting a handbook of spectra (Thompson and Walsh 1986). Inductively coupled plasmas can also be used as ion sources in conjunction with mass spectrometers (ICP-MS) (Gray 1986). These complicated but efficient instruments considerably increase the sensitivity of ICP-AES. Not much experience has been gained so far with its application towards the practical analysis of wine and related samples and with interference from matrix elements. ICP-MS certainly has good prospects of becoming a multi-element method for ultra-trace element analysis. It is certainly not a replacement of ICP-AES but a sensible extension and supplement to it. Fluorescence measurements (Van Loon 1981) in the visual range of the spectrum have found little application. They can be carried out not only in solutions (molecules, color reactions of metals) but also by using techniques employed in AAS (atomic fluorescence spectrometry, AFS). So far, however, this method has not seen wide use, in spite of initial successes and high sensitivities for some elements.

X-ray fluorescence analysis, which is the classical form of X-ray emission largely used today, belongs in fact to the fluorescence methods. Here, the – most often solid – sample is irradiated with intense (monochromatic/characteristic) X-rays and the emitted characteristic fluorescence radiation of the K, L and sometimes M shells is measured in a similar fashion as with X-ray diffraction methods ("wavelength dispersive"). Fluorescence radiation can also be measured in an energy-dispersive fashion with suitable semi-conductors as detectors and without spectral dispersion. X-ray fluorescence analysis is a multi-element method with moderate sensitivity. The detection limit is around 1–10 ppm (1 ng g^{-1}), depending on the excitation source used and the element determined. The detection limit can be improved by a preconcentration step by which the ratio of the concentration of the trace element to that of the macrocomponent (matrix) is increased. Such preconcentration procedures can be: ion exchange on filter papers, co-precipitation with suitable precipitates, extraction of metal chelates, electrodeposition, and so forth. Determinations down to the ppb range can be carried out by using these combination methods which have been developed especially for water analysis (Van Grieken 1982). XRA has been used to determine Fe, Cu, Zn, Mn, and Br in wine after preconcentration of these elements (Bergner and Land 1970). Ni, Cu, Zn, Pb, Br, Rb, Se, Cr, Mn, Fe, and others have been determined in the ppb range by XRA without preconcentration, after ashing 700–1000 mg of sample. (Zielkowsky and Bächmann 1978).

Excitation of secondary, characteristic X-rays can also be carried out by charged particles (particle-induced X-ray emission, PIXE) (Young et al. 1973). Some experience, especially with excitation by protons, has been reported, also with regard to practical applications (Campbell et al. 1975; Khan and Crumpton 1981; Noble et al. 1976; Zee et al. 1983). The protons are generated in a vacuum discharge tube. A thin film of the sample (1 μm–1 mm, 0.1–2.0 cm^2) is irradiated with the charged particles. Calibration is carried out by adding internal standards (e.g., Sr or Cd). The X-rays are measured in an energy-dispersive fashion. The advantage of PIXE is a considerably higher sensitivity compared to classical XRA. Most micro-components/trace elements with atomic numbers above 15 can be directly determined in wine.

3.3 Spectrophotometry/Colorimetry

In spite of the increased importance of instrumental methods in the last decennia (Crosby 1977), spectrophotometric methods are still used frequently for the determination of relatively high concentrations of trace elements in wine (Stöppler and Nürnberg 1984; see Sect. 1). They are still attractive methods of analysis since they are easy to carry out, require a relatively simple instrumentation, and are sufficiently sensitive for many elements (Marczenko 1981). After a chemical reaction, the elements are determined by measuring the selective absorption of light by the reaction products. The measurement is performed with filter photometers or at present most frequently with spectrophotometers. The quantitative determination is based on Beer's law. The chemical reactions are carried out with redox reagents or complexing agents, especially organic ones. The compounds thus formed can often be extracted into organic solvents ("extraction spectrophotometry"), a procedure by which the sensitivity and selectivity are increased. Selectivity can also be optimized by adjusting the pH value or adding complexing agents to mask interfering elements. Numerous reactions are known with which sensitive and selective determinations can be carried out (Koch and Koch-Dedic 1974; Lang and Vejdelek 1980; Sandell 1959; Sandell and Onishi 1978). As a rule, the sample must be mineralized prior to the spectrophotometric method, for example, by treating it with oxidizing acids or by dry ashing. Some elements can be determined directly in wine by extraction spectrophotometry. For instance, Cu can be reacted with zinc dibenzyldithiocarbamate to form brown copper dibenzyldithiocarbamate which is then extracted into CCl_4 from the acidified sample (Strunk and Andrease 1967). These direct colorimetric methods, however, can produce results which are too low if the complexing agents used are weaker than those naturally present in wine (Ivanov et al. 1972). Only absorbances down to 1×10^{-2} can be measured with the usual spectrophotometers. However, the detection limits can be improved by using sophisticated instrumentation (Harries 1982). It is questionable whether this increased sensitivity – absorbances down to 1×10^{-4} can be measured – will again result in an increased use of colorimetric methods, since the development of other instrumental methods has been extremely rapid.

3.4 Electrochemical Methods

Not only spectroscopic methods and activation analysis but also voltammetric methods are today powerful tools of trace analysis (Geissler 1980; Henze and Neeb 1986). A prerequisite is that the sample must be in solution. Hence, a digestion or another pretreatment is generally required. Dissolved species, even if not electrochemically active themselves, can affect the voltammetric determination, e.g., by influencing the reactions at the electrode. It is, therefore, often necessary to remove dissolved organic substances from solution. In many cases, a UV-photolysis prior to the actual determination will be sufficient if the sample does not contain too much organic material, as is the case with wine. Polarographic and voltammetric methods are superior to all other methods due to their favorable cost-performance ratio. Several elements can often be determined simultaneously ("oligo-element analysis"). Therefore, these methods have become extremely important in environmental analysis (Nürnberg 1982). Since they all respond to dissolved ionic species, they are also suitable for speciation studies, a feature which has been used especially in water analysis (Florence 1986). All present modern voltammetric methods are based on the direct current polarography that was developed by Heyrovsky, starting in 1923. A dropping mercury cathode is employed to record current-voltage curves which show step (wave) – like progressions with a plateau current (diffusion current) that is proportional to the concentration of the electrochemically active species. This current is caused by diffusion-controlled electrode reactions (reversible waves). In the case of slow electrode reactions (irreversible waves), the polarograms are not as well defined but can still be used analytically within certain limits.

Concentrations down to 10^{-5} mol l^{-1} can be determined by classical polarography. Smaller concentrations cannot be determined any longer due to charging and discharging phenomena at the electrode (electrochemical double layer). A large charging current is then superimposed on the diffusion current. The development of pulse techniques by Barker in 1953 considerably improved the possibilities of voltammetric methods. In these techniques, a rectangular potential pulse is applied on the dc-potential of the electrode and the current is measured at the end of the pulse when the charging current, in contrast to the diffusion current, has practically decayed to zero. Of the various methods based on this principle, especially differential pulse polarography (DPP) is widely used today in trace analysis. Concentrations in the range of 10^{-7} to 10^{-8} mol l^{-1} can be determined by DPP. Numerous ions and molecules can be determined by voltammetry[1] and polarography. These methods are especially important for the determination of heavy metals.

A further increase of sensitivity is obtained if "stripping" voltammetry is used (Neeb 1969; Wang 1985). In the simplest case, the enrichment is carried out by depositing metals electrochemically onto a stationary Hg-electrode. The metals are then dissolved again from the amalgam by anodic stripping, e.g., by differential pulse anodic stripping voltammetry (DPASV).

The signal obtained is higher than without deposition due to the fact that the concentration of the metals is higher in the amalgam than in the aqueous solution, depending on the conditions under which the deposition is carried out (e.g., stirring rate, deposition time, etc.). Cu, Sb, Bi, Tl, Sn, Pb, Cd, for instance, can be determined this way.

The application of stripping voltammetry can be extended by using other enrichment procedures. With cathodic stripping voltammetry (CSV), the enrichment takes place through anodic processes at the electrode, while stripping is car-

[1] The term "polarography" is used for techniques with a dropping mercury electrode, "voltammetry" for all others.

ried out in cathodic direction. An example is the determination of Se(IV) which is deposited as HgSe (A) onto a Hg-electrode from acid solution at a potential of −0.3 V. The determination is then carried out by cathodic stripping (B):

$$H_2SeO_3 + 4\,H^+ + 4\,e^- + Hg \rightarrow HgSe + 3\,H_2O \qquad (A)$$

$$HgSe + 2\,H^+ + 2\,e^- \rightarrow Hg + H_2Se\ (E_{peak}\text{: } -0.5 \text{ to } -0.8\text{ V}) \qquad (B).$$

In adsorption voltammetry, the ions, mostly cations, are reacted with suitable reagents to form stable, insoluble chelates or other complexes which can be adsorbed onto the electrode. After enrichment by adsorption under controlled conditions, the adsorbed species are determined by voltammetry. In this way, vanadium can be determined in the ppm to ppb range with catechol (Van den Berg 1984), nickel and cobalt with dimethylglyoxime (Pihlar et al. 1981).

Today, polarographic and voltammetric determinations can be carried out automatically by commercially available instruments with sample changers and computerized data processing (e.g., VA-Processor of Metrohm). The classical dropping mercury electrode and the old hanging mercury drop electrode (Kemula electrode) have been replaced by an automatically controlled mercury electrode. The mercury drops are no longer formed by gravity but by regulating the pressure in the mercury reservoir. Drop size and frequency are adjusted automatically by valves and timers. This multi-mode-electrode (MME) (Metrohm) can be used for polarographic as well as voltammetric stripping analysis.

While the usual polarographic and voltammetric methods all measure the current as a function of potential, potentiometric stripping analysis (PSA) is performed by measuring the potential with respect to time (Jagner 1982). After completion of the electrolytic enrichment at a mercury film electrode with subsequent formation of an amalgam, the potential of the electrode changes with time due to the reaction of an oxidizing agent (e.g., Hg^{2+}) in solution with the species deposited onto the electrode. At certain potentials, the resulting potential-time curves show "transition times" which are proportional to the concentration of the metal in solution and which can, therefore, be used to determine the metal concentration. An efficiency approaching that of voltammetric methods can be achieved only by computerized instrumentation which, however, is not available commercially.

Of the other electrochemical methods of analysis, only ion selective electrodes can be taken into consideration for the determination of small concentrations of ions (Cammann 1977). These electrodes respond directly to ionic species and should, therefore, be very suitable for speciation studies (see Sect. 4). However, their sensitivity and selectivity are not sufficient for the determination of microcomponents in wine and related samples (after their decomposition) so that this method has found hardly any application in this area.

Cu, Pb, and Zn in amounts of 0.03–0.08 ppm have been determined simultaneously by DPP in wine at a pH of 3.1 (potassium hydrogen phtalate) after concentrating the sample by evaporization and decomposing it with HNO_3/H_2O_2 (Bruno 1978). Higher sensitivities are obtained with stripping voltammetry, so that smaller sample volumes (1–10 ml) are required (Mannino 1982; Oehme and Lund 1979; Popko et al. 1978). Small amounts of cadmium can also be determined in this way. These elements can also be determined in wine by DPASV after

UV-irradiation of the acidified sample to which H_2O_2 has been added. (Golimowski et al. 1979). Ni and Co can be determined by adsorption voltammetry after the same treatment (Golimowski et al. 1980). A combination of methods is often necessary to determine extremely low concentrations of trace elements by voltammetry. Thallium, for instance, can be determined in the sub-ppb region in wine after a wet sample digestion and extraction of Tl(III)-bromide into ether (Eschnauer et al. 1984). Potentiometric stripping analysis has also been proposed to determine Pb, Cd, and Cu in wine (Jagner and Westerlund 1980). The determination can be carried out in slightly acidified wine without prior digestion.

Frequently, the results thus found for Cu are too low. This can, however, be avoided by heating the sample solution after acidification with nitric acid (1 h, 50 °C) McKinnon and Scollary 1986). Determinations without prior digestion can also be carried out in flow-through cells. The solution is changed between the electrolytic enrichment and the voltammetric determination so that the latter is carried out in a pure electrolyte solution without interfering matrix components. Pb has been determined in a flow-through cell by PSA (Anderson et al. 1982), Pb, Cd, and Cu by DPASV (Wahdat and Neeb 1987).

3.5 Isotope Methods

Since the 1950's, when nuclear reactors became available, neutron activation analysis has been developed into a reliable and efficient multi-element method of trace analysis, also for wine (Beridze 1969; Dedvarinani et al. 1973; Grimanis 1969; Siegmund and Bächmann 1977; May et al. 1982).

High neutron fluxes and cross-sections and favorable half-lives lead to low detection limits. The evaluation of an activation analysis, however, is not easy, due to a multitude of lines and half-lives. With germanium (lithium) detectors, which provide a good energy separation, numerous elements can be determined simultaneously by computerized instrumentation (instrumental neutron activation analysis, INAA). An example of this application is given in Chap. 5. A further refinement is obtained by carrying out an additional, radiochemical separation into single elements or a group of elements (radiochemical neutron activation analysis, RNAA).

The 14 rare earths have been identified in three Ingelheimer wines (vintage 1980) by radiochemical neutron activation analysis (RNAA). After separation by a fluoride-group precipitation each was determined quantitatively by using a C1-chondrite standard. The heavy rare earths with 0.52 µg Ce l^{-1} are significantly enriched in wine as compared to only 0.01 µg Eu l^{-1} for the lighter rare earths. The distribution pattern shows a slight negative anomaly for europium, which is rather remarkable for biologic material: Eu $<$ Tb $<$ Tm $<$ Ho $<$ Lu $<$ Sm $<$ Gd $<$ Dy $<$ Yb $<$ La $<$ Nd $<$ Ce (Schwellenbach 1983).

3.5.1 Isotope Dilution Analysis (Isotope dilution mass spectrometry, IDMS)

Here, a known spike of a suitable isotope of the element to be determined is added to an aliquot of the sample before or after digestion of the sample. Subsequently, the mass ratio is determined. This method has a high precision and reproducibility but it has not been applied to wine so far.

3.5.2 Radioactive Elements

The radio nuclides ^{14}C and ^{3}H can be of either natural or artificial origin. Natural ^{14}C and ^{3}H in wine are good model substances to examine atmospheric, meteorologic and hydrologic influences on vine growth.

Increased activities of artificial ^{14}C and ^{3}H were found at the beginning of the 1960's as a result of atomic bomb tests. After the test ban, the activities decreased again. It has also been possible to localize atomic explosions by measuring the activities of wines from the northern and southern hemispheres (Fischer et al. 1980).

Measurement of the ^{14}C content of ethanol can be utilized to distinguish between synthetic alcohol manufactured from fossile material and natural alcohol obtained by fermentation of recent material (brandy from potatoes and fruits) (Rauschenbach and Simon 1975). The ^{14}C content is measured with liquid scintillation spectrometers after distillation and purification of the alcohol. The procedure given has to be followed exactly (L'Orange and Zimen 1968).

3.5.3 Stable Isotopes

The natural variation of the isotopic forms of water, $H_2{}^{18}O$ and $H_2{}^{16}O$, can be used to characterize the origin and purity of wine. The measurement is preferably carried out by nuclear magnetic resonance (NMR) or mass spectrometry. The determination can be performed on wine without prior purification. CO_2 has been used as carrier gas (Förstel 1985).

3.6 Chromatographic Methods

Analytical chromatography is based on the fact that different substances pass a system consisting of a mobile and a stationary phase with different velocities. The stationary phase consists of an impregnated solid carrier material packed into suitable tubes (columns) of quartz, glass, or metal. The liquid phase is passed through the column, whereby the various components are separated. The determination is carried out in the eluate with a suitable detector.

In gas chromatography (GC), the mobile phase is an inert gas (nitrogen, hydrogen, or argon). Volatilization of the compound is achieved by injecting the sample into a heated entrance port.

In liquid chromatography the mobile phase is a liquid. High performance liquid chromatography (HPLC) is most commonly used. Working temperatures are relatively low so that sensitive substances can also be determined. Since inorganic compounds cannot generally be determined by chromatographic techniques, they must be transformed into suitable compounds by derivatization reactions (Nickless 1985). Metals are most often reacted to form chelates which can then be determined by HPLC or by GC in the case of substituted dithiocarbamates (Neeb

1982). Chromatographic methods are a combination of a separation and a determination technique, each of which is specific for a particular compound or element. Therefore, they are suitable for speciation studies (see Sect. 4), for example:

- to determine and quantify the valencies of an element in relation to its toxicity (As^{III}–As^{V} or Cr^{III}–Cr^{VI})
  or
- to determine the bonding in order to differentiate between inorganic and organic metal compounds of Hg, Pb, or Sn.

Multi-element analysis by ion chromatography (IC) with post chromatographic derivatization has been used for wine determinations. The trace elements Cu, Zn, Fe, Mn and the minerals Ca and Mg have been determined in ten drinks (wine and fruit juice) by this method and the results compared with AAS. IC was suitable and not subject to matrix interferences (Yan et al. 1986).

3.7 Mass-Spectrometry

Here, simultaneous multi-element trace analysis is most often performed by spark source mass spectrometry (SSMS). First, gaseous ions are produced thermally by a spark. Then, the ions are separated in a strong magnetic field by energy focusing. The sample must be conductive or be mixed with very pure graphite powder. The practical detection limit depends more on the purity of the graphite and on the preparation of the target than on the absolute detection limit which is around 10^{-9} to 10^{-12} g. It is absolutely required to use a control or reference material with exactly the same matrix composition. It is not easy to apply this method to wine or other biologic samples.

Thirteen elements have been determined simultaneously in several red and white wines from various European countries. The instrument used was a double focusing mass spectrometer (Model AEI, MS 702) with a spark source and electrical detection. The wines were ashed at 770° C during 24 h. The ash was then mixed with graphite and an internal standard. The error was high (20%) and could not be explained (Mannonen and Sihvonen 1983).

4 "Enologic" Speciation

The term "speciation" is applied to a much wider and more demanding field of trace analysis than the mere determination of total contents of trace elements. Speciation covers identification, separation, determination, and characterization of the chemical forms (species) of trace and ultra-trace elements. When applied to wine it could be coined "enologic" speciation. It includes methods which have been specifically developed for wine, methods for the determination of the elements in question, as well as separation methods. Examples of "enologic" speciation were known long before the term was applied to the field of trace analysis (Vogt 1979).

Already at the beginning of this century, two problems caused by iron in wines, especially white ones, were solved by "enologic" speciation. One of these was a persistent turbidity caused by a white or grey precipitate ("Bruch") of ferric phosphate, the other was caused by black ferric tannate. Once the cause of the turbidity was known, a way was also found to prevent it. By reducing Fe^{3+} to Fe^{2+}, the problem is only solved temporarily. The addition of ferro cyanide, on the other hand, has a lasting effect, since iron is precipitated as $Fe_4[Fe(CN)_6]_3$.

Towards the end of the 1950's, a method was found to prevent turbidity caused by copper. The insoluble copper (I) compounds, which cause the turbidity, are oxidized by air or H_2O_2 to copper (II) compounds, with the result that the tubidity disappears.

Tin can form colloidal, milky-white compounds and brownish precipitates of tin sulfide SnS with a hydrogen sulfide smell. Both compounds cause a turbidity in the wine.

Chromium is present in wine as Cr (III). Even Cr(VI), if added to wine, is reduced spontaneously to Cr(III), so that chromate ions cannot be detected any longer by their reaction with diphenylcarbazide (Beyermann and Eschnauer 1962).

As(V) is the most stable form of arsenic in the soil of vineyards. It is taken up by the vine and then reduced to the more toxic As(III) during fermentation of the must. In the wine finally, 60% of the total arsenic content is present as As(III), 40% as As(V). Methylarsonic acid (MAA) and dimethylarsinic acid (DMAA) can be determined by special methods of analysis, but their concentrations in wine are below the detection limit (Crecelius 1977).

The analysis of chemical species in red wine above the trace level is carried out by chromatographic or spectroscopic methods. Preliminary results can be obtained by simple methods of separation, such as ultra-filtration, liquid-solid extraction (questions regarding solubilities), and liquid-liquid extraction (especially with organic solvents). Separation of organically complexed elements can be achieved by a selective adsorption on different types of silica gels. Further investigations with respect to organic complexing agents of nickel showed that different amounts of nickel were distributed among the various fractions (Schwedt and Weber 1983).

5 "Enologic" Taxonomy (Pattern Recognition)

Wines can be categorized by numeric taxonomy (pattern recognition) according to location, vine variety, vintage year, climate, etc. Monovariate (single-element) or multivariate (multi-element) methods are used to determine the concentrations of minerals, trace and ultra-trace elements. Further information such as organic substances, physicochemical data, and sensory and enologic data can be included, but these will not be considered here.

The origin, i.e., the history, of a substance can be elucidated by determining the concentration of trace elements. Examples of this can be found in archeology, geology, and forensic science. Wine is a suitable matrix since a homogeneous sample of a large number of vines can be obtained from a large area. The term "enologic" taxonomy can then be used.

"Enologic" taxonomy is a typical two-step method. In the first step, as many elements as possible are determined by multi-element methods of analysis. These results are then used in the second step for an evaluation and characterization of the wine by mathematic methods. The work done so far shows that the geographic origin can be identified by multi-variate methods. It is possible to differentiate according to geologic formation, vineyard location and region of origin but less according to vine variety, vintage year, or climate. This characterization has been used to identify German wines from the Pfalz and Rheinhessen (Siegmund and Bächmann 1977, 1978), Pinot Noir wines from France and the USA (with a differentiation between wines from California and the Pacific Northwest) (Kwan et al. 1979; Frank and Kowalski 1984), Hungarian wines from three different regions of Transdanubia (Borszeki et al. 1983), and Spanish red wines from four regions of Catalonia (Rius et al. 1986).

"Enologic" taxonomy utilizes a variety of multi-element methods with which the concentration of a large number of elements can be determined simultaneously with sufficient reproducibility and sensitivity. Fifteen elements (Na, K, Sc, Cr, Fe, Co, Zn, Rb, Ag, Sb, Cs, Eu, Hf, Ta, Br) have been determined by instrumental neutron activation analysis (INAA) in 70 German wines, vintage 1970–1974 and produced from 11 different vine varieties in Pfalz, Rheinhessen and Franken. A sample of 200 ml of wine was ashed and irradiated for 48 (or 72) h in a quartz ampule in a research reactor of GfK in Karlsruhe with a neutron flux of 8×10^{13} cm^2 s^{-1}. Calibration was carried out with multi-element standards or with an iron monitor. The ratio of thermal to epithermal neutrons was 40 and the cross-sections of the elements in question were corrected accordingly. The measurements were carried out 5, 30, and 110 days after irradiation. In addition, Mn, Sr, and Y were determined in some samples by energy dispersive X-ray fluorescence analysis (XFA).

Seventeen elements (Cd, Mo, Mn, Ni, Cu, Al, Ba, Cr, Sr, Pb, B, Mg, Si, Na, Ca, P, K) have been determined in 40 Pinot Noir wines, *Vitis vinfera,* vintage 1966–1975 (14 from France, 9 from Washington and Oregon) by optical emission spectrometry (OES). For each element a standard was used (ACS grade chemicals). The samples were diluted with deionized, distilled water containing 12% ethanol. Four replicate measurements were carried out for each element and each sample.

The concentrations of Fe, Mg, Cr, P, B, Mn, Pb, Ni, Al, Ag, Ti, Co, and Zn have been determined in 38 Riesling wines of the same vintage year from three different regions of Transdanubia/Hungary (viz. Balatonfüred, Badascony, and Sopron) by optical emission spectrometry (OES). Two more elements have identified qualitatively.

In another publication, 105 Spanish red wines from four regions of Catalonia (viz. Conca de Barbera, Priorat, Terra Alta, and Tarragona) have been examined with regard to ten enologic parameters and seven element concentrations. Ca,

Mg, Sr, Mn, and Zn were determined by AAS; K, Mg, and Li by "flame"-OES. The second step of enologic taxonomy consists of a mathematical treatment of the results obtained by the multi-element methods of analysis mentioned above.

Possible connections can be detected by a monovariate examination of only one trace element concentration (key element) using a graphical representation (histogram) or the average value ($\bar{x}$) with its standard deviation (σ). The information thus obtained is, however, limited.

A more detail characterization is only possible with multivariate methods involving several elements. Several mathematical methods can be used. One of these is the cluster analysis (normalization of analysis results, choice of variables and alogarithm). The results are plotted in a dendrogram from which the relationships can be read off.

Several other mathematic methods have been developed and applied, e.g., partial least-squares regression models, AUTO-SCALE numerical method, WEIGHT correlation matrix, SELECT, K–L TRANSFORM, LEAST least-squares multilinear regression analysis, KNN K-nearest-neighbor analysis, SIMCA statistical isolinear multiple component analysis. Details of these methods should be obtained from the literature cited above. The error of these methods is surprisingly small, so that the origin of the wines can be determined with sufficient confidence.

References

Addink NWH (1971) DC Arc analysis Macmillan. London and Basingstoke

Anders U, Hailer G (1976) The substoichiometric extraction system MIBK/APDC and its application to the direct determination of lead and cadmium in wine. Fesenius Z Anal Chem 278:203–206

Anderson L, Jagner D, Josefson M (1982) Potentiometric stripping analysis in flow cells. Anal Chem 54:1371–1376

Bergner KG, Lang B (1970) Utilization of X-ray fluorescence spectroscopy in wine analysis (iron, copper, zinc, manganese and bromide). Dtsch Lebensm-Rundsch 66:157–164

Beridze GI, Macharashvili GR, Mosulishvili LM (1969) Quantitative determination of gold in some wines by neutron activation method. Radiokhimya Coden: RADKAU. Ser. 11.6. (1969) 726

Beyermann K, Eschnauer H (1962) Bestimmung von Chrom im Most and Wein. Z Lebens Unters Forsch 118:308–311

Blake CJ (1980) Sample preparation methods for the analysis of metals in foods by atomic absorption spectrometry – a literture review. Sci Tech Surv-Br Food Manuf Ind Res Assoc 122:57

Bock R (1979) A handbook of decomposition methods in analytical chemistry. Int Textbook Company

Borszeki J, Koltay L, Inczedy J, Gegus E (1983) Untersuchung der Mineralstoffzusammensetzung von Weinen aus Transdanubien und ihre Klassifikation nach Weingegenden. Z Lebensm Unters Forsch 177:15–18

Bruno P, Caselli M, Di Fano A, Fragale C (1978) Simultaneous determination of copper, lead and zinc in wine by differential-pulse polarography. Analyst 103:868–871

Cammann K (1977) Das Arbeiten mit ionenselektiven Elektroden. Springer, Berlin Heidelberg New York

Campbell JL, Orr BH, Herman AW, McNelles LA, Thomson JA, Cook WB (1975) Trace element analysis of fluids by proton-induced X-ray fluorescence spectrometry. Anal Chem 47:1542–1553

Cela R, Cabezon LM, Perez-Bustamente JA (1983) Application of the modified Fenton's reagent to the determination of metal traces in Sherry-type wines. An Quim 798:229–232

Crecelius EA (1977) Arsenite and arsenate levels in wine. Bull Environ Contam Tox 18:227–230

Cresser MS, Ebdon LC, McLeod CW, Burridge JC (1986) Atomic spectrometry update-environmental analysis. J Anal Atomic Spectrom 1:1R–27R

Crosby NT (1977) Determination of metals in foods. Analyst 102:225–268

Dedvariani TG, Beridze GI, Zakharov EA, Macharashvili GR, Daneliya GI (1973) Neutron activiation determination of rubidium in red wines of the georgian SSR. Soobshch Akad Nauk Gruz SSR. Coden: SAKNAH. Ser. 72. 1. 161

Dedvarinani TG, Beridze GI, Mosulishvili ML, Shoniya NI, Macharashvili GR, Daneliya GI, Osei YP (1973) Neutron activation analysis used for the determination of rubidium in wines. Vestn Sel'skokhoz Nauki (Moscou) Coden: VSNLAF 9:113

Dorten W, Valenta P, Nürnberg HW (1984) A new photodigestion device to decompose organic matter in water. Fres Z Anal Chem 317:264–272

Ebel S, Kamm K (1983) Statistical definition of the limit of determination. Fres Z Anal Chem 316:382–385

Eschnauer H (1972) Spurenelemente in Wein und anderen Getränken. Verlag Chemie, Weinheim

Eschnauer H (1986) Spurenelemente und Ultra-Spurenelemente in Wein. Naturwissenschaften 73:281–290

Eschnauer H, Gemmer-Colos V, Neeb R (1984) Thallium in wine trace-element-vinogram of thallium (in German). Z Lebensm Unters Forsch 178:453–460

Fassel VA (1986) Analytical inductivety coupled plasma spectroscopies-past, present, and future. Fres Z Anal Chem 324:511–518

Fischer E, Müller H, Rapp A, Steffan H (1980) Tritium- und Kohlenstoff-14-Gehalte von Weinen verschiedener Jahrgänge der nördlichen und südlichen Hemisphäre. Z Lebensm Unters Forsch 171:269–271

Florence TM (1986) Electrochemical approaches to trace element Speciation in Waters. Analyst 111:489–505

Förstel H (1985) Die natürliche Fraktionierung der stabilen Sauerstoff-Isotope als Indikator für Reinheit und Herkunft von Wein. Naturwissenschaften 72:449–455

Frank IE, Kowalski BR (1984) Prediction of wine quality and geographic origin from chemical measurements by oartual least-squares regression modeling. Analytica Chimica Acta 162:241–251

Galan L de (1986) New directions in optical atomic spectrometry. Anal Chem 58(6):697A-707A

Gallego R, Bernal JL, Del Nozal MJ (1983) Determination of iron, copper, manganese, zinc and lead in wines by atomic absorption spectrometry (in Span) An Bromatol Vol 1981, 33:175–190

Geißler M (1980) Polarographische Analyse. Akademische Verlagsgesellschaft, Leipzig

Golimowski J, Valenta P, Nürnberg HW (1979 a) Toxic trace metals in food I. A new voltammetric procedure for toxic trace metal control of wines (in German) Z Lebensm Unters Forsch 168:353–359

Golimowski J, Valenta P, Stoeppler M, Nürnberg HW (1979 b) Toxic trace metals in food II A comparative study of the levels of toxic trace metals in wine by differential pulse anodic stripping voltammetry and electrothermal atomicalsorption spectrometry. Z Lebensm Unters Forsch 168:439–443

Golimowski J, Nürnberg HW, Valenta P (1980) Die voltammetrische Bestimmung toxischer Spurenmetalle im Wein. Lebensmittelchem und Gerichtl Chem 34:116–120

Gorsuch TT (1970) The Destruction of organic matter. Pergamon Press, Oxford

Gray AL (1986) Ions or photons- an assessment of the relationship between emission and mass spectrometry with the ICP. Fres Z Anal Chem 324:561–570

Grimanis AP (1969) Simultaneous determination of arsenic and copper in wines and biological materials by neutron activation analysis. Proc 1968 Int Conf "Modern Trends in activation analysis" National Bureau of Standards. Special publication 312 1:197
Haller HE (1971) Determination of lead in wine by atomic absorption spectrometry (in German) Dtsch Lebensm Rundsch 71:430–431
Harnly JM (1986) Multielement atomic absorption with a continous source. Anal Chem 58(8):933A–943A
Harries TD (1982) High sensitivity spectrophotometry. Anal Chem 54(6):741A–750A
Henze G, Neeb R (1986) Elektrochemische Analytik, Springer, Berlin Heidelberg New York Tokyo
Interesse FS, Lamparelli F, Alloggio V (1984) Mineral contents of some southern italian wines. Z Lebensm Unters Forsch 178:271–278
Ivanov K, Maneva D, Popov D (1972) Determination of various forms of copper in Bulgarian white wines, brandy distillates, and brandies by ion or exchange chromatography (in Bulgarian). Nauchni Tr Vissh Inst Kharanit Vkusova Prom-st Plodiv 19:289–296
Jackwerth E, Gomiscek (1984) Acid pressure decomposition in trace element analysis. Pure and Appl Chem 56:479–489
Jagner D (1982) Potentiometrie stripping analysis-A Review. Analyst 107:593–599
Jagner D, Westerlunds (1980) Determination of lead, copper and cadmium in wine and beer by potentiometric stripping analysis. Anal Chim Acta 117:159–164
Jakob L (1987) Anwendung der ICP (Inductively coupled plasma atomic emission spectroscopy) in der Weinanalytik. In: 8. Int Önol Symp 28.–30. April 1987, Kapstadt, Republik Südafrika
Jenkins R, De Vries JL (1975) Practical x-Ray Spectrometry 2nd ed Phjilips Technical Library. Springer, New York Inc
Jenkins R, Gould RW, Gedcke D (1981) Quantitative X-ray spectrometry. Marcel Dekker Inc, New York and Basel Jenkius et al.
Junge C, Spadinger C (1987) Die Konsequenzen aus dem Einsatz moderner Analysenmethoden. In: 8. Int Önol Symp 28.–30. April 1987, Kapstadt, Republik Südafrika
Khan R, Crumpton D (1981) Proton induced X-ray emmission analysis Part I, II. Crit Rev Anal Chem 11:103–193
Knapp G (1984) Decomposition methods in elemental trace analysis. Trends Anal Chem 3:182–185
Knapp G (1985) Sample preparation techniques – an important part in trace element analysis for environmental research and monitoring. Int J Environ Anal Chem 22:71–83
Knapp G, Raptis SE, Kaiser G, Tölg G, Schramel P, Schreiber B (1981) A partially mechanized system for the combustion of organic samples in a stream of oxygen with quantitative recovery of the trace elements. Fres Z Anal Chem 308:97–103
Koch OG, Koch-Dedic GA (1974) Handbuch der Spurenanalyse Teil 1, 2. Springer, Berlin Heidelberg New York
Kwan WO, Kowalski BR, Skogerboe RK (1979) Pattern recognition analysis of elemental data. Wines of Vitis vinifera cv. Pinot Noir from France and the United States. Agricult Food Chem 27:1321–1326
Lang B, Vejdelek ZJ (1980) Photometrische Analyse. Verlag Chemie, Weinheim Deerfield-Beach Basel
Liebhafsky HA, Pfeiffer HG, Winslow EH, Zemany PD, Liebhafsky SS (collab. editor) (1972) X-rays, electrons and analytical chemistry. Wiley-Interscience, New York Canada Sydney Toronto
Mack D (1975) Determination of lead in wine and juices with flameless atomic absorption spectrometry (in German) Dtsch Lebensm-Rundsch 71:71–72, 430–431
Mankopff R, Friede G (1975) Grundlagen und Methoden der chemischen Emissionsspektralanalyse. Verlag Chemie, Weinheim
Mannino S (1982) Determination of lead, copper, and cadmium in wines using anode dissolution potentiometry (in Italian) Riv Vitic Enol 35:297–304
Mannonen R, Sihvonen M (1983) The use of spark source mass spectrometry in analyzing trace elements in wine. Proc Int Conf Heavy Metals in the Environment. Heidelberg, Sept. 1983, 1:261

Marczenko Z (1981) Spectrophotometric determination of trace elements. Crit Rev Anal Chem 11:195–260
May S, Leroy J, Piccot D, Pinte G (1982) Application de l'analyse par activation neutronique à'la détermination des vins provenant de différents crus. Possibilité d'identification d'un vignoble par la détermination des oligoélements. J Radioanal Chem 72:305–318
McKinnon A, Scollary G (1986) Determination of Copper in Wine by Potentiometric-stripping analysis. Analyst 111:589–591
Medina B, Sudraud P (1980) Levels of chromium and nickel in wines. Enrichment cause (in French). Connaiss Vigne Vin 14:79–96
Méranger JC, Somers E (1968) Determination of heavy metals in wines by atomic absorption spectrometry. J Ass Offic Anal Chem 51:922–955
Misselhorn K, Grafahrend W (1986) Identifizierung des Rohstoffs von hochrektifiziertem Alkohol. In: Spirituosen Jahrbuch. Versuchs- und Lehranstalt für Spiritusfabrikation und Fermentationstechnologie in Berlin, pp 448–452
Mitchel JW (1982) State of the art contamination control techniques for ultratrace elemental analysis. J Radioanal Chem 69:47–105
Neeb R (1969) Inverse Polarographie und Voltammetrie. Verlag Chemie, Weinheim
Neeb R (1982) Metal-chelate gas-chromatography for trace element analysis. Pure and Appl Chem 54:847–852
Nickless G (1985) Trace Metal Determination by Chromatography. J Chromat 313:129–159
Noble AC, Orr BH, Cook WB, Campbell JL (1976) Trace element analyses of wine by proton-induced X-ray fluorescence spectrometry. J Agric Food Chem 24:532–535
Nürnberg HW (1982) Voltammetric trace analysis in ecological chemistry of toxic metals. Pure and Appl Chem 54:853–878
Oehme M, Lund W (1979) Determination of cadmium, lead and copper in wine by differential pulse anodic stripping voltammetry. Fres Z Anal Chem 294:391–397
L'Orange R, Zimen KE (1968) C-14 aus Kernwaffenexplosionen Szintillations-Messungen des Äthanols naturreiner Weine. Naturwissenschaften 55:35–36
Pihlar B, Valenta P, Nürnberg HW (1981) New highperformance analytical procedure for the voltammetric determination of nickel in routine analysis of waters, biological materials and food. Fres Z Anal Chem 307:337–376
Popko RA, Pichugina IM, Petrov SI, Neiman EY (1978) Determination of heavy metals as microimpurities in wine by inverse voltammetry (in Russian). Zh Anal Khim 33:2108–2112
Rauschenbach P, Simon H (1975) Weitere Untersuchungen zum ^{14}C-Gehalt von Gärungsalkohol in Abhängigkeit von Wachstums-Zeit und Ort des Fermentationsmaterials. Z Lebensm Unters Forsch 157:143–146
Ristow R, Bernau M (1982) Reproducibility and comparability of various atomic absorption spectrophotometric methods to determine lead in wine (in German) Dtsch Lebensm-Rundschau 78:125–130
Rius FX, Franques MR, Ferre M, Larrechi MS (1986) Pattern recognition techniques applied to micro and trace constituents of Catalan wines. 10 ISM 86 abstracts Antwerp 25.–29.8.1986 Nr. 39
Sandell EG (1959) Colorimetric Determination of traces of metals, 3ed. Intenscience, New York
Sandell EB, Onishi H (1978) Photometric determination of traces of metals. John Wiley & Sons, New York Chichester Brisbane Toronto
Sansoni B, Kracke W (1968) Decomposition and ashing of organic substances by means of radicals in aqueons solution II. Wet-ashing of food stuffs and biological material by H_2O_2/Fe^{2+} (in German) Z Anal Chem 243:209–241
Schramel P, Klose BJ, Masse S (1982) Efficiency of ICP emission spectroscopy for the determination of trace elements in bio-medical and environmental samples (in German) Fres Z Anal Chem 310:209–216
Schramel P, Hasse S, Knapp G (1987) Application of the highpressure ashing system HPAV according to Knapp for voltammetric determinations of trace elements in biological material (in German) Fres Z Anal Chem 326:142–145

Schwedt G, Weber G (1983) Trennverfahren zur Analytik von Elementbindungsformen in Lebens- und Genußmitteln. Dtsch Lebensm Rundsch 79:213–221
Schwellenbach W (1983) Neutronenaktivierungsanalytische Untersuchungen über die Verteilung der Seltenen Erden und anderer Spurenelemente in Wein, Rebpflanzen und Böden. Diplomarbeit Universität Köln
Severin G, Schumacher E, Umland F (1982) Determination of cadmium by flameless atomic-absorption spectroscopy II. Modification of graphite furnace tubes by aluminium oxide (in German). Fres Z Anal Chem 34:205–208
Siegmund H, Bächmann K (1977) Die Lagezuordnung von Weinen durch Bestimmung des Spurenelementmusters. Z Lebensm Unters Forsch 164:1–7
Siegmund H, Bächmann K (1978) Anwendung der numerischen Taxonomie für die Klassifizierung von Weinen. Z Lebensm Unters Forsch 166:298–303
Siemer DD, Vitek RK, Koteel P, Houser WC (1977) Determination of arsenic in beverages and foods by hydride generation atomic absorption spectroscopy. Anal Lett 10:357–369
Stephen R (1980) Zeeman modulated AAS. Crit Rev Anal Chem 9:167–195
Stoeppler M, Nürnberg HW (1984) Analytik von Metallen und ihren Verbindungen. In: Merian E (ed) Metalle in der Umwelt. Verlag Chemie, Weinheim Deerfield Beach Basel, pp 45–104
Strunk D, Andreasen AA (1967) Collaborative study using a Zinc dibenzyldithiocarbamate colorimetric method for the determination of copper in alcoholic products. J Assoc Off Anal Chem 50:334–338
Thompson M, Walsh JN (1986) A handbook of inductively coupled plasma spectrometry. Blackie, Glasgow London
Tschöpel P (1982) Modern strategies in the determination of very low concentrations of elements in inorganic and organic materials. Pure and Appl Chem 54:913–925
Tschöpel P, Kotz L, Schulz W, Weber M, Tölg G (1980) Causes and elimination of systematic errors in the determination of elements in aqueous solutions in the ng/ml and pg/ml range (in German). Fres Z Anal Chem 302:1–14
Van den Berg CMG (1985) Direct determination of molybdenium in seawater by adsorption voltammetry. Anal Chem 57:1532–1536
Van den Berg CMG, Huang G (1984) Direct determination of dissolved vanadium in seawater by cathodic stripping voltammetry with the hanging mercury drop electrode. Anal Chem 56:2383–2386
Van Grieken R (1982) Preconcentration methods for the analysis of water by X-ray spectrometric techniques. Anal Chim Acta 143:3–34
Van Loon JC (1981) Atomic fluorescence spectrometry-present status and future prospects. Anal Chem 53:333A–361A
Van Loon JC (1985) Selected methods of trace metal analysis: biological and environmental samples. Wiley & Sons, New York Chichester Brisbane Toronto Singapore
Vogt E (1979) Der Wein, 8. Auflage. Verlag Eugen Ulmer, Stuttgart
Wahdat F, Neeb R (1987) Digestion – free determination of trace metals (Zn, Cd, Pb, Cu) in beverages (wine) by inverse voltammetry in flow-through cells. Fres Z Anal Chem (in press)
Wang J (1985) Stripping Analysis. Verlag Chemie, Weinheim
Weber G, Schwedt G (1982) Zur Analytik chemischer Bindungsformen von Nickelspuren in Kaffee, Tee und Rotwein mit chromatographischen und spektroskopischen Methoden. Anal Chim Acta 134:81
Welz B (1986) Atomic absorption spectroscopy. 2nd edn, Verlag Chemie, Weinheim New York
Winge RK, Fassel VA, Peterson VJ, Floyd MA (1985) Inductively coupled plasma – atomic emission spectroscopy – an atlas of spectral information. Elsevier Science Publishers, Amsterdam
Yan D, Stumpp E, Schwedt G (1985) Vergleich von Ionen-Chromatographie und Atomabsorptions-Spektrometrie zur Metallionen-Analytik in Wein und Obstsäften. Fres Z Anal Chem 322:474–479

Young FC, Roush ML, Berman PG (1973) Trace element analysis by proton – induced X-ray fluorescance. Int J Appl Rad Isotopes 24:153–163
Zander AT (1986) Atomic emission sources for solution spectrochemistry. Anal Chem 58(11):1139A–1149A
Zee JA, Szogmy IM, Tremblay JS (1983) Elemental analysis of Canadian, European and American wines by photon-induced X-ray fluorescence. Am J Enol Vitic 34:152
Zielkowski R, Bächmann K (1978) Instrumental multielement analysis of biological matrixes: a comparison of NAA, RFA and AAS (in German). Fres Z Anal Chem 290:143–144

Acids and Amino Acids in Grapes and Wines

C. S. Ough

1 Introduction

Acids in grapes and wines generally refer to the organic acids either found naturally in the grapes or wines produced by biochemical activity in the grapes or by yeast or bacteria during their growth and fermentation.

The inorganic acid anions which are found in grapes and wines are mainly phosphate, sulfate, nitrate and chloride. These are not thought of as acids in grapes or wines. Also, there are acids which are added to wines for fungistatic or other purposes – sorbic acid, sulfurous acid, benzoic acid, 5-nitrofurylacrylic acid and others. These also will not be covered.

The amino acids discussed are those which are produced by the grape or in certain instances by the yeast or bacteria. Peptides and proteins will not be covered in these discussions. The discussion will be restricted to the natural amino acids present.

2 Organic Acids

Radler (1975) reviewed the sources of organic acids found in grape juice and wines. Listed in Table 1 are those acids which can be measured effectively in most

Table 1. Acids which can be measured effectively under normal control

Source	Name of acid	Concentration $g\,l^{-1}$
Juice	Tartaric	0.5–5.0
	Malic	0.0–8.0
	Citric	0.0–1.0
	Fumaric	0.0–1.0
Yeast	Succinic	0.5–2.0
	Pyruvic	0.1–0.5
	α-Ketoglutaric	0.1–0.5
Bacteria	Lactic	0.0–5.0
	Acetic	0.0–1.0
Botrytis	Muccic	0.0–1.0
	Gluconic	0.0–1.0

laboratories with normal control and research capabilities and properly trained personnel.

There are many other organic acids identified in wine which can be measured. These include the fatty acid series, other hydroxy and α-keto acids, cyclic and aromatic series, and nitrogen- and sulfur-containing acids.

2.1 Total Titratable Acidity

Almost all grapes and wines have a "total" acidity determined for them. This measurement is one of the key determinants on when to harvest grapes and how to treat the harvested grapes and later the wines. Total acidity refers to the titratable acidity. In France, the titratable acidity is over the range of the natural pH of the grape or wine to pH 7.0. In most other countries, the end point is pH 8.2–8.6 or the inflection point in the base titration of the juice or wine. Because of the varying concentrations of the various mon-, di- and tribasic acids present and the buffering effects of the various proteins, amino acids and other organic and inorganic compounds, the answers obtained are arbitrary in the sense that they do not measure anything specific other than the amount of base required to shift the pH from one level in the grape juice or wine to another.

It has been known for many years that the titratable acidity could be related to the sum of the individual acids, the base materials, and phosphate and sulfur dioxide. In recent years, Amerine and Ough (1980), Ough et al. (1969) and Boulton (1980) have reported the relationship.

In France the titratable acidity is calculated as sulfuric acid as g l^{-1} and in the United States as tartaric acid as g 100 ml. The official Office International de la Vigne et du Vin method (Anonymous 1978) defines the French method. Amerine and Ough (1980) discussed and gave the method approved for use in the United States and many other countries.

Guymon and Ough (1962) identified the various ways to remove the possible interference of carbon dioxide and offered several alternatives. The end point can be determined either by phenolphthalein indicator or by a pH meter. The most efficient is by pH meter.

Method

Take a 5- or 10-ml aliquot of juice or wine and pipet it into a 250-ml vacuum flask. Subject the sample to a vacuum (water aspirator intensity) and shake for 10–20 s to remove the carbon dioxide. Wash the sample into a 200-ml beaker with about 100 ml of previously boiled and cooled (under nitrogen) distilled water adjusted to pH 8.4. Place the beaker on a magnetic stirrer, add a stir bar, immerse previously standardized pH electrodes into the diluted sample and stir. Titrate to pH 8.4 with standardized base. The total titratable acidity is calculated as g tartaric acid per 100 ml by this expression.

$$\frac{\text{Vol. NaOH} \times \text{Normality NaOH} \times 7.5}{\text{Volume of sample}} .$$

Errors can occur from changes in the standard NaOH, measurement errors, failure to use distilled water, poorly standardized pH meter. Errors in grape juice or wine can occur from super saturation with potassium hydrogen tartrate. When the juice or wine is chilled, losses in this compound occur which affect the titratable acidity.

The titration can be done with automatic titration units which eliminate some human error in determining the end point; however, the other factors must still be considered.

Application

The total acidity is determined on grapes for the purpose of deciding when to harvest. In addition, when the grapes are low in acidity, tartaric or other acids are added to increase the acidity and lower the pH. Wines from colder climates may have to be deacidified because of excess acidity. Wines which undergo malolactic fermentation may have the titratable acidity reduced to a level requiring the addition of acid. The reasons for addition of acids are to improve the microbiological stability and the sensory characteristics of a wine. Deacidification of grape juice or wines is to allow for better microbiological growth and to improve sensory characteristics and stability of components (Amerine et al. 1980; Winkler et al. 1974). Kliewer (1965) followed the total titratable acidity in grape flowers through overripe berries. The highest level per gram of fresh tissue was in the green berry juice at veraison. The changes in total acidity after veraison under field conditions were noted by Kliewer (1968 b) and for artificial conditions (Kliewer 1973). Leafroll virus caused increases in infected vines (Alley et al. 1963; Kliewer and Lider 1976).

Godinho et al. (1984) used the regular potentiometric and thermometric method with fair agreement. A second endpoint was discussed which was assumed to represent titration of phenol groups.

2.2 Paper, Thin Layer and HPTLC Chromatography

Scholten et al. (1983) describes the use of a computer-controlled continuous flow wine analyzer to measure different components at once; among these were tartaric, citric, malic, and lactic acids.

Paper and thin layers were popular methods to separate and estimate juice and wine acids. The report of Selmeci and Hanusz (1981) documents the usual analytic problems with visual estimates; ±20% error in the determinations. Bourzeix et al. (1970) used thin layer cellulose and photodensitometer to quantify wine acids. By careful work, Lin and Tanner (1985), using cellulose thin layer, color development with xylose-aniline reagent, and densitometer scan, reduced the coefficient of variation to about 3–6%. This method is presented below.

Method

Filter white wine and juice samples. Filter red wine samples with polyvinypolypyrrolidone to remove red pigments. For citric acid analysis dilute juice samples 1:12 with water. Use high performance thin layer chromatography (HPTLC) 10 × 10 cm precoated plates (such as Merck, without F). Spray on band 10 mm wide, 2 µl volume, keep bands 5 mm apart, use six samples/plate. Dry in cold air stream for 3 min, then place in chamber containing 3 ml of developing solvent.

Mix ethyl acetate, toluene, water and formic acid (60:20:20:15) in an appropriate separator funnel and let settle overnight. Take the upper clear layer for the developing solvent. Develop for 50 mm without any preconditioning (takes about 10 min). Dry the plate for 10 min at 140° C. After cooling to room temperature spray with equal volumes of freshly mixed staining solvent [a) 4 g xylose dissolved in 12 ml H_2O then diluted to 200 ml with methanol and (b) 4 ml aniline dissolved in 200 ml ethanol] and dry at 140° C for 5 to 10 min. Acid bands are brown on grey. Scan the plates at 546 nm with TLC scanner and integrate the areas. Relate the areas to standards prepared for gluconic acid 1.9 g l^{-1}, L(+) tartaric acid 2.6 g l^{-1}, citric acid 3.0 g l^{-1}, L(−) malic acid 4.9 g l^{-1}, glycolic acid 2.9 g l, α-ketoglutamic acid 1.3 g l^{-1}, succinic acid 3.9 g l^{-1} and DL-lactic acid 8.1 g l^{-1}. Dilute to appropriate levels and make standard curves using technique described above.

Applications

Dimotaki-Kourakou (1965) used paper chromatography to separate the organic acid formed with ^{14}C glucose during yeast fermentation and detected them by autoradiography. Beridze and Sikharulidze (1972) measured some old wines and found the usual acids with the exception of pyruvate. Souza (1971) used thin layer (silica gel) to chromatography acids in wines. Brun and Grau (1968) used polyamide flexible sheets to separate the organic acids in wines Webb et al. (1967) used paper chromatography to identify ethyl acid tartrate and ethyl acid malate. Paper chromatography can be used for determination of completion of malolactic fermentation. This will be covered in Section 2.8.

2.3 Gas Chromatography

The nonvolatile acids can be volatilized by forming methyl esters (Drummond and Shama 1982) in fermentation products and by Di Stefano and Bruno (1983) in wine. However, by far the most common derivatization method is that of silylation. The main reason is that silylation reacts with all the polar sites and the resultant esters are nonpolar and more easily volatilized and separated.

Fernandez-Flores et al. (1970) and Brunelle et al. (1967) indicated that silylation was practical for wine and juice analysis. Mattick et al. (1971) modified the method slightly, reducing sample size. Wagener et al. (1971) further modified the method. The work of Ryan and Dupont (1973) fully identified the silyl derivatives by thin layer and mass spectral analysis. Baker (1973) also used anion exchange isolation and silylation to separate and quantify organic acids. Bertrand (1974) made the methyl esters and then silylated the hydroxyl groups prior to gas chromatography. Ribereau-Gayon and Bertrand (1971) also reported the successful silylation of the organic acids of wine.

The methods more or less follow the procedure of isolation of the organic acids then silylation with one or another silylating agents.

The method chosen here is that of DeSmedt et al. (1981). The method includes the analyses of sugar as well, but seems to have the best features of all the variations. The Marcy and Carroll (1982) method is similar.

Method
Dilute sweetened samples as necessary. Mix equal volume of diluted sample and internal standard (1 $g\,l^{-1}$ of vanillic acid and 0.5 $g\,l^{-1}$ of α-methyl-D-mannoside in 20% v/v ethanol water solution). Take 50 μl and transfer to 2-ml sample vial, add 1 drop of aqueous ammonia and evaporate to dryness under nitrogen stream at room temperature. Dissolve the dry residue in 50 μl of anhydrous pyridine and 100 μl of silylating reagent [N,O-bis(trimethylsilyl)trifluoroacetamide (BSTFA) plus 1% trimethylchlorosilane (TMCS)]. Seal the vial with a Teflon-faced cap and heat for 1 h at 80° C. Inject 1 μl of the clear, single-phase liquid. Use a methyl silicone bonded capillary column 30 m in length and 0.25 mm inside diameter. Set the split on the column at 1:10. Program the temperature form 60° to 280° C at 4° C min^{-1}. Have the injector and the flame ionization detector at 290° C. Set the carrier gas at 3 ml min^{-1}.

Make a series of standards covering the range of anticipated organic acid levels. Prepare as described and chromatograph and contruct curves relating ratio of the acid area/int. std. area to the concentration of the acid. Relate the sample ratio of acid area/int. std. area to the standard prepared curve for each acid and determine concentration, taking into account any dilutions used.

Clean the glass injector insert if the peak areas of the two internal standards vary more than 1.55. Remove a coil or two from the column next to the injector if the problem of variable results persists. The coefficients of variation of the measurement of the organic acids are reasonable (2 to 16%).

Application
Mattick et al. 1971; Martin et al. 1971; Zanier and Tanner 1977; Tanner and Zanier 1976; Modi et al. 1976 measured either one or two samples of wines by the use of one form or another of a silylating agent. Some, for example Wagener et al. (1971), isolated the acids by anion exchange. Others used lead salts to precipitate them, then dried and silylated them. Either way some losses occur. Philip and Nelson (1973) document the losses by anion treatment. The use of columns varied, as did the acid separated and measured. Actual use of the method to measure metabolic changes has been limited to the paper of Cash et al. (1977), who showed changes in Concord grape acids during maturation, Kluba and Mattick (1978) showed similar changes in the fruit as well as changes during fermentation, and Wagener et al. (1971) demonstrated changes during fermentation and explained the loss of fumaric acid during fermentation.

2.4 High Performance Liquid Chromatography

HPLC methods of separation and detection can be divided into two major groups – anion or cation exchange separation and reverse phase column separation. The first group can be subdivided into the more or less classical elution with a suitable solvent and UV detection. The second and newly developed system is the ion exclusion system, where the eluting solvent ions are removed by one method or another (supressors) and the acid anions are detected by a very sensitive conductivity detector. This method is described in its latest form by Rocklin et al. (1986). Monk and Iland (1984a, b) used the method to detect malic and tartaric acids in grape juice and wine. They reported good agreement with standard chemical methods for tartaric acid but high values for malic acid compared to the enzymatic method. This is a developing method and will undoubtedly be of value in the future.

Direct separation by reverse phase generally is unsuccessful due to interferences from sugars and other components in the product to be measured. Shen et al. (1984) reported satisfactory recoveries from wine. Droz and Tanner (1982) used a pretreatment of cation exchange then C_{18} separation and UV detection. Bigliardi et al. (1979), Schneyder and Flak (1981), Flak and Pluhar (1983) analyzed juices or wines using an anion clean-up, then C_{18} separation and elution with either water-phosphate or methanol-phosphate buffer. Symonds (1978) pretreated with a strong cation resin, then used HPLC with anion resin column to separate, and detected acids using differential refractometry. Yoshida et al. (1985) used an analytical anion column and direct elution with disodiumphthalate and UV detection.

Several reports discussed first derivatization of the organic acids, then separation on C_{18} reverse phase columns. Mentasti et al. (1985) used phenacyl bromide, with 18-crown-6 as the catalyst, to make the derivatives. No clean-up was required, only a pH adjustment. Derivatization was reasonably complete. The p-nitrobenzyl esters were made by Steiner et al. (1984) using p-nitrobenzyl-N,N^1-diisopropylurea. Derivatization was simple and a Sep-Pak clean-up was all that was required after derivatization. Recoveries were adequate. Separation and detection were on C_{18} reverse phase column with elution gradient of methanol-H_2O and UV detection. Either of these two derivatization methods seem adequate.

Effective separation of organic acids by use of anion exchange resin columns was shown for grape juice by Palmer and List (1973). Shimazu and Watanabe (1976, 1981) also separated the organic ions by anion exchange but derivatized them post-column and detected in the visible range.

The method of choice for organic acid separations seems to be cation exchange columns which act as ion exclusion principle and allow the rapid elution of the acids. Early work by Rapp and Ziegler (1976, 1979) used anion exchange pretreatment then cation column for the HPLC separation. Woo and Benson (1984), Pfeiffer and Radler (1985), and McCord et al. (1984) have reported good success. The method detailed is that of McCord et al. (1984).

Method

Adjust sample pH in the range of 8 to 9 with concentrated NH_4OH. Draw 2 ml of the sample into a 5-ml syringe and inject into the specially prepared anion exchange resin column. [Bio-Rex 5 (Bio-Rad Lab.) 200–400 mesh packed in a special Econocolumn 7 mm ID × 8 cm modified to have male luer ends with a frit at the bottom end. Place a three-way valve above to luer on the top of the Econocolumn. Take 2 g of resin slurried in distilled water and pressure pack the Econocolumn. Wash the bed with an additional 5 ml distilled water. Keep resin wet at all times]. Wash the resin bed until 15 ml of water is collected (contains neutral compounds). Elute the acid fraction with 2 ml of 10% sulfuric acid followed by 10 ml of water. Collect 10 ml. Membrane filter (0.45 μm) a portion of the acid fraction and remove phenolic material by treatment with C_{18} Sep Pack (Water Associates). Inject 5 to 25 μl into the HPLC using an Aminex HPX-87H (Bio-Rad Lab) cation column. Operate at 65° C using 0.05 N H_2SO_4 as the mobile phase. Use either UV detection 210 nm or refractive index detector and base results either on peak height or, better, on area under the curves. Prepare standard curves of the acids and relate results to the standard curves. Coefficients of variation were from 0.5 to 6.0%.

2.5 Isotachophoresis

This method is relatively old, but has only recently been applied for separation and quantification of organic acids as well as inorganic acids (Kaiser and Hupf 1979) in wines. Wine samples are injected into a capillary system which has a leading electrolyte, which precedes the sample, and a terminal electrolyte, which follows the sample. Separation of the moving band(s) of organic acids depends on the mass and charge of the ions and the electric field imposed on the column. The lower weight organic acids should reach the detector first and be measured either by conductivity or UV detection. Chauvet and Sudraud 1983; Farkas and Koval (1982) discussed the principles and indicated the usefulness. The method has been applied to wine and must (Reijenga et al. 1982; Klein and Stettler 1984; Prusa and Smejkal 1983; Kakalikova 1985) with reasonable success. This also is a method with potential for future development.

2.6 Tartaric Acid

Tartaric acid is present in wines and in grapes from warmer, hot climates in the greatest amount of all the organic acids. The grape is postulated by Saito and Kasai (1978) to make tartaric acid from ascorbic acid in young berries, while sucrose and gluconate contribute more in the later stages of maturity. Ruffner and Rast (1974) had questioned some of the earlier work of Saito and Kasai (1969) suggesting the ascorbic acid route. They (Ruffner and Rast 1974) suggested gluconate and glycoaldehyde as the precursors. Further elucidation is necessary.

Usseglio-Tomasset (1973) reviewed the various methods for analysis of tartaric acid and concluded that the only satisfactory method was that used as the OIV reference method (Anon 1978). Ubigli (1981) came to the conclusion that the OIV reference method was also superior. However, this method is slow in that it involves time-consuming precipitations and weighings. Addeo and Musso (1973) suggested separation on anion exchange resin and oxidation with periodate to iodate and the excess periodate inactivated. I^- was added to form I_3^- and measured spectrophotometrically. Krasnova et al. (1986) suggested precipitation as potassium acid tartrate, then polarographic measurement. The most popular and useful methods appear to be variations of the method developed using vanadate salts to make colored derivatives. Matchett et al. (1944) first used the vanadate reaction. Red wines need to be decolorized. The process of decolorization removes some of the tartrate. Absorbing carbon can be presaturated with tartaric acid, but this is difficult to do properly.

Pilone (1977), using the metavanadate procedure, found anion collection of the tartaric acid prior to reaction successful, which eliminates the need for charcoal. Mattick and Rice (1981) used PVPP (polyvinylpolypyrrolidine) to remove the wine color and found comparable results with properly treated charcoal and PVPP. They pointed out that if the charcoal is supersaturated or undersaturated, as long as it is used to make the standard curve, the results are comparable. Rebelein (1961) used ammonium vanadate. Tanner and Sandoz (1972) praised the method. The method was updated (Rebelein 1973) by the use of silver nitrate to precipitate the excess vanadate and other wine components which might interfere. Tanner and Lipka (1973) and Lipka and Tanner (1974) found the method to be satisfactory, as did Ferenczi and Uray (1973) and Chauvet and Sudraud (1977). Nikova et al. (1980) found the method good for determination of tartrate in waste products. Kozub et al. (1983) recommended the Rebelein method. Battle et al. (1978a), using a dialysis step instead of carbon, automated the method.

Vidal and Blouin (1978) studied the method of Rebelein and had trouble with decoloration. They did not like nylon or PVPP and felt the carbon was too hard to use with reproducible results.

Hill and Caputi (1970) reviewed the work done, and determined that separation by anion exchange and color development with metavanadate gave the best results. The standard method of the OIV is very similar (Anon 1978).

Method

Prepare a strong anion exchange resin column with 20 ml of washed resin. Charge with 20 bed volumes of 30% acetic acid. Wash with 30 ml of 0.5% acetic acid then 50 ml of water all at 2.5 to 5 ml min^{-1}. Place 10 ml of sample on the column followed by 20 ml of 0.5% acetic acid and 50 ml of water. Using 0.5 M sodium sulfate elute off the acids. Lactic acid is eluted in the first 30 ml; then discard the next 15 ml and collect the next 30 ml containing the tartaric acid. Dilute this 30 ml to 100 ml with water. Place 20-ml portions in each of two 100-ml Erlenmeyer flasks. Label one "a" and one "b". To flask "a" add 2 ml of 2 N H_2SO_4, 5 ml of 0.1 N H_2SO_4 and 1 ml of 10% v/v glycerol. To flask "b" add 2 ml of 2 N H_2SO_4, 5 ml of 0.05 N periodic acid, let stand 15 min then add 1 ml of 10% glycerol to destroy periodic acid. Let stand for 2 min. Mix in 5 ml of vanadate solution (10 g of ammonium metavanadate in 150 ml of N sodium hydroxide in a 500-ml volumetric flask, add 200 ml of a 27% w/v sodium acetate solution and bring to volume with water), first flask "b" then in "a". Immediately start a timer and transfer the solutions to 10-mm cuvets. Before 1 min and 30 s set zero absorbance with the "b" blank. Then read "a" and record.

Prepare a standard solution that contains about 500 mg of tartaric acid in 6.66 ml of N sodium hydroxide in 500 ml of 7% of 0.5 M sodium sulfate. Take 10, 20, 30, 40 and 50 ml of this solution to 100-ml flasks and bring to volume with 7% 0.5 M sodium sulfate. Take each standard and treat as above, using both "a" and "b" flasks for each. Relate the results to the standard curve. Consider the dilution.

Application

Kliewer (1967) studied the amounts of tartaric acid in table grapes at early and late stages of maturity. Kliewer (1968b) also demonstrated temperature effects on the concentration of tartaric acid. These and many other reports on grapes and wines have been summarized in Winkler et al. (1974), Amerine et al. (1980), and in Ribereau-Gayon et al. (1975, 1976a, b).

2.7 Succinic Acid

Pires and Mohler (1970) developed an enzymatic method using succinic dehydrogenase and developed a color reagent measured at 420 nm.

As suggested by Michal et al. (1976), previous chemical methods were inadequate for the analysis of succinic acid. Their report gave a more acceptable enzymatic method for the determination in foods.

$$\text{succinate} + \text{CoA} + \text{GTP} \underset{\text{synthetase}}{\overset{\text{succinyl CoA}}{\rightleftharpoons}} = \text{succinyl CoA} + \text{Pi}$$

$$\text{GDP} + \text{PEP} \xrightarrow[\text{kinase}]{\text{pyruvate}} \text{pyruvate} + \text{GTP}$$

$$\text{pyruvate} + \text{NADH} + \text{H}^+ \xrightarrow{\text{lactate dehydrogenase}} \text{lactic acid} + \text{NAD}^+ .$$

The NADH absorbancy was measured. Recovery of added succinic was 98%. Joyeux and Lafon-Lafourcade (1979) recommended its use for wine. They used inosine-5-phosphate (ITP) instead of guanosine-5-phosphate (GTP).

Method

Prepare the following reagents: Buffer (dissolve 2.4 g glycineglycine, 600 mg $MgSO_4 \cdot (H_2O)$ in 50 ml of double distilled water and adjust pH to 8.4 and bring to 60 ml volume. Store at 4° C – good for 4 weeks); NADH (dissolve 45 mg $NADH \cdot Na_2$, 80 mg $NaHCO_3$ in 6 ml of distilled water); solution of CoA (ITP, PEP) (dissolve 60 mg $CoA \cdot Li_3$, 60 mg $ITP \cdot Na_3$ and 60 mg PEP in 6 ml distilled water and keep at 4° C – good for 2 weeks; pyruvate kinase and lactate dehydrogenase solution (suspend 3 mg pyruvate kinase and 1 mg lactate dehydrogenase in 1 ml H_2O (hold at 4° C – good for 1 year); succinyl-CoA synthetase (suspend 5 mg l^{-1} of the enzyme in distilled water).

Add to 10 mm curvets "a" and "b" 1 ml of buffer 0.1 ml NADH solution, 0.1 ml CoA/ITP/PEP solution, 0.05 ml of PK/LDH suspension. To "a" (the blank) add 1.5 ml H_2O and to "b" 1.4 ml H_2O and 0.1 ml of sample. Mix and after 5 min at 37° C determine the absorbancy of each at 340 nm. Call the blank E_1^b and the sample E_1^e. Add 0.02 ml of succinyl-CoA synthetase suspension, mix and after 5 min determine optical densities. Call the blank E_2^b and the sample E_2^e. The extinction coefficient for NADH is 6.3×10^{-6} cm^2 mol^{-1} for 340 nm reading.

Calculate concentration succinic acid = C = mg^{-1}

$$\Delta E = [E_2^e - E_1^e - (E_2^b - E_1^b)]$$

V = volume of test solution (2.77 ml)

v = volume of sample (0.1 ml)

F = dilution factor

$$C = \frac{(\Delta E)(118.09)(V)(F)}{(6.3)(v)} .$$

Sponholz and Dittrich (1977) used the method to successfully determine succinic acid in a large number of wines from various sources.

2.8 Malic Acid

Malic acid is of interest for several reasons. It respires in the fruit and changes dramatically during harvest, and malolactic bacteria and certain yeasts metabolize it during their growth in juice or wine.

The method of choice given by the OIV (Anonymous 1978) is the method of Rebelein (1964). This method requires ion exchanging of the juice to remove interfering substances and correction for lactic and tartaric acids. A color reaction with chromatropic acid under acid conditions forms the chromophore. Corradini and Pellegrini (1978) noted that the high values from the Rebelein method were due to glucuronic and galacturonic acid interferences. Brunner and Tanner (1979) improved the method by adding sodium boron hydrate to reduce the uronic acids and prevent their interference. This improved version agreed with results for the enzymatic method. Arranz et al. (1981) compared the OIV method to the enzymatic and a method using *Schizosaccharomyces* yeast to metabolize the malic acid and calculating the amount from change in the total acid. They determined that the OIV method was superior. Olschimke et al. (1969) found good correlation with enzymatic and the Rebelein method if they hydrolyzed the enzymatic sample first to free the esterfied acids.

Castino (1974) offered a fluorometric method where anion separation of the malic acid then reaction with orcinol in sulfuric acid form 7-hydroxy-5-methyl-coumarin, which is strongly fluorescent. No interferences were found. Huang and Dai (1985) stated that the excitation wave length was 369 nm and emission was at 445 nn. Ridomi and Pezza (1982) gave a method of calculating the malic acid, knowing total titratable acidity, pH, total tartaric acid and the proper pK's of the acids. At best it is an approximation. Gump et al. (1985) gave a rapid gas chromatographic method for malic acid using the butyl ester. Only fair agreement with the enzymatic method was achieved.

The enzymatic determination seems the standard method of choice for most work in grape tissue and in enological work with malolactic bacteria. The method given by Poux and Caillet (1969) shows that the presence of tartaric acid and ammonia can give false positive reactions. This method, however, has been used extensively.

$$\text{L-}(-)\text{ malate} + \text{NAD}^+ + \text{hydrazine} \xrightarrow[\text{pH 9.5}]{\text{malic hydrogenase}} \text{oxaloacetate hydrazone} + \text{NADH} + \text{H}^+.$$

This method was given in Amerine and Ough (1980) in detail. McCloskey (1980 b) has adapted a version of this method which used glutamate oxaloacetate transaminase (GOT) to remove the oxaloacetate and speed the reaction rather than the use of hydrazine to bind the oxaloacetate. His method is outlined below.

Method

Prepare solutions: Buffer (11.4 g glycine and 3.0 g glutamic acid dissolved in 250 ml distilled water. Adjust the pH to 9.8 with 5 *N* NaOH or KOH and bring volume to 300 ml. Store at 2° to 10° C – stable for 60–120 days). NAD^+ solution (add 1.0 g NAD^+ free acid to 6 ml of water. Store at 4° C – stable for 120 to 200 days). Enzymes (mix the enzymes together to make about 1 ml of suspension –

activities should be 1250 IU ml^{-1} for MDH and 400–500 IU ml^{-1} for GOT). Mix 3 ml of glycine glutamate buffer with 100 µl of NAD^+ solution. Add 25 µl of a sample to a 1-cm cuvet. Read absorbance at 340 nm and record as E_1. Then add 25 µl of enzyme mixture. Incubate for 5 to 10 min at 25–30° C and then determine absorbance E_2.

Calculate the concentration as $mg\,l^{-1}$

$\Delta E = E_2 - E_1$

V = Volume in cuvet (3.15 ml)

v = Volume of sample (0.1 ml)

F = Dilution factor

$$C = \frac{(\Delta E)(134.09)(V)F}{6.3 \times v}.$$

Battle et al. (1978 b) indicated that good replication could be achieved in an automated system using the usual enzymes and buffers and hydrazine. A novel counter dialysis system was suggested. Lonvaud-Funel et al. (1980) improved upon the automated system. Glutamate oxaloacetate transaminase was used and the NADH produced was used to reduce an organic dye to a highly colored compound in the visible range. Another dye was used as an intermediate to transport electrons. Excellent results were indicated. The automated results compared favorably with the gas chromatographic method of Bertrand (1974).

Several good methods are available for the quick measurement of the disappearance of malic acid to determine the progress of malolactic fermentation. The paper method most used is that of Kunkee (1968), which is simple and inexpensive. Improvements have been offered by Salgues and Andre (1977) using thin layer chromatography, which takes only 10 min from start to finish. Stamer et al. (1983) improved on the Kunkee method by using thin layer, but the time required was ¾ to 1 h.

Application

The enzyme method has been used by Corranza et al. 1981; Kliewer 1965 and many others follow changes in grape berries (see Winkler et al. 1974) and investigate the activity of various malolactic bacteria added or found in wine (Möhler and Looser 1969 a; Maccarrone et al. 1977; see Amerine et al. 1980 for further examples).

2.9 Lactic Acid

L(+) lactic acid comes from bacterial activity, whereas D(−) lactic acid comes from yeast action. The L(+) is present in relatively large amounts after the malolactic fermentation finishes in a wine. The D(−) form is there in relatively small amounts. Both can be independently measured enzymatically. A chemical or physical method is sometimes used to determine the total lactic acid and the specific enzymatic method to determine the L(+) portion. The D(−) form is then obtained by difference.

Mattick and Rice (1970) developed a quick gas chromatographic method. The lactic acid was extracted with ethanol as barium salt, a trimethylsilyl derivative made and chromatographed.

Guimberteau and Peynaud (1966) compared five chemical methods for lactic acid determination. Twenty five wines were measured by the five methods with eight replications each. The best method appeared to be fixing the acid on an anion column, eluting it and oxidizing it to acetaldehyde with cerric ions. The acetaldehyde is reacted with nitroprusside and piperidine and the colored compound measured spectrophotometrically, which is the OIV method (Anon 1978). Another chemical method developed by Pilone and Kunkee (1970) is relatively simple and accurate. It also oxidizes the lactic acid to acetaldehyde, but forms a color product with p-hydroxydiphenyl. Other acids, sugars, or sulfur dioxide are reported not to interfere. It is the chemical method given.

Method

Add water to sample to dilute the lactic acid to 10–100 µg/ml. Place 1 ml of sample in a 20 × 180 mm tube, add 1 ml 20% w/v $CuSO_4 \cdot 5H_2O$ solution and dilute to 10 ml. Add 1 g $Ca(OH)_2$ and mix and let stand for 30 min then centrifuge. Pipet 1 ml of the supernatant into a similar tube, add 0.05 ml 4% wt/v $CUSO_4 \cdot 5H_2O$ solution. While swirling the tube add 6 ml concentrated H_2SO_4 and then put tube into boiling water bath for 7 min, cool to 20° C or less in ice bath. Add 0.1 ml of a 1.5% p-hydroxydiphenyl in 0.5% NaOH stored in brown bottle. Shake, incubate for 15 min in 30° C bath, shake, continue to incubate for 15 min more. Place in boiling H_2O bath for 1.5 min to dissolve ppt. Cool under running H_2O. Read absorbance at 560 nm. Relate to a standard curve prepared with lactic acid conc. of 0, 10, 20, 30, 50 and 70 µg ml^{-1}. Consider dilutions.

The enzymatic procedure has been reported on most often. Mayer and Pause (1969) compared it favorably to the ion exchange method. Postel et al. (1973) and Bandion and Valenta (1977) found it was excellent and determined the amounts of each isomer enzymatically. Kanbe et al. (1977) gave a semi-automated method.

Method

Prepare solutions: Buffer pH 9.5 (7.5 g glycine, 5.2 g hydrazine sulfate, 0.2 g EDTA and 4 g NaOH dissolved in 150 ml H_2O. Stable for 1 week at 4° C); NAD^+ solution [177 mg NAD^+ dissolved in 100 ml distilled H_2O. Stable for 1 week at 4° C]; L(+) lactate standards [L(+) lactate(Li) at 0.1, 0.2, 0.3, 0.4 and 0.5 mg ml^{-1} levels]; L(+) lactate dehydrogenase solution [L(+) lactic dehydrogenase solution at 0.25 units μl^{-1} diluted five fold with buffer solution without hydrazine sulfate]. Take 0.1 ml sample solution, 1.5 ml buffer solution and 50 µl of LDH solution mixed in a 10 mm cuvet. After 1.5 min add 1.5 ml NAD^+ solution. Incubate at 37° C for 30 min. Prepare a blank by omitting the 50 µl of LDH solution to zero instrument. Read absorbance at 340 nm. The activity of each new batch of LDH could vary. A new standard curve should be prepared for each.

The concentration can be calculated from the standard curve or related to the calculations based on the extinction coefficient. D(−) lactic dehydrogenase can be used to measure the D(−) lactic acid.

Application
Lafon-Lafourcade et al. (1977) demonstrated that botritized wines which had D(−) lactic in excess of 200 mg l^{-1} were spoiled and that if the L(−) lactic was in excess of 100 ml l^{-1} malolactic fermentation had occurred. Both acetic bacteria and lactobacillus make acetic acid. Knowledge of which is present if the acetic acid is forming can be determined by knowledge of the amount of D(−) lactic formed. Acetobacter activity increases D(−) lactic acid (Sponholz et al. 1982).

2.10 Acetic Acid

Acetic acid, more commonly called volatile acids in wines, results from the activity of microorganisms. Yeast forms only a small amount under normal conditions. If the yeasts are stressed by cold temperatures and high sugar situations they can make much more than usual amounts (Lafon-Lafourcade et al. 1977). The acetic acid is a product of the fatty acid synthesis cycle. *Acetobacter* can directly oxidize ethanol to acetic acid in large amounts under certain conditions to form what is commonly known as vinegar. *Lactobacillus* or *Leuconostoc* can form acetic acid from sugar or from citric acid.

The standard method, approved by the AOAC (Horowitz 1980) consists of a specially designed still for the steam distillation of the acetic acid. It has a one-plate reflux. Conditions are defined and standardized. The distilled acids are titrated with standard base and any interfering components subtracted away. The method actually detects only a percentage of the volatile acids, but the amount in total is close to the amount of acetic acid found by gas chromatography (Ough and Amerine 1967). This is probably because of other volatile acids which partially distill and titrate (Pilone 1967). Pilone et al. (1972) suggested the addition of mercury salts to bind the sulfite and prevent its distillation, so that a special titration and correction would not have to be made. Pilone (1978) refereed an AOAC collaborative study and obtained approval. The method holds the sulfite but in routine work a build-up of metallic mercury occurs, which is unwanted for health reasons; hence, it is of a doubtful improvement. Schneyder and Pluhar (1977) suggest the oxidation of the sulfite with ferric ions prior to steam distillation. Dubernet (1976) assembled an automatic analyzer system where the excess sulfite was oxidized by hydrogen peroxide. The results still had to be corrected for the presence of sorbic acid which also steam distills. Instead of titrating the distillate with base, Florica and Ghimicescu (1977) suggest the use of vanadate and determine colorimetrically.

The method and the necessary corrections are defined in detail in Amerine and Ough (1980) with all the corrections allowed.

Method
Pipet 10 ml of wine into the inner tube of an electric cash still (see Fig. 1). Add 0.5 ml of a 1% solution of red mercuric oxide (dissolve 1 g mercuric oxide in 10% H_2SO_4). Add deionized water to the outer body of the still. Turn on the still and distill exactly 100 ml into a 250 ml wide-mouth Erlenmeyer flask. The condenser should condense but not cool the distillate. Bring the distillate to an incipient boil, add several drops of a 1% phenolphthalein solution (1 g phenolphthalein dis-

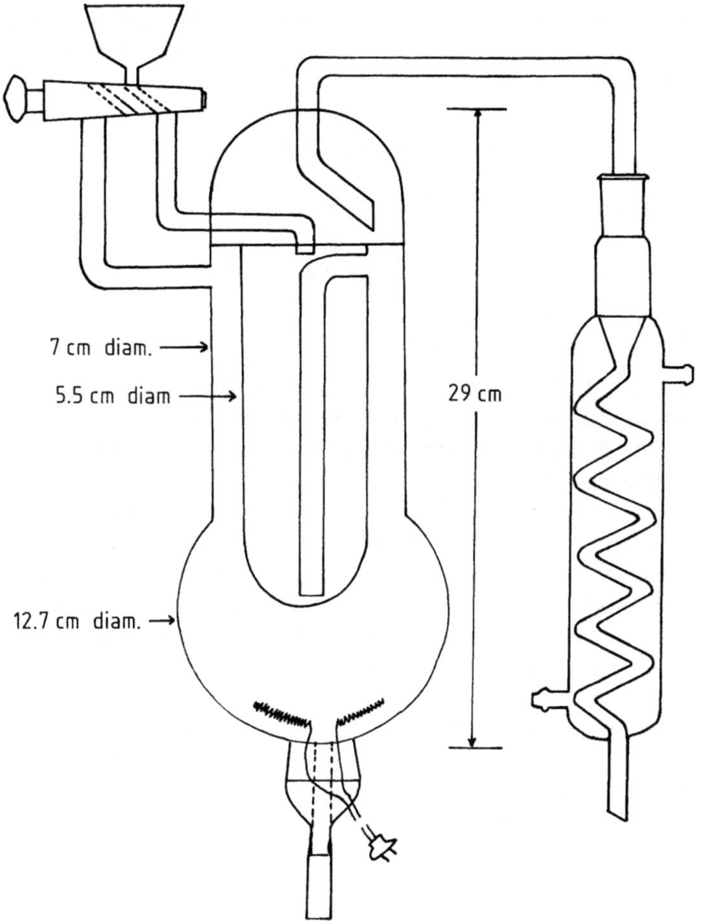

Fig. 1. Cash electric steam distillation apparatus for volatile acidity measurements of wines or juices

solved in 100 ml of 80% ethanol) and titrate to a slight pink end point. Calculate the volatile acids as acetic acid.

$$\text{acetic acid g/100 ml} = \frac{(\text{Vol. Std. Base})\ (N\ \text{Std. Base})\ (6.0)}{(\text{Volume of Sample})}.$$

Correct for sorbic acid by analyzing for it and subtracting equivalent. Each gram of sorbic acid is equivalent to 0.536 g acetic acid (Method for sorbic acid analysis: Steam distill 2 ml of wine as above. Collect about 100 ml add 0.5 ml of 0.1 N HCl and bring to 200 ml with water. Determine absorbance at 260 nm. Compare to a 200 mg l^{-1} acid acidified and steam distilled and calculate amount in wine).

McCloskey (1976a, b, 1980a) applied enzymatic analyses to the determination of acetate in juices and wines. He found it to be more specific, more accurate, and faster than the distillation methods. His final version is rather complex and requires 30 min development but is specific for acetate and more rapid than previous enzyme tests. The reactions are:

$$\text{acetate} + \text{ATP} \xrightleftharpoons{\text{acetate kinase}} \text{acetyl phosphate} + \text{ADP}$$

$$\text{acetyl phosphate} + \text{CoA} \xrightarrow{\text{phosphotransacetylase}} \text{acetyl-CoA} + \text{Pi}$$

$$\text{phosphoenolpyruvate} + \text{ADP} \xrightarrow{\text{pyruvate kinase}} \text{pyruvate} + \text{ATP}$$

$$\text{pyruvate} + \text{NADH} + \text{H}^+ \xrightarrow{\text{lactate dehydrogenase}} \text{lactate} + \text{NAD}^+ + \text{H}_2\text{O}.$$

The disappearance of NADH is followed at 340 nm. Polyvinylpyrrolidone is used to bind the tannins in red wines to protect the enzymes.

Bruer et al. (1985) found the enzymatic method superior to the distillation method for acetic acid determination.

Gas chromatographic determination for acetic acid by direct injection has been reported (Ough and Amerine 1967; Bertuccioli 1982). Trombella and Ribeiro (1980) used Carbopack C modified with 0.3% Carbowax 20 M and 0.1% phosphoric acid to attain excellent stability and sharp peaks. They used propionic acid as an internal standard. The carrier gas was saturated with formic acid. Further work on this method was done by Kupina et al. (1982).

Application

Kupina (1984) offered a HPLC method for acetate determination in grape juice applicable to use on a grape inspection station. Junge and Spadinger (1979) pointed out the need for rapid and accurate knowledge of acetic acid accumulation in order to protect wines from spoilage.

2.11 Citric Acid

Citric acid is a minor component of the acids in grapes. It is often added to wines at time of bottling to chelate iron and prevent precipitates as well as to balance wines lacking in acid.

There have been no quick effective methods described for the colorimetric determination of citric acid in the presence of other organic acids. The normal method of the OIV (Anon. 1978) is that of Rebelein (1967). Their reference method requires careful anion exchange separation, oxidation of the citric acid to acetone and then titration with standardized iodine. The normal method is time-consuming and with complicated washing and centrifuging of the precipitates, diazotation of the isolated citric acid to make a colored compound which is measured spectrophotometrically. Addeo (1972) has suggested anion exchange, then making the colored derivative after periodate oxidation and titration of the excess iodine with starch. Pyridine and acetic anhydride were used for derivatization. The method compared favorably to the enzyme method.

The enzymatic method seems to be the simplest to use and the most dependable. This method was suggested by Moellering and Gruber (1966). The reactions are as follows:

$$\text{Citrate} \xrightarrow{\text{citrate lyase}} \text{oxaloacetate} + \text{acetate}$$

$$\text{oxaloacetate} + \text{NADH} + \text{H}^+ \xrightarrow{\text{malic dehydrogenase}} \text{malate} + \text{NAD}^+$$

$$\text{oxaloacetate} \xrightarrow{\text{oxaloacetate decarboxylase}} \text{pyruvate} + \text{CO}_2$$

$$\text{pyruvate} + \text{NAD}^+ + \text{H}^+ \xrightarrow{\text{lactic dehydrogenase}} \text{lactate} + \text{NAD}^+ .$$

Bergner-Lang (1977) found that citrate lyase could produce Maillard reaction products giving high results. Also, if the citric acid esters have been hydrolyzed, false values are obtained.

Seppi and Sperandio (1983) compared the enzymatic method to the OIV standard method of Rebelein with excellent correlation and preferred the enzymatic method. Henniger and Mascaro (1985) did a collaborative study with recoveries of 12 individual laboratories averaging over 99% and a coefficient of variation of 2.3%. The method they outline is given below.

Method

Prepare solutions: Solution 1 (mix 11 ml of glycineglycine pH 7.8 buffer made at 0.51 mol l^{-1} with 0.6 m mol l^{-1} $ZnCl_2$ with 1 ml $NADH(Na_2)$ at 6 mg ml^{-1} and containing 12 mg ml^{-1} of $NaHCO_3$ and add 0.24 ml enzyme suspension containing 0.5 mg ml^{-1} malate dehydrogenase at 1200 U mg^{-1} and 2.5 mg ml^{-1} lactate dehydrogenase (from rabbit muscle) at 550 U mg^{-1} in $(NH_4)_2SO_4$ at 3.2 mol l^{-1} and pH 6–7. Solution is stable for 1 week at 4° C). Solution 2 (dissolve 50 mg lyophilized citrate lyase, about 12 units in 0.3 ml of redistilled H_2O. Solution is stable for 1 week at 4° C or 4 weeks at −20° C).

Dilute wine or grape juice sample to contain less than 0.4 g l^{-1} citric acid. Label 1 cm cuvets as "blank" and "sample". Pipet 1 ml solution 1 into each. Pipet 2.0 ml of redistilled H_2O into "blank" and 1.8 ml into "sample" cuvets. Add 0.2 ml wine sample to sample cuvet. Cover with Parafilm and mix and wait 5 min at 20 to 25° C. Determine absorbance of each cuvet at 340 nm and record as E_1^b and E_1^s. Add 0.2 ml of solution 2 to each cuvet. Mix and hold at 20 to 25° C about 5 min. Read absorbance on each again at 340 nm and record as E_2^b and E_2^s.

Calculate as follows:

$$\Delta E = (E_1^s - E_2^s) - (E_1^B - E_2^B)$$

V = Volume in cuvet (3.02)

v = sample volume (0.2)

F = dilution factor

$$C \text{ mg} \, l^{-1} = \frac{(V)(192.1)(\Delta E)(F)}{(6.3)(v)} .$$

Application

This enzymatic method has been used to measure citric acid in wines by Pachki (1974) and by Jeszenszky and Szalka (1976), as also by others in their study comparisons with the colorimetric method.

2.12 Uronic Acids

Normally the most common uronic acid in grapes and wines is galacturonic. This acid varied by year and cultivar from 1.5 to 2.0 g l^{-1} in Italian wines. The gluconic was much lower (Usseglio-Tomasset and Gabri 1975). Arndt and Thaler (1974) reported similar galacturonic acid levels in all wines. Bandion et al. (1980), measuring Austrian wines, found some very low in gluconic acids but others very high. Gluconic acid was related to the action of molds by the work of McCloskey (1974). Holbach and Woller (1978) verified this.

Sponholz and Dittrich (1984) used liquid chromatography to separate five uronic acids. Arndt and Thaller (1974) used thin layer chromatography of the ozazones of galacturonic acid and measured the spots photometrically. Souty et al. (1980) employed m-hydroxydiphenyl and sulfuric acid with orcinol to measure both galacturonic acid and arabinose in pectins with an autoanalyzer.

Gluconic acid was determined enzymatically by McCloskey (1974). The assay mechanism was:

$$\text{D-gluconate} + \text{ATP} \xrightarrow{\text{gluconate kinase}} \text{gluconate-6-phosphate} + \text{ADP}$$

$$\text{gluconate-6-phosphate} + \text{NADP}^+ \xrightleftharpoons{\text{6-phosphogluconate dehydrogenase}} \text{ribulose-5-phosphate} + \text{NADPH} + \text{H}^+ + \text{CO}_2 .$$

McCloskey measured NADPH absorption at 340 nm to quantify.

Gas chromatographic separation and quanification of the uronic acid was demonstrated (Bertrand and Dubourdieu 1978). This method will be outlined below.

Method

Dry methanol by refluxing with magnesium (4 g l^{-1}) then distill into a dry flask. Using laboratory grade bottled HCl (100%) bubble into the methanol, with cooling until it is about 0.5 *N* HCl. Dilute with dry methanol if necessary.

Take 5 ml of wine and add 25 ml of 95% v/v ethanol. After 1 h filter and wash the precipitate three times with 2 ml each of ethanol. Combine the filtrate and washes. Place onto an anion column (AG 2X 50–100 mesh) in the formate form. After washing with 150 ml water (300 ml for high alcohol or very sweet wines) elute the acids with 6 *N* formic acid. Collect in a 250-ml Erlenmeyer flask containing 5 ml of an erythritol internal standard (100 mg l^{-1}). Evaporate the formic acid away at 50° C. Transfer the residue with three portions of methanol to a 20-ml reaction flask. Evaporate off the methanol with a nitrogen stream at 50–60° C. Add 1 ml of the methanol-0.5 *N* HCl reagent, close the cap and hold at 80° C for 12 h. Eliminate the residual reagent by heating to 50–60° C under a stream of dry nitrogen. Add 0.2 ml of methylene chloride and 0.2 ml of trifluoroacetic anhydride. Close the cap and heat to 150° C for 5 min.

The liquid phase is QF1, 7% on chromosorb W H.M.D.S. 80–90 mesh. Tube length is 4 m ID 1/8 inch. Nitrogen flow is 10 ml min^{-1}. Program at 1° C min^{-1} from 120° to 200° C. Inject 5 μl sample. Construct standard curves that

$$\text{relate ratio of the } \frac{\text{areas}}{\text{int. std. area}} \text{ to concentrations.}$$

Effective separation and quantification of mucic, gluconic, glucuronic, and galacturonic acids are achieved.

2.13 α-Ketoacids

The importance of the α-ketoacids as precursors for amino acid synthesis is well known. Pyruvic and α-ketoglutaric are especially important in the initial synthetic pathways and in $-NH_2$ transport.

Novikova and Kudryashova (1981) suggested anion exchange separation, then to develop a color with 2,4-dinitrophenol hydrazine. Deibner and Cabibel-Hughes (1966) used a similar technique, but separated the derivatives on cellulose thin layer and determined the spot intensities photometrically.

Determination of the α-ketoacids is predominately done by enzymatic analysis. Pyruvate and α-ketoglutarate are most commonly detected this way (Rankine 1965; Lafon-Lafourcade and Peynaud 1965, 1966; Blouin and Peynaud 1963; Graham 1979; Sponholz et al. 1981 a, b, Möhler and Looser 1969 b).

Peynaud et al. (1966) reviewed the enzyme methods for the pyruvate and α-ketoglutarate. The method for pyruvate is outlined. If the reactions are carried out as outlined, no sample treatment is required, as was suggested by Delfini (1983). The enzyme reaction is:

$$\text{pyruvate} + \text{NADH} + \text{H}^+ \xrightleftharpoons{\text{lactate dehydrogenase}} \text{lactate} + \text{NAD}^+ .$$

Method

Prepare solutions: Buffer pH 6.8 (36.8 g $Na_2HPO_4 \cdot 12\,H_2O$, 13.6 g $KHPO_4$ and 7.4 g $EDTANa_2H_2$ diluted to 1 l with H_2O); NADH solution (2.0 mg ml^{-1} NaDH in 1% $NaHCO_3$); lactate dehydrogenase solution [(0.75 mg (protein) LDH in $(NH_4)_2SO_4$ l^{-1})] at pH 7.0 from a 5 mg ml^{-1} enzyme preparation].

Add successively to two 1-cm cuvet 2 ml buffer solution, 0.05 ml wine (diluted 10 to 40 times), 0.05 ml NADH solution. To the blank add 0.05 ml H_2O and to the sample add 0.05 ml enzyme solution. React for 6 min at room temperature. Determine the absorbance at 340 nm in each before the final addition of 0.05 ml of H_2O or enzyme; then again after the additions and calculate the ΔE.

$$\Delta E = [E_2^s - E_1^s - (E_2^b - E_1^b)]$$

V = volume test solution (2.15)

v = volume of sample (0.05)

F = dilution factor

$$\text{mg}\,l^{-1}\ \text{pyruvate} = \frac{(\Delta E)(88.06)(2.15)\,F}{(0.05)\,6.33}.$$

α-Ketoglutaric acid interferes only slightly. Glyoxylic acid reacts to the extent of about 80% but only a few mg l^{-1} are normally found.

The α-ketoglutarate enzymatic reactions are:

$$\alpha\text{-ketoglutarate} + NADH + NH_4^+ \xrightarrow{\text{glutamate dehydrogenase}} \text{glutamic acid} + NAD^+ + H_2O\,.$$

The method given is again that outlined by Peynaud et al. (1966).

Method

Prepare the following solutions: Buffer pH 7.6 (58.4 g $Na_2HPO_4 \cdot 12H_2O$, 3.8 g KH_2PO_4, 11.5 g $NH_4H_2PO_4$ and 7.4 g $EDTANa_2H_2$ diluted to 1 l); NADH solution (same as for pyruvate); glutamate dehydrogenase solution [2 mg protein from a 20 mg ml^{-1} enzyme preparation suspended in one ml pH 7.0 $(NH_4)_2SO_4$ solution (260 g l^{-1})].

Add successively to each of two 1-cm cuvets 2 ml buffer, 0.05 ml wine (diluted 10–40 times), 0.05 ml NADH solution. Read the absorbances at 340 nm then to the blank cuvet add 0.05 ml H_2O and to the sample cuvet 0.05 ml enzyme preparation and react for 9 min at room temperature and again read absorbances. Calculate △E.

$$\Delta E = [E_2^s - E_1^s - (E_2^b - E_1^b)]$$

V = volume of test solution (2.15)

v = volume of sample (0.05)

F = dilution factor

$$\alpha\text{-ketoglutarate mg}\,l^{-1} = \frac{(\Delta E)(146.1)(2.15)(F)}{(6.33)(0.05)}\,.$$

Interferences are 25% of the 2-ketovalerate and 7% from oxaloacetic acid.

2.14 Fatty Acids

The fatty acids in grapes and in wine are usually by-products of the fatty acid cycle. The straight chain acids mostly come in even carbon numbers. These acids are steam-distilled into spirits. Formic acid is found in grapes in the greatest amount. The others come mainly from yeast activity (Sponholz et al. 1981 a, b). Recently, Lafon-Lafourcade et al. (1984) indicated that C_8–C_{10} acids were inhibitors to yeast vitality. The usual way of determining these acids is by gas chromatography. Shinohara (1985) neutralized 10 ml wine with KOH, concentrated to dryness after adding internal standards, then added 5 ml *N* H_2SO_4 and extracted with 2 ml of ethyl acetate-n-pentane (2:1 v/v) with 10 min of shaking and chromatographed the C_6, C_8, and C_{10} acids. Masuda et al. (1985) used distilled spirits directly with internal standards added to chromatograph the same three acids. Hess et al. (1977) determined C_8 through C_{18} acids by steam distillation separation, ether-pentane extraction, methylation, and gas chromatographic analysis. Crowell and Guymon (1969) and Nykanen et al. (1968) also used gas chromatographic techniques to separate the fatty acids in distilled spirits.

2.15 Ascorbic Acid

Ascorbic acid is sometimes added to wines as an antioxidant. It is present in fresh grapes, but fairly rapidly disappears with exposure to oxygen.

As recently as 1982, it was recognized that the specific measurement of ascorbic acid in biological material was difficult (Sauberlich et al. 1982). Successful high-speed liquid chromatography separation and determination of ascorbic acid on an anion exchange column with a UV detector was reported (Williams et al. 1973). The chemical methods appearing for the measurement of this vitamin are extensive. However, relatively few have been applied to grape juice or wine. A few of the more recent methods which probably could or have been applied to grape juice or wine: HPLC fluorometric detection involving anion column separation, post column derivatization with o-phenylenediamine (Vandersplice and Higgs 1984; Kacem et al. 1986); polarographic (Lau et al. 1985); coulometric (Karlsoon 1975); HPLC using a (Waters Assoc.) carbohydrate column (Geigert et al. 1981); immobilized ascorbate oxidase on calcium alginate gel (Esaka et al. 1985); HPLC using anion exchange and UV detection (Ashoor et al. 1984).

The AOAC (Horowitz 1980) does not give an official method for grapes or wine, although they do give a general titration method for vegetable and fruit juices involving indolphenol. Sulfite interferes. The OIV manual (Anonymous 1978) gave a method to measure total ascorbic acid. The ascorbic acid is oxidized by iodine to dehydroascorbic acid and then precipitated as 2,4-dinitrophenylhydrazone. The derivative is separated by paper chromatography, eluted with acetic acid, the red color measured photometrically. Thaler and Gieger (1967) reported thoroughly on this method. The rapid method given for ascorbic acid for wine and juice by the OIV (Anonymous 1978) is given below. It measures only ascorbic acid, not dehydroascorbic acid.

Method

Prepare solutions: acetaldehyde (distill sufficient paraldehyde, in the presence of sulfuric acid, to give about 7 to 8 g acetaldehyde. Dilute to 500 ml and remove an aliquot and standardize with sodium sulfite. Adjust to 6.9 $g\,l^{-1}$ with water. 1 ml will combine with 10 mg SO_2); sulfuric acid 0.1 N (5.23 g H_2SO_4 and bring to 1 l carefully with H_2O); starch indicator (dissolve 5 g of soluble starch in 1 l H_2O); iodine 0.05 *N* (dissolve 12.5 potassium iodide and 6.35 g iodine in H_2O and bring to 1 l with H_2O. Standardize as necessary with standard thiosulfate).

Place 50 ml of juice or wine in a 300-ml Erlenmeyer flask, add 5 ml acetaldehyde solution. Let stand for 30 min, add 3 ml 0.1 *N* sulfuric acid, 5 ml of starch solution and titrate with standardized iodine to blue starch endpoint. 1 ml of 0.05 $N\,I_3^-$ will oxidize 4.4 mg ascorbic acid. (Generally wine or juice will consume about 0.2 to 0.3 ml of the iodine for other reducing substances).

$$\text{mg ascorbic acid } l^{-1} = (88)\,(\text{ml } 0.05\,N\,I_3^- - 0.3).$$

The starch end point with highly colored red wines is difficult. This problem can be averted by using an oxidation-reduction electrode.

2.16 Other Acids

There are many other organic acids reported in grapes and wines. Specific analytical methods are available, but not done in any sense as routine analysis. Such compounds as indoleacetic acid, abscisic acid, oxalic acid, agaric acid, salicylic acid and a whole series of phenolic acids are some examples. Many other acids are present in trace amounts.

3 Amino Acids

Grapes are usually a very good source of amino acids. There is always sufficient to support microbial growth. However, there are times when the levels are low enough to cause low cell numbers, hence, incomplete fermentations. For this reason and others discussed later, considerable amino acid analyses have been made on grapes and their products. Also, evidence of amino acid deficiencies in the grapes is a good indicator of vineyard problems.

There are numerous good books and reviews which describe the details of the biosynthesis of amino acids in plants (Lea and Wallsgrove 1985; Miflin et al. 1979; Bryan 1976). Without going through the intermediate compounds, families of the various amino acids are shown in Fig. 2 with their precursor compounds. There are three main families – glutamate, aspartate, and pyruvate with 3-phosphoglycerate furnishing the starting point for histidine and an alternate path for formation of serine, glycine, and cysteine. There are other secondary routes for the yeast or bacteria to form these amino acids.

The ammonia is taken up by amination via the glutamate synthase cycle. This involves the glutamate-forming glutamine, which then transfers the amino group to glutarate to form glutamate. The glutamic acid in turn can transaminate other α-keto acids to form the amino acid can return the glutarate back to the cycle (Lea and Wallsgrove 1985). This has been demonstrated in grape tissue by Roubelakis-Angelakis and Kliewer (1983a, b).

It is generally assumed that the formation of the amino acids follows the same routes in grapes as in other plants. Relatively incomplete evidence is available to prove this. Roubelakis and Kliewer (1978a, b) found that the enzymes of the Krebs-Henseleit cycle were operative in grape seedlings and in berry tissue. They also noted (Roubelakis and Kliewer 1978c, d) that there was ornithine transcarbamylase and arginase activity in grape berries during maturation, and further demonstrated (Roubelakis-Angelakis and Kliewer 1981) the effect of field nitrogen fertilization on these enzymes.

The uptake and formation of amino acids by yeast or bacteria are reasonably well described by good biochemistry or microbiology textbooks. The difference between the systems for yeast and bacteria and that for plants is much less than one might imagine.

Ough and Bustos (1969) reviewed the analytical methods available for grapes and wine and detailed their application through about 1967. The emphasis on methodology will be from this time on. A few important earlier references will be included.

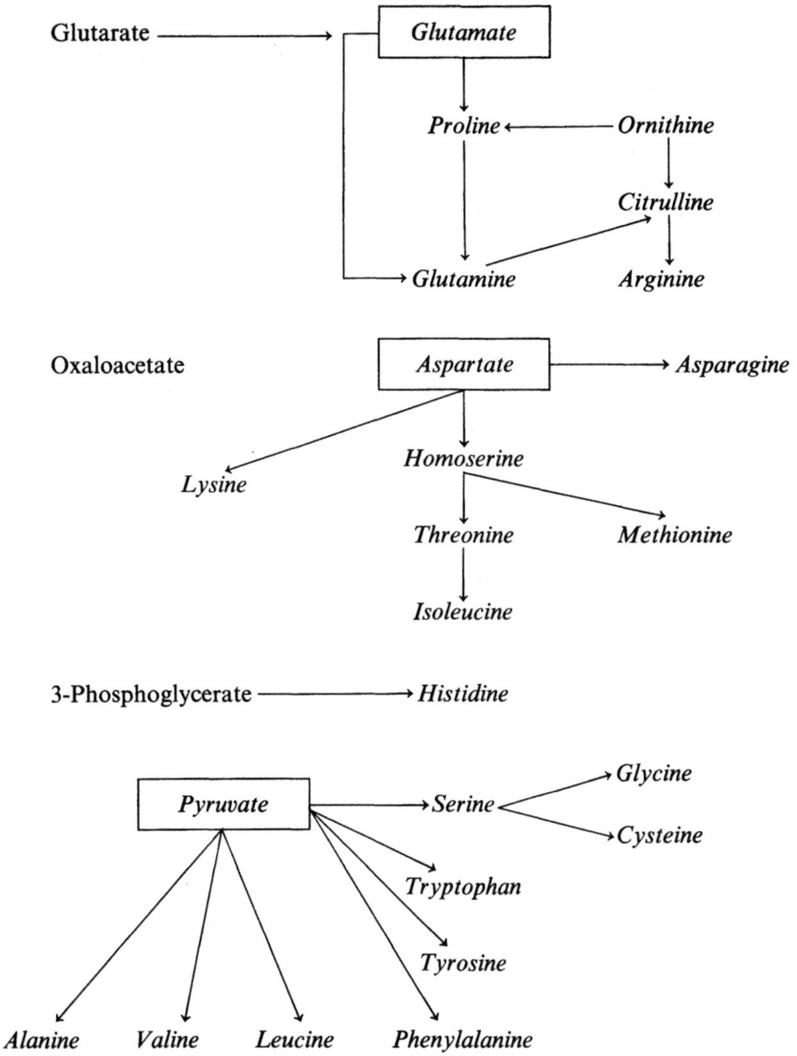

Fig. 2. Precursor compounds and abbreviated pathway relationships of the natural amino acids in grape and wines

3.1 Total α-Amino Nitrogen

Often it is valuable or necessary to know the status of the total α-amino nitrogen in grapes or in wine. It is an excellent measure of the possibility of a successful or a slow fermentation or of the risk of microbial instability of a wine.

The formol titration as described by Amerine and Ough (1980) was the choice for many years. The addition of formaldehyde converts the relatively weak α-amino acids into relatively strong acids. The before- and after-addition titrations with base between the proper set pH values allow the calculation of the α-amino acids.

The 2,4,6-trinitrobenzene sulfonic acid (TNBS) method for total α-amino acid measurement was first suggested by Satake et al. (1960). It reacts with $-NH_2$ groups and is not specific for α-amino acids. Lie (1972 a, b) compared the method, in beer and wort, to the ninhydrin method and found the TNBS method less reproducible. Crowell et al. (1985) developed the method for grape juice and wines. Pretreatment of the juice was necessary to remove interfering proteins and peptides. Wines with the much depleted α-amino acid content were more difficult to measure accurately. Beveridge and Harrison (1985) also used this method, with no pre-removal of interfering compounds to measure α-amino acid in fruit drinks. They compared it to the formol titration method and obtained good correlation of results.

The ninhydrin method (Rosen 1957) for total α-amino acid given below is similar to that published by the European Brewing Convention (1975). The method is successful on grapes and wines (Ough et al. 1968 b).

Method

Prepare a sample of juice or wine as described by Nassar and Kliewer (1966). Apply a sample of ethanol extract or wine equal to 10 g of fresh weight or 10 ml to a Dowex 50W–X8 (H^+ form) column 2 × 20 cm. Thoroughly wash with neutral water. Elute with 150 ml of 3 *N* NH_4OH and wash with three bed volumes of distilled H_2O. Dry the eluate and washings at 45 to 50° C with a stream of filtered air. Dissolve the residue in 5 ml of 10% isopropanol.

Take an aliquot of 1 ml containing about 0.20–0.40 mol of amino acids and determine α-amino acids according to Rosen (1957). Add ½ ml of cyanide-acetate buffer mixture and ½ ml of 3% ninhydrin in methyl cellosolve. Heat for 15 min in 100° C water bath. Maximum color is obtained in 10–12 min. Immediately remove from the water bath and add 5 ml of isopropanol; water mixture (1 : 1) and mix. Rapidly cool to room temperature and read at 570 nm in spectrophotometer. If reading is over 0.8 optical density, dilute with isopropanol : water mixture until it is less than 0.8.

Make up a sodium cyanide stock solution of 0.01 M [490 mg l^{-1} in acetate buffer (270 g sodium acetate trihydrate, 50 ml glacial acetic acid and 0.2 l water and dilute to 0.75 l with water)] and mix 20 ml of the 0.01 M sodium cyanide solution with sufficient concentrated acetate buffer to give 1 l.

The response to most α-amino acids is near 100% that of leucine. Proline and hydroxyproline do not respond significantly. Use leucine to prepare a standard curve.

Application

Ough et al. (1968 a) used this method to measure the α-amino nitrogen in ten different varieties of grapes, fertilized and nonfertilized, on two different rootstocks. Differences were attributable to fertilization and to rootstocks. Also, there were related fermentation and sensory effects (Ough et al. 1968 b). More recently (Vos et al. 1979; Vos and Gray 1979; Ingeldew and Kunkee 1985), using this method, demonstrated the effect of α-amino acids on fermentation rates. Catalina et al. (1981, 1982) showed the change in α-amino acids in the grape with maturity.

Ough and Nagaoka (1984) used the TNBS method to correlate α-amino nitrogen in the juice to the fermentation rate in red grapes. The total α-amino acids in grapes has been reported from 16% to be as high as 50% of the total nitrogen (Toth 1982; Astabatsyan 1980).

Microbiological

The realization that *Lactobacillus* or *Leuconostoc* sps. could be found that were dependent on specific amino acids for growth led to the development of assays for essential amino acids. Castor (1953a) developed the method for use in musts and wines for 14 different amino acids. Castor (1953b) showed the relationship of the rapid loss of the various amino acids to the growth of the yeast in normal grape juice fermentation. Ammonia added to the must increased cell numbers but did not spare the amino acids present from assimilation by the yeast (Castor and Archer 1959). Castor and Archer (1956) found that proline was not taken up by the yeast. They used the method of Moore and Stein (1951) to separate proline from threonine and serine, which eluted as one; then used the microbiological assay to quantify the three amino acids. Peynaud and Maurie (1953) and Lafon-Lafourcade and Guimberteau (1962) followed the amino acids during grape maturation. Lafon-Lafourcade and Peynaud (1959) measured amino acids for the red and white wines of Bordeaux. Peynaud and Lafon-Lafourcade (1962, 1963) confirmed that amino acids could be assimilated directly by the yeast *Saccharomyces cerevisiae* and also excretion of amino acids was noted and it was determined not to be confused with cell autolysis. Poux and Ournac (1970) compared the microbiological method with the automatic amino acid analyzer system and found that the microbiological method gave false positive results because of response to peptides.

Because of the work-intensiveness of the method and the lack of true specificity for individual free amino acids, the method is seldom used now.

3.2 Paper Chromatographic

Nassar and Kliewer (1966) described a method applicable to grapes or wine. This method can be related back to general procedures partially described by others (Redifield 1953; Wiggins and Williams 1952; Block et al. 1955). It is a two-part system. First, the amino acids are extracted from the crushed berries with ethanol, isolated by application to a cation column, eluted and concentrated, and then chromatographed on paper.

Method

Apply a sample of ethanol extract or wine equal to 10 g of fresh weight or 10 ml to a Dowex 50W–XB (H^+ form) column 2 × 20 cm. Thoroughly wash with neutral water. Elute with 150 ml of 3 *N* NH_4OH and wash with three bed volumes of distilled H_2O. Dry the eluate and washings at 45 to 50° C with a stream of filtered air. Dissolve the residue in 5 ml of 10% isopropanol. Separate the amino acid by two-dimensional paper chromatography. The first solvent system is sec-butanol : t-butanol : 2-butanone : H_2O (4 : 4 : 8 : 5 v/v) with 0.5% NH_4OH. The second solvent system is n-butanol : acetic acid : water (4 : 1 : 5 v/v). For additional evidence for identification of amino acids is a third solvent system n-butanol : 2-butanone : 17 *N* NH_4OH : H_2O (5 : 3 : 3 : 1 v/v). Develop the spots, dip the dried chromatogram in 0.5 ninhydrin in acetone (w/v) then heat to 60° C for 30 min and store in dark for 24 h. Use 1 to 100 μl aliquots spotted on 18 × 22 (in) Whatman No. 3 MM chromatographic paper. Use a densitometer to estimate the amounts and refer to a standard curve.

Application

Others have used paper chromatography in a manner somewhat similar to that above (Luthi and Vetsch 1952; Colagrande 1957; Martinez Burges 1960; Formisano et al. 1967) with either similar or nearly similar results. Variations have been suggested by Feuillat and Bergeret (1967 b). Juhasz and Polyak (1976) used a strong cation exchange resin attached to a plastic film to separate up to 16 amino acids. Pretreatment was necessary. van Wyk and Venter (1965) used high voltage electrophoresis to separate the acidic and basic amino acid, but resorted to two-dimensional paper chromatography to separate the neutral amino acids. The ninhydrin spots were intensified by $CdSO_4$ treatment.

Kliewer et al. (1966) surveyed the major free amino acids in the genus *Vitis*. Individual *V. vinifera* varieties of table grapes (Kliewer 1969) and wine grapes (Kliewer 1970) were measured for these eight amino acids. Generally, arginine was predominant in table grapes and proline in wine grapes in California. Feuillat (1974 a) and Feuillat and Bergeret (1967 b) reported similar results for grape juice of Bourgogne. Togawa and Takezawa (1978) measured grapes from Japan and found results similar to those of Kliewer. Gallander et al. (1969) found a different pattern of amino acids in eastern American grapes.

The changes in amino acids during grape maturation were reported (Kliewer 1968 a; Juhasz and Torley 1979; Khachidze and Matikashvili 1973). Proline showed the most dramatic increase over the normal ripening period.

Kliewer (1967) followed the annual cyclic changes in the concentration of the various amino acids in the main roots, lateral roots, cane wood, trunk wood and buds. He found main changes at bud burst, especially in arginine, aspartic acid, glutamic acid, and proline.

Low light intensity resulted in lower proline and increased arginine in berry juice (Kliewer and Lider 1970).

Increased nitrogen application increased free amino acids in leaves, grape yields and shoots. Further increases were caused when molybdenum was added (Veliksar 1977).

Lepadatu and Tanase (1967) followed changes in the amino acids during skin contact of red wine fermentations – some decreased, others increased.

Nastic et al. (1967, 1968) reported on the amino acids in wines in Yugoslavia. Dorer and Malnersic (1978) measured amino acids in red wines and grapes, and concluded that proline was not a good index for identification of variety. The method used is similar to the one described by Nassar and Kliewer (1966). Sachde et al. (1979) measured the amino acids in Egyptian wines and reported that amino acids made up most of the total nitrogen, with proline the predominant amino acid.

Erdelyi et al. (1967) indicated that heating of juice from 85 to 100° C, besides other damage, decreases the amino acid content. Goren'kova and Nechaev (1972) also indicated that juice pasteurization at 100° C for 24 h decreases amino acids by melanoid formation. Grechko and Emel'yanova (1984) heated concentrate to 110° C for 30–60 min and lost 40–70% of the amino acids. Contrary to this, Feuillat and Bergeret (1967 a) showed no amino acid decrease when heating crushed grapes to 80° C. Relatively short periods (5 days) of heat at 65° C cause no changes in sherry, according to Kurganova et al. (1974).

By several periodic aerations and filtrations during fermentation, Martakov et al. (1972) found sufficient assimilation of the free amino acids to prevent further yeast growth, resulting in semi-sweet stable wine.

Kozub et al. (1981) determined the free amino acid in musts, wines and sherries. In musts and wines the free amino acid to peptide ratios were high and in sherry production the peptide increased and the free amino acids decreased.

Mndzhoian et al. (1971) demonstrated the extraction of amino acid from *Quercus* oak. Petrosian (1971) measured various aged brandies, from 2 to 26 years in oak, and found increasing amounts of amino acids with age.

3.3 Automatic Amino Acid Analyses

The original method of amino acid analysis with post-column derivatization published by Spackman et al. (1958), with improvements over the years, is without doubt the most acceptable method of quantitative determination of amino acids. Hare et al. (1985), Perret (1985) and Rosenthal (1985) have effectively reviewed the methods available and discussed the pros and cons of various derivatizing reagents and detectors. The most discussed derivatizing agents are the classical ninhydrin, o-phthalaldehyde (with 2-mercaptoethanol), fluorescamine, and dansyl chloride. It has been suggested (Fujita et al. 1979; Ishida et al. 1981) that NaOCl be continuously mixed in with the buffers in order to detect secondary amines at a single wavelength using the OPT method without base line variations. Yokotsuka and Kushida (1983), apparently unaware of the work cited above, proposed similar treatments for use with grape juice and wine. Stein et al. (1973) and Felix and Terkelsen (1973) suggested fluorescamine as derivatization agent, as it is more sensitive and allows determination of all amino acids with one channel detector. Discussions in Section 3.4 also can apply to this section as far as derivatization and detection problems.

There are numerous commercial systems available from very simple skeleton set-ups to very expensive models fully computerized. Figure 3 shows a scheme of the equipment and arrangement for post-derivatization following cation exchange.

The method described is one that is in use in a commercial analytical laboratory to measure amino acids in grape juice and wine as well as in the authors' laboratory. The column and post derivatization equipment is similar to that described by Henshall et al. (1983).

Method

Make sample of juice or wine approximately 2.5 pH by diluting 1 to 4 with 0.25 *N* buffer 2.2 pH. Filter sample through 0.45 μm membrane and inject a 20 μl sample. Use a 3 mm × 250 mm, 10 μm cation exchange type column charged with lithium ion. Keep column oven temperature at 40° C and flow rate at 0.3 ml min^{-1}. The lithium gradient system is as follows: 0–20 min, 100% pH 2.75 (0.24 *N*); 20–70 min gradient from 100% pH 2.5 to 65% pH 2.75 and 35% pH 7.5; 70–100 min gradient from previous to 100% pH 7.5 (0.64 *N*); 100–151 min 100% pH 7.5; 151–185 min gradient from previous to 94% pH 7.5 and 6% regeneration solution (0.3 *N* Li ion with 2 m*N* EDTA); 185–186 min gradient 100% pH 2.75 and 186–210 min 100% pH 2.75.

For the post-column derivatization combine the ninhydrin solution (such as Trione, 17 g l^{-1} ninhydrin, 0.3 g l^{-1} hydrindantin made up in pH 5.5 lithium ace-

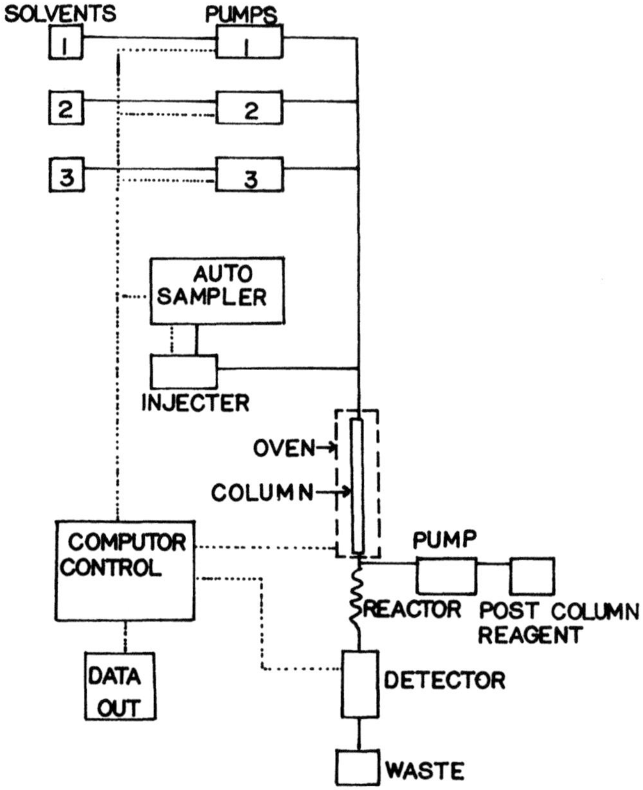

Fig. 3. Schematic of a typical HPLC system with post column derivatization set up for automated amino acid analysis

tate and sulfalane) with the column effluent at a 1 : 1 ratio. Heat mixture to 100° C for approximately 1 min in a continuous flow reactor. Determine the chromophere at 570 nm.

Figure 4 is an example of a chromatogram by this method.

Application

Rapp and Reuther (1971) measured the amino acid composition in sound and rotted grapes and noted generally amino acid decreases in the rotted fruit, with the exception of lysine, which increased. They also noted a minimum amino acid requirement for completion of fermentation. Bravdo et al. (1984) found relatively small differences in amino acids in grapes by cluster-thinning the fruit. Ough and Anelli (1979) compared the amount of grape cropping for amino acid levels by thinning and pruning, and also noted only small differences. Lotti and Anelli (1971) followed the changes in amino acids during grape maturation with variable results. Further work (Lotti et al. 1972) showed similar variability. Juhasz et al. (1984) found increases in arginine during grape maturation and that field nitrogen treatments increased arginine content of the grape. Juhasz and Torley (1985) suggested that arginine adversely affects the sensory quality of the wine

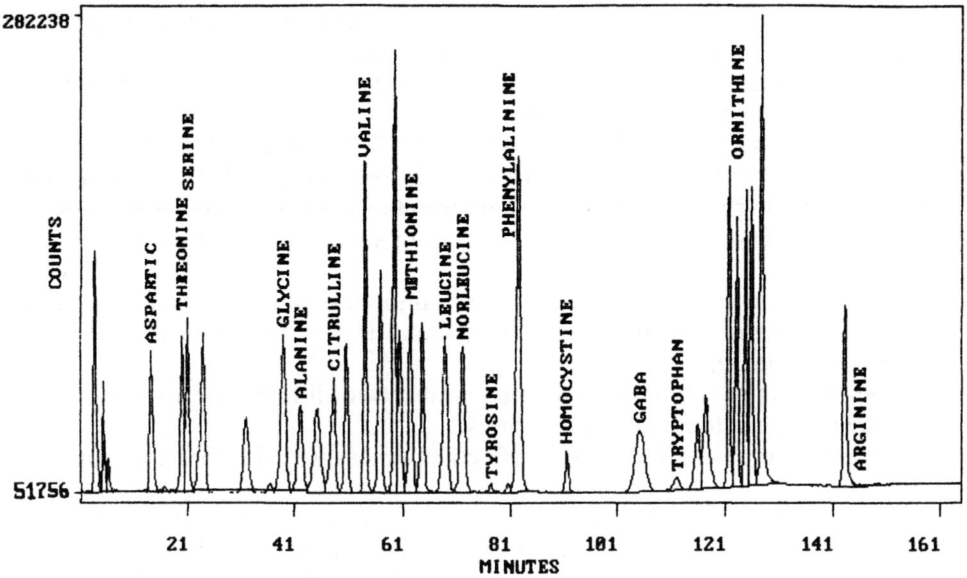

Fig. 4. Chromatogram of a standard amino acid run using an automated HPLC post column ninhydrin derivatization system

and that nitrogen fertilization may damage general wine quality. This is opposite to what was demonstrated by Bell et al. (1979). Many factors are involved beyond simple nitrogen levels.

Several species of wine yeasts were tested by Anelli et al. (1972) to determine if differences in residual amino acids resulted from their use to ferment grape juice. They found significant differences. Polo and Llaguno (1974 a, b) found that proline, alanine, and glycine remained in the wine in the greatest amount after various *Saccharomyces* sp. fermentation. Proline, arginine and glutamic acid were present in the largest quantities before fermentation. Kozub et al. (1977) found high amounts of proline and γ-aminobutyric acid in wines.

Salagoity et al. (1985) determined that the amino acid content of the juice of two white varieties did not change appreciably after 48 h of juice-skin contact. Nanitashvili (1973) found the treatment of various wines with a proteolytic enzyme increased the free amino acid content considerably. Zamorani and Nicolosi (1978) found greater amounts of amino acids in the juices with excessive pressing of the crushed grapes. Poux et al. (1974) and Leone et al. (1976) showed that heating juice and skins did not change amino acid concentrations appreciably.

Tsatsaronis et al. 1977; Mayer et al. 1973; Lotti et al. 1972; Cerutti et al. 1978; Temperli and Kuensch (1976) reported on the changes during vinification in the amino acid content. Pekur et al. (1981) noted that under aerobic conditions of grape juice fermentation, about 97% of the amino acids were metabolized, while under anaerobic conditions and CO_2 pressure (Pekur 1978), only about half were assimilated. Zamorani et al. (1973) found that fermenting juice with bentonite present decreased the amino acid content. This in turn did not appreciably effect the amyl alcohol concentration in the wines (Zamorani and Nocolosi 1974).

Wucherpfenning and Millies (1976) found that electrodialysis of wines did not change the amino acid concentrations.

Ooghe et al. (1981) and Ooghe and Kasteljin (1984) were able to differentiate wines from the various areas of France by their free and hydrolyzable amino acid composition. On the other hand, Casagrande et al. (1975) were not able to make a significant separation of areas by wines using amino acids and other chemical measurements.

Sciancalepora et al. (1983) measured the amount of amino acids in lees (yeast residue from fermentation of grape) and found it very similar to that of yeast grown on n-paraffin.

Rodopulo et al. (1969) found some amino acid differences in sherry wines made by two yeast fermentation methods.

Bergner and Haller (1969) very effectively measured the changes in amino acids during sparkling wine production from starting with the grapes to the final product. Suarez et al. (1979) found most changes in the first 3 months of storage. Margheri et al. (1982) noted little change in free amino acids during sparkling wine production or during autolysis. Feuillat (1974b) found significant uptake of amino acids, then a gradual release during production and aging. Colagrande et al. (1984) noted wide variation in amino acids in champagne due to treatments. Margheri et al. (1983, 1984a, b, c) further confirmed that liberation of amino acids by yeast autolysis was negligible, but did find an assimilation of amino acids under sparkling wine conditions. Spettoli and Zamorani (1982) found proline content to increase on aging more than the other amino acids. Scopigno et al (1983) showed a decrease in amino acids in Asti sparkling wines by the charmat process. Bergner and Wagner (1965) found no qualitative or quantitative difference in amino acids in sparkling wines made from the bottle process or the tank process.

Carbonic maceration and normal vinifications investigated by Andre et al. (1981) showed some differences in the catabolism of the amino acids. Flanzy et al. (1977) found that proline, arginine, alanine, and γ-aminobutyric acid predominate in the polypeptides that are broken down during carbonic maceration. The general amounts of free amino acids remain constant during the treatment.

The amino acid content of raisins from Thompson Seedless grapes was reported by Bolin and Petrucci (1985).

The amino acid content before and after the fermentation of *Vitis labrusca* cultivars was reported by Kluba et al. (1978a, b). Changes were similar to those found in *V. vinifera*. The concentration of amino acids was different, however, from normal *vinifera* varieties. Marcy et al. (1981) also showed some variations from *vinifera* amino acid composition when they studied *V. rotundifolia*. These were mainly relatively low proline values.

The malolactic fermentation of wine causes an expected decrease in all measured amino acids (Spettoli et al. 1984). Weiller and Radler (1976) reported great variation in the ability of malolactic bacteria to metabolize the various amino acids. Kuensch et al. (1974) also showed variations in changes in amino acid consumption by various types of malolactic bacteria. Of particular interest, they demonstrated the ability of *Leuconostoc oenos* to metabolize arginine to ornithine, indicating a strong arginase activity. Feuillat et al. (1985) further verified the need for arginine as a substrate for this malolactic bacteria.

3.4 High Performance Liquid Chromatography

Evans (1983) briefly reviewed the use of High Performance Liquid Chromatography (HPLC) in enology. It is difficult to separate HPLC methods with post-derivatization and use of ion-exchange columns from the classical automatic amino acid analyses, since they function in essentially the same manner. Therefore that type will be covered under the Automatic Amino Acid Analyzer section.

The pre-column derivatization of amino acids with o-phthaldialdehyde (OPT) (Roth 1971) and separation (Roth and Hampai 1973) of the fluorescent derivatives on reverse phase columns and fluorescent detection using HPLC systems has developed considerably in the last few years (Lindroth and Mopper 1979; Larsen and West 1981; Price et al. 1984; Kan and Shipe 1981; Hodgin 1979; Benson and Hare 1975; Benson 1974). Fleury and Ashley (1983) compared the results obtained by pre-column automated OPT derivatization to the results by an amino acid analyzer. With the exception of glutamine, the coefficients of variation for the HPLC system were twofold or greater. A Resolve 5 µm (Water Assoc.) (3.9 × 15-cm) was used without mercaptoethanol. Tyrosine, α-aminobutyric acid, proline, hydroxyproline, and cysteine were not quantified. Lee and Dreschert (1979) considered the way to improve cysteine detection by oxidation prior to OPT-mercaptoethanol treatment. Complete measurement of all the amino acids was obtained by Umagat et al. (1982) by oxidation of cysteine and cystine by performic acid to cysteic acid which will react with the OPT and derivatization of proline and hydroxyproline to fluorescent products with 4-chloro-7-nitrobenzofurazan. Svedas et al. (1980) studied the kinetics of the OPT, amino acid, 2-mercaptoethanol reactions and gave optimum concentrations.

Two other methods have been mentioned as possible derivatives for amino acid dansylation and phenylthiohydantoin formation. Bayer et al. (1976) showed reasonable results and a twofold increase in sensitivity over the ninhydrin method by use of dansyl chloride. Szokan (1982) improved the elution solvent system. Rapid and complete separation of amino acid phenylthiohydantoins was achieved by Bledsoe and Pisano (1981), who gave some useful hints in improving derivative separation. Noyes (1983) also gave methods for optimizing separations. Black and Coon (1982), using a C_8 column, also gave ways to optimize separations and show excellent runs of only 15 min length.

Martin et al. (1980) used dansylation to derivatize and separate 19 amino acids in wine. No quantification was given. Martin et al. (1984) gave further details on the methodology and considered the method satisfactory for juice, wine, and vinegar, although no replication data or analytical values for the grape products are shown. Casoli and Colagrande (1982) used dansylation also. They report very satisfactory relative standard deviations of 1.5 to 6.75% and separation of all but aspartic, glutamic, and cysteine.

Sanders and Ough (1985) used a C_{18} column and OPT with 2-mercaptoethanol to derivatize amino acids remaining in wine. The arginine and γ-aminobutyric acid derivatives were not resolved completely and cysteine was not found. Tryptophan recovery was low. Dual peaks from some of the amino acids are possible.

While this method and the dansyl method of Casoli and Colagrande (1982) both seem adequate, neither is perfect. We will detail the OPT method because there appears to be more work going on with this method, and because of the authors' familiarity with it.

Method

Take 1 ml of wine or grape juice which has been appropriately diluted (depending on concentration of amino acids) and add 2 ml of 0.1% trifluoracetic acid (TFA) in 70:30 water:methanol. Activate a C_{18} Sep-Pak (Waters Assoc.) attached to a 10-ml syringe by successively washing with 20 ml methanol, 20 ml of 0.1% TFA in water and 10 ml of 0.1% TFA in 80:20 water:methanol. Attach to the outlet end of a 0.22 µm 29-mm filter unit (Millipore) connected to a 2-ml syringe. Pass the diluted test sample through the Sep-Pak and filter and follow with a 1.0 ml wash of 0.1% TFA 70:30 water:methanol solution. Store in the refrigerator until ready to use.

To a 0.3-ml glass micro vial add 20 µl of the diluted prepared sample (or standard) and mix with 100 µl of the OPT derivatizing agent. [Prepare this reagent 1 day prior to use by dissolving 64.8 mg OPT in 1.2 ml ethanol (anhydrous) in a 12-ml graduated tube. Add 9 ml of 0.4 M borate buffer (pH 9.5) and 48 µl of 2-mercaptoethanol and mix. Filter the reagent through a 0.45 µm filter and store in the dark]. Allow the reaction to proceed for 2 min, then fill the 20 µl injection loop and inject.

All reagents should be HPLC quality or redistilled and cleaned as required. Reagent blank should give base line in the areas of the amino acid OPT derivatives.

Select a C_{18} reverse phase analytical column such as a Resolve, 15 × 0.39 cm ID, 5 µm particle size. Precede the column with a guard column such as Bondapak C_{18}/corasil 37–50 µm (Waters Assoc.). A mixture 0.035 M acetate buffer (pH 6.5), methanol and tetrahydrofuran (THF) (91:5:4) is solvent A and solvent B is a mixture of methanol and 0.01 M acetate buffer (pH 6.5) (60:40). Apply a multiple-step mobile phase gradient (0–5 min, solvent A 100%; 5–15 min, solvent A 86% B 14%; 15–31 min, solvent A 50% B 5%; 31–35 min, solvent B 100%; 45–55 min solvent A 100%. Linear ramping to next solvent mixture from 1–5 min, 10–15 min, 19–31, and 31–45 min. Next injection at 55 min). Set flow for 1 ml min^{-1} and oven temperature at 30° C. Inject a fresh sample every 55 min.

The fluorometric detector should read at 370 nm and the excitation emission filter should have a 418–700 cut-off. Different instruments or columns will need slightly different conditions.

Application

No applications other than those in the papers discussed. The complexity of the equipment needed and the lack of good definitive separations and quantifications have discouraged the use of this method.

3.5 Gas Chromatography

Development of the gaschromatographic analysis of amino acids goes back to the early 1960's (Johnson et al. 1965; Ruehlmann and Giesecke 1961; Zomzely et al. 1962). However, Lampkin and Gehrke (1965a, b) were perhaps the first to quantitate the amino acids, along with Johnson and Goodson (1965). The former (Lampkin and Gehrke) used the n-butyl N-trifluoroacetyl esters, while Johnson and Goodson used the methyl-N-trifluoroacetyl esters. The latter also demonstrated its use on biological material. In a series of papers (Stalling and Gehrke 1966; Gehrke and Shahrokhi 1966; Stalling et al. 1967), details for the method were worked out and summarized by Gehrke and Stalling (1967). Gehrke et al. (1968) gave details for macro- to micromethods for biological substances and the use of internal standards. An improved revision of the method (Zumwalt et al. 1971) uses 4-(aminoethyl)-cyclohexanecarboxylic acid as the internal standard and takes only 30 min to prepare the derivatives by eliminating the inter-esterification step. Ussary (1973) was enthusiastic over the improved version. Bengtsson et al. (1981) gave a variation of the esterification method using glass capillary columns and either electron capture or single-ion monitoring detection, both extremely sensitive. Kolb (1974) reviewed the various gas chromatographic derivitization and detection schemes. He pointed out some of the inherent errors in quantification. Engel and Hare (1985) reviewed the use of gas liquid chromatography and concluded that its best use is the identification of amino acids and determination of purity in conjunction with mass spectrometry and recommended that it be used to compliment quantitative ion-exchange measurements.

From a practical point of view the method described by Gehrke and Leimer (1971) and used for grapes and wines by Fantozzi and Montedoro (1973), and revised slightly by Ough and Tabacman (1979), is rapid and, while limited in that certain amino acids are not measured, is reasonably accurate for those amino acids measured. It gives results in agreement with classical ion exchange methods.

Method

Put 9.3 cm Dowex 50W × 8 (50–100 mesh) cation resin into a column (63 × 2 cm glass with sintered disc). Prewash the resin with water, ethanol, acid and base and charge with 5*N* HCl (112 ml at 2 ml min^{-1}) and rinse with H_2O until eluate free of Cl^- to Ag^+ test. Keep the resin covered with liquid at all times.

Take a 10-ml centrifuged and filtered wine or must sample and put onto the column and load at the rate of 1 ml min^{-1}. Decolorize wine samples with activated charcoal prior to ion-exchanging. Wash resin with 200 ml of distilled H_2O for must samples and 350 ml for wine, to remove glycerol and other substances which formed during fermentation and will silylate, at 2 ml min^{-1}. The silylated glycerol co-chromatographs with α-aminobutyric acid. Elute the amino acids with 200 ml of 3*N* NH_4OH at the rate of 1–2 ml min^{-1} into a round-bottom flask. Evaporate off the liquid to near dryness on a rotary evaporator at 45–50° C at 12 torr. Add 50 ml of H_2O to the flask and again take to dryness. Repeat until the odor of ammonia is gone. Volumetrically take up the sample to 10 ml with H_2O. Pipet 1 ml of this into a 2 ml screw cap vial and dried at 70° C with a stream of nitrogen. Add 1 ml dichloromethane and redry. Repeat this last step two more times to remove the H_2O. Close the vial with a Teflon-lined rubber septum and add 500 μl of bis-trimethylsilyl-trifluoroacetamide (BSTFA) and 500 μl of acetonitrile (with the internal standard decanoic acid, included). Sonicate the vials for 3 min then heat in oil bath at 145–150° C for 3 h.

Pack a 2-mm ID, 10-ft glass column, with Suplecoport (100–200 mesh) coated with 10% OV-11. Plug each end with silanized glass wool. Cure the column for 48 h at 300° C with N_2 gas flow at 20 ml min^{-1}. Set the injector temperature at 250° C and the detector temperature at 300° C. Inject 3 μl of the derivatized amino acids. Keep column at 110° C for 8 min then program 4° C min^{-1} to 250° C. Use a flame ionization detector.

Prepare a standard solution of 18 amino acids each at 100 mg l^{-1} concentration in water. To make standards add 0.02, 0.4, 0.6, 0.8 and 1.0 ml to 2 ml vials and proceed as above. Prepare the decanoic acid internal standard by adding 33 mg to 100 ml of acetonitrile.

The coefficient of variations varied from 4 to 12% for the standard curve data except for methionine and glutamic acid which were much higher. Fantozzi and Montedoro had no trouble with methione and glutamic acid. Their coefficients of variation were slightly lower, but similar.

Figure 5 shows the chromatogram for a grape juice sample.

Neves and Vasconcelos (1983) used a complex derivitization combining reaction of ethyl chloroformate and BSTFA and tested several different capillary columns with moderate success.

The use of capillary columns for better separations is obvious. However, there is sometimes a problem with replication because of the small volume used. The use of an internal standard is mandatory.

Applications

Lhuguenot et al. (1979) made the isobutyl-N-heptafluorobutyric ester and acyl derivatives of the amino acids to determine effects of different yeasts on the residual amino acids. Histidine was not measured by the method. Vasconcelos and Neves (1984, 1985) used capillary columns (OV-101); N,O-heptafluorobutyryl amino acid isopropyl esters to measure the amino acids released into a nitrogen-deficient media by growing yeast and thus characterize yeast. The standard deviations in the results were extremely small. Tunblad-Johansson (1977) used n-propyl-N-acetyl esters to look at the free amino content of yeast.

Drawert et al. (1977, 1978) used a trimethylsilyl derivatization method to determine amino acids in grape juice. Bertrand et al. (1982), using esterification methods, determined changes in amino acids in fermenting red wines. Miguel et al. (1985) used silylation methods to follow amino acid changes during grape maturation. Drawert et al. (1976) also demonstrated the trimethylsilylation method for use with grapes and wine. Campos and Severin (1970), and Feuillat and Demaimay (1972) used the n-butyl-N-trifluoroacetyl derivatives to identify amino acids in wine. Sedova and Kahler (1980), using the isobutyl-N-heptafluorobutyl derivatives, had less than 3% deviations in replicate samples.

Ough and Bell (1980) and Leone et al. (1980) demonstrated the effect of juice precursor amino acids on the concentration of n-propanol, isobutanol, and active and isoamyl alcohols.

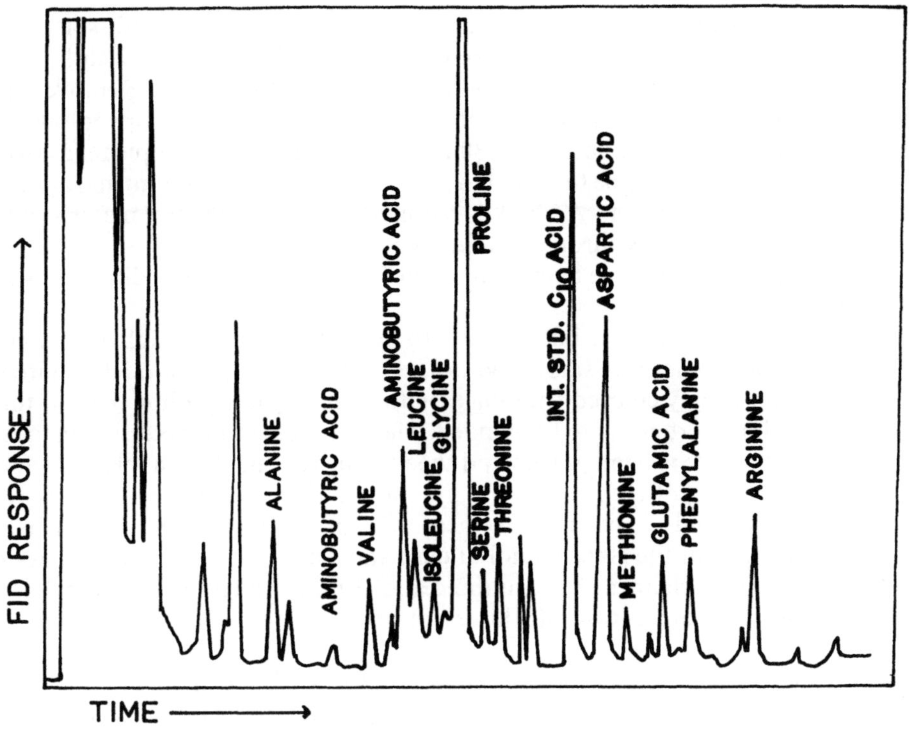

Fig. 5. Gas chromatogram of amino acids from a grape juice sample separated by anion exchange, concentrated and silylated with BSTFA

3.6 Proline

Proline, except for a few instances, makes up the majority of the free amino acids in wine and is the largest amount found in fresh grapes.

Proline and hydroxyproline are the only two amino acids which do not have a free $-NH_2$ group. Hydroxyproline is either not present or present in less than 10 mg l^{-1} levels in wines (Rougereau 1974). These two amino acids can be measured roughly by the neutral ninhydrin method (Rosen 1957) by reading the reaction products at 440 nm rather than the usual 570 nm for the blue color developed by the rest of the amino acids. Microbiological methods are available (Castor 1953 a). However, a good chemical method for proline was desirable, considering it is one of the most abundant amino acids in grapes and wines. A method suggested by Chinard (1952) was adapted to grapes and wines by Ough (1969 a). The method is applicable to most biological materials unless lysine, hydroxyproline, tryptophan and glutamine greatly exceed the concentration of proline. With slight variations, this method has been accepted for use in grape and wine research.

Method

Dilute the sample of grape juice or grape part extract or wine with water so that it contains between 0.05 and 0.50 mol ml^{-1} of proline. Add 0.5 ml of the diluted sample, 0.25 ml of formic acid and 1 ml of 3% wt/v ninhydrin (dissolved in methyl cellosolve) to 15 × 130 mm screw top tubes. Mix and close tightly with screw caps. Place the tubes into a boiling water bath for 15 min. Remove and place into a 20° C bath and immediately add 5 ml of 1:1 isopropanol: water diluent. Read after 5 min, but before 30 min, at 517 nm. If the sample reads over 0.8 optical density, it may be further diluted by pipeting out an aliquot and adding a known amount of diluent. A blank is required. Instead of the test material add 0.5 ml of water and carry through all the steps.

Prepare a standard curve by weighing out 57 mg of proline and diluting to 1 l (to obtain 0.5 mol ml^{-1}).

Wallrauch (1976) used n-butylacetate to extract the colored ninhydrin complex. The color then is more stable and can be kept for a longer period if required. Also with very deeply colored fruit juices with relatively low proline content this allows the color interference to be avoided. The extra step is not necessary in most instances if the juice is from grapes and the readings are made in a relatively few minutes. Betz (1980) offered a variation in solutions and read at an alternate wave length (492 nm) and claims improvement. Cavasino and Forsaci (1975) used n-butanol instead of the isopropanol diluent and claimed better results. Giaccio (1981) erroneously claimed that ornithine and citrulline interfere with the determination. At wine concentrations this is not possible.

Application

Ough (1968) measured the proline content of juice, seeds, skins, and pulp of berries of several varieties, and the proline in grape juice and wines, and related it to the total nitrogen content. Di Stefano (1979) suggested that the pulp was rich in proline. Modi and Guerrini (1975) noted that most wines with higher proline were better wines, but failed to realize that most of the quality varieties have a high level of proline. A number of Italian investigators (Cerutti et al. 1975; Mincione et al. 1976; Colagrande et al. 1975; Carrubba et al. 1979; Nicolosi and Lanza 1976) measured the proline content of the fruit and wines. Pallotti et al. (1976) stated that it was possible to distinguish between regions of Italy and suggested free proline, total and amino nitrogen as the criteria. Minguzzi and Amati (1973) noted the adulturated wines contained much less proline than standard wines. Salzedo et al. (1976) suggested that since variations occur from year to year, between and in varieties it would be impossible to detect adulterators. Colagrande et al. (1976) agreed. Ough and Stashak (1974), while showing differences in varieties, also pointed out maturity changes and the potential for disease effects on the proline. Ough and Singleton (1968) found proline increased with grape maturity and accounted for about 90% of the nitrogen left in Cabernet Sauvignon wines. Giaccio (1979) found that analysis for proline was not a proper measurement to detect adulteration. His data showing the formation of proline by yeast from arginine is what would be expected if arginine were the only source for yeast growth. Skin contact also increased the proline content (Ough 1969 b). The concentration ranges that are found for each variety are quite large, for example, see Ethiraj and Suresh (1982).

Some yeasts will metabolize proline and others will not (Bertolini and Paronetto 1976). It was determined by Ough and Winger (1982) that the *Saccharomyces cerevisiae* strain Montrachet, used commonly, could partially metabolize proline and used more at increasing temperature compared to *Saccharomyces bayanus* and *S. fermentati*. Farris et al. (1978) noted that five strains of *Saccharomyces bayanus* and five of *S. prostoserdovii* completely metabolized proline in either fermenting musts or during flor sherry production. Di Stefano (1977) also found variations in proline metabolism depending on yeast and conditions.

The explanation offered for the lack of proline uptake is that the transport system for proline into the cell is inhibited by aromatic amino acids as well as by anaerobic conditions and internal inhibition of the transport when the proline content inside the cell is sufficient. The work of Horak et al. (1977) indicated the inhibiting tendencies of some amino acids on proline transport.

Heat treatments of grapes caused no increase in proline extracted (Di Stefano 1979; Bertolini and Paronetto 1976).

Colagrande et al. (1973) and Bertolini and Paronetto (1976) found removal of proline by bentonite treatment.

Yellow mosaic virus in grapes caused marked increases in the proline concentration in the leaves (Abrasheva 1975).

Betz and Schmitt (1981) found that proline declined with increasing yield. Zelleke and Kliewer (1979) verified the report of Ough et al. (1968 a) that proline varied with rootstocks. Buttrose et al. (1971) showed the affect of increased temperature on proline accumulation in the berry. Downton and Loveys (1981) and Coombe and Monk (1979) showed that salt stress and water stress respectively affect the abscisic acid concentration in the leaves and in the berries and cause increases in proline content of both.

3.7 Arginine

Arginine is one of the most abundant amino acids in grapes. It is an important nutrient for yeast and bacteria growth.

An arginine electrode has been suggested by Neubecker and Rechnitz (1972) involving the reaction of two enzymes to form ammonia from arginine.

$$\text{Arginine} \xrightarrow{\text{Arginase}} \text{Ornithine} + \text{Urea}$$

$$\text{Urea} + \text{Water} \xrightarrow{\text{Urease}} HCO_3^- + NH_4^+ .$$

The ammonia is then detected by an ammonia electrode such as that produced by Orion. Ammonia already present in juice would interfere. Also, as pointed out by the authors, it is a slow reaction and not at the optimum pH for either enzyme. Valle-Vega et al. (1980) pursued the electrode use for peanut maturity using immobilized enzymes with some success.

Kliewer (1974 personal communication) applied the method of Gilboe and Williams (1956) to grape juice.

Method

Dilute clear grape juice to about 1:50 with water. In most cases this will result in diluting to 1 to 8 μg of arginine ml^{-1}. Put all reagents into an ice bath. Prepare a set of standards from an original stock arginine solution of 500 μg ml^{-1} (60.5 mg arginine hydrochloride per 100 ml of water). Dilute the stock solution with water to give 10 µg ml^{-1}. To a set of five test tubes 22–25 mm in diameter add 0, 1, 2, 3, and 4 ml of the 10 µg ml^{-1} solution to each respectively and bring each to 5 ml volume with water. Add 5 ml of the diluted juice(s) and place all tubes into an ice bath. To each tube add 1 ml of 0.02% 8-hydroxyquinoline and 1 ml of 10% sodium hydroxide, mix thoroughly and replace into the ice bath. Add quickly 0.2 ml of 1% sodium hypobromate (prepare by diluting 1 g of liquid bromine with 100 ml of 5% sodium hydroxide), shake and within 15 s add 1 ml of 40% urea, mix and within 1 min add 5 ml of cold distilled water and mix. Read absorption of color at 500 nm within 5 min of color development. Use a reagent blank to zero the instrument. Determine the juice values by comparison to the standard curve.

Application

Kliewer and Cook (1971, 1974) and Kliewer (1977) investigated the relationship of arginine to the status of the grapevine and further related this to the nitrogen status of the vineyard soil and the need for nitrogen additions. They found the arginine was a better indicator than proline, though both were well correlated to nitrogen deficiency. It was shown by Kliewer and Ough (1970) that defoliation decreased the amount of arginine in the juice. Juhasz et al. (1984) found significantly increased arginine in Hungarian vineyards treated with nitrogen. Bell et al. (1979) found threefold increases in arginine in the juices after extreme nitrogen treatments of the soil. There was a 20-fold difference in the arginine concentrations after fermentation with the wines as low as 24 mg l^{-1} and as high as 489 mg l^{-1}.

Kliewer and Lider (1976) found only small effects on the arginine levels of the juice of berries of vines heavily infected with leafroll virus. Pool et al. (1972) measured the effect of girdling and growth regulators on the arginine content of the resulting grapes in storage.

References

Abrasheva P (1975) Effect of the yellow mosaic on the free amino acid composition in grape leaves. Gradinar Lozar Nauka 12(5):123–127

Addeo F (1972) Determinazione colorimetrica dell'acido citrico nei mosti e nei vini. Sci Technol Aliment 2:87–92

Addeo F, Musso SS (1973) Determinazione colorimetrica dell'acido tartarico nei mosti e nei vini. Ind Agr 11(1):13–18

Alley CJ, Goheen AC, Olmo HP, Koyama AT (1963) The effect of virus infections on vines, fruits and wines of Ruby Cabernet. Am J Enol Vitric 14:164–170

Amerine MA, Ough CS (1980) Methods for analysis of musts and wines. Wiley, New York, p 341

Amerine MA, Berg HW, Kunkee RE, Ough CS, Singleton VL, Webb AD (1980) The technology of wine making. 4th edn, Avi Publishing, Westport Conn, p 794

Andre P, Benard P, Flanzy C, Tacchini M (1981) Wine making the carbonic maceration. Evolution of free and bound amino acids present in musts and wines. Sci Aliment 1(1):27–53

Anelli G, Lotti G, Lepidi AA (1972) Gli amminoacidi liberi nei vini prodotti con lieviti diversi. Ind Agr 10:205–210

Anonymous (1978) Recueil des methodes internationales d'analyse des vins. 4th edn. Office International de la Vigne et du Vin, Paris

Arndt W, Thaler H (1974) Zur Bestimmung der Galakturonsäure in Wein und Fruchtsäften. Mitt Höheren Bundeslehr Versuchsanst Wein Obstbau Klosterneuburg 24:325–340

Arranz M, Tienda P, Mareca I (1981) Estudio critico de los methodes de analisis de acido malico aplicados a mostos y vinos. Sem Vitivinic 36(1.837–1.838):4341–4343

Ashoor SH, Monte WC, Welty J (1984) Liquid chromatographic determination of ascorbic acid in foods. J Assoc Off Anal Chem 67:78–80

Astabatsyan GA (1980) Free amino acids in grapes during storage. Sadovod Vinograd Vinodel Mold 35(7):59–60

Baker DW (1973) Analysis of organic acids in fruit products by anion exchange isolation and gas chromatographic determination. J Assoc Off Anal Chem 56:1257–1263

Bandion F, Valenta M (1977) Zur Beurteilung des D(−) und L(+) Milchsäuregehaltes in Wein. Mitt Höheren Bundeslehr Versuchsanst Wein Obstbau Klosterneuburg 27:4–10

Bandion F, Roth I, Mayr E, Valenta M (1980) Zur Beurteilung der Gluconsäuregehalte bei Wein im Hinblick auf mögliche Veränderungen während der Lagerung. Mitt Rebe Wein Klosterneuburg 30:32–36

Battle JL, Joubert R, Collon Y, Jouret C (1978a) Utilisation du flux continu pour le dosage colorimetrique de l'acide tartrique dans les mouts de raisin et les vins par la method Rebelein. Ann Falsif Expert Chim 71:155–158

Battle JL, Joubert R, Collon Y, Jouret C (1978b) Dosage enzymatique en flux continu du L(−) malate et du L(+) lactate dans les mouts de raisin et les vins. Ann Falsif Expert Chim 71:223–227

Bayer E, Grom E, Kaltenegger B, Uhmann R (1976) Separation of amino acids by high performance liquid chromatography. Anal Chem 48:1106–1109

Bell AA, Ough CS, Kliewer WM (1979) Effects on must and wine composition, rates of fermentation, and wine quality of nitrogen fertilization of *Vitis vinifera* var. Thompson Seedless grapevines. Am J Enol Vitic 30:124–129

Bengtsson G, Odham G, Westerdahl G (1981) Glass capillary gas chromatographic analysis of free amino acids in biological microenvironments using electron capture or selected ion monitoring detection. Anal Biochem 111:163–175

Benson JR, Hare PE (1975) o-Phthalaldehyde: fluorogenic detection of primary amines in the picomole range. Comparison with fluorescamine and ninhydrin. Proc Natl Acad Sci USA 72:619–622

Benson JV (1974) Single column versus two column chromatographic system for amino acid analysis. Analysis 1(2):1–7

Bergner KG, Haller HE (1969) Das Verhalten der freien Aminosäuren von Weisswein im Verlauf der Gärung bei Ausbau, Lagerung und Umgärung. Mitt Rebe Wein Obstbau Früchteverwert (Klosterneuburg) 19:264–268

Bergner KG, Wagner H (1965) Die freien Aminosäuren während der Flaschen- und Tankgärung von Sekt. Mitt Rebe Wein Ser A (Klosterneuburg) 15:181–198

Bergner-Lang B (1977) Bemerkungen zur enzymatischen Citronensäure- und L-Apfelsäurebestimmung im Wein. Dtsch Lebensm Rundsch 73:279–280

Beridze GI, Sikharulidze TG (1972) Composition of organic acids of complex wines. Vinodel Vinograd SSSR 32(3):54–55

Bertolini C, Paronetto L (1976) Variazioni del contenuto in prolina del vino. Vignevini 3(5):13–18

Bertrand A (1974) Dosage des principaux acides du vin par chromatographie en phase gazeuse. Ann Falsif Expert Chim 67:253–274

Bertrand A, Dubourdieu D (1978) Dosage des acides dérivés des sucres, dans les mouts et dans le vin, par chromatographie en phase gazeuse. Ann Falsif Expert Chim 71:303–312

Bertrand A, Medina B, Chevallier JP (1982) Evolution des acides amines de vins rouges en fonction de la durée de maceration. Connaiss Vigne Vin 16:111–123

Bertuccioli M (1982) Direct gas chromatographic determination of some volatile compounds in wine. Vini Ital 138:149–156

Betz R (1980) Beitrag zur Bestimmung von Prolin in „weißen" Traubenmosten und Weißweinen. Wein-Wiss 35:211–213

Betz R, Schmitt A (1981) Über den Einfluß unterschiedlicher Ertragsleistungen der Rebe auf Most- und Weininhaltsstoffe. Wein-Wiss 36:126–134
Beveridge T, Harrison JE (1985) Amino nitrogen in fruit juice, juice concentrates and fruit drinks determined with 2,4,6-trinitrobenzenesulfonic acid. Can Inst Food Sci Technol J 18:259–262
Bigliardi D, Gherardi S, Poli M (1979) Determinazione degli acidi tartarico, malico e citriso nei succhi di frutta mediante cromatografia liquida ad alta pressione. Ind Conserve 54:209–212
Black SD, Coon MJ (1982) Simple, rapid, and highly efficient separation of amino acid phenylthiohdantoins by reversed-phase high performance liquid chromatography. Anal Biochem 121:281–285
Bledsoe M, Pisano JJ (1981) Rapid analysis of amino acid phenylthiohydantions by high performance liquid chromatography: stepwise elution with a 15-CM column. In: Liuty TY, Schechter AN, Heinrikson RL, Concliffe PG (eds) Chemical synthesis and sequencing of peptides and proteins. Elsevier North Holland Amsterdam, pp 245–249
Block R, Durrum E, Zweig G (1955) A manual of paper chromatography and paper electrophoresis. Academic Press New York, p 132
Blouin J, Peynaud E (1963) Présence constante des acides pyruvique et α-cetoglutaricque dans les mouts des raisins et les vins. CR Acad Sci (Paris) Ser D 256:4521–4522
Bolin HR, Petrucci V (1985) Amino acids in raisins. J Food Sci 50:1507
Boulton R (1980) The relationship between total acidity, titratable acidity and pH in wine. Am J Enol Vitic 31:76–80
Bourzeix M, Guitraud J, Champagnol F, Heredia N (1970) Identification des acides organiques et évaluation de leurs teneurs individuelles dans les jus de raisin et les vins par chromatographie et photodensitometrie. J Chromatogr 50:83–91
Bravdo B, Hepner Y, Loinger C, Cohen S, Tabacman H (1984) Effect of crop level on growth, yield and wine quality of a high yielding Carignane vineyard. Am J Enol Vitic 35:247–252
Bruer DRG, Iland PG, Navin JD (1985) Comparison of methods of analysis for ethanol, volatile acidity, and reducing sugars in wines. Aust Grapegrower Winemaker (186):8–9
Brun S, Grau C (1968) Caractérisation de acides organiques dans les vins par chromatographie en couches minces sur feuilles souples „chromatogram Kodak“ Qual Plant Mater Veg 16:197–199
Brunelle RL, Shoenman RL, Martin GE (1967) Quantitative determination of fixed acids in wines by gas-liquid chromatographic separation of trimethylsilylated derivates. J Ass Off Anal Chem 50:329–333
Brunner HR, Tanner H (1979) Chemische Apfelsäure-Bestimmung in Fruchtsäften und Weinen. Schweiz Z Obst Weinbau 115:249–259
Bryan JK (1976) Amino acid biosynthesis and its regulation. In: Bonner J, Varner JE (eds) Plant biochemistry. Academic Press, NY, pp 525–560
Buttrose MS, Hale CR, Kliewer WM (1971) Effect of temperature on the composition of "Cabernet Sauvignon" berries. Am J Enol Vitic 22:71–75
Campos L, Severin M (1970) Analise dos aminoacidos livres dos vinhos por cromatografia em fase gasosa. De Vinea et Vino Porugaliae Documenta Serie II 5(2):1–32
Carrubba E, Pastena B, D'agostino S, Alagna C (1979) Contenuto in prolina di uve e di vini Siciliani. Riv Vitic Enol 32:104–111
Casagrande S, Sperandio A, Seppi A (1975) Studio su alcuni parametri chimi per il controllo della genuinita dei vini. Riv Soc Ital Sci Aliment 4:331–353
Cash JN, Sistrunk WA, Stutte CA (1977) Changes in nonvolatile acids of Concord grapes during maturation. J Food Sci 42:543–544
Casoli A, Colagrande O (1982) Use of high performance liquid chromatography for the determination of amino acids in sparkling wines. Am J Enol Vitic 33:135–139 [see also Ind Bevande (2):29–34 (1982)]
Castino M (1974) Determinazione fluorimetrica dell'acido malico nei mosti e nei vini. Vini Ital 16:43–47
Castor JGB (1953a) The free amino acids of musts and wines. I. Microbiological estimation of fourteen amino acids in California grape musts. Food Res 18:139–145

Castor JGB (1953b) The free amino acids of musts and wines. II The fate of amino acids of must during alcoholic fermentation. Food Res 18:146–151
Castor JGB, Archer TE (1956) Amino acids in must and wines, proline, serine and threonine. Am J Enol Vitic 7:19–25
Castor JGB, Archer TE (1959) The free amino acids of musts and wines. III. Effect of added ammonia and of fermentation temperature on the fate of amino acids during fermentation. Food Res 24:167–175
Catalina L, Sarmiento R, Romero R, Valpuesta V, Mazuelos C (1981) Estudi de la fertilizacion diferenciada en la vid. I. Evolucion del nitrogeno total, nitrogeno proteico, amino acidos libres y prolina. An Edafol Agrobiol 40:667–675
Catalina L, Mazuelos C, Romero R, Sarmiento R (1982) Cambios metabolicos durante el proceso de maduracion de la uva (*Vitis vinifera* L. var. Palomino) en la zona del marco de jerez de la frontera (Cadiz). An Edafol Agrobiol 41:1503–1511
Cavasino G, Forsaci L (1975) Contenuto in prolina nei vini prodotti nella provincia di Trapani nelle annate 1973/74 e 1974/75 determinato col metodo di Ough modificato. Riv Sci Technol Aliment Nutr Um 5:113–114
Cerutti G, Mazzolini C, Ziliotto R (1975) Sul contenuto di prolina di vini nazionali. Riv Sci Technol Alimenti Nutr Um 5:309–311
Cerutti G, Gelati R, Zappavignia R (1978) Fermentazione del mosto d'uva in presenza di aminoacidi. Riv Vitic Enol 31:249–257
Chauvet S, Sudraud P (1977) Evaluation rapide de l'acide tartrique dans les mouts. Rev Fr Oenol 15:58–59
Chauvet S, Sudraud P (1983) Application de l'isotachophorese au dosage des acides des vins. Analusis 11:243–250
Chinard FP (1952) Photometric estimation of proline and ornithine. J Biol Chem 199:91–95
Colagrande O (1957) Ricera chromotrografica degli amminoacidi nei mosti e nei vini. Ist Ind Agrar Univ Cattol Sacro Cuore (Milan) 63:84–94
Colagrande O, Grielli F, Del Re AA (1973) Étude des phénomènes d'échange lors de l'emploi des bentonites oenologiques I. Les échanges de sodium et de proline. Connaiss Vigne Vin 7:93–106
Colagrande O, Mazzoleni V, Del Re A (1975) La prolina nei vini. Ann Fac Agrar Univ Cattol Sacro Cuore (Milan) 15(1–3):11–21
Colagrande O, Mazzoleni V, Del Re A (1976) La proline dans les vins. Connaiss Vigne Vin 10:23–32
Colagrande O, Silva A, Casoli A (1984) Acides amines dans les vins mousseux. Connaiss Vigne Vin 18:27–48
Coombe BG, Monk PR (1979) Proline and abscisic acid content of the juice of ripe Riesling grape berries: effect of irrigation during harvest. Am J Enol Vitric 30:64–67
Corradini F, Pellegrini R (1978) Determinazione dell'acido L(−) malico per via enzimatica nei mosti e nei vini confronto con il metado di Rebelein. Vini Ital 20:225–230
Corranza MCP, Iglesias JIM, Marino JIM, Camacho JR, Cortes IM (1981) Aportacion al estudio de los acidos L(−) malico, D(−) y L(+) lactico en mostos y vinos de „Tierra de Barros". Sem Vitivinic 36(1.821–1.822): 2515, 2517, 2519, 2521–2522
Crowell EA, Guymon JF (1969) Studies of caprylic, capric, lauric, and other free fatty acids in brandies by gas chromatography. Am J Enol Vitic 20:155–163
Crowell EA, Ough CS, Bakalinsky A (1985) Determination of alpha amino nitrogen in musts and wines by TNBS method. Am J Enol Vitic 36:175–177
Deibner L, Cabibel-Hughes M (1966) Dosage éléctrophotométrique des acides pyruvique et alpha-cetoglutarique dans les jus de raisin et les vins après séparation de leur dinitrophenylhydrazones au moyen de la chromagraphique sur couche mince de cellulose. Ann Technol Agric 15:127–134
Delfini C (1983) Sull'asserita efficienza dei metodi enzimatici applicati sui vini tal quali per la delerminazione dell-acido piruvico e 2-cheto-glutarico. Riv Vitic Enol 36:307–315
DeSmedt P, Liddle PAP, Cresto B, Bossard A (1981) Analysis of non-volatile constituents of wine by glass capillary gas chromatography. J Inst Brew 87:349–351 [see also Ann Falsif Expert Chim Toxicol 72:633–642 (1979)]

Dimotaki-Kourakou V (1965) Contribution de la autoradiographie a l'étude des acides organiques produits au cours de la fermentation alcoolique. Proceedings Journées Helbènes de séparation immédiate et de chromatographie, Athens 19–24 Sept 1965, Union of Greek Chemists, pp 231–235
Di Stefano R (1977) Fattori che influenzano il contenuto in prolina dei vini. Vini Ital 19:89–93
Di Stefano R (1979) Variazione del contenuto in prolina dei mosti durante la fermentazione. Riv Vitic Enol 32:112–119
Di Stefano R, Bruno C (1983) Applicazione della gascromatografia-spettrometria di massa all'analisi degli esteri metilici degli acidi fissi del vino. Vignevini 10(9):43–45
Dorer M, Malnersic R (1976) Proste amino kisline v rdecih vinih primorskokraskega vinorodnega rajona. Farm Vestn 29:165–175
Downton WJS, Loveys BR (1981) Abscisic acid content and osmotic relations of salt-stressed grapevine leaves. Aust J Plant Physiol 8:443–452
Drawert F, Lessing V, Leupold G (1976) Die Gruppentrennung von organischen Säuren, Kohlenhydraten und Aminosäuren mit Ionenaustauschern und quantitative gas-chromatographische Bestimmung der Einzelsubstanzen. Chromatographia 9:373–379
Drawert F, Leupold G, Lessing V (1978) Gaschromatographische Bestimmung der Inhaltsstoffe von Gärungsgetränken. XII. Mitt Gaschromatographische Bestimmung von organischen Säuren, Neutralstoffen (Kohlenhydraten), Aminosäuren von phenolischen Verbindungen in Weinen. Wein-Wiss 33:54–70
Drawert von F, Leupold G, Lessing V (1977) Gaschromatographische Bestimmung der Inhaltsstoffe von Gärungsgetränken. XI. Mitt Quantitative gaschromatographische Bestimmmung von organischen Säuren, Neutralstoffen (Kohlenhydraten) und Aminosäuren in Traubenpressäften. Wein-Wiss 32:122–133
Droz C, Tanner H (1982) Über die Trennung und quantitative Bestimmung der organischen Säuren in Fruchtsäften und Weinen mittels HPLC. Schweiz Z Obst-Weinbau 118:434–438
Drumond IW, Shama G (1982) A rapid gas chromatographic method for the analysis of acidic fermentation products. Chromatographia 15:180–182
Dubernet M (1976) Dosage automatique de l'acidité volatile dans les vins. Connaiss Vigne Vin 10:297–309
Engel MH, Hare PE (1985) Gas liquid chromatographic separation of amino acids and their derivatives. In: Barret GC (ed) Chemistry and biochemistry of amino acids. 1st edn Chapman and Hall, New York, pp 462–479
Erdelyi E, Dworschak E, Vas K, Linder K, Telegdy-Kovats M, Szoke-Szotyori K (1967) Die Wirkung der Hitzebehandlung auf die organoleptische Qualität und einige Inhaltsstoffe der Fruchtsäfte. Fruchtsaft-Ind 12(2):54–73
Esaka M, Suzuki K, Kubota K (1985) Determination method for L-ascorbic acid in foods with immobilized ascorbate oxidase. Agric Biol Chem 49:2955–2960
Ethiraj S, Suresh ER (1982) The proline of some experimental wines made in India. Am J Enol Vitic 33:231–232
European Brewing Convention (1975) Free alpha amino nitrogen in worts and beers. Anal EBC (3rd edn) pp 61–62
Evans ME (1983) High performance liquid chromatography in oenology. J Liq Chromatogr 6:153–178
Fantozzi P, Montedoro G (1973) Determinazione degli amminoacidi liberi presenti nei mosti e nei vini per cromatografia in fase gassosa. Sci Technol Alimenti 3:53–54 [see also Am J Enol Vitic 25:151–156 (1974)]
Farkas J, Koval M (1982) Vyuzitie izotachoforezy na identifikaciu a stanovenie kyselin vo vine. Kvasny Prum 28:256–260
Farris GA, Fatichenti F, Deiana P, Madau G (1978) L'effetto dei lieviti flor sulla prolina. Riv Vitic Enol 31:431–439
Felix AM, Terkelsen G (1973) Total fluorometric amino acid analysis using fluorescamine. Arch Biochem Biophys 157:177–182
Ferenczi S, Uray G (1973) Comparative study of methods of tartrate determination. Borgazdasag 21(2):77–79

Fernandez-Flores E, Kline DA, Johnson AR (1970) GLC determination of organic acids in fruits as their trimethylsilyl derivatives. J Assoc Off Anal Chem 53:17–20
Feuillat M (1974a) Les constituants azotes ru raisin et du vin. Ann Sci Univ Reims (Assoc Reg Etude Rech Sci) 11(3–4):37–46
Feuillat M (1974b) Les constituants azotes du raisin et du vin. Vigneron Champenois 95:228–236
Feuillat M, Bergeret J (1967a) Influence des traitements thermiques de la vendange sur l'extraction des constituants azotes. C R Acad Sci (Paris) Ser D 264:2520–2523
Feuillat M, Bergeret J (1967b) Identification et dosage des aminoacides dans les mouts et les vins de Bourgogne. C R Acad Sci (Paris) Ser D 264:1757–1759
Feuillat M, Demaimay M (1972) Separation des acides amines du mout de raisin par filtration sur gel de dextrane reticule (Sephadex) et analyse qualitative par chromatographie gazeuse. Ann Technol Agric 21:131–143
Feuillat M, Gouilloux-Benatier M, Gerbaux V (1985) Essais d'activation de la fermentation malolactique dans les vins. Sci Aliment 5:103–122
Flak W, Pluhar G (1983) Ergebnisse von Untersuchungen über die quantitative Bestimmung der Säurehauptkomponenten von Traubenweinen unterschiedlicher Herkunft, Sorte und Reife mit einer modifizierten hochdruckflüssigkeitschromatographischen Methode. Mitt Rebe Wein Klosterneuburg 33:60–68
Flanzy C, Andre P, Buret M, Chambroy Y, Flanzy M (1977) Evolution des substances azotées au cours du metabolisme anaerobie de la baie de raisin. Ann Technol Agric 25:175–190
Fleury MO, Ashley DV (1983) High performance liquid chromatographic analysis of amino acids in physiological fluids: on-line precolumn derivatization with ophthaldialdehyde. Anal Biochem 133:330–335
Florica D, Ghimicescu G (1977) La determination colorimétrique des acides volatils dans le vin. Ann Falsif Expert Chim 70:35–38
Formisano M, Minicone B, Coppola S (1967) Identificazione per via cromatografica delgi aminoacidi e dei composti volatili elaborati dai signoli lieviti nei vini prodotti con uve dell'Isola d'Ischia. Ann Fac Sci Agrar Uni Naples (Portici)2:119–140
Fujita K, Takeuchi S, Ganno S (1979) Fast separation of amino acids using ion exchange chromatography. In: Charalambous G (ed) Proc Symp Anal Foods Beverages 1:81–97
Gallander JF, Cahoon GA, Beelman RB (1969) Free amino acids in musts of eight eastern grape varietes. Am J Enol Vitic 20:140–145
Gehrke CW, Leimer K (1971) Trimethylsilyation of amino acids. Derivatization and chromatography. J Chromatogr 57:219–238
Gehrke CW, Shahrokhi F (1966) Chromatographic separation of n-butyl N-trifluoroacetyl esters of amino acids. Anal Biochem 15:97–108
Gehrke CW, Stalling DL (1967) Quantitative analysis of the twenty natural protein amino acids by gas liquid chromatography. Separation Sci 2:101–138
Gehrke W, Roach D, Zumwalt RW, Stalling DL, Wall LL (1968) Quantitative gas liquid chromatography of amino acids in proteins and biological substances. Macro, semimicro and micro methods. Analytical Biochemistry Laboratories, Columbia, Missouri, p 108
Geigert J, Hirano DS, Neidleman SL (1981) High performance liquid chromatographic method for the determination of L-ascorbic acid and D-isoascorbic acid. J Chromatogr 206:396–399
Giaccio M (1979) Sulla presenza di prolina nei vini. Riv Merceol 18:33–47
Giaccio M (1981) Sulla non validita' del dosaggio della prolina per il giudizio di genuinita' del vino. Riv Vitic Enol 34:34–36
Gilboe DD, Williams JN (1956) Evaluation of the Sakaguchi reaction for quantitative determination of arginine. Proc Soc Exp Biol Med 91:535–536
Godinho OES, Coelho JH, Chagas AP, Oleixo LM (1984) Investigation of the use of thermometric titrimetry for the determination of acidic substances in wine. Talanta 31:218–220

Goren'kova AN, Nechaev LN (1972) Loss of amino acids of grape juice. Vinodel Vinograd SSSR(5):216-222

Graham RA (1979) Influence of yeast strain and pH on pyruvic acid production during alcoholic fermentation. Am J Enol Vitic 30:318–320

Grechko NYa, Emel'yanova NA (1984) Change in the amino acid composition during thermal treatment of kvass must concentrate. Izv Vyssh Ucheb Zaved Pishch Tekhnol (3):52–53

Guimberteau G, Peynaud E (1966) Comparison de quelques méthodes de dosage de l'acide lactique dans les vins. Ann Technol Agric 15:303–309

Gump BH, Saguandeekul S, Murray G, Villar JT (1985) Determination of malic acid in wines by gas chromatography. Am J Enol Vitic 36:248–251

Guymon JF, Ough CS (1962) A uniform method for total acid determination in wines. Am J Enol Vitic 13:40–45

Hare PE, St John PA, Engel MH (1985) Ion separation of amino acids. In: Barret GC (ed) Chemistry and biochemistry of amino acids. 1st edn Chapman and Hall, New York, pp 415–425

Henniger G, Mascaro L (1985) Enzymatic-ultraviolet determination of L-citric acid in wine: collaborative study. J Ass Off Anal Chem 68:1024–1027

Henshall A, Pickering MJ, Soto D (1983) Amino acid analysis with a ternary gradient HPLC system. Chromatogr Rev 9:8–10

Hess D, Schindler I, Mager L (1977) Zur Bestimmung höherer aliphatischer Säuren im Weinbrand. Branntweinwirtschaft 117:217–219

Hill G, Caputi A (1970) Colorimetri determination of tartaric acid in wine. Am J Enol Vitic 21:153–161

Hodgin JC (1979) The separation of pre-column o-phthalaldehyde derivatized amino acids by high performance liquid chromatography. J Liq Chromatogr 2:1047–1059

Holbach B, Woller R (1978) Der Gluconsäuregehalt von Wein und seine Beziehung zum Glyceringehalt. Wein-Wiss 33:114–126

Horak J, Kotyk A, Rihova L (1977) Specificity of trans-inhibition of amino acid transport in baker's yeast. Folia Microbiol 22:360–362

Horowitz W (1980) Official methods of analysis of the Association of Official Analytical Chemists. 13th edn. Assoc Offic Anal Chem Washington, DC, p 186

Huang Z, Dai R (1985) Quantitative determination of malic acid in wine by fluorescence spectrophotometry. Shipen Yu Fajiao Gongye (6):26–29

Ingeldew WM, Kunkee RE (1985) Factors influencing sluggish fermentations of grape juice. Am J Enol Vitic 36:65–76

Ishida Y, Fujita T, Asai K (1981) New detection and separation method for amino acids by high-performance liquid chromatography. J Chromatogr 204:143–148

Jeszenszky Z, Szalka P (1976) Borols cetromsavtartalmanak meghatarozasa. Borgazdasag 24:136–139

Johnson DE, Goodson T (1965) Research on analysis of amino acids by gas chromatography. Southwest Res Inst, Aerospace Med Res Lab Wright-Patterson Air Force Base, Ohio

Johnson DE, Scott SJ, Meister A (1965) Gas-liquid chromatography of amino acid derivatives. Anal Chem 33:669–673

Joyeux A, Lafon-Lafourcade S (1979) Dosage de l'acide succinique dans les vins par méthode enzymatique. Ann Falsif Expert Chim 72:317–320

Juhasz O, Polyak D (1976) Study of free amino acids of grape berries on fixion 50 X 8 layer containing cation exchanging resin. Acta Agron Acad Sci Hung 25:299–308

Juhasz O, Torley D (1979) A szolobogyo szabad aminosavainak es oldhato feherjeinek vizsgalata. Borgazdasag 27(2):64–68

Juhasz O, Torley D (1985) Effect of free amino acids of the grape on the development of organoleptic properties of wine. Acta Aliment 14:101–112

Juhasz O, Kozma P, Polyak D (1984) Nitrogen status of grapevines as reflected by arginine content of the fruit. Acta Agron Acad Sci Hung 33:3–17

Junge C, Spadinger C (1979) Die flüchtigen Säuren des Weines. I. Mitteilung: Bildung und Abbau der Essigsäure im Verlauf der Gärung. Dtsch Lebensm Rundsch 75:12–15

Kacem B, Marshall MR, Matthews RF, Gregory JF (1986) Simultaneous analysis of ascorbic and dehydroascorbic acid by high-performance liquid chromatography with post column derivatization and UV absorbance. J Agric Food Chem 34:271–274
Kaiser KP, Hupf H (1979) Isotachophorese in der Lebenmittelanalytik. Dtsch Lebensm Rundsch 75:346–349
Kakalikova L (1985) Stanovenie kyseliny vinnej, joblcnej a mliecnej v hroznovych vinach (1). Vinohrad 23:16–17, 38–41
Kan TA, Shipe WF (1981) Modification and evaluation of a reverse phase high performance liquid chromatography method for amino acid analysis. J Food Sci 47:338–341
Kanbe C, Ozawa Y, Sohasai T (1977) Automated measurement of D(−), L(+) lactate in soy sauce and wine. Agric Biol Chem 41:863–867
Karlsson R (1975) Iodometric determination of ascorbic acid by controlled potential coulometry. Talanta 22:989–993
Khachidze OT, Matikashvili IA (1973) Change in grape juice amino acids during ripening. Sadovod Vinograd Vinodel Mold 28(1):26–28
Klein H, Stettler HB (1984) Zur Säuredifferenzierung Rheingauer Moste und Weine des Jahres 1982 mittels Isotachophorese. Wein-Wiss 39:51–66
Kliewer WM (1965) Changes in the concentration of malates, tartrates and total free acids in flowers and berries of berries of *Vitis vinifera.* Am J Enol Vitic 16:91–100
Kliewer WM (1967) Annual cyclic changes in the concentration of free amino acids in grapevines. Am J Enol Vitic 18:126–137
Kliewer WM (1968 a) Changes in the concentration of free amino acids in grape berries during maturation. Am J Enol Vitic 19:166–174
Kliewer WM (1968 b) Effect of temperature on the composition of grapes grown under field and controlled conditions. Am Soc Hort Sci 93:797–805
Kliewer WM (1969) Free amino acids and other nitrogeneous substances of table grape varieties. J Food Sci 34:274–278
Kliewer WM (1970) Free amino acids and other nitrogeneous fractions in wine grapes. J Food Sci 35:17–21
Kliewer WM (1973) Berry composition of *Vitis vinifera* cultivars as influenced by photo- and nycto-temperatures during maturation. Am Soc Hortic Sci 98:153–159
Kliewer WM (1977) Arginine – a new indicator for estimating nitrogen needs of vineyards. South Afr Soc Enol Vitic Proc (Capetown) pp 119–143
Kliewer WM, Cook JA (1971) Arginine and total free amino acid as indicators of the nitrogen status of grapevines. J Am Soc Hortic Sci 96:581–587
Kliewer WM, Cook JA (1974) Arginine levels in grape canes and fruits as indicators of nitrogen status of vineyards. Am J Enol Vitic 25:111–118
Kliewer WM, Lider LA (1970) Effects of day temperature and light intensity on growth and composition of Vitis vinefera fruits. J Am Soc Hortic Sci 95:766–769
Kliewer WM, Lider LA (1976) influence of leafroll virus on composition of Burger fruits. Am J Enol Vitic 27:118–124
Kliewer WM, Ough CS (1970) The effect of leaf area and cropt level on the concentration of amino acids and total nitrogen in „Thompson Seedless“ grapes. Vitis 9:196–206
Kliewer WM, Nassar AR, Olmo HP (1966) A general survey of the free amino acids in the genus *Vitis.* Am J Enol Vitic 17:112–117
Kluba RM, Mattick LR (1978) Changes in the nonvolatile acids and other chemical constituents of New York grapes and wines during maturation and fermentation. J Food Sci 43:717–720
Kluba RM, Mattick LR, Hackler LR (1978 a) Changes in the free and total amino acid composition of several *Vitis labruscana* grape varieties during maturation. Am J Enol Vitic 29:102–111
Kluba RM, Mattick LR, Hackler LR (1978 b) Changes in concentration of free and total amino acids of several native American grape cultivars during fermentation. Am J Enol Vitic 29:181–186
Kolb B (1974) Gas chromatography of amino acids and peptides. In: Korte F (ed) Analytical methods. Academic Press, New York, pp 1020–1034 (methodicum chimicum vol 1 part B)

Kozub GI, Balanutza AP, Furtune LA (1977) Use of lithium citrate buffer solutions for determining amino acids in grape must and wine. Sadovod Vinograd Vinodel Mold 32(8):26–28
Kozub GI, Koreisha MA, Furtune LA, Belousova VN, Tkachuk LA (1981) Changes in peptide amino acid composition at different stages of sherry production. Sadovod Vinograd Vinodel Mold 36(10):33–37
Kozub GI, Pershina LZ, Koreisha MA (1983) Colorimetric determination of tartaric acid in wines and related products. Sadovod Vinograd Vinodel Mold 38(7):43–45
Krasnova NS, Gershkovich IA, Kozub GI, Babich VV (1986) Polarographic determination of tartaric and malic acids in must and wine. Sadovod Vinograd Vinodel Mold (2):31–32
Kuensch U, Temperli A, Mayer K (1974) Conversion of arginine to ornithine during malolactic fermentation of red Swiss wines. Am J Enol Vitic 25:191–193
Kunkee RE (1968) Simplified chromatographic procedure for detection of malolactic fermentation. Wines & Vines 49(3):23–24
Kupina SA (1984) Simultaneous quantitation of glycerol, acetic acid and ethanol in grape juice by high performance liquid chromatography. Am J Enol Vitic 35:59–62
Kupina SA, Kutschinski JL, Williams RD, DeSoto RT (1982) A refined gas chromatographic procedure for the measurement of acetic acid in wines and its comparison with the distillation method. Am J Enol Vitic 32:67–74
Kurganova GV, Saenko NF, Ivanova NN (1974) Change in nitrogenous substances during heat treatment of sherry blends. Vinodel Vinograd SSSR 2:15–17
Lafon-Lafourcade S, Guimberteau G (1962) Évolution des amino-acides au cours de la maturation des raisins. Vitis 3:130–135
Lafon-Lafourcade S, Peynaud E (1959) Dosage microbiologique des acides amines des mouts de raisin et de vins. Vitis 2:45–46
Lafon-Lafourcade S, Peynaud E (1965) Sur l'évolution des acides pyruvique et α-cétoglutarique au cours de la fermentation alcoolique. CR Acad Sci Paris Ser D 261:1778–1780
Lafon-Lafourcade S, Peynaud E (1966) Sur les taux des acides citoniques formés au cours de la fermentation alcoolique. Ann Inst Pasteur 110:766–778
Lafon-Lafourcade S, Ribereau-Gayon P, Joyeux A (1977) Origines de l'acidite-volatile des grands vins liquoreux. CR Seances Acad Agric Fr 63:551–558
Lafon-Lafourcade S, Geniex C, Ribereau-Gayon P (1984) Inhibition of alcoholic fermentation of grape musts by fatty acids produced by yeasts and their elimination by yeast ghosts. Appl Environ Microbiol 47:1246–1249
Lampkin WM, Gehrke CW (1965a) Quantitative gas chromatography of amino acids. Preparation of n-butyl N-trifluoroacetyl esters. Anal Chem 3:383–389
Lampkin WM, Gehrke CW (1965b) Quantitative gas chromatography of amino acids. Anal Chem 37:383–389
Larsen BR, West FG (1981) A method for quantitative amino acid analysis using precolumn o-phthalaldehyde derivatization and high performance liquid chromatography. J Chromatogr Sci 19:259–265
Lau OW, Shiu KK, Chang St (1985) Determination of ascorbic acid in vegetables and fruits by differential pulse polarography. J Sci Food Agric 36:733–739
Lea PJ, Wallsgrove RM (1985) The biosynthesis of amino acids in plants. In: Barrett GC (ed) Chemistry and biochemistry of amino acids. Chapman and Hall, New York, pp 197–226
Lee KS, Dreschert DG (1979) Derivatization of cysteine and cystine for fluorescence amino acid analysis with the o-phthaldialdehyde-2-mercaptoethanol reagent. J Biol Chem 254:6248–6251
Leone AM, La Notte F, Lamparelli F (1976) Gli amminoacidi liberi e gli alcoli superiori nella vinificazione per macerazione. Note I. Gli amminoacidi liberi. Vini Ital 18:391–401
Leone AM, La Notte E, Santoro M (1980) Gli amminiacidi liberi a gli alcoli superiori nella vinificazione per macerazione Nota II. Gli alcoli superiori e la loro formazione. Vini Ital 22:231–239

Lepadatu V, Tanase I (1967) Recherches sur l'évolution de la teneur en glucides et en acides amines dans la première phase de la vinification en rouge. Ann Technol Agric 16:321–331

Lhuguenot JC, Ude L, Dymarsky E, Baron C (1979) Analyse des acides amines libres dans les vins et dans les mouts de raisin en cours de fermentation alcoolique par chromatographie gaz-liquide. Ann Falsif Expert Chim 72:275–286

Lie S (1972a) Die EBC-Ninhydrin-Methode zur Bestimmung des freien α-Amino-Stickstoffs. Brauwissenschaft 25:250–253

Lie S (1972b) Die EBC-Ninhydrin-Methode zur Bestimmung des freien Alpha-Amino-Stickstoffes. Mitt Versuchssta Gärungsgewerbe Wein 26:141–145

Lin L, Tanner H (1985) Quantitative HPTLC analysis of carboxylic acids in wine and juice. J High Resolut Chromatogr Commun 8:125–131

Lindroth P, Mopper K (1979) High performance liquid chromatographic determination of subpicomole amounts of amino acids by precolumn fluorescence derivatization with o-phthaldialdehyde. Anal Chem 51:1667–1674

Lipka Z, Tanner H (1974) Une nouvelle methode de dosage rapide de l'acide tartrique dans les mouts, les vins et autres boissons (selon Rebelein). Rev Suisse Viticult Arboricult Hortic 6:5–10

Lonvaud-Funel A, Doneche B, Bleuze D (1980) Automatisation en flux continu de dosage enzymatique de l'acide malique des vins. Connaiss Vigne Vin 14:207–217

Lotti G, Anelli G (1971) Variazione degli amminoacidi liberi durante la maturatzione delle uve. Riv Sci Technol Aliment 1:25–32

Lotti G, Anelli G, Pellegrini A (1972) L'evoluzione degli amminoacidi liberi nel corso della vinificazione. Riv Vitic Enol 25:459–474

Luthi H, Vetsch U (1952) Papierchromatische Bestimmung von Aminosäuren in Weinen. Schweiz Z Obst- Weinbau 61:390–394, 405–408

Maccarrone A, Asmundo CN, Lupo MCC (1977) Il contenuto in acidi malico e lattico del vini dell'Etna. Vignevini 4(2):13–15

Marcy JE, Carroll DE (1982) A rapid method for the simultaneous determination of major organic acids and sugars in grape musts. Am J Enol Vitic 33:176–177

Marcy JE, Carroll DE, Young CT (1981) Changes in free amino acids and total nitrogen concentrations during maturation of muscadine grapes (*V. rotundifolia*). J Food Sci 46:543–547

Margheri G, Gianotti L, Mattarei C, Pellegrini R (1982) Evoluzione degli aminoacidi liberi nel corso della elaborazione dei vini spumanti. Vignevini 9(11):19–30

Margheri G, Gianotti L, Mattarei C, Tonon D, Pellegrini R (1983) Evoluzione degli aminoacidi liberi nel corso della elaborazione dei vini spumanti di qualita del trentino. Vignevini 10(7–8):27–32

Margheri G, Versini G, Gianotti L (1984a) Vini spumanti di qualita'metodo champenois. Vini Ital 26:51–59

Margheri G, Gianotti L, Pellegrin R, Mattarei C (1984b) Vini spumanti di qualita' metodo champenois. Nota II. Assimilazione di aminoacidi liberi e di ammoniaca nel corso della presa di spuma. Vini Ital 26:21–26

Margheri G, Versini G, Daal Serra A, Gianotti L, Pellegrini R, Mattarei C (1984c) L'ausolisi del lievito in enologia. Vignevini 11(5):25–28

Martakov AA, Ermachkova LT, Potoroko GV (1972) Amino acid assimilation by wine yeast during aerated fermentation of grape must. Prikl Biokhim Mikrobiol 8:932–938 [see App Biochem Microbiol 8:623–628 (1972)]

Martin GE, Sullo JG, Schoeneman RL (1971) Determination of fixed acids in commercial wines by gas-liquid chromatography. J Agric Food Chem 19:995–998

Martin P, Suarez A, Polo C, Cabezudo D, Dabrio MV (1980) Analisis de 19 aminoacidos por H.P.L.C. Anal Bromatol 32:289–293

Martin P, Polo C, Cabezudo MD, Dabrio MV (1984) Dansyl amino acids behaviour on a Radial Pak C18 column. Derivatization of grape wine musts, wine and wine vinegars. J Liq Chromatogr 7:539–558

Martinez Burges R (1960) Cromatografia de algunos vinos navarros (aminoacidos libres) Anal Bromatol 12:19–26

Masuda M, Yamamoto M, Asakura Y (1985) Direct gas chromatographic analysis of fusel oils, fatty acids and esters of distilled alcoholic beverages. J Food Sci 50:264–265
Matchett JR, Legault RR, Nimmo CC, Notter GK (1944) Tartrates from grape wastes. Ind Eng Chem 36:851–857
Mattick LR, Rice AC (1970) Quantitative determination of lactic acid and glycerol in wines by gas chromatography. Am J Enol Vittic 21:205–212
Mattick LR, Rice AC (1981) The use of PVPP for decolorizing wine in the determination of tartrate by the metavanadate method. Am J Enol Vitic 32:297–298
Mattick LR, Rice AC, Moyer JC (1971) Determination of fixed acids in musts and wines by gas chromatography. Am J Enol Vitic 21:179–183
Mayer K, Pause G (1969) Enzymatische Milchsäurebestimmung in Weinen. Mitt Geb Lebensmittelunters Hyg 60–230–233
Mayer K, Pause G, Vetsch U, Kuensch U, Temperli A (1973) Aminosäurengehalte im Verlauf der Vinifikation einiger Rotweine. Mitt Rebe Wein Ser A Klosterneuburg 23:331–340
McCloskey LP (1974) Gluconic acid in California wines. Am J Enol Vitic 25:198–201
McCloskey LP (1976a) An acetic acid assay for wines using enzymes. Am J Enol Vitic 27:176–180
McCloskey LP (1976b) An enzymic assay for acetate fruit juices and wines. J Agric Food Chem 24:523–526
McCloskey LP (1980a) An improved enzymic assay for acetate in juice and wine. Am J Enol Vitic 31:170–173
McCloskey LP (1980b) Enzymatic assay for malic acid and malo-lactic acid fermentations. Am J Enol Vitic 31:212–215
McCord JD, Trousdale E, Ryu DDY (1984) An improved sample preparation procedure for the analysis of major organic components in grape must and wine by high performance liquid chromatography. Am J Enol Vitic 35:28–29
Mentasti E, Gennaro MC, Sarzanini C, Baiocchi C, Savigliano M (1985) Derivatization identification and separation of carboxylic acids in wines and beverages by high-performance liquid chromatography. J Chromatogr 322:177–189
Michal G, Beutler HO, Lang G, Guentner U (1976) Enzymatic determination of succinic acid in foodstuff. Fresenius Z Anal Chem 279:137–138
Miflin BJ, Bright WJ, Davies HM, Shewry PR, Lea PJ (1977) Amino acids derived from asparate; their biosynthesis and its regulation in plants. In: Hewett EJ, Cutting CV (eds) Nitrogen assimilation of plants. Academic Press, London, pp 335:358
Miguel C, Mesias JL, Maynar JI (1985) Évolution des acides amines pendant la maturation des raisins des variétés Cayetana et Macabeo (*Vitis vinifera*). Sci Aliment 5:599–605
Mincione B, Spagna Musso S, Coppola V (1976) Il contenuto in prolina dei vini di produzione Campana. Riv Vitic Enol 29:492–515
Minguzzi A, Amati A (1973) Il contenuto in prolina dei vini in relazione alla loro genuinita. Sci Technol Aliment 3:371–372
Mndzhoian EL, Saakian RG, Akhnazarian FA (1971) The conversion of nitrogeneous substances from the wood of *Quercus* in an alcoholic medium [wine manufacture]. Vinodel Vinograd SSSR (5):19–20
Modi G, Guerrini M (1975) Determinazioni di prolina effettuate su vini di varia origine. Boll Chim Unione Ital Lab Prov 26:269–275
Modi G, Guernini M, Simiani G (1976) Sul contenuto di acidi fissi liberi in alcuni vini nazionali. Boll Chim Unione Ital Lab Prov 27:183–189
Möhler K, Looser S (1969) Enzymatische Bestimmung von Säuren in Wein. I. Mitteilung Apfelsäure, Milchsäure, Citronensäure. Z Lebensm Unters Forsch 140:94–100
Möhler K, Looser S (1969) Enzymatische Bestimmung von Säuren, Kohlenhydraten und Glycerin in Wein. Mitt. II. Z Lebensm Unters Forsch 140:149–154
Moellering H, Gruber W (1966) Determination of citrate with citrate lyase. Anal Biochem 17:369–376
Monk PR, Iland PG (1984a) Ion-exclusion chromatography of carboxylic acids with conductimetric estimation. 1. Methodology. Food Technol Aust 36:16–17

Monk PR, Iland PG (1984 b) Ion-exclusion chromatography of carboxylic acids with conductimetric estimation. 2. Application to fruit juice and wine. Food Technol Aust 36(1):18–20

Moore S, Stein WH (1951) Chromatography of amino acids on sulfonated polystyrene resins. J Biol Chem 192:663–681

Nanitashvili TS (1973) Amino acid composition of wines prepared using the proteolytic enzyme preparation protawamorin G10 x, 78-2. Prikl Biokhim Mikrobiol 9:102–105 [for transl see Appl Biochem Microbiol 9:85–87 (1974)]

Nassar AR, Kliewer WM (1966) Free amino acids in various parts of *Vitis vinifera* at different stages of developement. Proc Am Soc Hortic Sci 89:281–294

Nastic D, Stanimirovic D, Stanimirovic SG, Vuckovic VM (1968) Saccharides organic, and amino acids in wines from Vrsac vineyards. Glas Khem Drush 33:495–510

Nastic DR, Stanimirovic DL, Vuckovic VM, Stanimirovic SG (1967) Saccaharides, organic acids and amino acids in grapes from the Vrsak vineyards. Glas Khem Drush 32:491–504

Neubecker TA, Rechnitz GA (1972) Arginine determination with ion selective membrane electrodes. Anal Lett 5:653–659

Neves HJC, Vascolcelos AMP (1983) Separação e analise quantitiva de aminoacidos por cromatografia gas-liquido, esteres n-etoxicarbonil isopropilicos, e esteres n-etoxicarbonil-O(S)-trimetilsilil isopropilicos. Rev Port Quim 25:184–192

Nicolosi AC, Lanza CM (1976) Il contenuto in prolina dei vini dell'Etna. Riv Sci Technol Aliment Nutr Um 6:185–186

Nikova Z, Simov N, Moldovanska I (1980) Methods for determination of tartaric acid levels in secondary raw materials. Lozar Vinar 29(8):24–27

Novikova VN, Kudryashova NA (1981) Determination of keto acids in wine. Vinodel Vinograd SSSR (3):22–24

Noyes CM (1983) Optimization of complex separations in high performance liquid chromatography. J Chromatogr 266:451–460

Nykanen L, Paputti E, Soumalainen H (1968) Volatile fatty acids in some brands of whisky, cognac and rum. J Food Sci 33:88–92

Olschimke D, Niesner W, Junge CH (1969) Bestimmung der Apfelsäure in Weinen und Traubensäften. Dtsch Lebensm Rundsch 65:383–384

Ooghe W, Kastelijn H (1984) Détermination de l'authenticité d'un beaujolais primeur a l'aide du spectre des acides amines. Ann Falsif Expert Chim Toxicol 77:467–477

Ooghe W, Kastelijin H, De Waele A (1981) Détermination de l'origine d'un vin rouge a l'aide du spectre des acides amines. Ann Falsif Expert Chem Toxicol 74:381–408

Ough CS (1968) Proline content of grapes and wines. Vitis 7:321–331

Ough CS (1969 a) Rapid determination of proline in grapes and wines. J Food Sci 34:228–230

Ough CS (1969 b) Substances extracted during skin contact with white musts. General wine composition and quality changes with contact time. Am J Enol Vitic 20:93–100

Ough CS, Amerine MA (1967) Studies with controlled fermentations. X. Effects of fermentation temperature on some volatile compounds. Am J Enol Vitic 18:157–164

Ough CS, Anelli G (1979) Zinfandel grape juice protein fractions and their amino acid makeup as affected by crop level. Am J Enol Vitic 30:8–10

Ough CS, Bell AA (1980) Effects of nitrogen fertilization of grapevines on amino acid metabolism and higher-alcohol formation during grape juice fermentation. Am J Enol Vitic 31:122–123

Ough CS, Bustos O (1969) A review of amino acid analytical methods and their application to grapes and wine. Wines & Vines 50(4):50–58

Ough CS, Nagaoka R (1984) Effect of cluster thinning and vineyard yields on grape and wine composition and wine quality of Cabernet Sauvignon. Am J Enol Vitic 35:30–40

Ough CS, Singleton VL (1968) Wine quality prediction from juice brix/acid ratio and associated compositional changes for "White Riesling" and "Cabernet Sauvignon". Am J Enol Vitic 19:129–138

Ough CS, Stashak RM (1974) Further studies on proline concentrations in grapes and wines. Am J Enol Vitic 25:7–12

Ough CS, Tabacman H (1979) Gas chromatographic determinations of amino acid differences in Cabernet Sauvignon grapes and wines as affected by rootstocks. Am J Enol Vitic 30:306–311
Ough CS, Winger CL (1982) Changes in non-volatile compounds and extracts of wines due to yeast species and fermentation temperature. S Afr J Enol Vitic 3:17–21
Ough CS, Lider LA, Cook JA (1968a) Rootstock scion interactions concerning winemaking. I. Juice composition changes and effects on fermentation rates with St. George and 99-R rootstocks at two nitrogen fertilizer levels. Am J Enol Vitic 19:213–227
Ough CS, Cook JA, Lider LA (1968b) Rootstock scion interactions concerning wine making. II. Wine compositional and sensory changes attributed to rootstock and fertilizer level differences. Am J Enol Vitic 19:254–265
Ough CS, Amerine MA, Sparks TC (1969) Studies with controlled fermentations. XI. Fermentation temperature effects on acidity and pH. Am J Enol Vitic 20:127–139
Pachki A (1974) Citric acid content in wines. Lozar Vinar 23(5):46–48
Pallotti G, Bencivenga B, Brighina G, Palmioli A (1976) I composti azotati quali indici di caratterizzazione del vino Frascati. Riv Soc Ital Sci Aliment 5:303–305
Palmer JK, List DM (1973) Determination of organic acids in foods by liquid chromatography. J Agric Food Chem 21:903–906
Pekur GN (1978) Study of the effect of excessive CO_2 pressure on the enzyme activity in nitrogen metabolism of wine yeast [*Saccharomyces cerevisae*]. Prikl Biokhim Mikrobiol 14:615–620
Pekur GN, Bur'yan NI, Pavlenko NM (1981) Characteristics of nitrogen metabolism of wine yeast in different fermentation conditions. Prikl Biokhim Mikrobiol 17:342–347
Perret D (1985) Liquid chromatography of amino acids and their derivatives. In: Barret GC (ed) Chemistry and biochemistry of amino acids. 1st edn. Chapman and Hall, New York, pp 426–461
Petrosian TL (1971) Quantitative determination of the amino acid composition of brandy. Vinodel Vinograd SSSR (2):30–31
Peynaud E, Lafon-Lafourcade S (1962) Sur la nutrition azotee des leuvres de vins. Rev Ferment Ind Aliment (Brussells) 7(1):11–21
Peynaud E, Lafon-Lafourcade S (1963) Constitution azotée des leuvres en fonction des conditions de nutrition. Qual Plant Mater Veg 9:365–380
Peynaud E, Maurie A (1953) Sur l'évolution de l'azote dans les differences parties du raisin au cours de la maturation. Ann Technol Agric 2:15–25
Peynaud E, Blouin J, Lafon-Lafourcade S (1966) Review of applications of enzymatic methods to the determination of some organic acids in wines. Am J Enol Vitic 17:218–224
Pfeiffer P, Radler F (1985) Hochleistungsflüssigchromatographische Bestimmung von organischen Säuren, Zuckern, Glyerin und Alkohol im Wein an einer Kationenaustauschersäule. Z Lebensm Unters Forsch 181:24–27
Philip T, Nelson FE (1973) Procedure for quantitative estimation of malic and tartaric acids in grape juice. J Food Sci 38:18–20
Pilone GJ (1967) Effect of lactic acid on volatile acid determination of wines. Am J Enol Vitic 18:149–156
Pilone GJ (1977) Determination of tartaric acid in wine. Am J Enol Vitic 28:104–107
Pilone GJ (1978) Total volatile acidity in wines, exclusive of sulfur dioxide: collabroative study. J Assoc Off Anal Chem 61:292–295
Pilone GJ, Kunkee RE (1970) Colorimetric determination of total lactic acid in wine. Am J Enol Vitic 21:12–18
Pilone GJ, Rankine BC, Hatcher CJ (1972) Evaluation of an improved method for measuring volatile acid in wine. Aust Wine Brew Spirits Rev 91:62, 64, 66
Pires R, Mohler K (1970) Enzymatische Bestimmung von Bernsteinsäure in Wein. Z Lebensm Unters Forsch 143:96–99
Polo MC, Llaguno C (1974a) Évolution des acides amines libres au cours de la fermentation alcoolique de jus de raisin par les leuvres de fleur. II. Composition des vins et des cellules de leuvres. Connaiss Vigne Vin 8:321–343

Polo MC, Llaguno C (1974 b) Évolution des acides amines libres dans le mout de raisin sous l'action des leuvres de fleur. I. Étude qualitative et quantitative des acides amines libres des mouts de raisin. Connaiss Vigne Vin 8:81–90

Pool RM, Weaver RJ, Kliewer WM (1972) Effect of growth regulators on changes in fruits of Thompson seedless grapes during cold storage. J Am Soc Hortic Sci 97:67–70

Postel W, Drawert F, Hagen W (1973) Enzymatische Untersuchungen über den Gehalt von L(+) und D(−) Lactat in Weinen. Z Lebensm Unters Forsch 150:267–273

Poux C, Caillet M (1969) Dosage enzymatique de l'acide L(−) malique. Ann Technol Agric 18:359–366

Poux C, Ournac A (1970) Acides amines libres et polypeptidiques du vin. Ann Technol Agric 19:217–237

Poux C, Caillet M, Joubert R (1974) Chauffage de la vendange et composes azotes. Ind Aliment Agric 91(6):695–699

Price SJ, Palmer T, Griffin M (1984) High-speed assay of amino acids using reversed-phase liquid chromatography. Chromatographia 18:62–64

Prusa K, Smejkal O (1983) Pouziti izotachoforezy pro rozliseni kyselin ve vine. Kvasny Prum 29:7–9

Radler F (1975) Die organischen Säuren im Wein und ihr mikrobieller Stoffwechsel. Dtsch Lebensm Rundsch 71:20–26

Rankine BC (1965) Factors influencing the pyruvic acid content of wines. J Sci Food Agric 16:394–398

Rapp A, Reuther KH (1971) Der Gehalt an freien Aminosäuren in Traubenmosten von gesunden und edelfaulen Beeren verschiedener Rebsorten. Vitis 10:51–58

Rapp A, Ziegler R (1976) Trennung von Dicarbonsäuren von Traubenmost und Wein mit Hilfe der Hochdruckflüssigkeits-Chromatographie an einem Kationenaustauscher. Chromatographia 9:148–150

Rapp A, Ziegler R (1979) Bestimmung von Zuckern, Glyerin, Äthanol und Carbonsäuren im Traubenmost und Wein mit Hilfe der Hochdruck-Flüssigkeitschromatographie. Dtsch Lebensm Rundsch 75:396–398

Rebelein H (1961) Kolorimetrische Verfahren zur gleichzeitigen Bestimmung der Weinsäure und Milchsäure in Wein und Most. Dtsch Lebensm Rundsch 57:36–41

Rebelein H (1964) Kolorimetrische Bestimmung der Apfelsäure in Verbindung mit der gleichzeitigen Bestimmung der Wein- und Milchsäure in Most und Wein. Dtsch Lebensm Rundsch 60:140–144

Rebelein H (1967) Vereinfachtes Verfahren zur kolorimetrischen Bestimmung der Citronensäure in Wein und Traubenmost. Dtsch Lebensm Runsch 63:337–340

Rebelein H (1973) Verfahren zur genauen serienmäßigen Bestimmung der Wein- und Milchsäure in Wein und ähnlichen Getränken. Chem Mikrobiol Technol Lebensm 2:33–38

Redifield R (1953) Two-dimensional paper chromatographic system with high resolution power of amino acids. Biochim Biophys Acta 10:344–353

Reijenga JC, Verheggen TPEM, Everaerts FM (1982) Simultaneous determination of organic and inorganic acids and additives by capillary isotachophoresis using UV and a.c. conductivity detection. J Chromatogr 245:120–125

Ribereau-Gayon P, Bertrand A (1971) Dosage simultane dans le vin des acides, organiques, des polyalcools et des sucres. Applications. CR Acad Sci (Paris) Ser D 273:1761–1762

Ribereau-Gayon J, Peynaud E, Ribereau-Gayon P, Sudraud P (1975) Traité d'oenologie sciences et techniques du vin, vol 2. Charactères des vins maturation du raisin, levures et Bacteries. Dunod, Paris, p 556

Ribereau-Gayon J, Peynaud E, Sudraud P, Ribereau-Gayon P (1976 a) Traité d'oenologie scienes et techniques du vin, vol 1. Analyse et contrôle des vins. Dunod, Paris, p 671

Ribereau-Gayon J, Peynaud E, Ribereau-Gayon P, Sudraud P (1976 b) Traité d'oenologie sciences et techniques du vin, vol 3. Vinifications transformations du vin. Dunod, Paris, p 719

Ridomi A, Pezza L (1982) Bilancio acidimetrico del mosto con determinazione indiretta dell'acido malico. Riv Vitic Enol 35:3–12

Rocklin RD, Slingsby RW, Pohl CA (1986) Separation and detection of carboxylic acids by ion chromatography. J Liq Chromatog 9:757–775
Rodopulo AK, Egorov IA, Pisarnitski AF, Martakov AA, Levchenko TN (1969) Amino acid composition of sherries. Prikl Biokhim Mikrobiol 5:186–188
Rosen H (1957) A modified ninhydrin colorimetric analysis for amino acids. Arch Biochem Biophys 67:10–15
Rosenthal GA (1985) Colorimetric and fluorimetric detection of amino acids. In: Barrett BC (ed) Chemistry and biochemistry of amino acids. Chapman and Hall, New York, pp 573–590
Roth M (1971) Fluorescence reaction for amino acids. Anal Chem 43:880–882
Roth M, Hampai A (1973) Column chromatography of amino acids with fluorescence detection. J Chromatogr 83:353–356
Roubelakis KA, Kliewer WM (1978a) Enzymes of the Krebs-Henseleit cycle in *Vitis vinifera* L. I. Orthinine carbamoyltransferase isolation and some properties. Plant Physiol 62:337–339
Roubelakis KA, Kliewer WM (1978b) Enzymes of the Krebs-Henseleit cycle in *Vitis vinifera* L. II. Arginosuccinate synthetase and lyase. Plant Physiol 62:340–343
Roubelakis KA, Kleiwer WM (1978c) Changes in the activities of ornithine transcarbamylase and ariginase, and concentrations of nitrogeneous substances during germination and seeding development of *Vitis vinifera* L. Vitis 17:377–385
Roubelakis KA, Kliewer WM (1978d) Enzymes of Krebs-Henseleit cycle in *Vitis vinifera*. III. In vivo and in vitro studies of arginase. Plant Physiol 62:344–347
Roubelakis-Angelakis KA, Kliewer WM (1981) Influence of nitrogen fertilization on activities of ornithine transcarbamoylase and arginase in Chenin blanc berries at different stages of development. Vitis 20:130–135
Roubelakis-Angelakis KA, Kliewer WM (1983a) Ammonia assimilation in *Vitis vinifera* L. I.: Isolation and properties of leaf and root and root glutamate dehydrogenase. Vitis 22:202–210
Roubelakis-Angelakis KA, Kliewer WM (1983b) Ammonia assimilation in *Vitis vinifera* L.: II. Leaf and root glutamine synthetase. Vitis 22:299–305
Rougereau A (1974) Teneur en hydroxyproline des vins. Rev Fr Oenol 55:27–30
Ruehlmann K, Giesecke W (1961) Gaschromatographie silylierter Aminosäuren. Angew Chem 73:113
Ruffner H, Rast D (1974) Die Biogenese von Tartrat in der Weinrebe. Z Pflanzenphysiol 73:45–55
Ryan JJ, Dupont JA (1973) Identification and analysis of major acids from fruit juices and wines. J Agric Food Chem 21:45–49
Sachde AG, El-Zalaki EM, El-Tabey AM, Abo-Donia SA (1979) Study on Egyptian fresh and aged wines. II. Total nitrogen and amino acids content of wines of three vintages. Am J Enol Vitic 30:272–274
Saito K, Kasai Z (1969) Tartaric acid synthesis from L-ascorbic acid-1-^{14}C in grape berries. Phytochemistry 8:2177–2182
Saito K, Kasai Z (1978) Conversion of labeled substrates to sugars, cell wall polysaccharides, and tartaric acid in grape berries. Plant Physiol 62:215–219
Salagoity MH, Bertrand A, Tricard C (1985) Dosage des acides amines libres du vin par CLHP. Rapport des activités de recherches 1983–1984. Inst d'Oenololgie, Talence France, pp 124–127
Salgues M, Andre J (1977) Chromatographie immediate des acides maliques et lactiques des vins. Vignes Vins (261):36–37
Salzedo A, Vian P, Mattaeri C (1976) Ricerche sul contenuto in prolina nei vini della regione Trentino-Alto Adige. Vini Ital 18:415–421
Sanders EM, Ough CS (1985) Determination of free amino acids in wine by HPLC. Am J Enol Vitic 36:43–45
Satake K, Okuyama T, Ohashi M, Shinoda T (1960) The spectrophotometric determination of amine, amino acid and peptides with 2,4,6-trinitrobenzene-l-sulfonic acid. J Biochem 47:654–660

Sauberlich HE, Green MD, Omaye ST (1982) Determination of ascorbic acid and dehydroascorbic acid. In: Seib PA, Tolbert BM (eds) Ascorbic acid. Adv Chem Ser, Am Chem Soc, Wash DC, pp 199–219

Schneyder J, Flak W (1981) Quantitative Bestimmung der Säure-Hauptkomponenten von Weinen mittels Hochdruckflüssigkeitschromatographie. Mitt Rebe Wein Klosterneuburg 31:57–61

Schneyder J, Pluhar G (1977) Beitrag zur Bestimmung der flüchtigen Acidität von Weinen. Vorschlag eines durch Oxydation der schwefligen Säure vor der Wasserdampfdestillation vereinfachten Verfahrens. Mitt Höheren Bundeslehr-Versuchsanst Wein- Obstbau Klosterneuburg 27:14–17

Scholten G, Woller R, Steinmetz E (1983) Automatisierte Weinanalyse, 2. Mitteilung. Wein-Wiss 38:397–426

Sciancalepora V, De Goglio A, Pallavicini C (1983) Détermination de quelques composants des lies de vin. Ind Aliment Agric 100:365–367

Scopigno PT, Bonafaccia G, Mezzasoma A (1983) Studio sulla variabilita dei contenuti in aminoacidi liberi ed azoto totale, conseguente alla tecnologia di spumantizzazione dei vini. Riv Soc Ital Sci Aliment 12:267–272

Sedova H, Kahler M (1980) Pivovarstvi a sladarsti. Stanoveni aminokyselin plynovou chromatografii. Kvasny Prum 26:193–197

Selmeci G, Hanusz B (1981) Rapid determination of malic and lactic acid in wines. Elelmiszervizsgalati Kozl 27:135–138

Seppi A, Sperandio A (1983) L'acido citrico nei vine. Determinazione con metodo enzimatico e con metodo chimico ufficiale. Riv Soc Ital Sci Aliment 12:479–482

Shen G, Wang R, Ying M (1984) HPLC with direct injection method for the determination of organic acids in wine. Shipen Yu Fajiao Gongye (6):14–20, 40

Shimazu Y, Watanabe M (1976) Qualitative Analyse der organischen Säuren der Weine und Moste mit dem Carboxylsäureanalysator. Wein-Wiss 31:45–53

Shimazu Y, Watanabe M (1981) Determination of organic acids in grape must and wine by high-performance liquid chromatography. Nippon Jozo Kyokai Zasshi 76:418–423

Shinohara T (1985) Gas chromatographic analysis of volatile fatty acids in wines. Agric Biol Chem 49:2211–2212

Souty M, Lapize F, Breuils L (1980) Possibilité de dosage simultane sur autonalayse de l'acide galacturonique et des oses neutres lors de la détermination des substances pectiques. Ann Technol Agric 29:89–98

Souza AHD (1971) Caracterização dos acidos organicos dos vinhos pela cromatografia em camada delgada. Rev Brasil Farm 52:51–60

Spackman DH, Stein WH, Moore S (1958) Automatic recording for use in the chromatography of amino acids. Anal Chem 30:1190–1206

Spettoli P, Zamorani A (1982) Gli alcoli superiori e gli aminoacidi liberi nella spumantizzazione in bottiglia dei vini. Technol Aliment 5(5):21–25

Spettoli P, Nuti MP, Dal A, Peruffo B, Zamorani A (1984) Malolactic fermentation and secondary product formation in wine by *Leuconstoc oenos* cells immobilized in a continuous-flow reactor. Ann NY Acad Sci 434:461–464

Sponholz WR, Dittrich HH (1977) Enzymatische Bestimmung von Bernsteinsäure in Mosten und Weinen. Wein-Wiss 32:38–47

Sponholz WR, Dittrich HH (1984) Über das Vorkommen von Galacturon- und Glucuronsäure sowie von 2- und 5-Oxo-Gluconsäure in Weinen, Sherries, Obst- und Dessertweinen. Vitis 23:214–224

Sponholz WR, Wunsch B, Dittrich HH (1981 a) Enzymatische Bestimmung von (R)-2-Hydroxyglutarsäure in Mosten, Weinen und anderen Gärungsgetränken. Z Lebensm Unters Forsch 172:264–268

Sponholz WR, Dittrich HH, Haas F, Wunsch B (1981 b) Die Bildung von flüchtigen Fettsäuren durch *Saccharomyces*-Hefen während der Vergärung von Traubenmost. Z Lebensm Unters Forsch 173:297–300

Sponholz WR, Dittrich HH, Barth A (1982) Über die Zusammensetzung essigstichiger Weine. Dtsch Lebensm Rundsch 78:423–428

Stalling DL, Gehrke CW (1966) Quantitative analysis of amino acids by gas chromatography; Acylation of arginine. Biochem Biophys Res Commun 22:329–335
Stalling DL, Gille G, Gehrke CW (1967) Quantitative gas chromatography of amino acids. Anal Biochem 18:118–125
Stamer JR, Weirs LD, Mattick LR (1983) Thin layer chromatographic (TLC) analysis of malic and lactic acids. Food Chem 10:235–238
Stein S, Bohlen J, Stone J, Dairman W, Undenfriend S (1973) Amino acid analysis with fluorescamine at the picomole level. Arch Biochem Biophys 155:202–212
Steiner W, Muller E, Frohlich D, Battaglia R (1984) HPLC-Bestimmung von Carbonsäure als p-Nitrobenzylester in Fruchtsäften und Wein. Mitt Geb Lebensmittelunters Hyg 75:37–50
Suarez MA, Polo MC, Llaguno C (1979) Étude de la composition de vins mousseux pendant la prise de mousse au cours du vieillissement en bouteiles. I. Étude des acides amines libres, du glycerol des sucres et des acides organiques. Connaiss Vigne Vin 13:199–217
Svedas VJ, Galaev IJ, Borisov IL, Berezin IV (1980) The interaction of amino acids with o-phthaldialdehyde: a kinetic study and spectrophotometric assay of the reaction product. Anal Biochem 101:188–195
Symonds P (1978) Application de la chromatographie liquide haute performance au dosage de quelque acides organiques du vin. Ann Nutr Aliment 32:957–968
Szokan G (1982) Pre-column derivatization in HPLC of amino acids and peptides. J Liq Chromatogr 5:1493–1498
Tanner H, Lipka Z (1973) Eine neue Schnellbestimmungsmethode für Weinsäure in Mosten, Weinen und anderen Getränken (nach Rebelein). Schweiz Z Obst- Weinbau 109:684–692
Tanner H, Sandoz M (1972) Vergleichende Weinsäurebestimmung mit verschiedenartigen Methoden. Schweiz Z Obst- Weinbau 108:251–254
Tanner H, Zanier C (1976) über die qualitative und quantitative gaschromatographische Bestimmung der organischen Säuren und der Zucker verschiedener Getränke. Schweiz Z Obst- Weinbau 112:453–460
Temperli A, Kuensch U (1976) Die Veränderungen der Gehalte an freien Aminosäuren während der Vinifikation. Qual Plant Plant Foods Human Nutr 26:141–148
Thaler H, Gieger U (1967) Zur Bestimmung der Gesamt-Ascorbinsäure in Wein. Mitt Geb Lebensmittelunters Hyg 58:473–495
Togawa H, Takezawa Y (1978) Material grapes for winemaking. II. Detection of free amino acids in grapes and wine. Nippon Jozo Kyokai Zasshi 73:469–472
Toth G (1982) Hazai mustok, borok nitrogentartlmu anyagainak es feherjeosszetelenek jellemzese. Borgazdasag 30:65–70
Trombella B, Ribeiro A (1980) Improved gas chromatographic determination of acetic acid in wines. Am J Enol Vitic 31:294–297
Tsatsaronis GC, Pegiadou SA, Manoussopoulos ChI (1977) A study on the amino acids contained in several Greek wines. Chim Chronika 6:461–469
Tunblad-Johansson I (1977) Quantitative determination of free amino acids by gas-liquid chromatography with special reference to yeasts. Acta Pathol Mikrobiol Scand (suppl B) 259:17–24
Ubigli M (1981) Conffronto fra alcuvi metodi di determinazione dell'acido tartarico. Vini Ital 23:54–63
Umagat H, Kucera P, Wen LF (1982) Total amino acid analysis using pre-column fluorescence derivatization. J Chromatogr 239:463–474
Ussary JP (1973) Quantitative determination of amino acids by gas liquid chromatography. Food Prod Dev 7(7):84, 86, 88
Usseglio-Tomasset L (1973) Osservazioni sui metodi di determinazione dell'acido tartarico nei vini. Riv Vitic Enol 26:375–384
Usseglio-Tomasset L, Gabri G (1975) Gli acidi uronici nei mosti e nei vini. Atti Accad Ital Vino 27:151–172
Valle-Vega P, Young CT, Swaisgood HE (1980) Arginase urease electrode for determination of arginine and peanut maturity. J Food Sci 45:1026–1030

Vandersplice JT, Higgs DJ (1984) HPLC analysis with fluorometric detection of vitamin C in food samples. J Chromatogr Sci 22:485–489

van Wyk CJ, Venter PJ (1965) The determination of free amino acids in musts and wines by means of high voltage paper electrophoresis and paper chromatography. S Afr J Agric Sci 8:57–67, 69–71

Vasconcelos AMP, Neves HJC (1984) Estudo comparative ce castas vinicolas pela composicao de vinhos elementares em del aminoacidos livres e oligopeptidos. Proceedings Biotechnology Symposium (Porto) Feb 84 Portugal, pp 203–215

Vasconcelos AMP, Neves HJC (1985) Characterization of *Saccaromyces* sp. through determination of amino acid profiles by capillary gas chromatography. J High Resolut Chromatogr Commun 8:547–550

Veliksar SG (1977) Change in the amino acids composition of grape shoots under the effect of molybdenum. Sadovod Vinograd Vinodel Mold 32(5):25–27

Vidal M, Blouin J (1978) Dosage colorimétrique rapide de l'acide tartrique dans les mouts et les vins. Rev Fr Oenol 16(70):39–46

Vos PGA, Gray RS (1979) The origin and control of hydrogen sulfide during fermentation of grape must. Am J Enol Vitic 30:187–197

Vos PGA, Zeeman W, Heymann H (1979) The effects on wine quality of diammonium phosphate additions to musts. S Afr Soc Enol Vitic Proceedings (Stellenbosch) Nov 1978, pp 87–104 (1978)

Wagener WWD, Ough CS, Amerine MA (1971) The fate of some organic acids added to grape juice prior to fermentation. Am J Enol Vitic 22:167–171

Wallrauch S (1976) Prolinbestimmung in Fruchtsäften, Bedeutung für die Beurteilung. Flüss Obst 43:430, 435–437

Webb AD, Kepner RE, Maggoria L (1967) Identification of ethyl acid tartrate and one isomer of ethyl acid malate in California flor sherry. J Agric Food Chem 15:334–339

Weiller HG, Radler F (1976) Über den Aminosäurestoffwechsel von Milchsäurebakterien aus Wein. Z Lebensm Unters Forsch 161:259–266

Wiggins LF, Williams H (1952) Use of butanol-formic acid-water mixture in paper chromatography of amino acids and sugars. Nature 170:279–280

Williams RC, Baker DR, Schmit JA (1973) Analysis of water soluble vitamins by high-speed ion-exchange chromatography. J Chromatogr Sci 11:616–624

Winkler AJ, Cook JA, Kliewer WM, Lider LA (1974) General viticulture. University of California Press, pp 556–560

Woo DJ, Benson JR (1984) Organic acid determination using polymeric columns. Am Lab 16(1)50, 52–54

Wucherpfenning K, Millies KD (1976) Über den Einfluß der Elektrodialysebehandlung zum Zwecke der Weinsteinstabilisierung auf die Konzentration der Aminosäuren im Wein. Mitt Rebe Wein Obstb Früchteverwert Höheren Bundeslehr Versuchsanst 26:13–26

Yokotuska K, Kushida T (1983) An improved method for fluorometric amino acid analysis including proline and hydroxyproline. J Ferment Technol 61:1–6

Yoshida I, Hayakawa K, Miyazaki M (1985) Simultaneous determination of carboxylic acids and inorganic anions by photometric ion chromatography applied to liquid foods. Eisei Kagaku 31:317–323

Zamorani A, Nicolosi AC (1974) Sulla fermentazione dei mosti refrigerati e aggiunti di bentonite. Nota II. Gli alcoli Superiori. Riv Vitic Enol 27:274–279

Zamorani A, Nicolosi AC (1978) Tecnologia di produzione del mosto e qualita' dei vini bianchi della zonz dell'etna. Riv Vitic Enol 31:484–490

Zamorani A, Nicolosi AC, Maccarone A (1973) Sulla fermentazione dei mosti refrigerati e aggiunti di bentonite. Nota I. Le sostanze azotate e gli amminoacidi. Riv Vitic Enol 26:462–478

Zanier C, Tanner H (1977) Determinazione qualitativa e quantitativa degli acidi organici e degli zuccheri nei vini e nei mosti per mezzo della gaschromatografia. Vignevini 4(1):19–22

Zelleke A, Kliewer WM (1979) Influence of root temperature and rootstock on budbreak, shoot growth and fruit composition of Cabernet Sauvignon grapevines grown under controlled conditions. Am J Enol Vitic 30:312–317

Zomzely C, Marco G, Emery C (1962) Gas chromatography of n-butyl-N-trifluoroacetyl derivatives of amino acids. Anal Chem 34:1414–1417

Zumwalt RW, Juo K, Gehrke CW (1971) Applications of gas-liquid chromatographic method for amino acid analysis. A system for analysis of nanogram amounts. J Chromatogr 55:267–280

Alcohols Derived from Sugars and Other Sources and Fullbodiedness of Wines

W. R. SPONHOLZ

1 Introduction

When the sugar of a fruit juice is fermented to alcohol, the resulting beverage is called a wine. The main fruit juice used for fermentation is that of the grape *Vitis vinifera,* which is planted throughout the world in some thousand different varieties with very different flavor. Juices from apples, pears, plums, cherries and others are also wines when they are fermented with the same technology and, when it is a pure yeast fermentation, mainly with the yeast *Saccharomyces cerevisiae* (Dittrich 1977; Troost 1974; Würdig and Woller in prep). During fermentation not only the sugars of the grape juice – glucose and fructose – are fermented to ethanol, but also some minor components are formed by the yeast during sugar metabolism.

Directly related to sugar metabolism is the formation of glycerol, a component formed by different fungi in higher quantities. Growing yeast cells also require a nitrogen source, which is available in grape juice as a mixture of amino acids. These are assimilated and transformed during yeast growth. During this metabolism the higher alcohols are formed either from sugar (Thoukis 1958) or by degradation of the corresponding amino acids by the Ehrlich mechanism (Ehrlich 1907). Directly related to the amino acid synthesis is also the formation of the 2,3 butandiols.

Methanol is also a minor component always present in wines. In red wines the amounts are always slightly higher than in whites. This is caused by the fermentation on skins of the red mash for better color extraction, which also causes a more intensive breakdown of the pectins with which the methanol is esterified.

In wines made from healthy grapes, ethanol, glycerol, meso- and levo-2,3 butandiol, the higher alcohols and methanol are the main alcoholic constituents. Small amounts of terpeneols and other substances with hydroxy groups are very important for the wine flavor. Myo-inositol can also be included here, being substance ubiquitous in all plant tissues and juices (Loewus and Loewus 1980; Loewus and Dickinson 1982).

Wines of the Auslese type cannot be produced from normal sound grapes. These wines, originally from Austria, Germany, Sauterne (France), but nowadays also from Australia, California and Japan, always require infection by *Botrytis cinerea.* When this mold infects ripe berries and opens up the skin, it metabolizes the ingredients of the grape by itself and changes the composition of the resulting juice (Dittrich 1977). Molds like *Botrytis cinerea,* but also others, are well known to produce high amounts of polyalcohols like glycerol, arabitol, and mannitol (Blumenthal 1976). When the skin of the grape is then opened by the mold, other microorganisms like yeast and bacteria have the chance to develop in the exuding juice. Mainly yeasts are then present on the grape skin in extrordinarily high numbers. Especially those which are osmotolerant or have an osmophilic metabolism are involved in changing the composition of the must (Spencer and Spencer 1980). More glycerol, erythritol, arabitol, and mannitol are formed by the yeasts abundant on grapes. These yeasts belong to the species *Kloeckera apiculata, Candida stellata,* and *Metschnikowia pulcherrima.* The two latter were found in high numbers on infected grapes, *Candida stellata* 10^7 cells ml^{-1}, *Metschnikowia pulcherrima* 10^5 cells ml^{-1} (Fleet et al. 1984). Both synthesize sorbitol in high amounts (Sponholz et al. 1986), a substance previously seen as a marker of falsification of wines by the addition of pome fruit juice (Anonymous 1970; Junge 1970; Patschky 1974).

In high quality wines alditols are present in unexpectedly high quantities and belong to the main constituents of the sugar-free extract (Sponholz and Dittrich 1985). They must be seen in these wines, like glycerol, as substances causing fullbodiedness. The formation of polyols by microorganisms cannot always be regarded as an enhancement of wine quality. Heterofermentative lactic acid bacteria like *Leuconostoc* and *Lactobacillus* strains can produce very high amounts of mannitol from fructose, but this formation of mannitol is always connected with contemporary formation of acetic acid. This is well known to cause severe faults in wines (Wood 1961; Dittrich 1977; Sponholz et al. 1982; Sponholz in preparation).

During malolactic fermentation, a certain amount of glycerol is also fermented by lactic acid bacteria (Dittrich et al. 1980). These strains of *Lactobacillus* reduce the dehydration product of glycerol 3-hydroxypropionaldehyde to propandiol 1,3 (Schütz and Radler 1984). These strains are also able to reduce butandiol 2,3 to butanol-2 (Hieke and Vollbrecht 1974; Bertrand and Suzuta 1976; Tittel and Radler 1979).

A synthetic polyol, diethylenglycol, was used in Austria for falsification of wines. This substance was added to wines of lower quality for their enhancement. This means that the amount of sugar-free extract, which gives more fullbodiedness to the wine, was raised (Bandion et al. 1985). This was done because diethylenglycol has a similarity to glycerol and could not be detected so easily. Nowadays it is possible to analyze diethylenglycol by gas liquid chromatography (Bandion et al. 1985; Wagner and Kreuzer 1985; Führling and Wollenberg 1985; Littmann 1985; Haase-Aschoff and Haase-Aschoff 1985; Rapp et al. 1986) or by high pressure liquid chromatography (HPLC) (Pfeiffer and Radler 1985; Barka and Heidger 1986). This can be done together with ethylenglycol, meso- and levobutandiol 2,3 and also glycerol (see Conte and Minguzzi, this Vol.)

2 Analytical Methods

For qualitative and quantitative determination of alcoholics and polyalcoholics, a wide range of analytical methods is available. The special method used for analysis depends very much on the nature of the substance(s) to be analyzed, the matrix the substance is in and the amount of substance present or expected in the sample. For example, the ethanol present in high amounts in wine can be determined by jodometric methods (Rebelein 1971; Jakob 1973), the enzymatic method (Boehringer 1986), by different gas liquid chromatographic methods, as direct injection or headspace technique (Hachenberg and Schmidt 1977), or by high pressure liquid chromatography, or gravimetrically (Schmitt 1975).

The so-called higher alcohols which are present in all fermentation broths in the mg l^{-1} range can be analyzed with direct injection by gas chromatography on packed or capillary columns. The use of a stainless steel capillary column coated with Carbowax 400 is used in the wine and spirit industry to measure these fusel-oils including methanol. These columns are able to separate 2- and 3-methylbutanol(1), which are very valuable for the determination of the "alcohols'" origin. For alcohols with higher boiling points, such as hexanol(1), meso- and levo-butandiol 2,3- and 2-phenylethylethanol, a similar column coated with Carbowax 20 M is very useful. To analyze trace amounts of aroma components a concentration step is necessary (Rapp in prep.). These concentrates often contain several hundred components. The separation of these very complex mixtures needs columns with very high efficiency, such as glass or fused silica capillary columns. Because the mixtures contain substances with very different polarity (acids, alcohols, esters and hydrocarbons), have a boiling point range from 50 to 350° C and the concentrations vary from 10^{-2} to 10^{-10}, often also bidimensional gas chromatography is used.

For polyols, Bieleski (1982) and Tejedor and Santa-Maria (1984) reviewed the methods available for analysis. Chemical, microbiological, enzymatic, and chromatographic (ion exchange, gas chromatography, and high pressure chromatography) methods were used. Werder (1929) analyzed sorbitol in pome fruit wines by precipitation with benzaldehyde. This method was improved by Litterscheid (1931) and Vogt (1934) by using o-chlorbenzaldehyde for better precipitation. Because this method is not selective for sorbitol, the content of other polyols, mainly mannitol, must be known. The only microbiological method was used by Peynaud and Lafourcade (1955) for the evaluation of the myo-inositol content in musts and wines. They used the requirement of a *Kloeckera apiculata* yeast strain for the mesurement. This method is very time-consuming but also very selective for this particular substance. Commercially available methods for polyol determination were developed for the measurement of sorbitol and xylitol only (Junge 1970; Beutler and Becker 1977). The enzyme from sheep liver has the same activity on both polyols, but by the hexokinase method, which only phosphorylates the resulting fructose, a specifity for sorbitol is given. While xylitol only is found in small amounts in wine, there is a slight interference with glycerol, which causes a baseline drift. Glycerol is the other polyol that is very quickly and accurately determined in wine by enzymatic assay (Boehringer 1986).

Paper and thin layer chromatography were used very often (Tejedor and Santa-Maria 1984), but it is not possible to analyze wines directly by these methods. High concentrations of sugars, acids, and other substances interfere with the determination of the comparably small amounts of polyols in wine. The elimination of interfering substances is principally possible by passing the wine through an ion exchange column. Different ion exchangers were used for separation of polyols from acids and reducing sugars, including cation exchangers in Li^+, Rb^+, Na^+, Mg^{2+}, Ba^{2+}, Ca^{2+} and anion exchangers in Cl^-, HCO^{3-}, CH_3COO^-, OH^-, and $BO_3{}^{3-}$.

The borate method was used commercially by Biotronik for sugar analysis with an automatic buffer program. After separation on an anion exchanger, the eluting sugars and polyols were oxidized by periodate and determined photometrically. This method was valuable for dry wines, but failed with wines of high residual sugar content. The method was also only practicable with inert stainless steel pumps under special precautions. Gas chromatography is often used for the identification and quantification of sugars and polyols in biological material. These methods require the transformation of these nonvolatile substances to stable and volatile derivatives. The methods most used are acetylation with acetanhydryde or trifluoracetanhydryde and the silylation to the trimethylsilyl-derivativs. For the determination of sugars it is often useful to transform these to their oximes prior to acetylation or silylation to avoid more than one peak from one substance (Knapp 1976; Blau and King 1977; Pierce 1982). For the separation of complex biological matrices the use of capillary columns is recommended. With these columns, separation of acids, sugars and polyols is possible also in

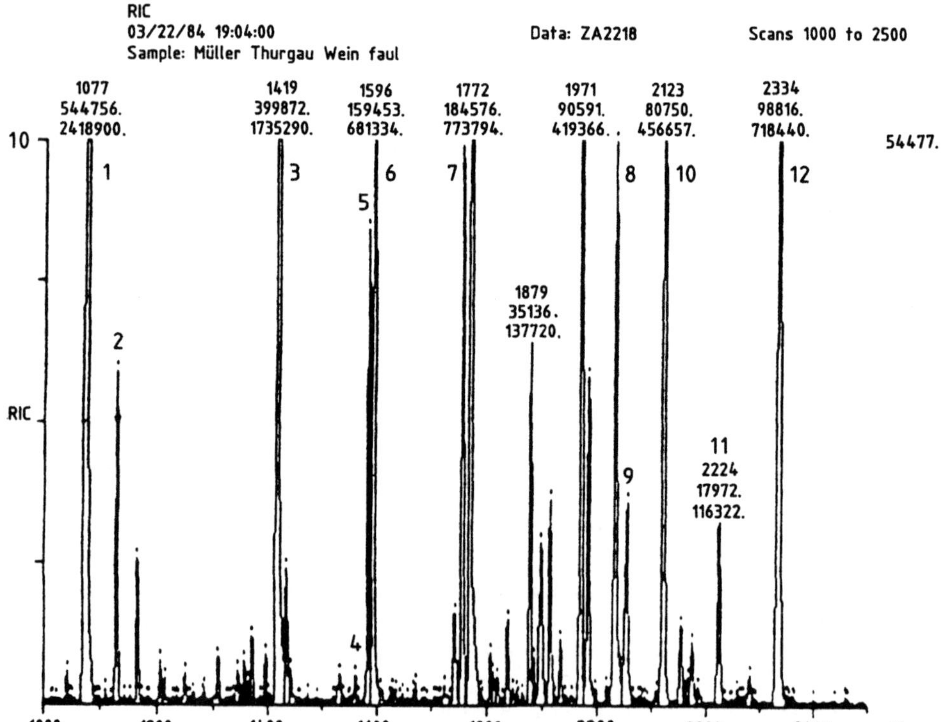

Fig. 1. Chromatogram of sugar alcohols and sugar acids as their trimethylsilyl derivatives. *1* malic acid; *2* erythritol; *3* tartaric acid; *4* xylitol; *5* arabitol; *6* ribitol (internal standard); *7* citric acid; *8* mannitol; *9* sorbitol; *10* gluconic acid; *11* mucic acid; *12* myo-inositol

musts and wines containing high sugar quantities, without prior separation by ion exchange. The coupling of a gas chromatograph to a mass spectrometer as a detector allows a quick and sure determination of the polyols also in trace amounts. The columns used were 30 m OV 1 on Pyrex glass (De Smedt et al. 1979) and a 25 m BP5 fused silica capillary (Sponholz and Dittrich 1985). The separation of the acids, sugars, and alditols of a wine are shown in Fig. 1.

The use of high pressure liquid chromatography for the separation of polyols in musts and wines is limited by the low sensitivity of the detectors used and also by the small amounts in which the polyols are often present in wines. The separation power of the available columns is also not efficient enough to separate all the polyols, sugars, and acids present. The use of this method for wine analysis is valuable for the separation of the main acids, the sugars, glycerol and ethanol and also for the determination of inositol (Flak 1981). It was also used for the determination of diethylenglycol in Austrian wines, as described earlier.

3 Biochemistry of Alcohol and Polyol Formation, and Their Occurrence in Wines

3.1 Methanol

Methanol is a breakdown product from grape pectin by the activity of pectin-methyl-esterase, which liberates methanol by hydrolysis. The grapes' own enzyme is set free during pressing. Because the skin contact is often prolonged in red wine-making, these wines therefore show higher amounts of methanol, as do wines from spicy varieties which sometimes also have longer skin contact for aroma extraction. In noble rot (*Botrytis cinerea*-infected) grapes, the mold's enzyme can also liberate methanol from pectin. The use of pectolytic enzymes for better color yield in red wine production or for better pressing can also enhance the methanol content of wines. White wines from sound grapes have 40–120 mg l^{-1}, those from red grapes show 120–250 mg l^{-1} of methanol. Wines from *Botrytis*-infected grapes can often have up to 364 mg l^{-1} of methanol. The methanol content of wines is very low compared to that of some spirits, and therefore it will not contribute so much to the fullobodiedness of wines; but as a part of the methylesters of wine it is involved in aroma formation (Nykänen and Suomalainen 1983).

3.2 Ethanol

Ethanol in wines is normally the product of the glycolytic pathway of sugar catabolism by the yeast *Saccharomyces cerevisiae*. Under anaerobic fermentation conditions it should follow the equation:

$$\begin{array}{llll} 1\ \text{Hexose} & \longrightarrow & 2\ \text{Ethanol} + & 2\ CO_2 \\ 1\ \text{mol} & \longrightarrow & 2\ \text{mol} & 2\ \text{mol} \\ 180.15\ \text{g} & \longrightarrow & 92.1\ \text{g} & 88\ \text{g} \\ \text{Theoretical yield} & & 51.1\% & 48.9\% \end{array}$$

Under wine-marking conditions, the practical yield of ethanol is only 47% (Trauth and Bässler 1936; Jakob 1974). This means that a part of the sugar is not fermented to ethanol by the yeast, but is transformed to other fermentation products. These can be very important for the taste and flavor of fermentation products, also for wine (Dittrich 1977). Pasteur (1858) showed that 100 g l^{-1} was fermented to 48.6 g l^{-1} alcohol, 46.6 g l^{-1} CO_2, 3.3 g l^{-1} glycerol, 0.6 g l^{-1} succinic acid and 1.2 g l^{-1} of other products. This means that the production of side products by the yeast during fermentation is very important for the flavor and taste of wines. Mainly the glycerol and meso- and levo-2,3 butandiols as produced polyhydric alcohols contribute ot the fullbodiedness of wines. Flavor compounds such as ethylesters and acetylesters are directly involved in wine flavor composition. So a high alcohol content of a wine not only gives the required body by itself to an alcoholic beverage; the higher amounts of byproducts like glycerol, which is formed in higher quantities when more sugar is fermented are of primary importance (Vogt and Bieber 1970). So the chaptalization, (the addition of sucrose to a must before fermentation), is not only done for a higher alcohol content of the wine, but also for body enhancement by the higher amount of byproducts formed.

3.3 Alcohols Related to the Amino Acid Metabolism

An important group of alcohols found in fermented beverages such as wine is related to the amino acid metabolism of the yeast. These alcohols, which can be present in amounts from 1–500 mg l^{-1}, are the main aroma constituents of wine or parent substances for a high number of esters, so that they not only contribute to the fullbodiedness of wines, but also to its flavor. The metabolism of some of them is shown in Fig. 2. Amino acids were proposed by Ehrlich (1907) as the parent substances for the formation of higher alcohols, as these alcohols are normally called. He fermented mashes rich in leucine and obtained a high yield of 3-methylbutanol(1). Other amino acids can also act as parent substances for higher alcohols.

Threonin	⟶	Propanol (1)
Valin	⟶	2-Methylpropanol (1)
Isoleucin	⟶	2-Methylbutanol (1)
2-Phenylalanin	⟶	2-Phenylethanol (1)
Tyrosin	⟶	Tyrosol
Tryptophan	⟶	Tryptophol
Methionin	⟶	Methionol

All these alcohols are present in normal wines. 3-Methylbutanol is always the main component. In spontaneous fermentations, the yield of 2-phenylethanol is often much higher than in fermentations with pure yeast cultures of *Saccharomyces cerevisiae* (Sponholz and Dittrich 1974). Thus the more aerobic spontaneous yeast flora present, the more of this alcohol forms. This can be of importance for the wine flavor. Wines from spontaneous fermentations, compared to those from pure yeast culture fermentations, are often described as more fullbodied.

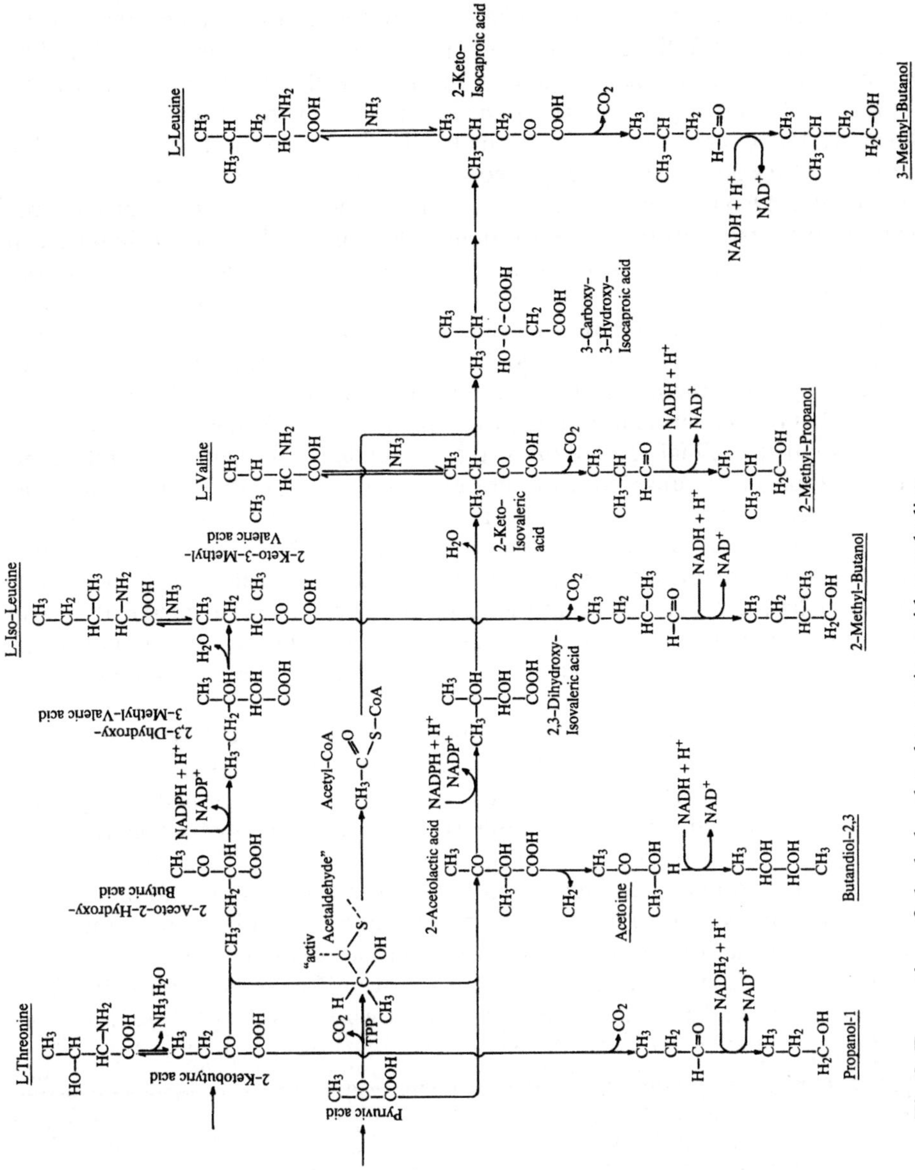

Fig. 2. Formation of alcohols related to amino acid metabolism

In the brandy industry also, 2-phenylethanol is seen as a substance causing wine character and fullbodiedness. An important role in wine aroma is played by the acetic ester of the 2-phenylethanol, a substance with a roselike flavor. The role of methionol as an important part of the aroma of red wines from California was shown by Muller et al. (1971). From the Ehrlich break-down of amino acids theoretically also ethanol can be expected from alanine, ethylenglycol from serine, and thioglycol from cysteine. According to the amino acid content of musts, especially the precursor substances for the higher alcohols, the Ehrlich mechanism can be only a minor part of fuseloil formation. Castor and Guymon (1952) showed that these amino acids were taken up by the yeast in the early process of fermentation, but the fuseloils were produced during the whole fermentation time. Genevois and Lafon (1956) showed that 3-methylbutanol(1) was also formed from ^{14}C acetate, when leucine was present in the fermentation broth. Thoukis (1958) showed that during fermentation the fusel alcohols were formed from sugar.

The synthesis of higher alcohols is strongly related to fermentation and amino acid metabolism. Because there are not enough of the specific amino acids that the yeast needs for protein synthesis, they have to be formed by it. Precursor for these syntheses is pyruvate from sugar metabolism. 2-Acetolactate and 2-aceto-2-ketobutyrate are formed as precursors for the needed 2-ketoacids for transamination to the corresponding amino acids valine, leucine, and isoleucine. Pyruvate decarboxylase, with an activity to all 2-ketoacids up to a C_6 chain, can now form the corresponding aldehydes. These are then reduced by the unspecific yeast alcoholdehydrogenase to alcohols.

Thus the fusel alcohols are by-products of the amino acid synthesis of the yeast. If a nitrogen donor is present in a very high amount, for example leucine, the corresponding 2-ketoacid will be decarboxylated and the formed aldehyde reduced to high amounts of 3-methylbutanol(1), as in Ehrlich's experiments.

If a yeast lacks a nitrogen source, as often occurs in juices from grapes grown in hot climates, often much higher amounts of these alcohols are found. This is caused by a high synthesis of the 2-ketoacids, which cannot be transaminated to the amino acids and then undergo decarboxylation followed by reduction to alcohol, as shown in juices lacking sufficient nitrogen (Rapp and Franck 1971).

The two isomer 2,3 butandiols which are present in wines up to 3.3 g l^{-1} are also related to amino acid metabolism. Guymon and Crowell (1967) reported a range from 420 to 1160 mg l^{-1} where the meso isomer was 20–38%. Castino (1969, 1970) found the same range in Italian wines. Dittrich (1964) found 420–1840 mg l^{-1} in German top quality wines, the same range was found in German reds (Lemperle and Kerner 1968). Patschky (1973) made a survey of Romanian wines. In 175 wines he found 310–1660 mg l^{-1}. A Trockenbeerenauslese showed 2689 mg l^{-1} and an Edelauslese 3300 mg l^{-1}, exeptionally high amounts. In French wines 220–400 mg l^{-1} were measured with 33.7% meso-2,3-butandiol (DeSmedt et al. 1979). Although there is no acetolactate decarboxcylase in yeast (Magee and De-Robin-Szulmajster 1968; Chuang and Collins 1968), a part of the 3-ketoacid, 2-acetolactate, can decompose spontaneously to acetoin and is reduced by yeast alcoholdehydrogenase or butandioldehydrogenase to the 2,3-butandiols. Most of the 2,3-butandiols can be formed by an acetoin-forming system (Tittel and Radler 1979). Acetoin is than reduced to the alcohol by acetoinreductase. During sherry production this substances are not affected by the flor yeasts growing on the wine (Ough and Amerine 1972). This indicates that the activity of the acetoinreductase in yeast must be very weak. Also homofermentative lactic acid bacteria can form both isomers of 2,3 butandiol. They possess a diacetylreductase (Strecker and Harary 1954) and an 2-acetolactate decarboxylase system to form acetoin. This is then reduced by stereo-specific 2,3 butandiol dehydrogenases which are responsible for the formation of the specific pattern by different bacteria strains (Taylor and Juni 1960; Höhn-Bentz and Radler 1978). Lactic acid bacteria are only involved in 2,3-butandiol formation when the malolactic fermentation occurs, or when they are fermenting sugar. The latter is uncommon in wine-making and mostly results in wine faults caused by high amounts of diacetyl. This can be "repaired" by a subsequent pure yeast culture fermentation, where diacetyl is reduced via acetoin to 2,3 butandiol. The normally small amount of these alcohols makes only a small contribution to the fullbodiedness of the wine by itself, but all the small amounts of the various substances are additive. At the other hand, 2,3-butandiols are precursors of butanol(2), an alcohol formed by *Lactobacillus brevis* (Hieke and Vollbrecht 1974; Bertrand and Suzuta 1976) and spoiling wines.

3.4 Other Aliphatic Alcohols

Some minor constituents with alcoholic character are sometimes found originating from wine-spoiling bacteria. At higher pH, that means in already spoiled wines, butanol(1) can be found in higher quantities than 1 mg l^{-1}, indicating the growth of *Clostridium*. This is also indicated when propanol(2) is present in wine. Hexanol(1) is also a minor constituent in wine, normally not exceeding 4 mg l^{-1}. In higher amounts, this alcohol with its grassy taste and smell indicates unripeness of the grapes. Its origin is the breakdown of linoleic and linolic acid to hexen-2-al and hexanal(1) during pressing and their reduction by fermenting yeast to hexanol(1) (Drawert and Rapp 1966). All the other aliphatic alcohols such as octanol(1), octanol(3), and octenol(3), which are mushroomlike, are connected with the *Botrytis* flavor, and all the alcoholic substances with terpenoid structure are present in such small amounts that they will not directly contribute to the fullbodiedness of wines. They are very important flavor substances in wines. to study this field see Nykänen and Suomalainen (1983); van Straten et al. (1977).

3.5 Glycerol

Glycerol as a constituent in wines has been known for a very long time. This polyhydric alcohol is miscible with water in all proportions and has a sweet taste, which compared to glucose is 50–80% (v. Rymon-Lipinski and Lück 1975). Very important for the mouth-filling character of this substance is also its viscosity. The glycerol content of musts from sound grapes is very low and does not exceed 1 g l^{-1}. Amounts of 0–0.4 g l^{-1} were analyzed in such musts (Dittrich 1964; Holbach and Woller 1976). During fermentation, yeasts produce glycerol from sugar. This production is dependent on the yeast strain used for fermentation (Mayer and Pause 1970) and the fermentation temperature (Ough et al. 1972; Rankine and Bridson 1971). If osmotolerant yeast strains are present during fermentation, higher amounts of glycerol can be formed, and also when higher amounts of oxygen are available (Spencer and Spencer 1980; Spencer and Sallans 1956). So the "wild" yeasts involved in wine-making, like *Kloeckera apiculata, Metschnikowia pulcherrima, Candida stellata* and others, can contribute to the glycerol content of a wine (Sponholz et al. 1986). The concentrations of glycerol measured are 4.1–9.9 g l^{-1} in 100 German wines (Rebelein 1957), 6.3–10.1 g l^{-1} for German red wines (Lemperle and Kerner 1968). Castino (1969, 1970) reported 6.5–13.7 g l^{-1} for Italian wines. The same amounts were reported from Rankine and Bridson (1971) for Australian wines (4.1–9.9 g l^{-1}) and from Ough et al. (1972) for Californian wines (1.86–8.2 g l^{-1}), where the 1.86 g l^{-1} seems to be an artificial, diluted, or heavily spoiled wine. All these data show the wide range of glycerol production by yeast and fermentation technology used for wine-making. Since Müller-Thurgau (1888), it is known that musts from grapes infected by noble rot *(Botrytis cinerea)* contain much higher concentrations of glycerol. Kielhöfer and Würdig (1961) found up to 21 g l^{-1} in musts from Trockenbeeren from 1959, Dittrich (1964) analyzed 3.8–9.2 g l^{-1} in musts from grapes with noble rot. A comparison of musts from sound grapes with those from grapes with noble rot showed an increase of glycerol ranging from 4.4 to 15 g l^{-1} (Dittrich et al. 1974, 1975). Holbach and Woller (1976, 1978) reported glycerol contents in noble rot grapes in combination with gluconic acid in these musts. For the glycerol present in musts and not synthesized during fermentation, the name Mostglycerin (must glycerol) was invented to identify it. Although wines produced from musts with high glycerol content show much more of this substance than those fermented from musts of sound grapes, the glycerol from yeast fermentation is always additional (Dittrich et al. 1974). Franconian Auslese has 9.7–20.0 g l^{-1}, Beerenauslese 13.1–25.1 g l^{-1}, Trockenbeerenauslese 24–26.7 g l^{-1} (Wagner and Kreutzer 1977). Wines from noble rot musts 7.5–22.4 g l^{-1} (Dittrich et al. 1974) or 7.4–42.4 g l^{-1} (Holbach and Woller 1976). Hungarian Tokay wines, which are also made from noble rot grapes, also have very high glycerol contents, 8.25–35.8 g l^{-1} (Drawert et al. 1976). The high glycerol contents in high quality wines can be caused by different reasons:

1. Synthesis by *Botrytis cinerea.*
2. Synthesis by „wild" yeasts.
3. By concentration in drying out grapes.
4. By synthesis during fermentation with *Saccharomyces cerevisiae.*

1. Synthesis of glycerol by *Botrytis cinerea*

Glycerol is synthesized by *Botrytis cinerea* from the sugars of the must. Mainly glucose is metabolized by this mold, while fructose is nearly unaffected. The metabolic pathway for glycerol synthesis is the glucose-monophosphate pathway (Gentile 1954), where glycerolaldehyde 3-phosphate is produced, the precursor for glycerol formation. Comparison of glycerol formation by *Botrytis cinerea* in musts with increasing sugar content showed that the highest amounts are formed under sugar stress. Up to 24% of sugar in the must, a maximum of 3.7 g l^{-1} glycerol was formed. When the sugar content was increased to 29%, 5.6 g l^{-1} were synthesized. Similar results were reported by Spencer and Spencer (1980) with the osmotolerant yeast *Zygosaccaromyces rouxii.*

2. Synthesis of glycerol by "wild" yeasts

Some wild yeasts are associated with *Botrytis cinerea.* This means that on grapes opened by this mold, yeasts which are always present have the chance to multiply and to metabolize the now available sugar. Mainly osmotolerant or osmophilic yeasts can be found on these grapes. High cell numbers of *Kloeckera apiculata,*

Metschnikowia pulcherrima, Candida stellata and others are present. Compared to *Saccharomyces cerevisiae,* these yeasts are much better glycerol producers (Sponholz et al. 1986). *Candida stellata* is present in such fermentations very often in much higher cell numbers than *Saccharomyces cerevisiae.* It can still be present when the fermentation is finished or stopped (Fleet et al. 1984).

3. The concentration effect in drying out berries
When the grape berries are infected by *Botrytis cinerea,* the skin is opened and the juice is available for the mold and the yeasts living on it. Sugars, amino acids, and other substances are metabolized and the composition of the juice is changed drastically (Dittrich and Sponholz 1975; Dittrich et al. 1974). During the whole of this time, the berries are shrinking by losing water by active uptake from the microorganisms and by evaporation. This is causing a concentrating of the grapes' ingredients and also a stimulation for the mold and the yeasts to overcome water stress by a higher production of polyols. Table 1 shows the concentration effect calculated from the weight of 100 berries from five different varieties. The concentration reaches sevenfold. The concentration of sugar in the variety *Pinot blanc* has risen from 182 g l^{-}, 184 g l^{-1}, 204 g l^{-1}, 311 g l^{-1} to 478 g l^{-1}, representing all the wine qualities from Kabinett to Trockenbeerenauslese type. The increase of the sugar content is only 2.5 times, all the other sugar is metabolized by the mold and the yeasts.

Table 2 shows the glycerol content in the different concentration stages. The sound uninfected berries show the known glycerol content of 0.4 g l^{-1}. In the first infection stage where no shrinking and only a very small concentration occurs,

Table 1. Weight of 100 grape berries (g) of increasing infection stages by *Botrytis cinerea* and the resulting concentration factor (C_f)

Stage	Pinot blanc		Pinot gris		Pinot noir		Osteiner		Optima	
	g	c_f	g	c_f	g	c_f	g	c_f	g	c_f
0	209	1.00	209	1.00	181	1.00	174	1.00	142	1.00
1	156	1.34	175	1.20	151	1.20	162	1.07	116	1.22
2	130	1.61	143	1.46	103	1.76	138	1.26	100	1.42
3	73	2.86	85	2.46	52	3.48	71	2.45	54	2.63
4	30	6.97	36	5.81	27	6.70	41	4.24	34	4.18

Table 2. Glycerol content of grape musts from berries with increasing infection by *Botrytis cinerea* (mg l^{-1})

Stage	Pinot blanc	Pinot gris	Pinot noir	Osteiner	Optima
0	370	91	68	46	55
1	1464	826	1227	415	1182
2	4010	3221	5250	2312	3312
3	15350	8006	17270	10580	8670
4	28420	20670	32890	22650	18160

the glycerol content has risen to 1 g l^{-1}. In the second stage, where the concentration factor is near 2, the glycerol content is nearly 4 g l^{-1}, rising then in the next stage to 17 g l^{-1} and in the 4th stage to 33 g l^{-1}. These are amounts often found also in wines of very high qualities, like Trockenbeerenauslese.

Combining these results, the glycerol content in high quality wines is a combination of the metabolism of *Botrytis cinerea,* some "wild" yeasts, the shrinking of the berries and the additional glycerol formed during fermentation by *Saccharomyces cerevisiae*. This forming of a metabolite and its concentration is common to most of the grapes ingredients, resulting in wines with increasing body. in certain circumstances, glycerol must not be a product of *Botrytis* or yeasts. Sometimes must glycerol is a product of *Penicillium* strains developing on the grape, causing Grünfäule (green rot) (Mühlberger and Grohmann 1962). Glycerol can be fermented by lactic acid bacteria during malolactic fermentation (Dittrich et al. 1980). this is also reported for apple and perry wines (Whiting 1961). *Leuconostoc mesenterioides* (Hirano et al. 1962), *Lactobacillus brevis* and *L. buchneri* (Schütz and Radler 1984) can cause this. A vitamin B_{12}-dependent glycerol dehydratase forms 3-hydroxypropanal (Smiley and Sobolow 1962), which normally is reduced to 1,3-propandiol by 1,3-propandiol dehydrogenase. Because of a proposed isomerisation of 3-hydroxypropanal to 2-hydroxypropanal (lactaldehyde) also 1,2-propandiol is formed. The unspecifity of this enzyme and its reaction with the resulting 1,2-propandiol forms propanol(1), which is often found in higher concentrations in wines with malolactic fermentation (Schütz and Radler 1984). Lacking glucose, 3-hydroxypropionic acid (hydracrylic acid) can be formed (Sobolow and Smiley 1960). Under acid conditions or on heating such mashes for distillation, the 3-hydroxypropanal formed loses one molecule of water-forming acrolein (Wilharm and Holz 1951). A condensation of acrolein or 3-hydroxypropanal with anthocyanins or other phenolic bodies in the wine can cause bitterness (Rentschler and Tanner 1951). Glycerol can also be metabolized by acetic acid bacteria. The resulting dihydroxyaceton can be found in wines which have acquired an acetic acid taint after fermentation by these bacteria (Sponholz et al. 1982). In vinegar, this substance is seen as a quality marker, indicating that the vinegar was made from a fermentation product like wine.

3.6 Polyols in Musts and Wines

3.6.1 General

The polyols present in musts and wines are numerous. Erythritol, arabitol, xylitol, mannitol and sorbitol were described (Dubernet et al. 1974; Bertrand and Pissard 1976; Drawert et al. 1976; De Smedt et al. 1979; Sponholz and Dittrich 1985). There is, however, also much literature on single polyols which will be discussed in the sections dealing with the occurrence and importance of these polyols for wine. The polyols are either the natural contents of the grape, or formed by molds or yeasts on the grapes. They can also be fermentation products by yeasts of a different kind, and they can be products formed during malolactic fermentation or spoiling of wines by lactic acid bacteria. There is also evidence that acetic acid bacteria can use these polyols as a source of carbon (Kersters et al. 1965). The metabolic pathway for polyol formation is mainly the glucosemonophosphate cycle (Holligan and Jennings 1972a), because the amount of arabitol synthesized is highly dependent on its activity. Furthermore these authors, were able to demonstrate that mannitol was partly formed by a "direct pathway", partly from a hexosephosphate cycling the hexosemonophosphate pathway. Its activity was highest with nitrate as a nitrogen source (Holligan and Jennings 1972b). Additional indications of a linkage between the hexosemonophosphate pathway and polyol formation comes from genetic experiments with mutants of *Aspergillus nidulans* (Hankinson 1974). Yeasts have also been shown to possess numerous polyol dehydrogenases, some of which are necessary for the final stages of the formation of the various polyols formed, such as the polyol dehydrogenase of *Candida utilis* (Chakravorty et al. 1962), which can be either NAD^+- or $NADP^+$-linked, or the pentitol dehydrogenases from *(Zygo)saccharomyces rouxii* (Ingram and Wood 1965). Both dehydrogenases react with the free sugars. A scheme for arabitol formation was outlined by Weimberg and Orton (1962).

$$\text{Glucose} \xrightarrow{\text{ATP}} \text{glucose 6-phosphate} \xrightarrow{\text{NADP}^+\text{-NADPH}}$$

$$\text{gluconic acid-6-phosphate} \xrightarrow{\text{NADP}^+\text{-NADPH}} \text{D-ribose-5-phosphate}$$

$$\text{Ribulose} + P_i \longrightarrow \text{NADPH} \longrightarrow \text{NADP}^+ \longrightarrow \text{D-arabitol.}$$

Weimberg also included the transaldolase-transketolase system in sugar phosphate formation. Other workers found a reductase in *Saccharomyces rouxii* with xylulose as substrate for arabitol formation (Blakeley and Spencer 1962); but in crude extracts both sugars wre reduced, indicating that both enzymes were present and active. In brewers' yeast *(Saccharomyces cerevisiae)* mannitol, sorbitol, galactitol, and erythritol were oxdized by a dehydrogenase (Wilhelmsen 1961). An enzyme forming arabitol from xylose was isolated by Onishi and Saito (1962). This enzyme also catalyzed the oxidation of mannitol and sorbitol to fructose, xylitol to xylulose, erythritol to erythrulose, and glycerol to dihydroxyaceton. Most of the polyols are formed from the free sugars by reduction and the sugars are formed in the pentosephosphate cycle as phosphorylated sugars. Phosphatases are therefore necessary to obtain the free sugar moiety. Such enzymes with a broad spectrum are described from *Saccharomyces mellis* (Weimberg and Orton 1963) and *Zygosaccharomyces rouxii*. Lactic acid bacteria, like *Leuconostoc dextranicum, Lactobacillus pentosaceus,* and *L. brevis,* form mannitol from fructose during growth in wine (Wood 1961). *Lactobacillus brevis* reduces D-fructose directly to D-mannitol with a NADH-dependent dehydrogenase (Martinez et al. 1962).

3.6.2 Erythritol

Erythritol can be formed by various enzymes from erythrulose. This must be synthesized from erythrose-4-phosphate, which is formed in the pentosephosphate cycle via transaldolase-transketolase reactions, by an isomerase. The resulting sugar phosphate must then be dephosphorilated.

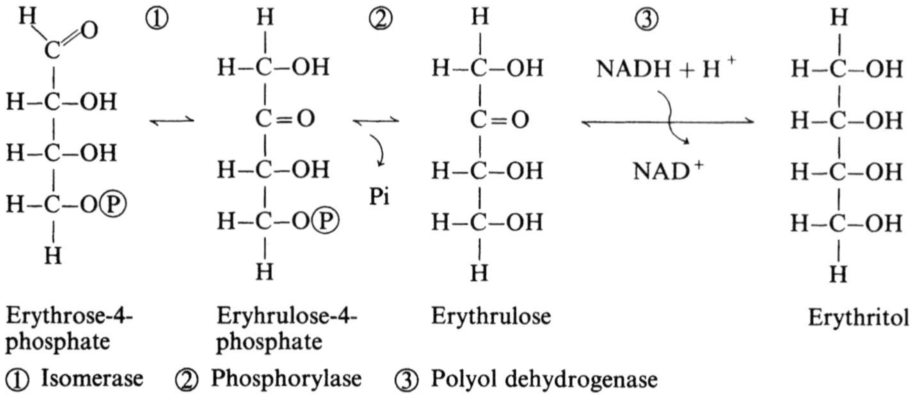

Erythritol in wine was described by Dubernet et al. (1974) in wines from the Bordeaux region. They found 33–100 mg l^{-1} in whites, 64–116 mg l^{-1} in reds and 160–272 mg l^{-1} in reds from infected grapes. A culture of *Botrytis cinerea* on a must initially containing 33 mg l^{-1} showed no increase within 2 months. The same amounts from 22–208 mg l^{-1} were found by De Smedt et al. (1979), whereas Drawert et al. (1976) did not report this polyol in their investigations of Tokay wines. In wines of increasing quality from Q. b. A. (quality wine) to Trockenbeerenauslese, increasing amounts of erythritol were reported (Sponholz and Dittrich 1985), ranging from 34–692 mg l^{-1} in whites, whereas in German and Rioja red wines the amounts found ranged from 53–182 mg l^{-1} and were similar to those reported for Bordeaux wines. Also in sherries, 43–105 mg l^{-1} were found. The high amounts in high quality wines may be dependent on the growth of *Botrytis cinerea* on the grapes (it forms up to 86 mg l^{-1}), the growth of yeasts on the infected grapes, or the concentration of the grape ingredients by drying-out of the grapes. The drying-out effect, which is very prominent with glycerol, exists here also, but the initial content of the grapes was too poor to contribute to the high content of erythritol in high quality wines. Investigations of the ability of yeasts related to grapes showed that these yeasts are able to form high amounts of erythritol during fermentation (Sponholz et al. 1986). While normally fermenting *Saccharomyces cerevisiae* strains form only up to 114 mg l^{-1}, a weakly fermenting but osmotolerant strain of *S. c.* (formerly *S. italicus*) formed up to 900 mg l^{-1} erythritol. Strains of *Kloeckera apiculata* formed up to 530 mg l^{-1}, those of *Candida stellata* 200 mg l^{-1} in normal musts. This indicates that high amounts of erythritol in wines may be formed by the yeasts associated with *Botrytis cinerea*-infected grapes and by those fermenting these musts.

3.6.3 Xylitol

Xylitol, which can be formed by the same pathway as erythritol but from xylulose, was measured in wines from 0–95 mg l^{-1} (Sponholz and Dittrich 1985). Some yeasts, like *Zygosaccharomyces rouxii, Kloeckera apiculata* and *Candida stellata,* show very weak formation during must fermentation (Sponholz et al. 1986). While there was also no production by *Botrytis cinerea,* the higher amounts found in high quality wines seem to be dependent on must concentration by shrinking of the berries.

3.6.4 Arabitol

Arabitol is reported as a main polyol formed in sugar metabolism via mannitol in molds (Blumenthal 1976) or yeast (Spencer and Spencer 1980). A NADH-dependent polyol dehydrogenase from *Candida utilis* (Chakravorty et al. 1962) forms it from xylulose, one which is NADPH-dependent from *Zygosaccharomyces rouxii* uses D-ribulose (Ingram and Wood 1965). In wine, arabitol was first reported by Dubernet et al. (1974). They found 13–59 mg l^{-1} in whites, 32–111 mg l^{-1} in reds and 165–359 mg l^{-1} in wines from infected grapes. They also showed that *Botrytis cinerea* forms this polyol during growth on must, resulting in an increase from 36 to 570 mg l^{-1} in a 2-month period. Much higher concentrations of arabitol, mainly in high quality wines, were found by Sponholz and Dittrich (1985). They reported in wines of normal quality like Q. b. A. and Kabinett 121–216 mg l^{-1} and increasing amounts in higher qualities. Spätlese wines showed 9–265 mg l^{-1}, Auslese wines 13–370 mg l^{-1}, Beerenauslese wines 175–2344 mg l^{-1} and Trockenbeerenauslese wines 396–2353 mg l^{-1}, indicating the enormous influence of molds and yeasts on these musts and wines. Whereas German reds showed 14–315 mg l^{-1}, Rioja reds showed 49–148 mg l^{-1} and sherries only 4–35 mg l^{-1}.

Growing *Botrytis cinerea* on must with increasing sugar content did not result in the high arabitol concentrations reported (Dubernet et al. 1974). The strain used formed only 246 mg l^{-1}. Investigations on the arabitol synthesis of yeasts showed that some species have a high ability for arabitol production (Spencer and Sallans 1956). *Zygosaccharomyces rouxii* formed 3034 mg l^{-1}, *Candida krusei* 2321 mg l^{-1} and *Metschnikowia pulcherrima,* which is always related to noble rot grapes, 1552 mg l^{-1} (Sponholz et al. 1986). Thus the arabitol found in musts from infected grapes is mainly formed by yeasts and may also be from *Botrytis.* During shrinking of the berries it is concentrated to the high amounts found in high quality wines.

3.6.5 Mannitol

Mannitol is formed by molds (Blumenthal 1976), yeasts (Spencer and Spencer 1980), and also by lactic acid bacteria (Wood 1961). The synthesis of mannitol can be a direct reduction:

$$\text{D-fructose} + NADH + H^+ \rightarrow \text{mannitol} + NAD^+$$

or a reduction of the phosphorilated sugar:

$$\text{D-fructose-6-P} + NADH + H^+ \rightarrow \text{mannitol-1-P} + NAD^+$$

followed by a dephosphorilation:

$$\text{mannitol-1-P} \rightarrow \text{mannitol} + P_i.$$

The direct reduction of D-fructose appears to be more abundant (Chakravorty et al. 1962; Blumenthal 1976; Spencer and Spencer 1980).

In wine, mannitol was known as a product of spoiling bacteria (Frerejaque 1935), and was found up to 35 g l^{-1}.

As a normal constituent of wine it was first described by Thaler and Lippke (1971); Thaler et al. (1970). They found 151–1401 mg l^{-1} in German and French wines. They also reported mannitol as a naturally occurring substance in fresh pressed grape juice. Mannitol was also found in Bordeaux wines. White contained 84–323 mg l^{-1}, reds 90–394 mg l^{-1} and reds from infected grapes showed 452–735 mg l^{-1}. During a 2-month growth of *Botrytis cinerea* on a grape must, the content rose from initially 82 mg l^{-1} to 536 mg l^{-1}. This indicates the ability of this mold to form this substance also on grapes (Dubernet et al. 1974). It seems therefore unusual that only such a small amount as 60 mg l^{-1} was found in increasing qualities of Hungarian Tokay wines, which are also made from heavily infected noble rot grapes (Drawert et al. 1976). Investigations on German white wines with increasing quality from Q. b. A. to Trockenbeerenauslese showed drastically increasing amounts of mannitol (Sponholz and Dittrich 1985). All these wines were wines without faults, excluding the formation of mannitol by spoiling lactic acid bacteria. Whereas Q. b. A. wines showed normal amounts, as reported previously, from 110–552 mg l^{-}, the other qualities showed much higher amounts, as can be seen in Table 3.

Mainly in the high quality wines, the sweet-tasting mannitol contributes to the fullbodiedness of the wines and to their extract. The sweetening factor for mannitol compared to sucrose is 0.7 (Von Rymon-Lipinski and Lück 1975). German red wines had 142–952 mg l^{-1} and Rioja red wines 172–422 mg l^{-1}, while sherries contained 10–1117 mg l^{-1} of mannitol. Investigations of musts from grapes infected by *Botrytis cinerea* showed increasing amounts of mannitol, which could not be dependent on shrinking. There was active synthesis by the mold or the associated yeast cells. Similar to the results of Dubernet et al. (1974), *Botrytis* formed large amounts of mannitol when growing on musts. Without any concentration effect, a maximum of 2.6 g l^{-1} was formed within 8 days. An evaluation of the mannitol-forming power of yeasts associated with the mold on grapes showed that only *Metschnikowia pulcherrima* produced 1.3 g l^{-1} of mannitol. So the mannitol found in high quality wines can be formed by either *Botrytis cinerea* or by *Metschnikowia pulcherrima* and it is then concentrated on the amounts found in grape juices and wines by shrinking of the berries. This production of mannitol is not associated with wine-spoiling microorganisms or other substances formed by them at the same time, like acetic acid, so it must be seen as a positive enhancement of the wine quality. On the other hand, there is production of man-

Table 3. Amounts of mannitol in German white wines of increasing quality (mg l^{-1})

Q.b.A.	110– 552
Kabinett	3– 998
Spätlese	158– 1672
Auslese	377– 2183
Beerenauslese	836–10639
Trockenbeerenauslese	4365–12884

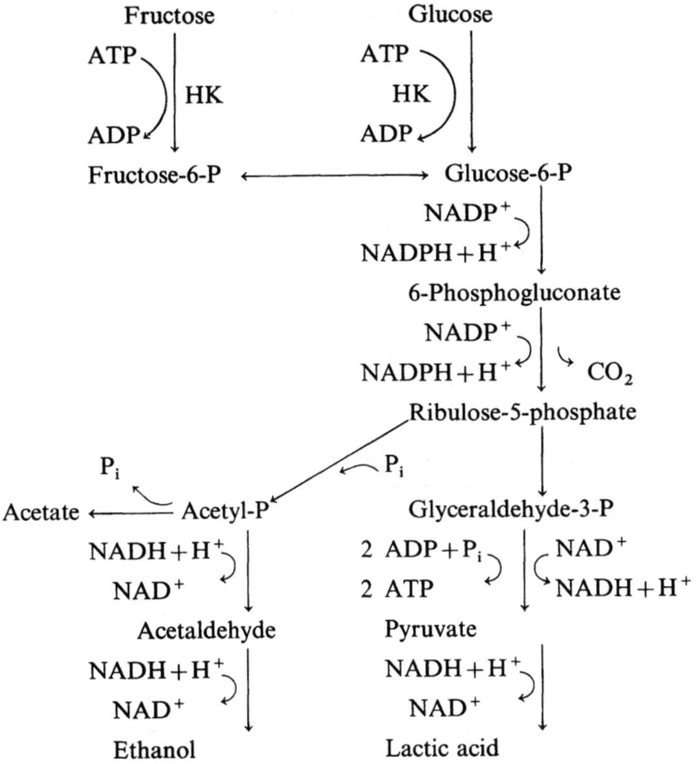

Fig. 3. Formation of mannitol from fructose by heterofermentative lactic acid bacteria

nitol by lactic acid bacteria, which is always associated with the formation of acetic acid. The resulting wines are then spoiled by a vinegar taint. Glucose and fructose are both fermented by heterofermentative lactic acid bacteria. When fructose is present in the medium, this is reduced to mannitol, causing that the C_2 moiety from the breakdown products from xylulose-5-phosphate glyceraldehyde 3-phosphate and acetylphosphate is only partly reduced to ethanol. The other part is dephosphorilated to acetic acid (Fig. 3). The strains of *Leuconostoc dextranicum, Lactobacilluc pentosaceus, L. brevis* and also *Leuconostoc oenos,* which is used for malolactic fermentation in wine, can produce mannitol from fructose. This can be seen in Table 4 (Wood 1961).

From the calculation of *Lactobacillus brevis* a stoichiometry of:

3-fructose → 2-mannitol + lactic acid + acetic acid + CO_2

results (Martinez et al. 1962). The other bacteria have slightly different stoichiometries. Because wines with a high mannitol content can be normal wines, there must be other parameters found to identify the wines spoiled by heterofermentative lactic acid bacteria. Also an additional high content of acetic acid is not enough to separate them analytically from normal wines, while acetic acid can also be formed by acetic acid bacteria and yeasts in wines with a natural high

Table 4. The products of glucose and fructose metabolism by heterofermentative lactic acid bacteria (Wood 1961)

	Leuconostoc dextranicum		*Lactobacillus pentoaceticus*		*Lactobacillus brevis*	
Glucose	+		+		+	
Fructose		+		+		+
Lactic acid	84	53	91	44	83	33
Ethanol	81	51	61	12	74	1
Acetic acid	11	35	35	32	15	40
CO_2	87	78	86	58	81	45
Glycerol	24	2	0	0	33	4
Mannitol	0	30	0	47	0	62

Table 5. Wines spoiled by heterofermentative lactic acid bacteria (mg l^{-1})

Mannitol	Acetic acid	D-Lactic acid	Sorbitol	Taste
9173	3039	3177	170	Ester, vinegar
1353	12801	2044	95	Ester, vinegar

polyol content (Sponholz et al. 1982). These wines thus also show high amounts of D-lactic acid, the indicator that sugar is fermented by lactic acid bacteria; the sorbitol content should also be low. Table 5 shows such wines with spoiling by heterofermentative lactic acid bacteria.

3.6.6 Sorbitol

Sorbitol is a very well-known ingredient of pome fruit juices and was described by Werder (1929). For the sorbitol content in some Rosaceae fruits and juices, see Mattioni and Valentinis (1971); Weiss and Sämann (1979); Bieleski (1982). In fresh grapes no sorbitol was detected, but some was found in raisins (Reif 1934). Therefore sorbitol in wines was seen as a marker for an addition of pome fruit juice for falsification. After improvement of the sorbitol detection method by Litterscheid (1931) and Vogt (1934), sorbitol was found also in wines. The highest sorbitol content then seen as natural was 75 mg l^{-1} (Vogt 1935; Schätzlein and Sailer 1935; Reichhard 1938). A further improvement of the detection method was developed by Tanner and Dupperex (1967, 1968), who used thin layer chromatography for the separation of sorbitol and mannitol, and by the enzymatic method (Junge 1970; Beutler and Becker 1977), as also by Thaler and Lippke (1971), who used a combination of enzymatic and polarographic methods.

Junge (1970) measured in 66% 50–100 mg l^{-1}, in 27% 100–150 mg l^{-1} and in 7% 150–200 mg l^{-1} sorbitol in a total of 200 wines. Nearly the same amounts were detected by Thaler and Lippke (1970). Similar results were also reported by Patschky (1974), who evaluated 553 wines. In French wines, the same sorbitol contents were reported by De Smedt (1979), but slighly higher concentrations up

Table 6. Concentration of sorbitol in German white wines ($mg\,l^{-1}$)

Q.b.A.	30– 97
Kabinett	14–220
Spätlese	17–202
Auslese	84–342
Beerenauslese	131–989
Trockenbeerenauslese	650–827

to 300 $mg\,l^{-1}$ in sweet wines by Bertrand and Pissard (1976). Drawert et al. (1976) reported carbohydrates in Hungarian Tokay wines and found 490 $mg\,l^{-1}$ of sorbitol in an "Essenz" from 1972, a wine made from the free run of extremely noble rot-infected grapes. As for the other polyols in wines, it was shown by Sponholz and Dittrich (1985) that high amounts of sorbitol can be natural in high quality German white wines. The concentrations of sorbitol in these wines are given in Table 6.

In German red wines, 4–181 $mg\,l^{-1}$, in Rioja red wines 57–228 $mg\,l^{-1}$, and in sherries 2–72 $mg\,l^{-1}$ were measured. Similar values in sherries were reported by Olano (1983). The very high concentrations in German white wines show that the maximum allowed amount of 200 $mg\,l^{-1}$ was already exceeded with some Kabinett wines. This also indicated that there must be a synthesis of sorbitol by the mold *Botrytis cinerea* or by the associated yeasts on the grape berries. Mainly therefore white wines of Auslese character only can be produced from infected grapes. The mold did not form any sorbitol when growing on grape juice, neither did *Penicillium expansum* nor *Aspergillus niger*. Sorbitol formation was detected on grape must by some yeasts (Sponholz et al. 1986), mainly the yeasts always present in high numbers on grapes infected by *Botrytis cinerea* (Fleet et al. 1984), showing an evident synthesis of sorbitol on grape must (Sponholz et al. 1986).

Kloeckera apiculata strains produced up to 120 $mg\,l^{-1}$, *Candida stellata* strains up to 439 $mg\,l^{-1}$ and *Metschnikowia pulcherrima* 257 $mg\,l^{-1}$. These yeasts are thus responsible for the high amounts of this polyol found in wines in reference to the high concentration factor of nearly 6 in high quality wines. The polyol sorbitol thus changed from an indicator of falsification with pome fruit juice (Junge 1970) to an indicator of quality in high quality white wines. Also in normal white and red wines, mainly from northern wine-growing regions, sorbitol can no longer be seen as an indicator for falsification, because a very great part of all grapes harvested in these areas can be heavily infected by noble rot and yeasts.

3.7 Inositol

Two of the five naturally occurring isomers of inositol have been detected in wine so far. The most common is myo-inositol, which is described as a ubiquitous substance present in all plants (Loewus and Loewus 1980), and therefore also in all

Table 7. Concentrations of myo-inositol in increasing qualities of German white wines (mg l^{-1})

Q.b.A.	270– 790
Kabinett	220–1 376
Spätlese	102–1 436
Auslese	443–1 399
Beerenauslese	290–2 483
Trockenbeerenauslese	1 640–2 066

plant juices. The other is scyllo-inositol, found in amounts of 10–143 mg l^{-1} (De Smedt et al. 1979).

Myo-inositol was measured by the growth of *Kloeckera apiculata*, an auxothrophic yeast for myo-inositol; Peynaud and Lafourcade (1955) detected in red Bordeaux musts 400–640 mg l^{-1}, which remained constant even after fermentation with *Saccharomyces cerevisiae*. They also reported an increase of it during grape maturation. In commercial wines 220–730 mg l^{-1} of inositol were measured. By gas chromatographic analysis 79–1044 mg l^{-1} were reported by De Smedt et al. (1979). In sherries 36–741 mg l^{-1} (Olano 1983; Santa-Maria et al. 1985) and 55–553 mg l^{-1} (Sponholz and Dittrich 1985) were reported.

Increasing amounts with increasing quality were found in Hungarian Tokay wines. The normal Furmint (a grape variety) wine had (450 mg l^{-1}) a normal inositol content, 5 butt Aszu showed 660 mg l^{-1}, 6 butt Aszu 1060 mg l^{-1}, and the "essence" 2560 mg l^{-1} of inositol (Drawert et al. 1976). This indicates a concentration or synthesis by microorganisms. The concentrations in German white wines can be seen in Table 7 (Sponholz and Dittrich 1985).

In German red wines, 309–805 mg l^{-1} were found, in Rioja red wines 137–619 mg l^{-1}. The concentrations of myo-inositol in German white wines are very similar to those reported for Hungarian Tokay wines (Drawert et al. 1976).

Investigations of grapes infected with *Botrytis cinerea* showd that there is a concentration of myo-inositol by shrinking of the grape berries. There is also a use of myo-inositol by the mold and the other microorganisms living on the grape. Thus *Botrytis cinerea* used 430 mg l^{-1} of a total of 630 mg l^{-1} when growing on a must with 310 g l^{-1} sugar content. Also all yeasts associated with the mold on grapes use myo-inositol during growth and fermentation, *Kloeckera apiculata* 400 mg l^{-1}, *Metschnikowia pulcherrima* 150 mg l^{-1} and *Candida stellata* 300 mg l^{-1}, when growing on grape juice. Also all strains of *Saccharomyces cerevisiae* used myo-inositol from 297–544 mg l^{-1} during growth and fermentation (Sponholz et al. 1986). Thus myo-inositol in wine is a natural grape ingredient and not formed by microorganisms related to wine-making. The often very high concentrations in high quality wines indicate the shrinking process in grapes infected by *Botrytis cinerea*.

4 Conclusion

The number and chemical structure of alcoholic compounds found in wines is enormous. They can be natural ingredients of the grape, like terpeneols as aroma compounds, or myo-inositol and xylitol. They can be formed during growth of *Botrytis cinerea* and the associated yeasts *Kloeckera apiculata, Metschnikowia pulcherrima* and *Candida stellata* on the grapes, like most of the polyols and glycerol. They can be fermentation products from glucose and fructose, like ethanol and the fusel alcohols, which are related to the amino acid synthesis of the yeast. They can also be products formed during detoriation and spoiling of the wine by lactic acid bacteria, like mannitol when associated with acetic acid and D-lactic acid. However, they can also be added by criminals as diethylenglycol for the enhancement of wine quality; this shows in fact, that polyols present in wines give more body, more smoothness to the product. So the ethanol from fermentation, the higher alcohols formed during fermentation, the production of glycerol in the infected grape and during fermentation, and the polyols from noble rot and the associated yeast cells give extract and body to the wine. This is constantly increased by the additional shrinking of the berries and the loss of water, resulting in highly concentrated musts with high extracts like Beerenauslese and Trockenbeerenauslese wines and similar products like Tokay and Sauternes.

References

Anonymous (1970) Methodes d'analyse et éléments constitutifs des vins. Bull OIV 43:767–798

Bandion F, Valenta M, Kohlmann H (1985) Zum Nachweis extrakterhöhender Zusätze zu Wein. Mitt Klosterneuburg 35:89–92

Barka G, Heidger V (1986) Eine effektive Bestimmung von Diethylenglycol in Wein und Ergebnisse eines Ringversuches. Supplement Lebensmittelchemie 2/86 57–60 GIT Verlag Darmstadt

Bertrand A, Pissard R (1976) Dosage du sorbitol dans les vins par chromatographie en phase gazeuse de son dérivé acétylé. Ann Falsif Expert Chim 69:571–579

Bertrand A, Suzuta K (1976) Formation du butanol-2 par les bactéries lactiques isolées du vin. Connaiss Vigne Vin 10:409–426

Beutler HO, Becker J (1977) Enzymatische Bestimmung von D-Sorbit und Xylit in Lebensmitteln. Dtsch Lebensm Rundsch 73:182–187

Bieleski RL (1982) Sugar Alcohols. In: Pirson A, Zimmermann MH (eds) Enzyclopedia of plant physiology. Vol 13A Loewus FA, Tanner W (eds) Plant carbohydrates I intercellular carbohydrates. Springer, Berlin Heidelberg New York

Blakeley ER, Spencer JFT (1962) Studies on the formation of D-arabitol by osmophilic yeasts. Can J Biochem Physiol 40:1737–1748

Blau K, King GS (eds) (1977) Handbook of derivatives for chromatography. Heyden, London

Blumenthal HJ (1976) Reserve carbohydrates in fungi. In: Smith JE, Berry DR (eds) The filamentous fungi, vol 2 biosynthesis and metabolism. Edward Arnold, London, pp 292–307

Boehringer (1986) Methoden der biochemischen Analytik und Lebensmittel-Analytik. Boehringer Mannheim (Firmenschrift)

Castino M (1969) Gli acidi 2-metilmalico, 2,3 diidrossiisovalerianico e 2,3 diidrossi 2-metilbutirico nei vini. Nota I: Loro correlazione con gli altri prodotti della fermentazione alcolica. Riv Vitic Enol 22:197–207

Castino M (1970) L'acido succinico nei vini. Nota II: Fattori che ne condiziano la formazione. Vini d'Italia 12:289–297

Castor JGB, Guymon JF (1952) On the mechanism of formation of higher alcohols during *alcoholic fermentation*. Science 115:147–149

Chakravorty, Veiga LA, Bacilla M, Horecker BL (1962) Pentose metabolism in *Candida* II. The diphosphopyridine nucleotide-specific polyol dehydrogenase of *Candida utilis*. J Biol Chem 237:1014–1020

Chuang LF, Collins EB (1968) Biosynthesis in bacteria and yeast. J Bacteriol 95:2083–2089

De Smedt P, Liddle PAP, Cresto B, Bossard A (1979) Application de la CPG sur colonne capillaire a l'analyse des composé fixes du vins. Ann Falsif Expert Chim 72:633–642

Dittrich HH (1964) Über die Glycerinbildung von *Botrytis cinerea* auf Traubenbeeren und Traubenmosten sowie über den Glyceringehalt von Beeren- und Trockenbeerenausleseweinen. Wein Wiss 19:12–20

Dittrich HH (1977) Mikrobiologie des Weines. Ulmer, Stuttgart; 2. Aufl. 1987

Dittrich HH, Kerner E (1964) Diacetyl als Weinfehler, Ursache und Beseitigung des „Milchsäuretones". Wein Wiss 19:528–535

Dittrich HH, Sponholz WR (1975) Die Aminosäureabnahme in *Botrytis*-infizierten Traubenbeeren und die Bildung von höheren Alkoholen bei ihrer Vergärung. Wein Wiss 30:188–210

Dittrich HH, Sponholz WR, Kast W (1974) Vergleichende Untersuchungen von Mosten und Weinen aus gesunden und Botrytis-infizierten Traubenbeeren. I. Säurestoffwechsel, Zuckerstoffwechselprodukte, Leucoanthocyangehalte. Vitis 13:36–49

Dittrich HH, Sponholz WR, Göbel HG (1975) Vergleichende Untersuchungen von Mosten und Weinen aus gesunden und aus *Botrytis*-infizierten Traubenbeeren II. Modellversuche zur Veränderung des Mostes durch *Botrytis*-Infektion und ihre Konsequenzen für die Nebenproduktbildung bei der Gärung. Vitis 13:336–347

Dittrich HH, Sponholz WR, Wünsch B, Wipfler M (1980) Zur Veränderung des Weines durch den bakteriellen Säureabbau. Wein Wiss 35:421–429

Drawert F, Rapp A (1966) Über die Inhaltsstoffe von Mosten und Weinen VII. Vitis 5:351–376

Drawert F, Leupold G, Lessing V, Kerenyi Z (1976) Gaschromatographische Bestimmung der Inhaltsstoffe von Gärungsgetränken. VI Quantitative gaschromatographische Bestimmung von Neutralstoffen (Kohlenhydraten) und phenolischen Verbindungen in Tokayer Wein. Z Lebensm Unters Forsch 162:407–414

Dubernet MO, Bertrand A, Ribereau-Gayon P (1974) Présence constante dans les vins d'erythritol, d'arabitol et de mannitol. CR Acad Sci Paris Serie D 279:1561–1564

Ehrlich F (1907) Über die Bedingungen der Fuselölbildung und über ihre Zusammensetzung mit dem Eiweißaufbau der Hefe. Ber Dtsch Chem Ges 40:1027–1047

Flak W (1981) Die quantitative Bestimmung von Sacchariden und Zuckeralkoholen in Wein mittels Hochdruckflüssigkeitschromatographie (HPLC). Mitt Klosterneuburg 31:204–208

Fleet GH, Lafon-Lafourcade S, Ribereau-Gayon P (1984) Evolution of yeast and lactic acid bacteria during fermentation and storage of Bordeaux wines. Appl Environ Microbiol 48:1034–1038

Frerejaque M (1935) Dosage polarimétrique du mannitol. CR Acad Sci Paris 200:1410–1412

Führling D, Wollenberg H (1985) Zur Bestimmung kleiner Mengen Diethylenglykol im Wein. Dtsch Lebensm Rundsch 81:325–328

Genevois L, Lafon M (1956) Transformation de l'acétate marqué par la levure en fermentation anaérobic. Bull Soc Chim Biol 38:89

Gentile AC (1954) Carbohydrate metabolism and oxalic acid synthesis by *Botrytis cinerea*. Plant Physiol 29:257–261

Guymon JF, Crowell EA (1967) Direct gaschromatographic determination of levo- und meso-2,3 butandiols in wines and factors affecting their formation. Am J Enol Vitic 18:200–209

Haase-Aschoff K, Haase-Aschoff I (1985) Bestimmung von Diethylenglykol (DEG) durch Kappilargaschromatographie. Weinwirtschaft-Technik, 121:266

Hachenberg H, Schmidt AP (1977) Gas chromatographic headspace analysis. Heyden, London

Hankinson O (1974) Mutants of the pentose phosphate pathway in *Aspergillus nidulans*. J Bacteriol 117:1121–1130

Hieke E, Vollbrecht D (1974) Zur Bildung von 2-Butanol durch Lactobacillen und Hefen. Arch Mikrobiol 99:345–351

Hirano S, Imamura T, Uchijama K, Yu TL (1962) Formation of acrolein by *Clostridium perfringens*. Kagoshima 4:239–264, zit Chem Abstr 59:3096c

Höhn-Bentz H, Radler F (1978) Bacterial 2,3 butandiol dehydrogenases. Arch Mikrobiol 116:197–203

Holbach B, Woller R (1976) Über den Zusammenhang zwischen Botrytisbefall von Trauben und den Glycerin- sowie Gluconsäuregehalt von Wein. Wein Wiss 31:202–214

Holbach B, Woller R (1978) Der Gluconsäuregehalt von Wein und seine Beziehung zum Glyceringehalt. Wein Wiss 33:114–126

Holligan PM, Jennings DH (1972a) Carbohydrate metabolism in the fungus *Dendryphiella salina* II. The influence of different carbon and nitrogen sources on the accumulation of mannitol and arabitol. New Phytol 71:583–594

Holligan PM, Jennings DH (1972b) Carbohydrate metabolism in the fungus *Dendryphiella salina* III. The effect of the nitrogen source on the metabolism of ($1-^{14}C$) and ($6-^{14}C$) glucose. New Phytol 71:1119–1133

Ingram JM, Wood WA (1961) Pentitol dehydrogenase of *Saccharomyces rouxii* J Bact 89:1186

Jakob L (1973) Die kombinierte Alkohol-Zucker Bestimmung, ein Weg zur Vereinfachung der Weinanalyse. Allg Weinfachztg 29:780–783

Jakob L (1974) Qualität und Spezialität - objektiv betrachtet. Allg Weinfachztg 110:1253–1255

Junge Ch (1970) Aktuelle Probleme der Weinanalytik. Dtsch Lebensm Rundsch 66:374–379

Kersters K, Wood WA, De Ley J (1965) Polyol dehydrogenases of *Gluconobacter oxydans*. J Biol Chem 240:965–974

Kielhöfer E, Würdig G (1961) Bildung von Glycerin und 2,3 Butylenglykol bei der Weingärung. Z Lebensm Unters Forsch 115:418–428

Knapp DR (1976) Handbook of analytical derivatization reactions John Wiley, New York

Lemperle E, Kerner E (1968) Untersuchungen zur Farbe von Rotwein. 1. Mitteilung: Versuch zur Objektivierung der Farbmessung. Wein Wiss 23:281–284

Litterscheid F (1931) Über ein neues „Sorbit-Verfahren“ zum Nachweis von Obstwein in Traubenwein. Z Lebensm Unters Forsch 62:653–657

Littmann S (1985) Zur Bestimmung kleiner Mengen Diethylenglykol in Wein. Dtsch Lebensm Rundsch 81:328–329

Loewus FA, Loewus MA (1980) Myo-Inositol:Biosynthesis and metabolism. In: Preiss J (ed) The biochemistry of plants, vol 3. Academic Press, New York, pp 43–76

Loewus FA, Dickinson DB (1982) In: Pirson A, Zimmermann MH (eds) Encyclop plant physiology New Series, Vol 13A Loewus FA, Tanner W (eds) Plant carbohydrates I Intracellular carbohydrates. Springer, Berlin Heidelberg New York

Magee PT, De-Robin-Szulmajster H (1968) Yeast:Ile, Val-biosynthesis. The regulation of the Ile-Val-biosynthesis in *S. cerevisiae*. 2. Identification and characterisation of mutants lacking the acetohydroxyacid-synthetase. Eur J Biochem 3:502–506

Martinez G, Barker HA, Horecker BL (1962) A specific mannitol dehydrogenase from *Lactobacillus brevis*. J Biol Chem 238:1598–1603

Mattioni R, Valentinis G (1971) Ricerca del sorbitolo nel succo di lampone con cromatographia su stratto sottile bidimensionale con corsa a tempo per l'identificazione delle adulterazioni con succo di armarena, di ciliegie, di melle e analoghi. Boll Lab Chim Prov 22:789–796
Mayer K, Pause G (1970) Überprüfung einiger Weinhefen auf Alkohol- und Glycerinbildung. Schweiz Z Obst- Weinbau 106:490–492
Mühlberger FH, Grohmann H (1962) Über das Glycerin in Traubenmosten und Weinen. Dtsch Lebensm Rundsch 58:65–69
Müller-Thurgau H (1888) Die Edelfäule der Trauben. Landwirtsch Jahrb Schweiz 17:83–100
Muller CJ, Kepner RE, Webb AD (1971) Identification of 3(Methylthio)-Propanol as an aroma constituent in Cabernet Sauvignon and Ruby Cabernet Wines. Am J Enol Vitic 22:156–160
Nykänen L, Suomalainen H (1983) Aroma of beer, wine and distilled alcoholic beverages. In: Rothe M (ed) Handbuch der Aromaforschung. Akademie Verlag, Berlin
Olano A (1983) Presence of trehalose and sugar alcohols in sherry. Am J Enol Vitic 34:148–151
Onishi H, Saito T (1962) Polyalcohol production by *Pichia miso* in a jar fermentor. Agric Biol Chem 26:804–808
Ough CS, Amerine MA (1972) Further studies with submerged flor sherry. Am J Enol Vitic 23:128–131
Ough CS, Fong D, Amerine MA (1972) Glycerol in wine; determination and some factors affecting. Am J Enol Vitic 23:1–5
Pasteur L (1858) Production constante de glycérine dans la fermentation alcoolique. CR Hebd Seances Acad Sci 46:857–859
Patschky A (1973) Über die chemische Zusammensetzung eingeführter rumänischer Weine. Mitt Klosterneuburg 23:101–116
Patschky A (1974) Der Sorbitgehalt von Weinen. Allg Weinztg 57:1336–1337
Peynaud E, Lafourcade S (1955) L'inositol dans les raisins et dans les vins. Son dosage microbiologique. Ann Technol 4:381–396
Pfeiffer P, Radler F (1985) Bestimmung von Diethylenglykol mit der HPLC-Methode. Weinwirtschaft Technik 121:234–235
Pierce AE (1982) Silylation of organic compounds, 4th edn. Pierce Chemical Company, Rockford, Illinois
Rankine BC, Bridson DA (1971) Glycerol in Australien wines and factors influencing its formation. Am J Enol Vitic 22:6–12
Rapp A (1988) Der Traubenmost. Aromastoffe (in preparation) In: Würdig G, Woller R (eds) Chemie des Weines. Ulmer, Stuttgart
Rapp A, Franck H (1971) Über die Bildung von Äthanol und einigen Aromastoffen bei Modellgärversuchen in Abhängigkeit von der Aminosäurekonzentration. Vitis 9:299–311
Rapp A, Engel L, Ullemeyer H (1986) Zur Bestimmung von Mono- und Diethylenglykol in Wein durch zweidimensionale Gaschromatographie. Z Lebensm Unters Forsch 182:498–500
Rebelein H (1957) Unterscheidung von naturreinen und gezuckerten Weinen und Bestimmung des natürlichen Alkoholgehaltes. Z Lebensm Unters Forsch 105:403–420
Rebelein H (1971) 5-Minuten-Methoden zur genauen Bestimmung des Alkohol-Zucker und Gesamt SO_2-Gehaltes (durch Destillation) in Weinen und Fruchtsäften. Allg Weinfachztg 24:590–594
Reichard O (1938) Wein 2. Teil. Analytischer Teil. Überwachung und Verkehr. In: Juckenack A (ed) Handbuch der Lebensmittelchemie, Bd VII. Alkoholische Getränke, Nachweis von Sorbit. Springer, Berlin, p 361
Reif G (1934) Über den Nachweis von Sorbit in Obsterzeugnissen. Z Lebensm Unters Forsch 68:179–186
Rentschler H, Tanner H (1951) Das Bitterwerden der Rotweine. Mitt Lebensm Unters Hyg 42:463–475

Santa-Maria G, Olano A, Tejedor M (1985) Quantitative determination of trehalose and inositol in white and red wines by gas liquid chromatography. Chem Mikrobiol Technol Lebensm 9:123–126

Schätzlein CH, Seiler E (1935) Sorbit in reinen Traubenweinen. Z Lebensm Unters Forsch 70:484–488

Schmitt A (1975) Aktuelle Weinanalytik. Heller Chemie- und Verwaltungsgesellschaft mbH, Schwäbisch Hall

Schütz H, Radler F (1984) Anaerobic reduction of glycerol to propandiol 1,3 by *Lactobacillus brevis* and *Lactobacillus buchneri*. System Appl Microbiol 5:169–178

Smiley KL, Sobolow M (1960) Metabolism of glycerol by an acrolein forming *Lactobacillus*. J Bacteriol 79:261–266

Smiley KL, Sobolow M (1962) A cobamide requiring glycerol dehydrase from an acrolein-forming lactobacillus. Archives Biochem Biophys 97:538–543

Spencer JFT, Sallans HR (1956) Production of polyhydric alcohols by osmophilic yeasts. Can J Microbiol 2:72–79

Spencer JFT, Spencer DM (1980) Production of polyhydric alcohols by osmotolerant yeasts. In. Preiss J (ed) The biochemistry of plants, Vol 3. Academic Press, New York, pp 393–425

Sponholz WR (1988) 4. Der Wein 4.3. Fehlerhafte und unerwünschte Erscheinungen im Wein. In: Würdig G, Woller R (eds) Chemie des Weines. Ulmer, Stuttgart

Sponholz WR, Dittrich HH (1974) Die Bildung von SO_2-bindenden Gärungs-Nebenprodukten, höheren Alkoholen und Estern bei einigen Reinzuchthefestämmen und einigen für die Weinbereitung wichtigen „wilden" Hefen. Wein Wiss 29:301–314

Sponholz WR, Dittrich HH (1985) Zuckeralkohole und myo-Inosit in Weinen und Sherries. Vitis 24:97–105

Sponholz WR, Dittrich HH, Barth A (1982) Über die Zusammensetzung essigstichiger Weine. Dtsche Lebensm Rundsch 78:423–428

Sponholz WR, Lacher M, Dittrich HH (1986) Die Bildung von Alditolen durch die Hefen des Weines. Chem Mikrobiol Technol Lebensm 10:19–24

Strecker HJ, Harary I (1954) Bacterial butylenglykol dehydrogenase und diacetylreductase. J Biol Chem 211:263–270

Tanner H, Dupperex M (1967) Über den dünnschichtchromatographischen Nachweis von Sorbit, Mannit und anderen Polyhydroxy-Verbindungen. Schweiz Z Obst- Weinbau 103:610–616

Tanner H, Dupperex M (1968) Über die Bestimmung von Sorbit in sortenreinen Birnen-, Apfel- und Kirschensäften. Schweiz Z Obst- Weinbau 104:508–511

Taylor MB, Juni E (1960) Stereoisomeric specifities of 2,3 butandiol-dehydrogenase. Biochem Biophys Acta 39:448–457

Tejedor M, Santa-Maria JG (1984) Métodos de análisis di polialcoholes. Sem Vinivinicola 39:2349–2357

Thaler H, Lippke G (1971) Zur Bestimmung von Sorbit und Mannit in Wein. Mitt Klosterneuburg 21:19–31

Thaler H, Lippke G, Lemelson D (1970) Über das Vorkommen von Mannit in Wein. Mitt Geb Lebensmittelunters Hyg 61:372–377

Thoukis G (1958) The mechanism of isoamyl alcohol formation using tracer techniques. Am J Enol Vitic 9:161–166

Tittel D, Radler F (1979) Über die Bildung von 2,3 Butandiol bei *Saccharomyces cerevisiae* durch Acetoinreductase. Monatsschr Brau 32:260–266

Trauth F, Bässler K (1936) Ein Beitrag zur Frage der Beziehung zwischen Mostgewicht und Alkoholgehalt und deren Nutzanwendung bei der Verbesserung der Moste. Z Lebensm Unters Forsch 72:476–498

Troost G (1974) Technologie des Weines, 4. Aufl. Ulmer, Stuttgart

Van Straten S, De Vrijer F, De Beauveser JC (1977) Volatile compounds in food. 4th edn Central Institute for Nutrition and Food Research TNO, Zeist

Vogt E (1934) Das Sorbitverfahren zum Nachweis von Obstwein in Traubenwein. Z Lebensm Unters Forsch 67:407–425

Vogt E (1935) Über das Vorkommen von Sorbit in reinen Traubenweinen. Z Lebensm Unters Forsch 69:587–591
Vogt E, Bieber H (1970) Weinchemie und Weinanalyse, 3. Aufl. Ulmer, Stuttgart
Von Rymon-Lipinski GW, Lück E (1975) Süßstoffe – Entwicklungen und Tendenzen. Chem unserer Zeit 9:142–145
Wagner K, Kreutzer P (1977) Zusammensetzung und Beurteilung von Auslesen, Beeren- und Trockenbeerenauslesen am Beispiel fränkischer Erzeugnisse unter besonderer Berücksichtigung des Jahrganges 1975. Weinwirtschaft 113:272–275
Wagner K, Kreutzer P (1985) Gaschromatographische Bestimmung von Diethylenglykol in Wein. Weinwirt Tech 121:213
Weimberg R (1962) Mode of formation of D-arabitol by *Saccharomyces mellis*. Biochem Biophys Res Commun 8:442–445
Weimberg R, Orton LL (1963) Repressible acid phosphomonoesterase and constitutive pyrophosphatase of *Saccharomyces mellis*. J Bacteriol 86:805–813
Weiss J, Sämann H (1979) Ergebnisse von Untersuchungen über die D-Sorbitgehalte von Fruchtsäften. Mitt Klosterneuburg 29:81–84
Werder J (1929) Zum Nachweis von Obstsaft (Obstwein) in Traubenwein. Mitt Geb Lebensmittelunters Hyg 20:7–14
Whiting GC (1961) "Non-fermentable" substances in ciders and perries. Rep Long Ashton Res Stat 1960, 1961:135–139
Wilharm G, Holz G (1951) Beitrag zur Kenntniss des Acroleins in Obstbränden, Maischen und Mosten. Arch Mikrobiol 15:403–413
Wilhelmsen JB (1961) Polyol dehydrogenases in yeast. Enzymologia 23:259–265
Wood WA (1961) The bacteria, vol 2, Academic Press, New York
Würdig G, Woller R (1988) (eds) Chemie des Weines. Ulmer, Stuttgart

Wine Phenols[1]

V. L. Singleton

1 Introduction

1.1 General

The term "phenols" has not been widely used for constituents of wine or other foods for at least two reasons. Phenol itself is thought of as a toxic disinfectant inimical to food and the adjective phenolic or the term polyphenol (most but not all natural phenols have more than one phenolic hydroxyl) have been substituted. Furthermore, owing to widely differing special attributes, natural phenols have usually been considered by separate groups as red pigments, tannins, browning substrates, antioxidants, etc., ignoring their phenolic commonality. It is useful, however, to consider these compounds together as natural phenols since that produces more order and perspective regarding their chemical reactivity and analysis. Now that their biochemical origins are understood, with phenylalanine the predominant precursor in food plants, this practice should become common.

Phenols are crucial components in wines. If all the natural phenols and their derivatives were absent from wines, wines would lack the differences that make them so interesting and diverse in color, astringency, bitterness, oxidation level, and related characteristics. Of course, wines denuded of phenols would still differ in sugar, acid, alcohol, odor, and carbon dioxide. Phenols apparently make little direct contribution to the odor of most wines, but may contribute pungency and, especially in wines aged in small barrels, may supply vanillin and smoky odor notes (Singleton and Noble 1976). Much of the differences among classes and types of wine, starting with white versus red, is due to the kinds and amounts of phenols present and the reactions they have undergone. After alcohol, acids, residual sugars, and perhaps proline, phenols are the constituents major in amount in many wines, especially red ones. They cover in wines as a whole one of the largest ranges in total content (at least 30-fold) of any constituent group.

Phenols in wines are, however, much more than just direct contributors to appearance and flavor. They participate in and serve as useful indicators of many reactions. They reflect the grape variety and sometimes the growing conditions in useful ways. They are indicators of the degree of berry maceration during winemaking. They are the substrates for enzymic browning of musts and are the major substrate for nonenzymic autoxidation in wines. They usually change in response to processing, stabilization, oxidation, and aging of a wine, depending on the conditions and the duration of the treatment. Differences in phenol composition among wines are sometimes qualitative, reflecting the presence or absence of certain phenols (red versus white wines lacking anthocyanins is an obvious example). More often, differences reflect ratios among common constituents. For example, two young wines with similar anthocyanin color may be quite different in astringent tannin content. Significant quantitative differences in phenol content are very likely between any two wines and may cover a wide range of content in total phenols or that of any one group of phenols (light pink through dark red wines, for example). Analyses of wines for these differences is obviously important and useful.

It is not possible in the space allotted to review exhaustively the reasons one might wish to analyze for grape or wine phenols. Much further detail on these topics can be found in recent papers and in books by Paronetto (1977), Ribéreau-Gayon (1972), and Singleton and Esau (1969). Rather it is intended to give as thorough a summary of the state of wine phenol analysis as possible in order to serve plant product researchers and analysts. Although the focus is on wine, a few references on related products will also be cited. In recent years almost every publication presenting detailed data on wine composition includes at least one and sometimes several different phenol analyses. Few of these papers can be cited. The three books just cited have considerable discussion of phenol analysis and particularly that by Paronetto (1977) gives detailed procedures. Amerine and Ough (1980), Ribéreau-Gayon et al. (1976), and several others give general coverage of wine analysis including procedures for phenols. A limited number of excellent references for phenol analysis, not specifically on wines, deserve mention at least as examples (King and White 1956; Pridham 1964; Seikel 1964; Mabry et al. 1970; Schanderl 1970; Markham 1982).

How to organize this summary presented some difficulty. One might use the application viewpoint such as analyses used during growing, ripening, harvesting, process control, quality control of the product, or legal standardization. The official methods with legal standing may not be the best available, but have had sufficient validation to achieve their standing (see for example Office International de la Vigne et du Vin 1978; Association of Official Analytical Chemists 1984). Methods could be considered by country or research teams. The fact that these often differ greatly indicates that there is little consensus. Easy, straightforward, unequivocal analyses are not yet the rule in this field. One might consider wine phenol analyses from the property-procedural viewpoint such as acidimetry, chromatography, colorimetry, fluorimetry, gravimetry, oxidation methods, polarography, or spectrophotometry. To some extent this approach is inevitable. The organization to be followed here will be to consider the methods considered useful for individual phenols first, then for classes of phenols, and finally for total phenols, with appropriate discussion from other viewpoints when possible.

1.2 Kinds, Amounts, and Sources of Phenols to Expect in Wines

Wines differ so much in phenol content that only rough approximations will be given as typical, but some gross comparisons seem useful as background. Unless otherwise specified in this article, wines from *Vitis vinifera* grapes are meant. This species is the overwhelming source of the world's wines. Wines from other grapes or other fruit frequently differ appreciably in phenol content. The typical finding is that the phenols originally present, and usually their relative proportions, are set by the genetics of the particular variety of grape (or other fruit) within the broader but still specific capabilities of the species, genus, family and higher category of plant taxonomy. The total amount of phenols is greatly influenced by grape-growing and wine-making conditions.

All the significant phenols of young, unflavored wines originate in the grapes (or in other fruit of fruit wines), with the notable exception of tyrosol, which can be produced from sugar in synthetic media by wine yeasts. Of course, fermentation and processing may modify the grape phenols. Aged wines involve further modifications and sometimes additional phenols extracted from wooden cooperage. The extractable phenols of oak cooperage are not flavonoids and a large portion is hydrolyzable tannins not otherwise found in wine.

The lowest phenol content belongs to wines prepared from clarified juice drained immediately after the grapes are crushed. Disregarding the relatively few commercial varieties such as Alicante Bouschet, Salvador, or Rubired that have red juice and even fewer that may have appreciable tannin in the juice, such as Calzin or Refosco, wines from juice have very little flavonoid content, perhaps 30 mg l^{-1} as catechin or epicatechin is typical. The phenols of such wines and the nonflavonoid phenols of all wines are mostly phenolic acids, particularly derivatives of caffeic acid and total of the order of 150 mg l^{-1}, at least at the start of the wine-making process.

Anthocyanins of most red grapes are confined to the firmer skin tissue of the berry and little appears in the must or wine unless appreciable maceration releases them from the skin into the fluid. This is true of other flavonoids as well, even from white grapes. Grape seeds, if present, also contain large amounts of flavonoid derivatives and may contribute them to wines. Cluster stems are removed and leaves are excluded in normal practices. Firmer tissues, skins and seeds, may in the long run contribute further nonflavonoid components, but usually the nonflavonoid composition of wine reflects the composition of the easily expressed vacuolar juice and is fairly similar regardless of the degree of pomace extraction. The usual flavonoid content, on the other hand, can rise from near zero for a wine from juice to approach the maximum possible if all the phenols of the berry were extracted. Increased steeping time between grape crushing and fluid separation (especially if alcohol is also being produced), increased temperature during that time, and factors such as increased stirring or harder pressing will increase the flavonoid content in the final wine. Actually the maximum wine content of phenols is somewhat less than the total fruit content because an appreciable fraction is retained in the pomace and lees as the fluid is removed and clarified. Nevertheless, rich red wines prior to aging may contain 3–6 g l^{-1} of total phenols (Singleton and Esau 1969).

Wines with such high content of total phenols are considered too rough, astringent, and bitter for commercial acceptance and by modified conditions of fermentation, processing, or aging are generally limited to perhaps 2000 mg l^{-1} or less as the wines are readied for sale.

Many efforts have been made to delineate typical wine phenol content (e.g., Amerine 1954; Singleton and Esau 1969; Van Buren 1970; Singleton and Noble 1976; Singleton 1982), but they involve many approximations and compromises. Table 1 is one example, to be considered as a rough guide only. Any one value can vary by at least 50% in different wines. During the course of exposure to oxygen and aging, caftaric acid will disappear and anthocyanins are incorporated into polymers. The products may no longer be identifiable as such but are considered as derived from the original cinnamate or flavylium precursors.

Table 1. Gross estimated phenol content for typical young, light table wines ($mg\,l^{-1}$)

	White	Red
Nonflavonoids, total	165	200
Volatile phenols	1	5
Tyrosol	14	15
Gallic and other C_6C_1 acids	10	40
Caffeic acid and related cpds.	140	140
Hydrolyzable tannins	0 (no oak)	0 (+250 only after cooperage)
Flavonoids, total	35	1000
Catechins (flavan-3-ols)	25	75
Anthocyanins and derivative	0	400
Other monomeric flavonoids	Trace	25
Oligomeric flavonoids	5	500
Total phenols	200	1200

2 Determination of Specific Individual Phenolic Substances

2.1 General, Nonchromatographic Analyses

Wines, and other food products derived from plants, if they contain phenols, usually contain at least low levels of several classes of phenols and frequently several phenols in each of the classes. For example, most red grapes and young red wines contain glucosides and some acylated glucosides of the five anthocyanidins cyanidin, peonidin, delphinidin, petunidin and malvidin. In such a situation it is generally not possible to analyze for any one specific phenol unless it has been separated more or less completely from its fellows. Ordinarily, electronic spectra, colorimetric reactions, etc. applied to the unfractionated sample are not capable of determining individual phenols in such mixtures. Even modern computerized multicomponent spectral analyses are not adequate for the task. It is true that a few reactions, notably coupling with selected diazonium salts, can give different colored products from very similar phenols and could be used to determine individual phenols at least after partial segregation. No such analysis has been widely applied. For practical purposes, present analysis for one or more specific phenols in wines requires detailed separation usually by chromatography prior to the analysis itself. Any one phenol could be determined after an isolation procedure peculiar to it, but that could lead to separate techniques for each phenol.

2.2 High Performance Liquid Chromatography (HPLC) of Wine Phenols

Phenols, covering as they do a wide range of spectral characteristics for detection and of properties permitting separation, are nearly ideally suited to separation and analysis by HPLC. This technique, with the best of the available equipment, appears ultimately capable of determining every significant phenol, at least those below certain molecular weight limits, in a single "run" with a wine sample. By present thinking, this is likely to require microbore columns with high theoretical plate numbers, gradient elution, and simultaneous multiple detection such as complete diode array UV-visible spectra on each peak, coupled perhaps with additional detectors such as electrochemical ones. There are several formidable barriers to immediate exploitation of the full capabilities envisioned. Such equipment is expensive and touchy to operate, sample throughput would be slow, and data evaluation complex. A major obstacle is the fact that wine chromatograms show very large numbers of components, many not identified, and many no doubt not phenols. No one has yet published such a complete and detailed HPLC (or any other kind) of analysis of all the individual phenols of a wine. Reviews of HPLC application include those by Daigle and Conkerton (1983), Evans (1983), Hardin and Stutte (1984) and Nagel (1985).

HPLC, nevertheless, is very useful in analysis of wine phenols and promises to become much more so. It is the method of choice for individual nonvolatile phenols to be determined with minimal wine sample pretreatment. Unless problems of solubility, reactivity or lability intervene, it appears possible to design an HPLC procedure capable of separating and quantitating any single phenol among similar compounds. Even cis-trans positional and diastereomeric optical isomers of the same phenolic structure are generally easily separated and determined. For most reliable quantitation, a known standard sample of each phenol to be determined must be chromatographed for comparison under identical conditions.

Owing either to concentrations present of desired phenols or to numbers and amounts of undesired phenols or extraneous substances, a preliminary treatment of the sample is often necessary. The sample for injection needs to be free of particulate matter or the column will soon plug. Alcohol and other nonaqueous solvents (for reversed phase systems) need to be removed by low temperature vacuum or other suitable treatment, and the pH needs to be acidic to ensure that the phenols are in uniformly protonated rather than mixed ionic form. Conditions for HPLC of different phenols usually need optimization for analysis, at least of those eluting very near to each other or to potentially interfering substances. Retention times and completeness of close separations are easily affected even if chromatographic conditions are reproduced as closely as possible. As a consequence, it seems most useful to describe general features as a point of departure for HPLC of wine phenols. The column is most commonly packed with uniform, spherical, porous beadlets (usually 5 or sometimes 10 μm in diameter) of silica gel coated completely with a covalently bonded stationary phase such as C_{18} or C_8 alkyl chains linked via oxygen bridges to the silica. On this reversed-phase, hydrocarbon-like, stationary surface the phenols with the most hydrophobic portions are held the tightest and elute the slowest. Developing (mobile) solvent systems that are usually adequate begin with the sample and initial development in water to elute the most hydrophilic phenols. More hydrophobic phenols are eluted by increasing proportions (isocratic, gradient, or step-gradient) of methanol or acetonitrile.

The elution order is not necessarily that expected from overall molecular polarity because an area of the molecule may govern its hydrophobic affinity with the stationary C_{18} surface regardless of polar units in other portions of the molecule. It is not uncommon to find a phenolic aglycone eluting fairly near its glycoside and well away from another different aglycone-glycoside pair where the aglycones differ considerably in polarity. The sequence of elution should remain constant from a given type of column (stationary phase) with a single gradient pair of solvents in mid-range. It does not always do so with the two single solvents compared with each other or with another gradient pair. The degree of separation and order of elution for closely similar phenols may be changed in aqueous acetonitrile compared to aqueous methanol. Some optimization under your conditions is frequently necessary and for close separations may need to be updated if conditions change – such as a new column.

The eluted peaks from an analytical HPLC column (commonly about 4×100 mm) are ordinarily too dilute for collection and further operations. Analysis consists of detection and peak area measurement directly on the effluent stream usually by absorbance near the wavelength of the most unique maximum of the desired phenol. Comparisons with peak areas of the known substance at a series of appropriate dilutions under identical conditions enable reliable quantitation. Useful approximate detector wavelengths are 280 nm for flavanols and simple phenols without further conjugation, 290 nm for flavanonols, 320 nm for hydroxycinnamates, 340 nm for flavones, 360 nm for flavonols, and 520 nm for anthocyanins. The more wavelengths that can be monitored simultaneously, the more confidence one can have of purity and proper identification of each phenol (peak). Electrochemical detection can also be very useful since it can distinguish, for example, between phenols with free vicinal dihydroxyls, catechols, and other less readily oxidized substitution patterns (Roston and Kissinger 1981).

Other references might be cited that use HPLC to separate members of specific groups of phenols, but a few will suffice to give fairly general results covering some phenolic acids and some flavonoids. Among 124 or more peaks separated by HPLC from white wines seven phenols were identified (Singleton and Trousdale 1983). On three different reversed phase (one C_{18} and two C_8) columns developed with a 0.1% formic acid and a water to 65% methanol gradient, the elution sequence was the same but retention times differed. Coefficients of variation were 1–2% for retention time in repeated analyses and 1.2–8.9% for peak areas. As long as peaks were well separated and absorbance appreciable, the amounts calculated from absorbance at 280, 292 or 320 nm were fairly constant even if well away from the maximum absorbance for a compound. Wine phenol composition was most affected by pomace contact but also was considerably affected by grape variety and less so by vintage. Retention times on a C_8 column were about 11 min for gallic acid, 28 for caftaric, 33 catechin, 38 coutaric, 39 epicatechin, 57 dihydroquercetin rhamnoside, and 61 min for dihydrokaempferol rhamnoside. On a C_{18} column the first four were faster and epicatechin slower to elute. With a C_{18} column developed with 5% acetic acid in water, Wulf and Nagel (1976) found relative retention times of about 2 min for gallic acid, 3 for protocatechuic, 4 for 4-hydroxybenzoic, 6 for catechin, 8 min for caffeic and 24 min for ferulic acid. In a methanol-water-acetic acid system flavones were much retarded compared to flavanols with similar substitution. Quercetin-3-rhamnoside and myricetin were very close together, indicating that one more B-ring phenolic hydroxyl had about the same effect on effective polarity as did rhamnose attachment to the 3-position. Salagoity-Auguste and Bertrand (1984), with an unthermostated C_{18} column and a water at pH 2.5 to methanol gradient, analyzed ethyl acetate extracts of wines made at pH 7 for neutral phenols and at pH 2 for phenolic acids. Eight phenolic acids totaled 8–15 mg l^{-1} in three red wines, gallic and syringic being predominant. Tyrosol was about 4 mg l^{-1} and eight identified flavonoids totaled 45–97 mg l^{-1}. Extraction efficiency was 32–98%, the reproducibility coefficients of repeated analyses were 4–34%, depending on the compound, and recovery of the phenols added to wine was of the order of 20% or better.

Lea (1982) has shown that phenolic carboxy acids can be moved rapidly through the HPLC column with improved net separation of the other phenols by a brief initial period of pH above 5.5 obtained by a trace of ammonia in the eluant. Furthermore, a steep gradient moved polymeric procyanidins off the column whereas slow gradients were unsuccessful.

2.3 Paper Chromatography of Wine Phenols

Considered by some to be obsolete, paper chromatography is still occasionally quite useful for preparing large two-dimensional "maps" of the qualitative presence and approximate relative amounts of phenols in extracts from grapes or wines. It also can be useful to isolate a small amount of or check identity of a given phenol. It is not a very useful quantitative method because of the variability inherent in methods of visualization and densitometry of the "spots". Probably the most useful single system for wine phenols remains Whatman No. 1 or 2 paper and liquid-liquid partition development with the upper phase of freshly prepared (to avoid production of butyl acetate) *n*-butanol : acetic : water at volume proportions of 4 : 1 : 5. Separations depend on relative polarity with glycosides being slower than aglycones (Roberts 1956; Mabry et al. 1970).

The dried chromatogram of a single sample can usefully be developed in the second (90° rotation) dimension with dilute, usually 2 or 6%, acetic acid giving separations based on hydrogen bonding between the phenols and the cellulose. Aglycones generally move more slowly than glycosides and more slowly the more free phenolic hydroxyls or the less soluble in water. The most sensitive detection is usually done by spraying the dried chromatogram with a freshly mixed solution of equal parts 1% potassium ferricyanide and 1% ferric chloride in 0.1 N HCl. Prussian blue spots are produced from most phenolic and other easily oxidized substances with about 5 μg/spot or more. Washed with dilute hydrochloric acid to remove excess reagent and with water to remove the acid, the dried chromatograms are permanent. Other sprays are more selective and generally less sensitive by an order of magnitude. There are many useful data on qualitative phenol identification and semi-quantitative phenol content by such methods, see for examples Roberts (1956); Singleton et al. (1966); Markh and Zykina (1969); Ribéreau-Gayon (1972); and Nozaki and Yokotsuka (1985). If one is limited by available equipment, making preliminary experiments, making wide surveys, or comparing with earlier data, the technique has value.

2.4 Thin Layer Chromatography (TLC) for Wine Phenol Analysis

TLC is faster in development of the chromatograms, adaptable to many adsorptive supports other than cellulose, and usually gives more discrete spots than paper chromatography. Otherwise, it shares many of the disadvantages as a quantitative technique and is more demanding, especially if you prepare your own plates, and, especially if plates are purchased, more expensive. As examples of specific analyses of wines by TLC Cabezudo et al. (1971) identified salicylic and *p*-hydroxybenzoic acids added to wines as preservatives at 2 mg l^{-1} or more by TLC on polyamide layers and benzene with 4% acetic acid development. Miskov et al. (1970) quantitated a number of phenols from ethyl acetate extracts of wine by TLC on silica gel with solvents such as toluene:ethyl acetate:formic acid 5:4:1 with ±13.5% precision. Differences in grape leaves in phenolic acids and esculetin were studied by Revilla et al. (1986b). High performance TLC was used by Lea et al. (1979) to identify and characterize individual polymeric procyanidins.

2.5 Gas-Liquid Partition Chromatography (GLC) of Wine Phenols

At least for those laboratories well equipped for modern GLC, it is the method of choice for individual volatile phenols. The amount of easily volatile phenols in wine is small and their estimation has been included with general analysis of the volatile odorants of wines (see for examples Schreier and Drawert 1977; Nykänen and Suomalainen 1983; Williams et al. 1983).

Tyrosol and other phenols of lower volatility, including ferulic acid and ethyl caffeate have been identified in wine extracts and could be quantitated by high resolution capillary GLC programmed at 2° C min^{-1} linearly from 40 to 250° C (Güntert et al. 1986). With a coupled mass spectrometer, the compounds are positively identified as they are detected. On the other hand, there are no detection methods specific for phenols in GLC.

Even less-volatile phenols can be analyzed by GLC by converting them (and other compounds such as sugars) to trimethysilyl derivatives. Christensen and Caputi (1968) produced by this technique very complex chromatograms from red wines that displayed different patterns for wines from different grape varieties. They determined ten phenolic acids and five flavonoids in eight red wines. Valouiko et al. (1980) analyzed wines and distillates by a similar silation GLC method. Drawert and Leupold by GLC of silyl derivatives, successfully quantitated anthocyanins in red wines through malvidin-3,5-diglucoside (1976a) and other phenols in Tokay wines through kaempferol-3-rhamnoglucoside (1976b). The silation and GLC method has not been more generally accepted at least partly for fear that polyphenols might not easily and completely silate to yield a single peak from each original phenol.

Given the importance of nonvolatile polymers and glycosidic phenols in wines and the greater opportunity for more selective determination by HPLC, GLC seems likely to be preferred only in special instances.

2.6 Low Pressure Column Chromatography

Ordinarily there are insufficient theoretical plates in simple columns packed with adsorbents, gel filtration materials, partition supports, or ion exchange or other resins to easily separate individual phenols from complex mixtures such as wines. With sufficient experimentation and rechromatography, isolation and therefore estimation of a specific major phenol or group of phenols is usually possible. Such columns are often most useful for preliminary concentration and partial separation prior to analysis by HPLC or another technique.

Somers (1968) used dextran gel columns to produce a reproducible analytical profile of monomeric nonacylated, monomeric acylated, and polymeric pigments in red wine. Glories and Ribéreau-Gayon (1973) similarly studied the state of condensation of red wine tannins and applied further analyses to the fractions obtained. Yokotsuka et al. (1980) separated tyrosol, seven phenolic acids and a number of fractions of wine tannin oligomers by chromatography on a polyacrylamide gel. Bourzeix (1978) estimated the content of yellow, brown, and red polyphenols in musts and wines by a combination of TLC and column chromatography on polyvinylpolypyrrolidone.

Probably the most useful low pressure column packing for phenol is the dextran modified by hydroxypropylation, Sephadex LH-20. This material has the advantage of acting primarily as a hydrogen bond acceptor with phenols and in the range of molecular sizes for phenols found in wine has little exclusion effect. Wine phenols can be eluted usually quantitatively with increasing content of methanol or of acetone (to about 60% in water maximum or the gel is partly dissolved). Phenolic acids such as coutaric and caftaric acids can be separated including their cis-trans isomers and the separation is improved if 0.2% acetic acid is added to the developing water. This acidity suppresses carboxyl ionization, improves peak symmetry, and allows H-bonding by the carboxyl hydrogen (Singleton et al. 1977, 1978). Sephadex LH-20 holds flavonoids more strongly, generally in order of the number of phenolic hydroxyls per molecule, and can be used to concentrate them from wine (10–12% ethanol). Lea et al. (1979) used Sephadex LH-20 chromatography along with counter-current distribution, TLC, high performance TLC, paper chromatography, colorimetry, and hydrolytic breakdown to characterize and determine the flavanols and procyanidins of white wine. The largest oligomers eluted last with the highest solvent content.

Initial chromatography on Sephadex LH-20 with a water to methanol gradient can give useful segregation prior to HPLC and this is visualized as its greatest permanent usefulness. Peaks monitored by ultraviolet absorbance from such a chromatogram are broad and seldom a single substance, but they can be shown by rechromatography to be purified of earlier and later compounds. Application of this peak's group of compounds to reversed phase HPLC generally gives a much simpler family of HPLC peaks, but because the procedures operate by quite different mechanisms (H-bonding vs. hydrophobic attraction), still cover a wide range of HPLC retention times. Combinations of this type are envisioned as a time-saving compromise between ultimately complete direct analysis by HPLC and cruder determinations.

3 Estimation of Phenols of Wines by Groups, Phenols Other than Flavonoids

The distinction between methods of analysis of individual phenols, analysis by groups, or by total phenolic content can be rather arbitrary. For example, restriction of the detection wavelength for HPLC to the red (520 nm absorbance) pigments clearly is an individual analysis, but the group may be analyzed by summing the individual compounds. On the other hand, spectrophotometry without separation would give total red pigments expressed in some general units without specific information on any one pigment. A group separation procedure such as distillation to separate volatile phenols could be followed by a method for total phenols in the distillate and yet the determination would be for the group. Furthermore, the groupings can be of more than one type such as structural class (e.g., flavones versus flavonols), functional groupings (monophenols, vicinal diphenols, vicinal triphenols, meta diphenols, etc.), or property (visible, brownable, precipitable, etc.). The discussion in this section will generally follow the structural class as far as possible. Methods that might be but have not been significantly applied to grapes or wines generally will be omitted.

3.1 "Nonflavonoids" – Phenols Other than Flavonoid Derivatives in Wines (Fig. 1)

Aldehydes, especially formaldehyde, react with phenols to produce large, polymeric, insoluble resins, the basis of the first "Bakelite" plastics. If the pH is sufficiently low, only highly nucleophilic phenols react quickly in solution at room temperature. The meta-hydroxylated A-rings of flavonoids have doubly activated electron-rich centers at carbons 6 and 8. Formaldehyde forms first methylol substitution at these sites then cross-links to make a methylene bridge between two flavonoid A-rings. The reaction proceeds and soon the polymer precipitates from solution (Fig. 5). This is the basis for the Stiasny test to distinguish condensed (flavonoid) tannins from hydrolyzable ones (gallo- and ellagitannins). Hillis and Urbach (1959) demonstrated that with catechin only the A-ring reacted in strong acid. They attempted unsuccessfully to dry the precipitate for gravimetry.

Kramling and Singleton (1969) developed a successful quantitative method for the nonflavonoid phenols by determining total phenol content remaining after precipitating the flavonoids with acidic formaldehyde. The method was applied to estimating the phenols contributed to wine by oak, redwood or cork contact, since such contributions were essentially all nonflavonoid (Singleton et al. 1971). Further study verified that the conditions employed precipitated little or none of most phenols considered likely to occur in wine or other foods if they were not flavonoids and precipitated them essentially completely if they were flavonoids or other phloroglucinol derivatives (Singleton 1974). The method is quite useful, but not without difficulties. It has the inherent problems of the method used for total phenols. The Folin-Ciocalteu method (Singleton and Rossi 1965) has been used and is not affected by the aldehyde.

Two conditions interfere with flavonoid precipitation and lead to falsely high results: (1) in many white wines there is insufficient flavonoid for precipitation

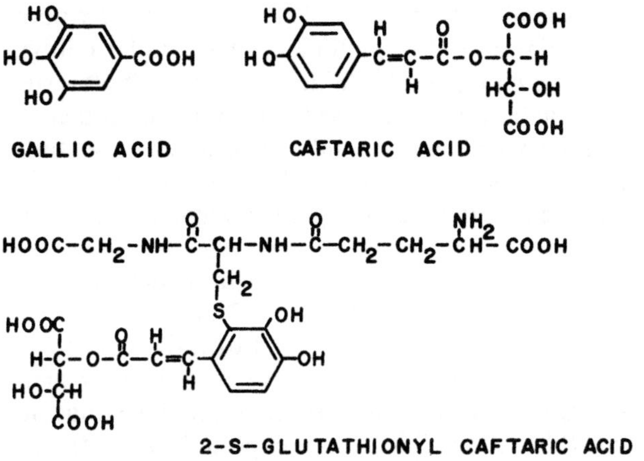

Fig. 1. Examples of important nonflavonoid phenols of wine. Coutaric acid = 3-deoxycaftaric acid; fertaric = 3-methoxy caftaric acid

and it remains suspended as a colloidal haze (2) anthocyanin glucosides react with the formaldehyde but, unless cross-linked with sufficient other flavonoids without sugars, do not become insoluble. Both problems have been attacked by adding phloroglucinol. It is essentially completely precipitated itself, so it does not affect the analysis but does provide sufficient bulk and cross-linking to ensure precipitation in the cases described. There usually remains a faint pink equivalent to perhaps 4 mg l^{-1} of anthocyanin from very young red wines with either low tannin or very high color. Formaldehyde precipitates from phloroglucinol or flavonoids have a slight solubility in the reagents, equivalent usually to about 5 mg l^{-1} gallic acid. If ethanol or formaldehyde content is too high, this solubility is increased and the precipitate can be dissolved in 95% ethanol. If too great an excess of phloroglucinol is added, a small amount of nonflavonoids can be adsorbed on the precipitate. If attention is paid to these considerations, the method has been reproducible, $\pm 2.5\%$, and informative with good recoveries and discriminations with knowns added to wines.

The precipitation is ordinarily carried out by mixing 10 ml of the wine with 2 ml of concentrated hydrochloric acid and 5 ml of 120 g l^{-1} aqueous formaldehyde solution. The solution is allowed to react under nitrogen at room temperature for 2–3 h and then 3 ml of 10 g l^{-1} phloroglucinol for red wines (1 ml for white wines plus 2 ml H_2O) is mixed in and the reaction continued under N_2 at room temperature for a total of about 24 h. The supernatant (centrifuged or decanted) is filtered through a 0.45 μ membrane filter if not absolutely clear. The total phenol content of the original wine also is determined (Singleton and Rossi 1965). Since all wines have about the same content of nonflavonoid phenols falling in the appropriate range, no dilutions are required, but calculation must allow for the one-half dilution by the reagents. The amount of phloroglucinol can be adjusted if it is unnecessary (old reds), insufficient (inky young reds) or excessive (light white), but the values stated have been satisfactory in nearly all cases.

Reaction with acetaldehyde instead of formaldehyde would avoid any additional hazard to the analyst by the latter and in preliminary tests molar equivalents of either aldehyde have been nearly equal. Acetaldehyde has not been tested as thoroughly, however. Slower reaction and increased solvent effect appear to be of some concern. Direct colorimetry with acidic vanillin is seen to be essentially the same reaction, but in contrast to formaldehyde vanillin does not react with deactivated A-rings like quercetin and is affected by degree of polymerization (see Sects. 4.3.2, 4.3.4).

The value obtained for the nonflavonoid phenols of wines had not been readily explainable in terms of the known phenols present (Myers and Singleton 1979). A major part of this confusion has now been clarified by the discovery that, unless thoroughly protected from enzymic oxidation, caftaric and coutaric acids convert to a high degree in most grapes during must processing to 2-S-glutathionyl caftaric acid (Singleton et al. 1984, 1985). That this compound can hydrolyze to a series of other derivatives and that other sulfhydryl compounds give similar products explains the disappearance of caftaric and coutaric acids, increased complexity of the chromatograms, and the production of very polar hydroxycinnamate derivatives (Cheynier et al. 1986).

3.1.1 C_6, C_6C_1–C_6C_3 Phenols of Wines, Other than Carboxylic Acids

Several phenols fitting this category have been identified in wines including vanillin, syringaldehyde, tyrosol, 4-vinylguaiacol, acetovanillone, eugenol, and 4-ethylphenol. In part they appear to be degradation products of other phenols and to increase in aged wines. They are easily extractable and relatively volatile. The volatile and steam-volatile phenols probably include some of the phenolic acids and particularly their ester derivatives. They total about 1–11 mg l^{-1} for white and up to about 40 mg l^{-1} for aged red wines (Singleton and Noble 1976). They are generally determined individually via gas chromatography as already outlined or by total phenol assay on the distillate. Individually, with the exception of tyrosol, they are well below 1 mg l^{-1} in a typical young red wine (Etievant 1981). Owing to the possibility of fragmentation, prolonged distillation may be producing artifacts.

3.1.2 Phenolic Acids and Their Derivatives

The group of nonflavonoid phenols major in amount in wines is the hydroxycinnamates, although there are also small amounts of gallic acid and a few other hydroxybenzoates. Esters and lactones (coumarins) from these compounds are more volatile and have been found in small amounts in some wines by chromatography, particularly GLC. The phenolic acids themselves can be separated as a group from phenols without carboxyl groups by differential extraction or chromatography. At pH about 5 the carboxyl is ionized and either remains in the water phase during extraction with solvents such as ethyl acetate or rapidly passes through chromatographic columns including HPLC and Sephadex LH-20 that depend on low polarity or protonated carboxyls for appreciable retention. Lowering the pH to 2 or lower reverses the situation and the phenolic acids can now be manipulated or analyzed in purified form by rechromatography or extraction and then chromatography. Conversely, raising the pH to about 9 would produce phenolates and allow such manipulation for phenols without carboxyls. In alkaline solution, however, phenols are so readily oxidized by air that this is rarely satisfactory as a preliminary to analysis.

The quantitatively most important members of this group of compounds, particularly caftaric, coutaric, and gallic acids, are generally determined individually by HPLC under conditions that allow separate determination of the cinnamates' cis-trans forms and derivatives. The hydroxycinnamates are emphasized by using detection near 320 nm and the benzoates near 280 nm (gallic 265 nm max). Examples, all using reversed phase HPLC similar to that already outlined, include Singleton et al. (1984, 1985) and Okamura and Watanabe (1981), who used direct injection of wines and musts and others who used some preliminary extraction procedure (Ong and Nagel 1978; Symonds 1978; Garcia Barosso et al. 1983, 1986; Romeyer et al. 1983; Estrella et al. 1986). The values for individual phenolic acids reported by Estrella et al. (1986) on sherries appear generally much too high compared to those by others and presumably represent mg l^{-1} of extract (1 ml per 100 ml wine) rather than the wine itself. Caffeic acid does, but caftaric acid, much less glutathione-substituted caftaric acid (Fig. 1), does not extract readily into ether or even ethyl acetate (Singleton et al. 1984).

A group analysis for hydroxycinnamates without separation can be based upon their similar absorbance maxima near 320 nm. Somers and Ziemelis (1985 b) report that all the phenols can be removed by treating 5 ml of juice or wine with 500 mg of powdered insoluble polyvinylpyrrolidone (PVP, Polyclar AT) for 30 min, centrifugation and membrane filtration. The solution is then considered to exhibit the spectrum of the nonphenolic interfering substances and its 320 nm absorbance can be subtracted from that of the untreated sample to give the hydroxycinnamate absorbance of the sample. On 230 commercial white wines the residual absorbance after PVP was 1.37 ± 0.31 and absorbance at 320 nm (measured in 1 mm cell to avoid dilution but calculated to 10 nm) minus 1.4 is considered a good measure of hydroxycinnamate content convertible via the extinction of caffeic acid (0.90 absorbance, 10 mm path at 10 mg l^{-1} in model wine) into apparent caffeic acid mg l^{-1}. Keeping in mind the assumptions (PVP adsorbs all of the phenols, it adsorbs no other species absorbing in the UV, the nonphenolic UV absorbance is constant, all hydroxycinnamates are the same in molar absorbance at 320, etc.) the method is a considerable improvement over previous ones. The commercial white wines had, on this basis, 17–130 mg l^{-1} as caffeic acid averaging 64 mg l^{-1}, or 110 mg l^{-1} as caftaric acid. This agrees well with the amounts determined elsewhere by HPLC on commercial musts and wines, subjected as they have been to variable enzymic oxidation.

4 Flavonoids in Grapes and Wines

Catechins (flavan-3-ols), condensed (flavonoid) tannins, flavonols, flavanones, flavanonols, flavenes, chalcones, and anthocyanins have all been reported in grapes or wines and their analyses will be summarized below. A well-validated method for determining these flavonoids as a general group is the difference between total phenols and nonflavonoid phenols after precipitation with acid formaldehyde (Kramling and Singleton 1969; Singleton and Rossi 1965). As a method for nonflavonoids this has been discussed and more will be said under methods for total phenols. Since the value for flavonoid content is the difference between the two analyses, nonflavonoids and total phenols, the coefficient of variability is increased to about 7%. The flavonoid content of wines from clarified juice is very low and after pomace fermentation can be very high as described earlier.

Somers and Ziemelis (1985b) proposed an ultraviolet absorbance method for flavonoids based on absorbance at 280 nm (10 mm cell) minus 4 (correction based on typical 280 nm absorbance after 10 g per 100 ml PVP treatment to remove all phenols) minus $^{2}/_{3}$ of the difference value absorbance at 320 nm minus 1.4 (correction for the contribution to 280 nm absorbance by hydroxycinnamates). On this basis wines from juice generally had no flavonoids and white wines ranged from 0 to 220 mg l^{-1} calculated as catechin. Of course, the procedure neglects phenolics such as tyrosol and gallic acid which are present in wines, absorb maximally near 280 nm and are not flavonoids.

4.1 Anthocyanins, the Most Important Phenols in Red Wines

The anthocyanin derivatives found in grapes include no unusual ones and all the common ones except pelargonidin, i.e., cyanidin, peonidin, delphinidin, petunidin, and malvidin (Singleton and Esau 1969; Ribéreau-Gayon 1982) (Fig. 2). The anthocyanidins do not occur free (and if produced by hydrolysis are unstable) but are glucosides in grapes or wines. *Vitis vinifera* grapes and wines have only the 3-glucosides, whereas most other species and their red varieties have in addition or instead the 3,5-diglucosides. Furthermore, with the exception of a few of the red varieties, notably Pinot noir, the pigments are partially acylated (on the 6-hydroxyl of the 3-linked glucose) by acetic acid, *p*-coumaric acid, or caffeic acid.

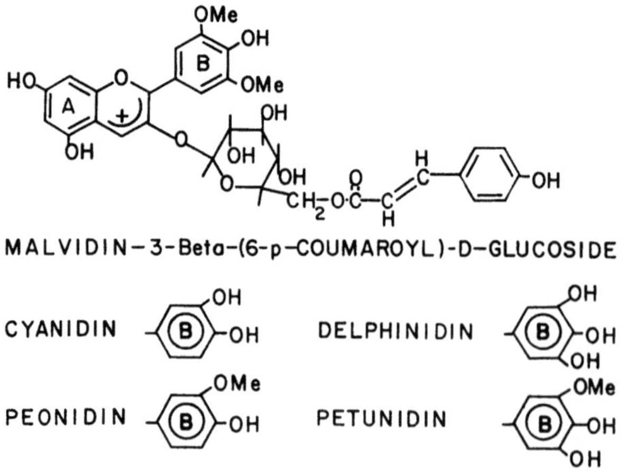

Fig. 2. Anthocyanins of *vinifera* grapes and their young red wines

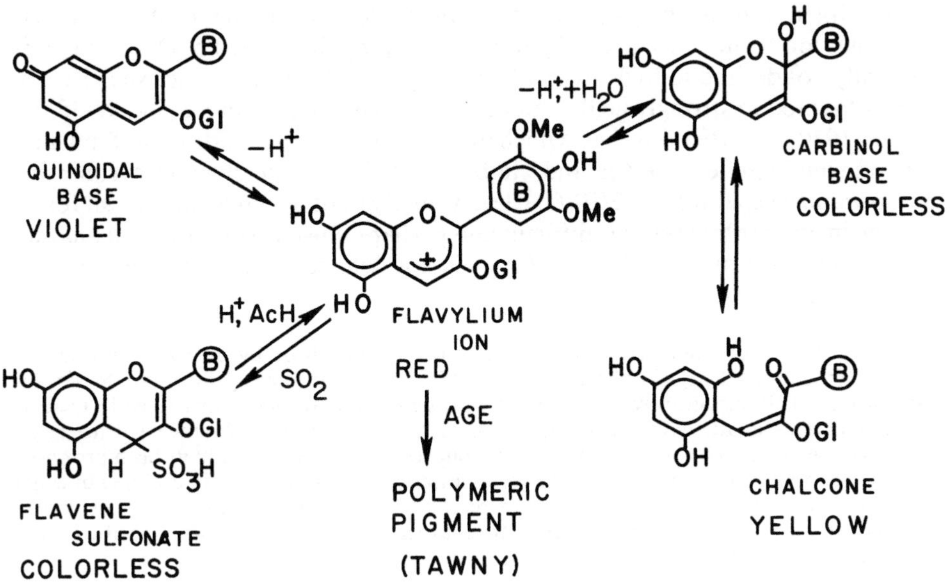

Fig. 3. Anthocyanin equilibria and reactions

If every pigment of a nomal red *vinifera* grape was present as the free glucoside and each of the three common acyl forms there would be a total of 20 anthocyanins present. *V. labrusca* and others with both mono- and diglucosides could have 40 or more individual pigments. Among individual grape varieties a wide range of differences exists, although the pattern for any one variety is relatively constant.

In wines at pH 3.5 or so the situation is complicated by the fact that the flavylium ion (red) forms of the anthocyanins are in equilibria with a little of the more purple anhydrobase and a lot of the colorless carbinol base. The carbinol is also in an equilibrium interconversion with the open-ring (yellow) chalcone form (Fig. 3). At wine pH only about ¼ or less of total potential anthocyanin is in the flavylium ion (red) form with a maximum near 520 nm. A second major complication in wines is that rather quickly during wine production and storage an increasing portion of the red pigment is incorporated into polymers. These polymers have greater absorbance near 420 nm than the monomeric anthocyanins and are relatively amber-red. The third major complication is that the color exhibited by anthocyanins in solution is affected by their own concentration, by temperature (Ohta et al. 1983), and by interaction and copigmentation with other compounds present (e.g., Timberlake and Bridle 1976, 1977). Sulfur dioxide bleaches them; catechin, certain other flavonoids, and acetaldehyde increase the color. Because of these factors the visible color of an undiluted wine and its description by tristimulus colorimetry or other means is another important area that cannot be discussed here. Only methods for the apparent content of individual, free, polymeric and total anthocyanins will be considered.

The intent during analysis of anthocyanins is to measure them entirely in their flavylium ion form. This is usually done by keeping the pH <0.5. Of course, it is generally not desired to hydrolyze off acyl or glucoside groups. Since such hydrolysis is favored in strongly acid solution, artifactual changes are a risk. Wulf and Nagel (1980) reported flavenes, which are capable of being oxidized to flavylium ions, in wine samples. Preston and Timberlake (1981) found chalcones from wine anthocyanins separable by HPLC. Clearly, analysis of anthocyanins in all forms in wine in an absolute sense is difficult and perhaps illusive, since the exact nature and permanence of the polymeric forms is not defined, and other equilibria are rapid and complex.

Reviews considering analyses for anthocanins include those by Bourzeix and Saquet (1974), Hrazdina (1979), Timberlake and Bridle (1980), Francis (1982) and Ribéreau-Gayon (1982). Polarography has been used to determine anthocyanins directly in wine without pretreatment, with results agreeing with other methods (Mareca Cortez et al. 1980). A few other techniques, such as electrophoresis or fluorescence, have been proposed but not adopted. For the most part, however, anthocyanins have been determined in wines by either spectrophotometry in the visible region without fractionation or by chromatographic separation, recently HPLC. Since these are monitored near 520 nm, specific for anthocyanin derivatives in wine, they become group analyses.

4.1.1 Anthocyanins by Chromatography

Fractions by low pressure chromatography give useful subgroupings and can be combined and repeated to give more complete separations. Hrazdina and Franzese (1974) and Hrazdina (1975) isolated and weighed up to 20 different pigments from *V. labrusca* varieties by combinations of columns of polyamide and polyvinylpyrrolidone. The procedure is too cumbersome for routine analysis, but useful data and spectral characteristics were obtained. Separation to indicate various proportions of free anthocyanins, acylated anthocyanins, or polymeric anthocyanins can be accomplished on dextran gels (Somers 1968), on polyvinylpyrrolidone (Bourzeix et al. 1982), polyamide (Berg 1963) and related adsorbents or combinations (Astegiano and Ciolfi 1974; Moutounet and Chudzikiewicz 1980; Zloch 1985). The best use of these methods appears to be to separate the monomeric anthocyanins from the polymeric. The monomeric can be readily eluted with dilute acidic aqueous alcohol from most adsorbents, whereas the polymeric forms do not elute under these conditions, if at all. The monomeric anthocyanins can then be determined by visible spectrophotometry. The polymeric forms will be further considered under tannins. After the first few months of a wine's life, the polymeric anthocyanins are likely to be more than $^2/_3$ of the total red pigment.

Earlier chromatography tended to list the separated anthocyanins as percentages of the recovered color or total absorbance. Astegiano and Ciolfi (1974), for example, showed via prior concentration and hydrolysis and TLC that very young red wines from different grape varieties could be grouped by their relative content of cyanidin versus delphinidin derivatives. Merlot gave about 50% malvidin, 15% each delphinidin and petunidin, 13% peonidin, and 8% cyanidin.

Earliest adaptations of HPLC showed that the method was capable of separating the grape and wine anthocyanins, progressed to percentages of peak area

Table 2. Anthocyanin analyses by HPLC on *vinifera* grapes or young wines

	Cabernet S. Wulf and Nagel (1978)	*Cavernet S.* Nagel and Wulf (1979)	*Merlot* Nagel and Wulf (1979)	*Cabernet S.* McCloskey and Yengoyan (1981)	*Sirah* Roggero et al. (1984)		*Souzao* Bakker and Timberlake (1985a, b)	
	Grape %	Wine $mg\,l^{-1}$		Wine %	Grapes %	Wine %	Grape (a) %	Wine (b) %
Cyanidin-3-glucoside	1.3	1.6	1.6	–	1.0	0.1	2	1
Peonidin-3-glucoside	5.3	10.7	11.7	6.0	8.3	3.5	6	7
Delphinidin-3-glucoside	10.0	30.8	19.8	5.6	5.2	1.2	8	8
Petunidin-3-glucoside	6.1	23.0	19.3	6.9	6.1	3.0	8	8
Malvidin-3-glucoside	42.6	119.6	105.4	43.1	36.0	62.0	57	64
Cya-3-gluc. acetate	0.1			–	0.2	–		
Peo-3-gluc. acetate	0.9	30.5	18.6	–	2.7	1.5		
Del-3-gluc. acetate	2.5			4.3	0.7	–		
Pet-3-gluc. acetate	2.2			tr	1.3	0.6		
Mal-3-gluc. acetate	20.5	63.0	29.4	25.0	9.4	16.5	4	5
Cya-3-gluc. coumarate	0.1			–	?	?		
Peo-3-gluc. coumarate	0.6	4.8	6.5	–	2.0	0.7		
Del-3-gluc. coumarate	0.5			–	1.0	0.2		
Pet-3-gluc. coumarate	0.4			–	–	–		
Mal-3-gluc. coumarate	6.4	9.7	8.7	9.1	21.9	10.2	15	8
Mal-3-gluc. caffeate	0.1			–	(+?) 2.4+	(+?)1.2		
Unknown	0.3			–	?	?		
Total %	99.8			100	99.2	100.9	96	96
Total $mg\,kg^{-1}$ or $mg\,l^{-1}$	–	294	221	–	1020	–	–	718
Total $mg\,l^{-1}$ (spectral)	–	322	207	–	–	–	–	–

compared to the total area of 520 nm absorbing material recovered, and most recently to mg l^{-1} of anthocyanins calculated based on either malvidin-3-glucoside or malvidin-3,5-diglucoside. The choice of reference standard is not trivial because the maxima and absorptivity can vary appreciably, but useful relative values are obtained regardless. For example, molar absorptivity of petunidin-3-glucoside acylated with *p*-coumaric acid was 20700 at 540 nm in methanol and the analogous acylated petunidin-3,5-diglucoside derivative was 37200 at 539 nm (Hrazdina and Franzese 1974). Table 2 presents anthocyanin values obtained by various workers on selected *vinifera* grapes and young wines (Wulf and Nagel 1978; Nagel and Wulf 1979; McCloskey and Yengoyan 1981; Bakker and Timberlake 1985 ab; Roggero et al. 1984, 1986). All were by reversed phase HPLC and, although other conditions varied, it does not seem useful to detail them except to note that formic acid and acetic acid in the presence of HCl produced troublesome artifacts by acylation (Bakker and Timberlake 1985a). As an example and perhaps the least complicated system, Bakker and Timberlake (1985a, b) used a 5 × 100 mm reversed phase column packed with Spherisorb-Hexyl (5 μm particles) eluted at 1 ml min^{-1} and 35° C with step gradients beginning with 20% of methanol in 0.6% $HClO_4$ in water and ending with 95% methanol 5% by volume aqueous 0.6% perchloric acid. Port wines were filtered and injected directly as were grape extracts diluted to $^1/_5$ in 0.6% perchloric at 20 μl per sample.

4.1.2 Total Anthocyanins by Spectrophotometry

To estimate anthocyanins in grape extracts or red wine is deceptively simple. Make the pH low enough, 0.5 to 0.8, to ensure that all the anthocyanins are in the flavylium ion form, determine absorbance at the visible maximum, 520 nm or so in water and about 540 in alcohol, and calculate the pigment concentration from the molar absorptivity of the predominant pigment in the same solvent. Some of the complications have already been mentioned. Malvidin-3-glucoside would be the best comparison standard for *vinifera* wines but generally must be isolated by the researcher. Niketić-Aleksić and Hrazdina (1972) report for malividin-3-glucoside, mol. wt. 529, molar absorptivity of 28,000. Malvin (malvidin-3,5-diglucoside, mol. wt. 691) is more readily available commercially. It had a molar absorptivity of 37,700. Fortunately, the increased extinction is almost balanced by the increased molecular weight and mg l^{-1} of anthocyanin should remain similar with either as standard, provided the correct figures are used.

The major problem with direct spectrophotometry on all but fresh grape extracts is the high proportion of red pigment that is polymeric and not anthocyanin per se. The total apparent anthocyan can be estimated but actual recoverable anthocyanin may be considerably less. On a weight basis the polymeric pigment is a large factor, but Niketić-Aleksić and Hrazdina (1972) calculated that it should contribute at most 100 mg l^{-1} and ordinarily considerably less to visible anthocyanin content.

The red of anthocyanin-incorporating polymers and the interference from tailing absorbance from brown substances has been compensated for by reactions that destroy the absorbance of the free anthocyanins and, hopefully, do not affect the interfering substances. Oxidation with hydrogen peroxide (Swain and Hillis

1959), pH shift effects and decolorization with SO_2 (Ribéreau-Gayon and Stonestreet 1965) have been applied, the latter two widely. In the pH shift method two tubes are prepared with 1.0 ml of wine (diluted if necessary) plus 1.0 ml of 0.1% concentrated HCl in 95% ethanol. To one tube is added 10.0 ml of 2% (v/v) concentrated HCl, pH 0.6, and to the other 10.0 ml of pH 3.5 buffer (303.5 ml 0.2 M Na_2HPO_4 plus 696.5 ml 0.1 M citric acid). The difference in absorbance at 520 nm, 1 cm cell, water reference, of the two is compared to a standard curve prepared from 375.0 mg l^{-1} of a known anthocyanin in 0.1% (v/v) conc. HCl in ethanol substitued for the wine or is calculated from the original observation (Ribéreau-Gayon and Stonestreet 1965) that 100 mg l^{-1} crystalline grape anthocyanin gave 0.260 absorbance difference. The standard solution 1.0 ml, is mixed with 1.0 ml of 10% (v/v) aqueous alcohol containing 5 g l^{-1} tartaric acid $^1/_3$ neutralized to simulate the wine situation.

The sulfite method is similar, uses the same standard and yielded 0.117 absorbance difference at 520 nm of the two final solutions in the presence of 100 mg l^{-1} of anthocyanin in the standard or wine. The two solutions are 10.0 ml each of 1.0 ml of wine plus 1.0 ml of 0.1% (v/v) conc. HCl in ethanol plus 20.0 ml of 2% (v/v) conc. HCl. To one add 4.0 ml of distilled water and to the other 4.0 ml of 15% $NaHSO_3$. Rather close agreement was obtained between the pH shift and sulfite methods on anthocyanin content of the same wines (Ribéreau-Gayon and Stonestreet 1965; Ribéreau-Gayon and Nedeltchev 1965). Agreement was also rather good with simple acidic spectrophotometry, except with older wines. The tendency was for the bisulfite method to give higher values than the pH shift method. Roson et al. (1978) found on 13 wines with about 300 mg l^{-1} of total anthocyanin that the bisulfite values averaged 28% higher than the pH shift values, with two direct acidic absorbance methods generally intermediate. All the methods paralleled each other, giving comparable relative values on this series of wines, but the pH shift method was somewhat more erratic. This difference between the sulfite and pH shift methods is apparently the result of variable SO_2 content in the wine affecting the pH shift method, and agreement is better in the presence of some SO_2 (Glories and Augustin 1980–1981).

The spectral methods of Somers and Evans (1977) are based upon the effect of sulfite on the anthocyanins (monomers) and relative lack of effect on the red polymers. Their system is the most widely used for total anthocyanins and the system can be used to determine other details, such as the percentage of the anthocyanins that are ionized in the wine, the fraction bound with SO_2, and estimation of the proportion of polymeric anthocyanins. For total anthocyanins, the procedure as published determines in a 10-mm cell the absorbance at 520 nm after 3 to 4 h of 100 μl of wine added to 10.0 ml of 1 M HCl multiplied by 101 (dilution). Separately determine the absorbance at 520 nm in a 1-mm path cell after 1 min of a mixture of 0.33 ml of wine plus 5 μl of freshly prepared 20% sodium metabisulfite solution. The anthocyanin in mg l^{-1} is calculated by subtracting 1.67 times the absorbance calculated to a 10-mm cell of the sulfite sample (to correct for the acid effect on the polymeric forms) from the absorbance of the acid solution and multiplying the difference by 20 (based on $E_{1\,cm}^{1\%}$ of 500 for malvidin-3-glucoside). The correction value is admittedly arbitrary and wines do vary in the apparently correct factor.

Bakker et al. (1986b) found total free anthocyanin measured by summing the HPLC peak areas and comparing with peak area absorbance at 520 nm from pure malvidin-3-glucoside chloride was considerably lower than anthocyanin content in wines measured by Somers and Evans' (1977) spectral method. They attributed this to the use of a higher molar absorptivity (28,000 vs. 26,455) and uncorrected partial bleaching of polymeric pigment forms by sulfite. The polymeric pigments were proportionately higher in the wines than indicated by the spectral method and were important earlier than expected. A port wine that had at fortification 737 mg l^{-1} of anthocyanins by the spectral method and 627 mg l^{-1} by HPLC had only 62 and 24 mg l^{-1} respectively 46 weeks later. It would appear that direct spectrophotometry gives useful values more easily than HPLC, but they must be viewed as approximate rather than absolute values.

4.1.3 Anthocyanidin Diglucosides

It was originally shown by Ribéreau-Gayon (1959) that the European wine grape does not produce diglucosidic anthocyanins, whereas other species do. This has been verified to be a dominant characteristic in crosses and extended so that the presence of diglucoside pigments in a red wine has become legally prohibited in several European countries as evidence of blending with wines from grapes not permitted. There is nothing wrong with the other grapes, but their wine is forbidden in some places for various economic, political, authenticity and pest control reasons. The analysis for diglucoside pigments is therefore of regulatory interest primarily on a qualitative basis. Methods have generally depended on paper chromatography in comparison with known malvin (malvidin-3,5-diglucoside). The diglucoside moves ahead of other pigments in aqueous solvents such as 0.6% citric acid. In ultraviolet light malvin produces a characteristic brick-red fluorescence (Ribéreau-Gayon 1982). With care and spectrophotometry of the eluted malvin spot addition of about 4 mg l^{-1} can be detected in a *vinifera* wine (Bourzeix and Heredia 1967). This would detect blending with a small percentage of young wine from hybrid or non-*vinifera* grapes. Malvin at higher levels can be also detected by a characteristic green fluorescence after reaction with sodium nitrite. Mix in order 1 ml of wine, 1 drop 1 N HCl, 1 ml 1% $NaNO_2$, wait 2 min, add 10 ml MeOH containing 5% (v/v) NH_4OH, and illuminate with a Woods lamp (Jouret 1967).

4.2 Flavones, Flavanones, Flavonols, Flavanonols (Fig. 4)

These compounds all have a 4-keto group and are not major phenols in grape berries or wines. Flavones have been reported in leaves of some grapes but not in fruit or wines (Revilla et al. 1985). The flavanones naringin and hesperetin have been reported in a few parts per million in grapes and wines (Drawert et al. 1980). The flavanonols dihydroquercetin and dihydrokaempferol in the form of their 3-rhamnosides have been found in grapes and wines at low concentration (Singleton and Trousdale 1983; Trousdale and Singleton 1983). Flavonols, especially quercetin and kaempferol, occur in grapes as glycosides and in wines partly as

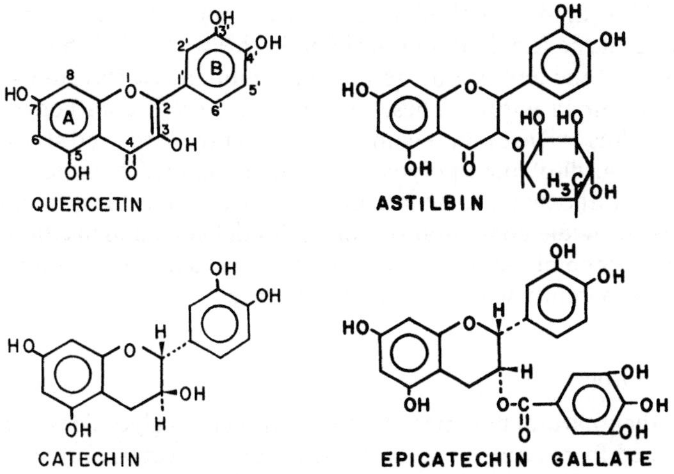

Fig. 4. Representative flavonoids from grapes and wines: flavonols (quercetin, kaempferol = 3′-deoxyquercetin, myricetin = 5′-hydroxyquercetin); flavan-3-ols [(+)-catechin, (−)-epicatechin-3-gallate]; flavanonols (astilbin = dihydroquercetin-3-rhamnoside, engeletin = 3′-deoxy astilbin)

aglycones only to perhaps 30 mg l^{-1}, but are relatively high in grape leaves (Revilla et al. 1985, 1986a). With all these substances, analyses depended upon chromatographic isolation and quantitation at 292–365 nm with a free 3-hydroxyl supplemented in the case of free flavonols with spectral shifts into the visible by aluminum chloride (Margheri et al. 1974; Somers and Ziemelis 1985a, b). The main analytical interest seems to be as means of detection of adulteration of wine with other fruit such as figs as indicated by flavones (Siewek et al. 1985) or contamination by leaves in mechanically harvested grapes sufficient to lead to flavonol hazes in the wine (Somers and Ziemelis 1985a).

4.3 Flavanols (Catechins) and Condensed Tannins (Fig. 4)

The flavan-3-ols of grapes include mainly (+)-catechin and (−)-epicatechin with usually small amounts of (+)-gallocatechin, (−)-epigallocatechin, and the 3-gallate esters of epicatechin and epigallocatechin (Singleton and Esau 1969; Czochanska et al. 1979). In wines, usually only the two main catechins are found in appreciable amount. Note that hydrolysis of the gallates is a source of free gallic acid in wine and that the gallocatechins are relatively unstable to oxidation.

Condensed, i.e., flavonoid, tannins, sometimes called flavolans, are dimers and largely made up of usually (−)-epicatechin (Fig. 4) repeating units linked covalently from their 4-carbon to the 8- or sometimes 6-carbon of another, usually terminating in a (+)-catechin unit. The resultant larger polyphenol is a tannin – precipitable with protein, capable of converting hide to leather – whereas the monomers are not. This polymeric structure when heated with acid breaks down to yield considerable cyanidin (dephinidin if gallocatechin units are present) and some catechins, at least the terminal unit. Originally this was called the leucoanthocyanin reaction, but the term is now reserved for monomeric flavan-3,4-diols that very seldom occur in nature. These tannins are therefore anthocyanogens or proanthocyanidins and are major components of grape seeds and skins. In the course of processing, ill-defined polymeric products lumped together as phlobaphenes, polymeric anthocyanins, etc. are produced from catechins, flavolans, and anthocyanins by acidic condensation, phenol-aldehyde condensation, oxidative polymerization, or combinations of these and other reactions, including some with proteins and apparently with large carbohydrates.

4.3.1 Catechins, Flavan-3-ols

These compounds and their natural polymers do not occur here as glycosides, are relatively readily extracted from aqueous solution by organic solvents, and have relatively nondescript ultraviolet absorbance spectra roughly equivalent to the sum of phloroglucinol and catechol, with a maximum near 280 nm. They are the major part of the "neutral" as opposed to carboxy acid, nonglycosidic, extractable phenols of grapes or wines. Their determination has been generally chromatographic or by colorimetry with acidic vanillin. Although the amounts of (+)-catechin and (−)-epicatechin are of the same order of magnitude, perhaps 300 mg kg^{-1} in fresh berries with seeds, the amounts reported in red wines by HPLC are usually about 100 mg l^{-1} or less each (e.g., Nagel and Wulf 1979) and in white wines from traces to about half that amount (e.g., Singleton and Trousdale 1983). After a fairly cumbersome polyamide and thin layer chromatographic separation, Berger and Herrmann (1971) determined content of individual catechins in vinegar stock wines by both direct spectrophotometry and by colorimetry with vanillin and obtained good agreement. Total monomeric catechins were about 40 mg l^{-1} in light white wines and 300 mg l^{-1} red. The difficulties of direct spectrophotometry in a mixture were pointed up by the fact that the catechin series had absorption maxima of 280 and the gallocatechins 271 nm, while the respective molar absorbances were 3600 and 1500 and for the gallates 14,000 and 11,500.

4.3.2 Colorimetry with Vanillin and Related-Aldehydes

Vanillin in mineral acid reacts with phloroglucinol and "undeactivated" flavonoids, i.e., those without a 4-keto group, to give a red color (Swain and Hillis 1959). Since catechins and their derivatives are the compounds in white wines that have this structure, the reaction has been used to estimate catechins (plus condensed tannins) in white wines (Rebelein 1965a, b; Pompei and Peri 1971). Note that the reaction is essentially the same as the formaldehyde precipitation of flavonoids, except that formaldehyde is more reactive and vanillin gives a colored product (Fig. 5). If the pH is not highly acidic additional phenols will react (Salem 1985). The reagent will react with resorcinol or pyrogallol but not with catechol or gallic acid (Goldstein and Swain 1963). It is not completely inert to 4-keto flavonoids (Sankar and Howarth 1976). It apparently reacts with oleanolic acid

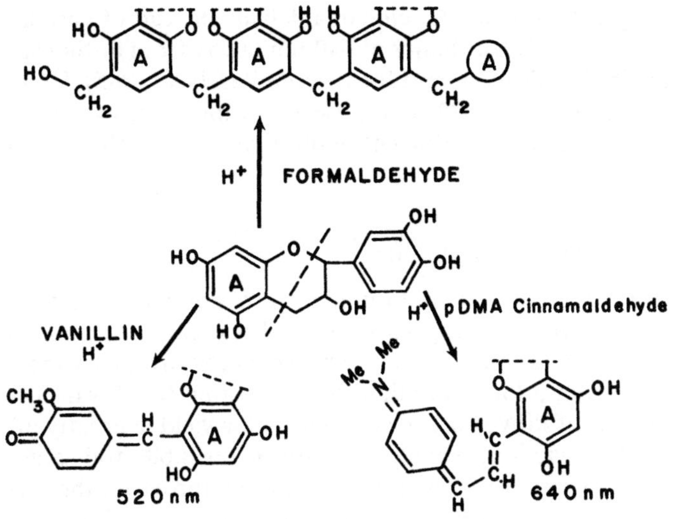

Fig. 5. In strongly acidic solution aldehydes react with the electron-rich 8- and 6-positions of the A-ring of flavans. Formaldehyde reacts with all and precipitates the polymerized flavonoids. Vanillin and *p*-dimethylaminocinnamaldehyde are less reactive (flavans without a 4-keto group react) and produce colored products

(Otsuka 1967); a terpenoidal nonphenolic and a similar reaction has been used to determine monoterpenes from muscat wines (Dimitriadis and Williams 1984).

Anthocyanins apparently react, at least the values are unrealistically high if applied without compensation to unfractionated red wines. This is confused by the fact that the vanillin reaction is measured at 490–500 nm and the acid used enhances the overlapping absorbance from anthocyanins. Furthermore, anthocyanogens may be partially converted to cyanidin by the acid and interfere at that wavelength.

The reaction can be useful, nevertheless, especially in white wines or after partial fractionation. The polymeric anthocyanogenic flavolans also react, but yield less color per catechin unit the larger the polymer. The most critical condition is the acid content of the reacting system – more acid gives more color and Beer's law is not followed in aqueous systems. Pompei and Peri (1971) recommend assay after preliminary extraction of the catechin with ethyl acetate (also applicable to red wines) or by diluting 1.0 ml of wine to 25.0 ml with absolute ethanol. (+)-Catechin standards at appropriate dilutions (0–300 mg l^{-1}) and 1% vanillin are freshly prepared in 96% ethanol. The alcoholic wine or standard solution, 10.0 ml, is mixed with 10.0 ml of 11.5 N HCl and 5.0 ml of vanillin reagent and the absorbance read at 500 nm versus the zero catechin blank after 30 min. Butler et al. (1982) recommend reaction in alcohol to emphasize polymeric flavolans over catechins because less color is produced from the monomers, but if the solvent is glacial acetic acid, oligomers react proportionally to the flavan-3-ol end groups. This has not yet been applied to wine.

A similar method using *p*-dimethylaminocinnamaldehyde instead of vanillin on beer analysis has advantages of absorbance at 640 nm (away from anthocyanins) and greater reproducibility (Delcour and Janssens de Varebeke 1985). However, it gave values that were only about 20% of those with vanillin, has not been reported on wines and the correlation coefficient in different beers with vanillin assays was only +0.6.

4.3.3 Anthocyanogenic Flavolans, Procyanidins

Since monomeric leucoanthocyanidins are not found in grapes or wine, it follows that the anthocyanogens are dimers or larger flavolans. These can be estimated by their conversion to cyanidin and comparison by absorbance with either known cyanidin or the color yield from a known anthocyanogen, often the flavan-3,4-diol synthesized by reduction of dihydroquercetin. The color yield is greatly affected by the conditions of the reaction, cyanidin is relatively unstable, and values obtained are necessarily somewhat arbitrary. It is difficult to improve on the earliest analyses of Masquelier et al. (1959), who subtracted pre-existing anthocyanin color and compared the color generated by heating in strong HCl with that produced by a given amount of pure leucocyanidin under the same conditions. They found 0–93 mg l^{-1} in white wines and 1100–4160 mg l^{-1} in red wines. Of this red wine value 17–66% was larger molecules (precipitable by saturation with NaCl) and 34–83% remained soluble with the higher proportion of soluble forms in older wines.

Ribéreau-Gayon and Stonestreet (1966) used essentially the same procedure and it has been widely applied to wine by others. Their version used as a standard a condensed tannin isolated from pine. Others have used synthetic 3′,4′,5,7-tetrahydroxyflavan-3,4-diol (leucocyanidin) and cyanidin itself with, of course, not strictly comparable results. They dilute red wine to $^1/_{50}$ adjusted as necessary and prepare the standard condensed tannin at appropriate dilutions of 1 g l^{-1} in model wine (10% v/v ethanol, 5 g l^{-1} tartaric acid partly neutralized with sodium hydroxide). The diluted wine or standard sample, 4.0 ml, is placed in two tubes with 2.0 ml H_2O and 6.0 ml conc. HCl. One tube is heated under reflux in boiling water for 45 min then cooled 10 min (protected from light considering cyanidin instability). The other tube is maintained at room temperature. To each tube 1.0 ml of ethanol is added to enhance color and absorbance is measured at 550 nm, 1 cm cuvettes. The heated tube's absorbance minus that of the unheated measures the absorbance increase from anthocyanogenic tannins and concentration is calculated from comparison with the standard.

Similar methods have been used before and after Ribéreau-Gayon and Stonestreet (1966), many with more or less consequential modifications. Some adsorb the anthocyanogens on polyamide or polyvinylpyrrolidone, wash it and then heat with acid, adsorbent and all, to generate the cyanidin for measurement. This can concentrate the anthocyanogens, help eliminate interference from sugars or free anthocyanins that contribute color, and lower the water content, which enhances the reaction to produce anthocyanidins. Others extract the cyanidin formed into a butanol layer with similar aims. Still others add prooxidants such as iron salts or benzoquinone to enhance the percentage conversion to cyanidin.

Pompei et al. (1971) investigated several such modifications and recommended a widely followed standard procedure for white wines. Agitate for 5 min 200 mg of polyvinylpyrrolidone (Polyclar AT) in 20.0 ml of wine diluted $^1/_{10}$ to $^1/_{20}$ in water. Recover the adsorbent by filtration and rinse it thoroughly with distilled water. Transfer it to a tube with 20 ml of a 1:1 (v/v) mixture of *n*-butanol:conc. HCl containing 150 mg l^{-1} of $FeSO_4$. Heat 50 min on a boiling water bath, cool, dilute to 25.0 ml with the same butanol solution, centrifuge until clear, and determine absorbance at 550 nm in a 10.0 mm cell. The absorbance is corrected with a blank made with the same amount of resin without any wine. The mg of cyanidin in 25.0 ml is calculated as 1.217 times the corrected absorbance.

Dadic (1974) applied to wine a similar method developed for beer to give a simultaneous estimation of catechins separately from anthocyanogens. After heating in butanol/HCl the absorbance was determined at 455 and at 545 nm. Catechins in mg $l^{-1} = 283\ A_{455} - 58.3\ A_{545}$ and anthocyanogens $= 5.5(20.5\ A_{545} - 5.16\ A_{455})$. The values obtained, respectively, for a white and red wine were catechins 183, 1087 and anthocyanogens 175, 1177 mg l^{-1}. Absorbances obtained with propanol rather than butanol and with polyamide rather than PVP were considerably different and the method does not appear to have had many users.

Flanzy et al. (1969) recommended precipitation of the anthocyanogens with $HgSO_4$ prior to cyanidin development, particularly for samples with high sugar content. Results were considered equivalent to or better than polyamide adsorption and gave values of 9–32 mg l^{-1} for white wine and 1315 mg l^{-1} for a red wine. Puisais et al. (1968) found that the total anthocyanogenic fraction could be determined from the linear relation in absorbance at 225 nm minus that at 235 nm at pH 3.2 in aqueous alcohol. Polymeric forms were separately determined by difference after precipitation by saturation with NaCl. Values obtained agreed on 12 wines averaging 1768 and 1769 for total anthocyanogens, 374 and 370 mg l^{-1} for polymeric forms, respectively, by this method and by that of Masquelier et al. (1959). Absorbances at these low wavelengths are very susceptible to interferences.

4.3.4 Tannins

Tannins are the astringent, protein-precipitating polyphenols in the range of about 500–5000 molecular weight. Larger units may not stay in solution nor bind as readily with proteins. Unless treated with tannic acid or aged with wood contact, wines have no hydrolyzable tannins. Grape tannins are flavanol polymers and include or are identical with anthocyanogens, and in wine range from dimers through about decamers. There is good evidence for polymerization-depolymerization reactions as well as for incorporation of anthocyanins into the oligomers, oxidation, and complex formation that modify the color, solubility, and other properties of the large-sized phenol fraction (Fig. 6).

These compounds can be estimated as anthocyanogens by the methods already outlined. To estimate molecular size or degree of polymerization (DP), methods have usually fallen into one of three major groups: ratios of two analyses, one affected by DP and one not, selective separation into size groups, and specific chromatographic study. The latter group is the least arbitrary and holds

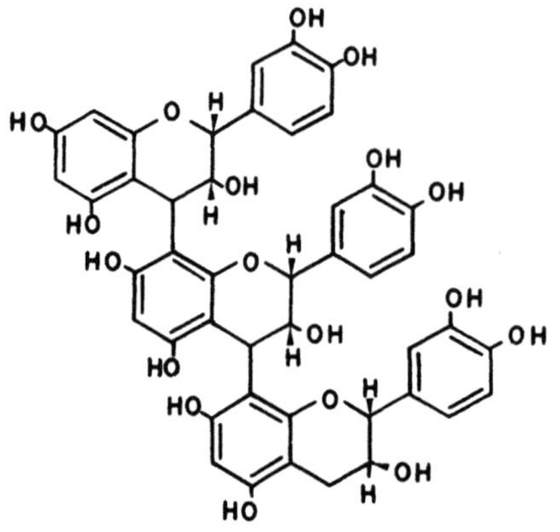

Fig. 6. A representative trimeric, anthocyanogenic condensed tannin made up of two (−)-epicatechin units linked to a terminal (+)-catechin unit via 4–8-bonds

the ultimate answer. However, it may well turn out that the difference in properties between a pentamer and a hexamer, say, and particularly the exact stereochemical linkages may not reflect differences important to a wine-maker or consumer. Estimation as a group may be more useful.

In probably the most complete study to date, Lea et al. (1979), by combinations of various counter-current and chromatographic techniques, showed the presence of, verified the structures of, and estimated the content of various members of this group in representative white wines and in seeds. A Müller-Thurgau wine prepared with considerable skin contact had about 22 mg l^{-1} of catechin plus epicatechin, 62 mg l^{-1} of mostly dimers with some trimers, 51 mg l^{-1} of tetramers plus pentamers and 186 mg l^{-1} of larger polymers and oxidized forms not extractable into ethyl acetate. Red wines generally have considerably more and analysis is affected by polymerization incorporating anthocyanins during aging.

The acidic vanillin reagent combines with the 6 or 8 positions of the flavanols which are exposed and not already substituted in the polymer. The vanillin value drops as polymerization proceeds. Italian workers have therefore called it an index of polymerizable flavanols. Total phenols by Folin-Ciocalteu or Folin-Denis (Swain and Hillis 1959) or total anthocyanogens (Ribéreau-Gayon and Stonestreet 1966) are not, or at least are much less, affected by the DP. Therefore the ratios (in mg l^{-1} of catechin or cyanidin) vanillin/Folin-Ciocalteu or vanillin/anthocyanogen decrease as polymer size is larger. Such data, for example, show that larger tannin polymers in wine are less extractable with ethyl acetate and more precipitable with gelatin than are smaller units.

Many different precipitation and extraction procedures have been used to capitalize on these differences. It seems inevitable, since the differences are subtle and continuous as polymerization increases, that the separations are not sharp and are influenced by conditions. Analysis by extraction or precipitation must be accompanied by a method for quantitation usually before and after the separation. Precipitates with lead and other metal salts generally include small phenols as well as polymers and are difficult to make quantitative (e.g., Mattick et al. 1969). Hide powder, gelatin and other proteins precipitate tannins, but the results are not entirely predictable, depending on the protein, the tannin, and the conditions (e.g., Asquith and Butler 1986). Polyamide and polyvinylpyrrolidone bind tannins by mechanisms similar to those of proteins and are more easily standardized, but often adsorb smaller phenols, depending on relative dosage.

Masquelier et al. (1959) precipitated the polymeric forms by saturation with NaCl and found they represented about 65% of the total anthocyanogens in young wines or roughly 1 g l^{-1}. Glories (1978) has made detailed studies of these compounds in red wines of different ages by a sequential method of separation. Raising the ethanol content precipitated tannin and carbohydrate or salt complexes which, in a young wine with 2340 mg l^{-1} total phenol and 531 mg l^{-1} anthocyanins, amounted to 18 mg l^{-1}. Then addition of methanol and chloroform precipitated large and very large polymeric tannins which, after elution of the large from PVP with formic acid (leaving the very large still bound) gave 13 mgl^{-1} large and 8 mg l^{-1} very large polymers. The supernatant, with the solvents removed and the phenols adsorbed on PVP, gave after elution of smaller phenols with water and acidic aqueous alcohols, a formic acid eluate that contained smaller tannins and polymers incorporating anthocyanins to the combined level of about 44 mg l^{-1}. Free anthocyanins totalled only 50 mg l^{-1}. There clearly was considerable discrepancy between the amounts accounted for and the total expected.

Mitjavila et al. (1971) used the Folin-Ciocalteu assay after separating the tannins from nontannins in wine by forming with the tannins a complex with soluble PVP (25,000–30,000 mol. wt.) and precipitating it with trichloracetic acid. A burgundy with 1830 mg l^{-1} total gave 1120 mg l^{-1} nontannic phenols and 690 mg l^{-1} astringent tannins by their test. Peri and Pompei (1971 a, b), defining tannins as those substances precipitating with the alkaloid cinchonine, used the Folin-Ciocalteu assay in combination with formaldehyde precipitation to separately determine nontannin flavonoids, simple phenolics, condensed tannins, and hydrolyzable tannins. With known mixtures, the calculated and determined values agreed well. White wine, 50.0 ml, is placed in a 100-ml centrifuge tube and neutralized with NaOH to pH 7.0. Add 25.0 ml of pH 7.9 phosphate buffer (1.36 g KH_2PO_4, 8.35 g $Na_2HPO_4 \cdot 12\,H_2O$, 12.50 g $NaHCO_3$, water to 500 ml), 12.5 ml of cinchonine (1.5 g base, 2 ml 1:3 H_2SO_4, water to 100 ml), mix, and after 20 min at room temperature, centrifuge. Wash the precipitate twice with 10 ml of 10% aqueous Na_2SO_4. The supernatant plus washings is acidified to pH 3.5 with HCl, brought to a known volume with water and assayed by the flavonoid-formaldehyde procedure for nontannin flavonoids and simple phenols. Similarly, the cinchonine precipitate is dissolved in ethanol containing 10% HCl and used to determine by the same means flavonoid tannins and, if present, hydrolyzable tannins. In gallic acid equivalents, a white wine had 81 mg l^{-1} total phenols, 33% nonflavonoid phenols, 25% nontannin flavonoids, 0% nonflavonoid tannins, and 42% condensed tannins.

Montedoro and Fantozzi (1974) used precipitation with methyl cellulose to quantitate the tannins in musts and wines. Phenol content was determined before and after the precipitation, with tannin being the difference. The coefficient of variation averaged 0.86% for musts and 2.95% for wines. By combination with the formaldehyde method for flavonoids, nontannin flavonoids can be separately determined. Into a 10 ml volumetric flask place 1–5 ml of sample containing 1–

1.5 mg of phenol, 1 ml of 0.4% methyl cellulose with a high degree of methylation, mix, add 2 ml saturated $(NH_4)_2SO_4$ and make to 10.0 ml with water. Centrifuge or filter off the precipitate after 20 min at room temperature and determine the phenol content on the supernatant and untreated wine by the Folin-Ciocalteu procedure. They compared the method with the cinchonine and PVP (Mitjavila et al. 1971) methods and obtained good agreement, 250–296 mg l^{-1}, on white wine and less satisfactory for red wine. Burkhardt (1976) compared the same three methods with red musts and wines and found all three methods were equally suitable, giving quite similar values. Cela et al. (1983) found substitution of sodium dodecylbenzenesulfonate for the ammonium sulfate produced more readily clarified supernatant.

5 Total Phenols

The determination of all phenols together in an ideal manner requires that every phenol but no other compounds present react or have the characteristic used for the determination. There is no such ideal method for phenols. Useful methods require understanding of their weaknesses. Of course, the medium in which the phenols occur also affects the methods which may be used. In red wine the situation is quite different than in white must, for example, the latter having much more sugar, much less phenol, and possibly more free sulfur dioxide.

It is important that the method for total phenol content give values that are directly or easily convertible to units that have meaning in relation to analyses for subcategories of phenols. For this reason, analyses that are expressed as "indexes" in ml of titrant, absorbance units, etc. leave much to be desired. It is very useful if the general method for total phenols is capable of easy application to separated fractions so that difference values and a "balance sheet" can be prepared. For studies of reactions molar concentrations are preferred, but to visualize content in wines we prefer mg l^{-1} in terms of standard substances. Total phenols calculated as gallic acid reach the order of 6000 mg l^{-1} in wines and about that order of magnitude in mg kg^{-1} in extracts of fresh grapes. If a wine had one millimolar content each of gallic acid, catechin and hexameric flavolan, the respective concentrations by weight would be 170, 290, and 1735 mg l^{-1}. These latter units can be more easily compared to total phenol content or sensory thresholds and applied in processing, such as fining of wines.

The methods for total phenols that have been widely applied to wines include oxidation with permanganate, with the Folin-Denis or Folin-Ciocalteu phosphotungstate-molybdate reagents, or with ferric reagents, and spectrophotometry in the ultraviolet. Other methods not widely applied, such as acidimetry, will be briefly considered separately. For reasons that will become apparent, comparisons among these main methods will be in reference to the Folin-Ciocalteu method when possible.

5.1 Assays Based upon Oxidation

Many assays for total phenols have been based upon the use of different reagents to oxidize phenols reproducibly. Problems with these methods arise mainly from either inclusion of other readily oxidized nonphenols or failure to oxidize all phenols equally or in a predictable manner.

5.1.1 Permanganate Oxidation

The earliest widely applied oxidation method for phenols in wine was the Löwenthal-Neubauer titration with $KMnO_4$ to an indigo-carmine end point (Singleton and Esau 1969). Differences between analyses before and after treatment with activated carbon and comparison with, usually, a tannic acid standard gave total phenols often called "tannin plus pigments." The difference value is necessary because permanganate can react with many substances including sugars and tartaric acid. Furthermore, some monophenols and meta-polyphenols do not react, some polyphenols give different values than expected from others, and reproducibility of the determination in the hands of different analysts was sometimes poor. The method has been considered superceded by the Folin-Denis-type procedures (Singleton and Esau 1969; Ribéreau-Gayon and Sartore 1970). Nevertheless, agreement between the two could be quite good, especially on red wines. Donko (1975) found the mean total phenol content in 32 red wines by Löwenthal-Neubauer was 1321 mg l^{-1} and by Folin-Ciocalteu 1260 mg l^{-1}. With 20 white wines, respective values were 313 and 203 mg l^{-1} with a considerably larger difference between duplicates for the permanganate method. Other comparisons (Gigliotti 1973, 1974; Munoz Alcon and Marine Font 1978) show similar results, with the permanganate method less preferred, slightly higher for red wines and slightly lower for white wines than the Folin-Ciocalteu.

5.1.2 The Folin-Ciocalteu Reagent and Related Methods

These methods determine phenols (and other easily oxidized substances) by producing a blue color from reducing yellow heteropolyphosphomolybdate-tungstate anions. The older formulation, Folin-Denis, was a reproducible method (Swain and Hillis 1959; Pridham 1964; Association of Official Agricultural Chemists 1984), but those who made direct comparisons recommend the Folin-Ciocalteu reagent as more sensitive, more reproducible and more trouble-free, although analytical values obtained on the same wine should be nearly identical (Singleton and Rossi 1965; Joslyn et al. 1968; Gigliotti 1973, 1974).

The colored products are evidently mixtures of the 1-, 2-, 4-, and 6-electron reduction products in the tungstate series, $P_2W_{18}O_{62}^{-7}$ to $H_4P_2W_{18}O_{62}^{-8}$, and the 2-, 4-, and 6-electron reduction products in the molybdate series $H_2P_2Mo_{18}O_{62}^{-6}$ to $H_6P_2Mo_{18}O_{62}^{-6}$ (Papaconstantinou and Pope 1970). All are blue with shorter wavelength maxima and higher extinction values with the greater reduction. Based upon the broad visible color maximum observed in the assay, several species are probably involved, but the 4-electron product with 760 nm maximum and about 19,000 extinction appears to predominate. The 1-

electron reduction product of the tungstate series is thought to help initiate the reaction with phenols (Singleton 1974).

The blue product evidently does not involve chelation or coordination complexing with the substance oxidized, and this helps explain why the color yield is reasonably predictable for different phenols. Except for very acidic phenols like picric acid or strongly internally hydrogen bonded ones like salicylic acid, color equivalent to about 13000 molar absorptivity is produced per reactive phenol group by most natural phenols (Singleton and Rossi 1965; Singleton 1974). Therefore it is possible to convert analytical values based on one standard substance into those for another. Phloroglucinol and most other meta-polyphenols react as monophenols; catechols and pyrogallols as diphenols. Catechin reacts as a triphenol, one for the A-ring and two for the B-ring. Gallic acid has most often been used as the standard and results expressed in $mg\,l^{-1}$ as gallic acid equivalents (GAE). To convert these values into, say, (+)-catechin $mg\,l^{-1}$, it is preferable to compare the two standard substances under your conditions of assay but good estimates can be made by multiplying gallic values by 1.14. To illustrate the method without using our own data, Table 3 presents some from Barna and Urdich (1975), who used the Folin-Denis procedure. By the same procedure a phenol determined specifically by some other technique can be readily calculated to its contribution to the total phenol gallic acid equivalent $mg\,l^{-1}$ or $mg\,kg^{-1}$. An average content of 145 $mg\,l^{-1}$ of caftaric acid in unoxidized *vinifera* juice would be equivalent via Folin-Ciocalteu (FC) analysis to $145 \times \frac{170}{312} = 79\ mg\,l^{-1}$ gallic acid equivalent (GAE).

The conditions affecting the reaction have been studied in considerable detail (Singleton and Rossi 1965; Joslyn et al. 1968; Blouin et al. 1972; Montreau 1972). Color development is faster but interference by sugar is larger at warmer temperatures. The blue is stable at 60° C for at least 48 min and longer at lower temperatures, but it fades and the assay becomes erratic if time or temperature are excessive. The maximum is broad and satisfactory measurement can be made over at least 740 to 770 nm, but for uniformity 765 nm is recommended. Unless there is reason to do otherwise, 2 h of color development at room temperature is considered suitable. Good results can be obtained between 1 h and about 12 h as long as standards are included with each assay. The method has been automated by

Table 3. Average content of total phenol in $mg\,l^{-1}$ in white wines

No. of wines	302	48	52	37	30
Kind of wines	Normal harvest	Late harvest	Auslese	Beeren-auslese	Trocken-beeren-auslese
Determined as gallic acid	341	384	414	485	566
Separately determined as (+)-catechin	365	424	470	570	681
Calculated as catechin from 1.14 × gallic acid	389	438	472	542	645

the air-segmented continuous flow method by developing the color in a 5-min coil at 55° C (Slinkard and Singleton 1977). Coefficients of variation averaged 5.8% for the manual and 2.2% for the automated method and the values obtained on dry wines or after correction for sugar on sweet wines agreed within the variation for manual duplicates.

The potentially interfering substances include aromatic amines (absent or very low in grapes or wines), sugars, certain nucleic acid fragments, and the readily oxidized substances ascorbic acid, sulfur dioxide, or ferrous ions. The potential for nucleic acid interference appears quite small in normal wines (Myers and Singleton 1979), as is that of iron. Ascorbic acid soon disappears in commercial processing of musts and, unless added, is absent from wines. If present, however, it interferes at about 1 mg of apparent gallic acid per mg of ascorbic acid and must be corrected for by separate determinations (Joslyn et al. 1968; Singleton 1974; Görtges 1981).

The interference by sugar in musts and sweet wines can be serious and is greater with higher temperature and higher phenol content and interacts to increase the interference by free SO_2 (Slinkard and Singleton 1977; Moutounet 1981). Fructose has a greater effect than glucose and for grapes invert sugar, 1 fructose:1 glucose, should be used for corrections. Donko and Phiniotis (1975) recommended 15% deduction from the phenol value if the sugar content was 20 to 25% decreasing to 3% at 1 to 2.5% sugar. Slinkard and Singleton (1977) recommended for best correction preparing standards at the same sugar content as the samples, but as a guide gave the data in Table 4. The corrections are higher at low phenol content and room temperature than indicated by Donko and Phiniotis (1975), but about the same at higher phenol content. Clearly the greatest problem is with juice and other samples low in phenol and high in sugar.

The mechanism of interference by sugar appears to be the production of endiol reductones from the sugar in alkaline solution. These compounds, like ascorbic acid, are very easily oxidized. Considering the high percentage of sugar present, reductones are produced in small proportions. They can interact with phenols by reducing quinoids in the early stages of oxidation by the FC reagent.

Sulfur dioxide and sulfites can also interfere (Singleton and Rossi 1965; Singleton 1974). The interference can be so large as to invalidate results if the

Table 4. Approximate corrections in mg GAE l^{-1} to be subtracted from the total phenol by FC to correct for sugar content

Apparent phenol content (mg GAE l^{-1})	Sugar (invert content)					
	2.5%		5%		10%	
	25° C 2 h	55° C 5 min	29° C 2 h	55° C 5 min	25° C 2 h	55° C 5 min
100	5	20	10	30	20	60
200	10	40	15	50	40	75
500	20	85	30	120	50	190
1000	30	105	60	150	100	250
2000	60	200	120	250	200	340

phenol content is low and the free SO_2 content high (Somers and Ziemelis 1980). This is not the typical situation, however. SO_2 in must is bound by the sugar and other carbonyls, by the fermentation, and soon bound if added later to wine in small amounts. Bound SO_2 has much less effect and in our unpublished tests (Orthofer R, Goldfield D, Singleton V 1986) did not interact so that a simple and generally small correction could be applied if the wine's total SO_2 is bound with acetaldehyde. Adding acetaldehyde or formaldehyde to most white wines did not lower the apparent phenol content since the SO_2 was already bound. With red wines no appreciable SO_2 effect has been noted not only because it is bound, but also because of the high level of phenols and consequent dilution of the sample and the relatively low total SO_2 in modern wines.

The mechanism seems to be similar to that of sugar in that they interact, and free SO_2 can especially augment the effect of a lower level of phenol on the reagent. This synergistic effect is related presumably to effective reduction of quinoid oxidation products enabling reoxidation by the reagent. When 100 mg l^{-1} of free SO_2 was added to different phenols so that there was 3.1 mol of SO_2 per mol of phenol, the color increase was 91 to 101 mg gallic acid equivalent (GAE) l^{-1} for gallic acid, caffeic acid or catechin, but only 40 mg GAE l^{-1} for *p*-coumaric and 16 mg GAE l^{-1} for resorcinol (Somers and Ziemelis 1980). SO_2 was only $^1/_3$ as effective in generating color on a molar basis as was gallic acid. Data from Joslyn et al. (1968) indicate about 5 mg of SO_2 alone is equivalent to 1 mg of catechin. They found a wine with 10 mg l^{-1} of tannin had only 0.5 to 1.5 mg l^{-1} added by the presence of 10 to 30 mg l^{-1} of SO_2.

With careful attention to avoid misuse, the Folin Ciocalteu assay remains quite valuable. It has been in use for some time and there are many comparative data. It is particularly good for dry wines with considerable pomace contact. It has been widely used to develop fractional analyses (flavonoid-nonflavonoid, cinchonine precipitation, etc.) that do not have ready alternatives. Above all it reacts predictable with essentially all natural phenols.

The reagent itself can be purchased, but is easily prepared. Dissolve 100 g of $Na_2WO_4 \cdot 2H_2O$ and 25 g of $Na_2MoO_4 \cdot 2\,H_2O$ in 700 ml of distilled water in a 2-l distillation flask. Add 50 ml of 85% H_3PO_4, 100 ml of conc. HCL and reflux 10 h (Singleton and Rossi 1965). Remove heat, rinse down the condenser with 50 ml distilled water, and when safely below boiling temperature add 150 g $Li_2SO_4 \cdot H_2O$ and (in ventilation hood) a few drops of Br_2 (or a few ml of 30% H_2O_2) and boil in the open for 15 min or until the Br_2 is gone. The resultant clear yellow solution is cooled and made to 1 l with distilled water.

It is recommended to pretreat the wine with excess acetaldehyde to ensure that all the SO_2 is bound, but not so high or so long as to cause precipitation. For wines with high free SO_2 we have used 1000 ppm acetaldehyde and 20-min reaction time plus appropriate correction for resultant dilution (Orthofer R, Goldfield D, Singleton V unpublished 1986). Mix appropriate dilutions of standard (such as 500 mg l^{-1} gallic acid and lower) or wine (white undiluted, red generally $^1/_5$) at 0.2 ml with 10.0 ml of $^1/_{10}$ diluted Folin-Ciocalteu reagent and 1.8 ml of water. After ½ to 8 min, mix in 7.0 ml of 115 g l^{-1} of Na_2CO_3, wait 2 h at room temperature and read absorbance versus a water blank at 765 nm (Singleton and Rossi 1965; Blouin et al. 1972; Montreau 1972; Slinkard and Singleton 1977).

With due consideration for sugar interference, a shorter time at a higher temperature may be used. Answers are usually expressed as mg l^{-1} gallic acid equivalents for must or wine and mg kg^{-1} fresh wt for grape extracts corrected for dilution as necessary.

5.1.3 Prussian Blue Assay

The production of Prussian blue as phenols are oxidized can be made a sensitive quantitative procedure and, as already noted, has been applied with grapes and wine as a spray for paper chromatograms. Monophenols react weakly and ascorbic acid interferes so strongly that it must be corrected for if present. Any substance capable of reducing ferric to ferrous ions will interfere, but since the reaction is acidic, carbohydrates should not, if the reaction time is limited. The procedure as recommended by Butler (1982) involves mixing 0.5 ml of sample or standard (catechin usually used) with 30.0 ml of water, 3.0 ml of 0.05 *M* $FeCl_3$ in 0.1 *N* HCl, and with timing, 3.0 ml of 0.008 *M* $K_3Fe(CN)_6$. At 20 min determine absorbance at 720 nm versus a suitable (similar alcohol content) blank. The relatively high $FeCl_3$ gave fast reaction and low ferricyanide gave low blanks. There is sevenfold decrease in color yield between quercetin and hydroquinone.

5.1.4 Jerumanis-Type Iron Colorimetry

After reviewing, for beer analysis, other methods for total phenols including Folin-Denis, permanganate, and 270 nm absorbance, De Clerck and Jerumanis (1967) proposed as an improvement colorimetry with ferric ammonium citrate. The method has been modified and become a standard for the European Brewery Convention (Bishop 1972). Degassed beer, 10.0 ml, is mixed with 8 ml of a solution of 1% low viscosity carboxymethylcellulose with 0.2% EDTA. A fresh 3.5% solution of ferric ammonium citrate (16% Fe) is added, 0.5 ml, mixed, 0.5 ml of a $^1/_3$ dilution of conc. NH_4OH added, mixed and made to 25.0 ml final volume. The absorbance at 600 nm is measured after 10 min (10 mm cell) against a blank similarly prepared without the ammonia. The color seems to result from a combination of oxidation and chelation.

The EBC or Jerumanis methods do not react or only very weakly with phloroglucinol or monophenols and give extra color with pyrogallol derivatives compared to catechol derivatives, in contrast to more uniform and predictable molar color yields from the FC reagent (Leupold and Drawert 1981). On the other hand, the Jerumanis methods are apparently less affected by SO_2, ascorbic acid or carbohydrates (Cela et al. 1982). On a series of beers the Folin-Ciocalteu assay gave much higher values especially on dark stout than did the EBC method (Leupold and Drawert 1981). This interference was removed and values brought into excellent agreement by adsorption of the phenols on PVP and elution with 50% dimethyl formamide. This interference with the FC values in dark beers appears to be from carbohydrates and melanoidins not present in wines. The total polyphenol content of a white wine was 335 mg l^{-1} by FC and 224 mg l^{-1} by ferric ammonium citrate (Möbius and Görtges 1974a). Fining caused approximately equal drops in phenol content by either method. A series of musts by FC and Jerumanis methods gave, respectively, 260, 225; 165, 120; 221, 143; and 238,

195 mg l^{-1} (Möbius and Görtges (1974b). The consistently lower value by the ferric method has been attributed to SO_2 (Cela et al. 1982), but the relative agreement and other factors make the blindness of the Jerumanis-type methods to monophenols and sensitivity to pyrogallols seem at least a contributing cause. This belief is reinforced by consistent but larger differences with five German red wines (Möbius and Görtges 1974a). Respective pairs by FC vs. Jerumanis methods ranged from 453, 189 to 877, 500 mg l^{-1}. Malvidin apparently does not react in the Jerumanis reaction, but contributes to the measured color (Cela et al. 1982). Although such application has not been made, the difference between FC and Jerumanis values should give an estimate of phenols not easily converted to quinoids.

5.2 Total Phenol Analysis by Ultraviolet Spectrophotometry

The ultraviolet absorption spectrum of red wines shows an intense maximum near 280 nm due in large part to phenols. White wines, as already discussed, are likely to have maxima near 280 nm and another or shoulders at longer wavelengths, 320 nm or so representing the hydroxycinnamates and perhaps 360 nm for the flavonols or certain quinoid oxidation products.

Margheri (1960) estimated phenols in wines by acidification and extraction with ethyl acetate, drying the extract, and absorbance at 275 nm in acidic methanol in comparison with tannic acid. The values obtained are fairly similar to those obtained by other methods such as FC. For example, white wines gave 200–320 mg l^{-1} a rosé 360 mg l^{-1} and red table wines 580–1520 mg l^{-1}. Ribéreau-Gayon and Sartore (1970) after comparative studies of the three, considered 280 nm absorbance of dry red wines diluted with water a better measure of total phenol than permanganate oxidation but not as good as Folin-Ciocalteu.

Blouin and Cordeau (1978) also rejected permanganate but found absorbance at 275 nm times water dilution almost identical with absorbance at 280 nm times dilution with 50% ethanol. Correlation coefficients were 0.95 between the two ultraviolet methods, but between the 275-nm method and permanganate 0.59 and with FC 0.42. They considered the FC method more complicated but a useful supplement to UV estimation. Margheri (1974) reported total phenol in mg/l with reference to tannic acid standards for a diluted but not extracted series of red wines by respectively, Folin-Ciocalteu and 280 nm absorbance as 1040, 500; 3560, 1540; 1860, 800; 1800, 770; 1750, 695; 1570, 695; 1100, 455; 1096, 465; and 1040, 500. Clearly relative correlation existed but concentration did not agree and a downward correction of absorbance would aggravate the difference.

Gigliotti (1973) found the standard deviation of 280 nm absorbance values on red wines lower than for permanganate or Folin-Denis methods, but slightly higher than FC. The ratios of values among the four methods were relatively constant. Castino (1979) found among 66 red wines that correlation between total phenol values by 280 nm absorbance and by Folin-Ciocalteu assay improved if the wines were divided into two groups. The differences were attributed to components separated by Sephadex LH-20 chromatography that gave similar response in the FC assay, but much different in the UV. The FC was preferred as more specific.

Somers and Evans (1977) and Somers and Ziemelis (1985 b) use (10 mm) E_{280} × dilution minus 4 as an index of total phenols in white or red wines. The −4 correction is, as discussed under total flavonoid analyses, intended to correct for constant UV absorbance by components presumably not phenols that did not adsorb on PVP. If sorbic acid is present it interferes and requires further correction. White wines gave values of 5–15 and, assuming 3600 molar absorptivity for catechin, 400–1200 mg l^{-1} of total catechin equivalents are indicated uncorrected for hydroxycinnamate contribution. If gallic acid molar absorptivity of near 7900 is used, about 100–300 mg l^{-1} are indicated in white wines and a range of about 500–2150 mg l^{-1} in red wines, values close to GAE expectations by FC.

Bakker et al. (1986 a) compared the Singleton and Rossi (1965) FC method with Somers and Evans (1977) 280 nm absorbance and found on a series of port wines a correlation coefficient of 0.956. The zero on the UV scale was 210 mg l^{-1} GAE on the FC scale, probably in part due to the uncorrected sugar effect of the ports. One absorbance unit was equivalent to 29.5 mg l^{-1} gallic acid by FC. By 7900 molar extinction it would be 21.5 mg l^{-1} per unit absorbance at maximum (about 270 nm for gallic acid). They also note that anthocyanins contribute 0.6 of their absorbance at 520 to the total phenols at 280 nm.

In sum, ultraviolet absorption spectral measurements are seen as important and relatively simple techniques for estimating phenols in wines. For the present, they are seen as supplementary to other techniques because the wavelengths and intensities of different phenol's maxima differ so much. As individual phenols are determined their contribution to the total can be calculated. More information is needed on the nature and variability of the nonphenolic substances interfering in grape extracts and wine. Just as with the FC and most other methods, the analyses are most questionable in the white wines very low in phenols. Ultraviolet absorbance cannot be used when UV-absorbing reagents are present as with cinchonine precipitation and formaldehyde or acetaldehyde to precipitate flavonoids or bind SO_2. High aldehyde in certain sherries and furfural derivatives in musts or wines subjected to heat would also absorb at 280 nm and prevent ultraviolet determination of phenols.

6 Phenol Grouping Analyses

6.1 Polarography

Polarographic methods have been applied to phenols and phenol fractions in wine, but not widely so (Singleton and Esau 1969). The equipment is specialized and can be costly. In aqueous solutions like wines multiple oxidations often give no sharp end points. Setting decomposition voltages to distinguish between vicinal dihydroxy and other phenols is a useful electrochemical detection system in HPLC (Roston and Kissinger 1981). Direct determination of anthocyanins in young red wines by polarography was linear over the range 50 to 200 mg l^{-1} as malvin and gave values between the SO_2 and pH-shift methods (Mareca Cortes et al. 1980; Davidovic et al. 1985). Pakhomova et al. (1980) via amperometric titration with vanillin and enotannin as standards found total phenols of white wines in the range 200 to 500 mg l^{-1} and of red 1170 to 1370 mg l^{-1}.

6.2 Titrimetry

Phenolic groups are weak acids as well as oxidizable and can be estimated by titrimetry. Khomenko and Oloinik (1968) titrated acidified wine with $KBr\text{-}KBrO_3$ to produce two potential jumps corresponding to pigments and tannins. The total phenol content agreed well with spectrophotometric and permanganate values on seven wines 550–1600 mg l^{-1}. Godinko et al. (1984) titrated 16 red wines with strong base following the titration reaction by measurement of the heat generated. The first inflection gave values agreeing well with total acidity determined by standard means and the second corresponded to the phenolic content as measured by permanganate index. Because of a rounded second break in the curve, a standardized method of calculating the end point for phenols was developed. It appears that only the first phenolic hydroxyl was titrated in gallic acid.

6.3 Inorganic Complex Colorimetry

It is sometimes useful to be able to further differentiate phenols on the basis of their phenolic group arrangement. Vicinal or ortho-diphenols are readily oxidizable to ortho-quinones. Mono or meta-polyphenols and methoxylated versions are not. Pyrogallol-like phenols (vicinal trihydroxyls) are especially oxidizable and chelate iron strongly.

Mono and meta-phenols react very little with the Prussian blue or Jerumanis-type ferric reagents and, as already mentioned, the difference from FC values where they do react could be used to estimate these phenols. Based on the original proposal by Mitchell (1923), essentially a specialized version of the later Jerumanis-type reaction can be used to estimate pyrogallol substitution patterns in the presence of catechol rings (King and White 1956). The regulation of pH is critical. Russian workers at one time frequently applied these techniques to wines, but both difficulties with the method and uncertainty of the utility of the results have limited their applications. The procedures must be meticulous and vary somewhat by product. For good details and examples see King and White (1956) and Inskip et al. (1973).

Another and more usually useful method is that of Quastel (1931) and Arnow (1937) which determines vicinal dihydroxy phenols via a reddish yellow color produced with molybdate. This is informative because, in the absence of 1,4-dihydroxybenzenes as is usual in grapes, wines, and other natural food products, the catechol and pyrogallol (included) derivatives are the easily oxidized forms. The method is simple. A suitably diluted sample or comparison standard solution, 1.0 ml, is mixed with 5.0 ml of phosphate buffer at pH 6.5 to 6.6, 1.0 ml of a water solution of 5 g/100 ml of Na_2MoO_4 is added and after 15 min the absorbance is measured at 420 nm vs. a blank identical except water replaces the molybdate. A higher molar color yield results from pyrogallol than from free catechol derivatives, but mono and metaphenols do not give the colored chelates with molybdate (Aubert 1970). SO_2 does not interfere and is added to inhibit oxidation. The method has been successfully automate (Aubert et al. 1972). The correlation between molybdate and Folin-Denis values, both in mg l^{-1} of gallic acid for all groups of pink and red wines was 0.967 or greater and with 280 nm absorbance was 0.713 for rosés rising to 0.997 for tannic red wines. The Folin-Denis values were usually higher than the molybdate values on the same wine, averaging about 120% on the lightest rosés and about 135% on the heaviest red, but showing considerable variation. With 20 wines ranging from 300 to 2960 total phenol as mg l^{-1} catechin by FC, the vicinal dihydroxy content as catechin was 100 to 1400 mg l^{-1} averaging 35% of the total (26 to 51% range) (Munoz Alcon and Marine Font 1978).

7 Concluding Comments and Multiple Comparisons

It would be desirable to know the exact identity and content of every phenolic derivative in grapes, musts, or wines. HPLC, especially, and other techniques are making that objective achievable. It probably will evolve, however, that we only need to know certain key phenols or groups of phenols to define environmental response, varietal characteristics, astringency, bitterness, color, oxidation status, etc. in particular scientific and practical contexts. Historically that has been the case out of necessity; in the future it may be so for the sake of convenience and simplicity.

A number of analytical procedures have been discussed that help bridge the past to the future. Many of them have flaws, but used intelligently they have given and will give much insight and useful information. It appears desirable to apply several complementary analyses to obtain as complete a picture of a wine's phenolic composition as may be necessary for a given purpose. On the one hand, analyses of wines for phenols by several methods may overlap in such a way that they can be seen to be measuring the same thing. With sufficient data, the better or easier method can be chosen and the others discarded. More comparisons over a wider range of wines are needed here along with thorough statistical evaluation. On the other hand, dichotomously branching analytical sequences are valuable such as dividing total phenols into flavonoid and nonflavonoid groups each subdivided into tannin polymers and not-tannins (Kramling and Singleton 1969; Montedoro and Fantozzi 1974; Peri and Pompei 1978a, b). Important applications of both of these approaches have been made, especially earlier by French and more recently by Italian researchers. Much of the Italian work has been with the intent of characterizing commercial wines of a given regional appellation to set standards and improve quality. It seems appropriate to close this discussion with a few recent illustrative examples of such combined analyses.

Carruba et al. (1981, 1982) analyzed 59 wines from four white grape varieties and 33 red wines from two Sicilian red varieties for many constituents including phenols. The averages and ranges for the phenol data are shown in Table 5. From

Table 5. Mean phenol content of 59 white and 33 red Sicilian wines of 1978 and 1979 vintages (Carruba et al. 1981, 1982)

	Total phenol (FC) mg l^{-1} GAE		Catechin mg l^{-1} (Vanillin)		Anthocyanin (pH Shift) mg l^{-1}		Leucoanthocyan (Pompei et al. 1971) mg l^{-1}	
	Mean	Range	Mean	Range	Mean	Range	Mean	Range
White wines	313	179– 542	67	26– 136	–	–	67	19– 15
Red wines	2388	1225–4280	688	274–1464	230	46–393	1772	740–316

Table 6. Examples of multiple phenol analyses (mg l^{-1}) of white wines

Analysis	Gattuso et al. (1986) 40 unfinished	Margheri and Tonon (1982) 6 white	Villa (1985) Pinot blanc
Total phenol (280 nm)	–	1197 (catechin)	–
Total phenol (FC)	672 (gallic)	529 (catechin)	275 (gallic)
Total flavonoid	489	–	100
Flavonoid tannin	216	–	39
Flavonoid monomers	208	–	61
Nonflavonoids	248	–	175
Ortho-diphenols	381	–	–
Catechin (vanillin)	561	43	14
Anthocyanogens (cyanidin)	567	34	97

Table 7. Examples of detailed phenol analyses (mg l^{-1}) of pink and red wines

Analysis	Margheri and Tonon (1982)		Ubigli and Barbero (1983) Barolo	LaNotte and Antonacci (1985) Castel del Monte	Gigliotti and Bucelli (1986) Rosé
	Schiava Gentile	Chianti			
Total phenol (280 nm) catechin	1625	2755	3280	3400	–
Total phenol (FC)	1100 (cat.)	1560 (cat.)	2840 (cat.)	2400 (cat.)	440 (GAE)
Total flavonoid	–	–	2980	–	230
Flavonoid tannin	–	–	–	–	80
Flavonoid not tannin	–	–	–	–	150
Non flavonoid	–	–	290	–	220
Anthocyanins mg l^{-1} (cyanidin)	136	80	36	85	44
Catechin (vanillin)	292	443	337	1192	87
Anthocyanogens (cyanidin) mg l^{-1}	446	490	1901	2366	63
Phenols extractable EtOAC (FC mg l^{-1})	464 (cat.)	520 (cat.)	–	–	176 (GAE)
Phenols ppt with NaCl	188	380	–	675	80
Phenols ppt with cinchonine	380	624	–	–	42

these data it is apparent that about half of the wines have had considerable pomace contact and would tend to be coarse and brown easily among the whites and be excessively tannic for the reds at least by finished wine standards. The values are not strictly comparable as presented. The simple sum of the fractions for the red wines is greater than the total phenol because the analyses are not exclusive and the bases of comparison are not the same. In FC mg GAE l^{-1} terms, the catechin value would convert to about 605 for the red wines and the leucoanthocyan value to roughly 1550 mg l^{-1}.

Selected data from a few other papers with detailed phenolic analyses are listed in Table 6 (white wines) and Table 7 (red and pink wines). Again these data bear out observations already made such as that nonflavonoid phenol content is relatively constant in wines. Although many more such papers could be cited, few use all the indicated analyses and there are usually gaps in the data presented. Each research group tends to use their own favorite methods of analysis for phenols and it is fairly frequent that the specific methodology is not certain from the meager details presented in a given paper.

Wider use, more comparisons, more standardization, and better understanding of advantages and pitfalls of each method are recommended.

References

Amerine MA (1954) Composition of wines. I. Organic constituents. Adv Food Res 5:353–510

Amerine MA, Ough CS (1980) Methods for analysis of musts and wines. Wiley-Interscience, New York

Arnow LE (1937) Colorimetric determination of the components of 2,4-dihydroxyphenylalanine-tyrosine mixtures. J Biol Chem 118:531–537

Asquith TS, Butler LG (1986) Interactions of condensed tannins with selected proteins. Phytochemistry 25:1591–1593

Association of Official Agricultural Chemists (1984) Official methods of analysis of the association of analytical chemists. 14th edn. AOAC, Washington, DC

Astegiano V, Ciolfi G (1974) Indagine sui costituenti antocianici dei vini rossi piemontesi. Riv Vitic Enol 27:473–479, 497–502

Aubert S (1970) Méthodes usuelles d'évaluation des anthocyanes et tannins dans les vins. Applications aux essais de definition des vins rosés. Ann Falsif Expert Chim 63(690):107–117

Aubert S, Ferry P, Lapize F (1972) Determination automatique des polyphénols totaux dans les vins. Ind Aliment Agric 89:1723–1730

Bakker J, Timberlake CF (1985a) The distribution of anthocanins in grape skin extracts of port wine cultivars as determined by high performance liquid chromatography. J Sci Food Agric 36:1315–1324

Bakker J, Timberlake CF (1985b) The distribution and content of anthocyanins in young port wines as determined by high performance liquid chromatography. J Sci Food Agric 36:1325–1333

Bakker J, Timberlake CF, Arnold GM (1986a) The colours, pigment and phenol contents of young port wines: effects of cultivar, season and site. Vitis 25:40–52

Bakker J, Preston NW, Timberlake CF (1986b) The determination of anthocyanins in aging red wines: comparison of HPLC and spectral methods. Am J Enol Vitic 37:121–126

Barna J, Urdich O (1975) Untersuchungen über den Polyphenolgehalt österreichischer Rot- und Weißweine. Mitt Klosterneuburg 25:379–386

Berg HW (1963) Stabilization des anthocyanes. Compartment de la coleur dans les vins rouges. Ann Technol Agric (Numero Hors) 12:247–259
Berger WG, Herrmann K (1971) Catechine und deren Abbauprodukte in Weintrauben, Trestern, Essigweinen und daraus hergestelltem Essig. Z Lebensm Unters Forsch 146:266–278
Bishop LR (1972) The measurement of total polyphenols in worts and beers. J Inst Brew 78:37–38
Blouin J, Cordeau J (1978) Mesure de la richesse potentielle des raisins en anthocyanes et polyphenols totaux. Bull Liaison Groupe Polyphenols 8:476–488
Blouin J, Llorca L, Montreau FR, Dufour JH (1972) Étude des conditions optimales pour la détermination des composés phénoliques totaux por le réactif de Folin-Ciocalteu. Connaiss Vigne Vin 6:405–413
Bourzeix M (1978) Estimation qualitative de la matiére colorande des moûts, des moûts concentrés et des vins. Mesure de son taux de polymères. Bull Liaison Groupe Polyphenols 8:459–467
Bourzeix M, Heredia N (1967) L'Isolement et le dosage des faibles quantités d'anthocyanes diglucosides dans les vins et les jus de raisin. Ann Technol Agric 16:357–364
Bourzeix M, Saquet H (eds) (1974) Les anthocyanes du raisin et du vin. Vignes Vins Spec No: 1–88 Institut Technique du Vin, Paris
Bourzeix M, Heredia N, Estrella Pedrola M-I (1982) Le dosage des anthocyanes des vins. Sci Aliment 2:71–82
Burkhardt R (1976) Analytische Bestimmung phenolischer Inhaltsstoffe von Mosten und Weinen. Mitteilungsbl GDCh-Fachgruppe Lebensmittelchem Gerichtl Chem 30:206–213
Butler LG (1982) Relative degree of polymerization of sorghum tannin during seed development and maturation. J Agric Food Chem 30:1090–1094
Butler LG, Price ML, Brotherton JE (1982) Vanillin assay for proanthocyanidins (condensed tannins): modification of the solvent for estimation of degree of polymerization. J Agric Food Chem 30:1087–1089
Cabezudo MD, De Gorostiza EF, Llaguno C (1971) Detección de conservadores en vinos por cromatografia en capa fina. Rev Agroquim Tecnol Aliment 11:526–540
Carruba E, D'Agostino S, Pastena B, Alagna C, Torina G (1981) Indagine analytica su vini Siciliani monovarietali. I. Vini bianchi. Riv Vitic Enol 34:359–400
Carruba E, D'Agostino S, Pastena B, Alagna C, Torina G (1982) Indagine analytica su vini Siciliani monovarietali. II. Vini rossi. Riv Vitic Enol 35:47–79
Castino M (1979) Sulla valutazione dei polifenoli totali nei vini rossi. Riv Vitic Enol 32:404–413
Cela R, Natera R, Pérez-Bustamante JA (1982) Determinacion de indices de polifenoles totales en mostos y vinos blancos. An Bromatol 34:207–217
Cela R, Buitago J, Pérez-Bustamante JA (1983) Modification of the Montedoro-Fantozzi method for routine determination of tannins by precipitation sensitized with methylcellulose and sodium dodecylbenzene sulfonate (trans). Quim Anal (Barcelona) 2:152–160
Cheynier VF, Trousdale EK, Singleton VL, Salgues MJ, Wylde R (1986) Characterization of 2-S-glutathionylcaftaric acid and its hydrolysis in relation to grape wines. J Agric Food Chem 34:217–221
Christensen EN, Caputi A Jr (1968) The quantitative analysis of flavonoids and related compounds in wine by gas-liquid chromatography. Am J Enol Vitic 19:238–245
Czochanska Z, Foo LY, Porter LJ (1979) Compositional changes in lower molecular weight flavans during grape maturation. Phytochemistry 18:1819–1822
Dadic M (1974) Spectrophotometric method for simultaneous determination of anthocyanogens and catechins (tanninogens). J Assoc Off Anal Chem 57:323–328
Daigle DJ, Conkerton EJ (1983) Analysis of flavonoids by HPLC. J Liq Chromatogr 6 (Suppl 1):105–118
Davidovic A, Michelic F, Tabakovic I, Davidovic D (1985) Polarographic determination of the anthocyanin malvin in wine (trans) Prehrambeno-Technol Rev 23:19–22; Chem Abst 104:87035

De Clerck J, Jerumanis J (1967) Dosage des polyphénols in brasserie. Bull Assoc R Anc Etud Brass Louv 63:137–161
Delcour JA, Janssens de Varebeke D (1985) A new colorimetric assay for flavanoids in Pilsner beers. J Inst Brew 91:37–40
Dimitriadis E, Williams PJ (1984) The development and use of a rapid analytical technique for estimation of free and potentially volatile monoterpene flavorants of grapes. Am J Enol Vitic 35:66–71
Donko E (1975) A borok összesfenol-meghatározasi módszereinek összehasonlítása. Szolesz Boraszat 1:343–355
Donko E, Phiniotis E (1975) A cukrok zavaró hatása az összesfenolmeghatározàsban Folin-Ciocalteu reagens hasznélata esetén. Szolesz Boraszat 1:357–366
Drawert F, Leupold G (1976a) Gaschromatographische Bestimmung der Inhaltsstoffe von Gärungsgetränken. 5. Quantitative gaschromatographische Bestimmung von Anthocyanen und Anthocyanidinen als TMS-Derivate in Rotweinen. Z Lebensm Unters Forsch 162:401–406
Drawert F, Leupold G (1976b) Gaschromatographische Bestimmung der Inhaltsstoffe von Gärungsgetränken. 6. Quantitative gaschromatographische Bestimmung von Neutralstoffen (Kohlenhydraten) und phenolischen Verbindungen in Tokajer Weinen. Z Lebensm Unters Forsch 162:407–414
Drawert F, Pivernetz H, Leupold G, Ziegler A (1980) Säulen- und dünnschichtchromatographische Trennung und spektralphotometrische Bestimmung von Rutin, Hesperidin und Naringin, besonders in Citrusfrüchten. Chem Mikrobiol Technol Lebensm 6:131–136
Estrella MI, Hernandez MT, Olano A (1986) Changes in polyalcohol and phenol compound contents in the aging of sherry wines. Food Chem 20:137–152
Etievant PX (1981) Volatile phenol determination in wine. J Agric Food Chem 29:65–67
Evans ME (1983) High-performance liquid chromatography in enology. J Liq Chromatogr 6 (Suppl 2):153–178
Flanzy M, Aubert S, Marinov M (1969) Nouvelle technique de dosage des tanins leucoanthocyaniques. Applications. Ann Technol Agric 18:327–338
Francis FJ (1982) Analysis of Anthocyanins. In: Markakis P (ed) Anthocyanins as food colors. Academic Press, New York
Garcia Barosso C, Cela Torrijos R, Perez-Bustamante JA (1983) HPLC separation of benzoic and hydroxycinnamic acids in wines. Chromatographia 17:249–252
Garcia Barosso C, Cela Torrijos R, Perez-Bustamante JA (1986) Evolution of phenolic acids and aldehydes during the different production process of "fino" sherry wine. Z Lebensm Unters Forsch 182:413–418
Gattuso AM, Indovina MC, Pirrone L (1986) Costituenti polifenolici di vini bianchi grezzi della Sicilia Occidentale. Vignevini 13(4):35–38
Gigliotti A (1973) La determinazione dei composti polifenolici nei vini rossi. Riv Vitic Enol 26:183–193
Gigliotti A (1974) La determinazione dei composti polifenolici nei vini bianchi. Riv Vitic Enol 27:206–214
Gigliotti A, Bucelli P (1986) Influenza della tecnologia di vinificazione adottata sulle caratteristiche organolettiche del vino rosato. Riv Vitic Enol 39:158–172
Glories Y (1978) Evolution des composés phénoliques au cours du vieillissement du vin. Ann Nutr Aliment 32:1163–1169
Glories Y, Augustin M (1980–1981) Influence du SO_2 sur le dosage des anthocyanes dans les vins. Inst Oenol Univ Bordeaux Rapport Act Rech, pp 88–92
Glories Y, Ribéreau-Gayon P (1973) Contribution a l'étude de la determination de l'etat de condensation des tanins des vins rouges. Connaiss Vigne Vin 7:15–38
Godinho OES, Coelho JA, Chagas AP, Aleixo LM (1984) Investigation of the use of thermometric titrimetry for the determination of acidic substances in wine. Talanta 31:218–220
Goldstein JL, Swain T (1963) Methods for determining the degree of polymerization of flavans. Nature 198:587–588

Görtges S (1981) Störfaktoren bei der Polyphenolbestimmung mit Folin-Ciocalteu-Reagenz. Flüss Obst 48:522–523
Güntert M, Rapp A, Takeoka GR, Jennings W (1986) HRGC and HRGC-MS applied to wine constituents of lower volatility. Z Lebensm Unters Forsch 182:200–204
Hardin JM, Stutte CA (1984) Chromatographic analyses of phenolic and flavonoid compounds. In: Lawrence JF (ed) Food constituents and food residues, their chromatographic determination. Marcel Dekker, New York
Hillis WE, Urbach G (1959) The reaction of (+)-catechin with formaldehyde. J Appl Chem (Lond) 9:474–482
Hrazdina G (1975) Anthocyanin composition of Concord grapes. Lebensm Wiss Technol 8:111–113
Hrazdina G (1979) Recent techniques in the analysis of anthocyanins in fruits and beverages. In: Charalambous G (ed) Liquid chromatographic analysis of food and beverages. Vol 1. Academic Press, New York
Hrazdina G, Franzese AJ (1974) Structure and properties of the acylated anthocyanins from *Vitis* species. Phytochemistry 13:225–229
Inskip EB, King P, Ziegler HW (1973) Spectrophotometric determination of gallotannins in beer. J Assoc Off Anal Chem 56:1362–1364
Joslyn MA, Morris M, Hugenberg G (1968) Die Bestimmung der Gerbsäure und verwandter Phenolsubstanzen mittels des Phosphormolybdat-Phosphorwolframat-Reagenz. Mitt Klosterneuburg 18:17–34
Jouret C (1967) Détection des diglucosides anthocyaniques dans les vins suivant la technique de Dorier et Verelle. Ann Technol Agric 16:373–377
Khomenko VA, Oloinik EG (1968) Metod potentsiometricheskogo titrovaniya s polyarizovannym elektrodom dla opredeleniya dybil'nykh i krasyashchikh veshchestv. Vinodel Vinograd 28(5):15–17
King HGC, White T (1956) The quantitative determination of specific nuclei and components of vegetable tannin extracts. In: The chemistry of vegetable tannins. Soc Leather Trades Chemists, London, pp 31–56
Kramling TE, Singleton VL (1969) An estimate of the nonflavonoid phenols in wines. Am J Enol Vitic 20:86–92
LaNotte E, Antonacci D (1985) Composizione polifenolica del vino rosso „Castel del Monte" nel corso dell' invecchiamento. Riv Vitic Enol 38:367–398
Lea AGH (1982) Reversed-phase high-performance liquid chromatography of procyanidins and other phenolics in fresh and oxidising apple juices using a pH shift technique. J Chromatogr 238:253–257
Lea AGH, Bridle P, Timberlake CF, Singleton VL (1979) The procyanidins of white grapes and wines. Am J Enol Vitic 30:289–300
Leupold G, Drawert F (1981) Zur Analytik phenolischer Verbindungen im Bier. Brauwissenschaft 34:205–210
Mabry TJ, Markham KR, Thomas MB (1970) The systematic identification of flavonoids. Springer, Berlin Heidelberg New York
Mareca-Cortez I, Mareca Lopez I, Tienda Priego P (1980) Une méthode polarographique pour l'analyse quantitative des anthocyanes. Bull Liaison Groupe Polyphenols 9:143
Margheri G (1960) Determinazione dei tannini nel vino mediante spettrofotometria nell'ultravioletto. Riv Vitic Enol 13:401–406
Margheri G (1974) Importanza tecnologica della conescenza della sostanze polifenoliche dei vini, con particolare riferimento alle fasi di stabilizzazione ed invecchiamento. Enotecnico 10(2):10–12, 14, 16–18
Margheri G, Tonon D (1982) Il significato tecnologico dei composti fenolici nei vini bianchi e rossi. Enotecnico 18(11):7–12
Margheri G, Avancini D, Romagnoli M, Martinelli C (1974) Recherches pour la caractérisation des vins à dénomination d'origine controlée de la région Trentino-Alto Adige, Caldaro (Kalterer) ou Lac de Caldaro (Kalterersee) Vini Ital 16(93):461–470
Markh AT, Zykina TF (1969) A chromatographic study of polyphenols of grapes and grape juice (trans). Prikl Biokhim Mikrobiol 5:189–194
Markham KR (1982) Technique of flavonoid identification. Academic Press, New York

Masquelier J, Vitte G, Ortega M (1959) Dosage colorimétrique des leucoanthocyannes dans vins rouges. Bull Soc Pharm Bordeaux 98:145–148
Mattick LR, Weirs LD, Robinson WB (1969) Effect of pH on the precipitation of the anthocyanins of wine as their lead salts. Am J Enol Vitic 20:206–212
McCloskey LP, Yengoyan LS (1981) Analysis of anthocyanins in *Vitis vinifera* wines and red color versus aging by HPLC and spectrophotometry. Am J Enol Vitic 32:257–261
Miskov O, Bourziex M, Heredia N (1970) Isolement individuel des acides-phénols et des catechins et évaluation densitometrique de leurs tenuers respectives dans les moûts et jus de raisin rouges et blancs et dans les vins blancs. Ind Aliment Agric 87:1515–1518
Mitchell CA (1923) The colorimetric estimation of pyrogallol, gallotannin and gallic acid. Analyst 48:2–15
Mitjavila S, Schiavon M, Derache R (1971) Teneur en tannins des boissons séparations et dosage colorimétrique de la fraction polyphénolique astringente. Ann Technol Agric 20:335–346
Möbius CH, Görtges S (1974a) Einfache Gerbstoffkontrolle in Wein und Most. Dtsch Weinbau (33–34):1236–1238
Möbius CH, Görtges S (1974b) Polyphenolbestimmung für die Praxis. Wein-Wiss 29:241–253
Montedoro G, Fantozzi P (1974) Dosage des tannins dans les moûts et les vins à l'aide de la méthylcellulose et évaluation d'autres fractions phénoliques. Lebensm Wiss Technol 7:155–161
Montreau FR (1972) Sur le dosage des composés phénoliques totaux dans les vins par la methode Folin-Ciocalteu. Connaiss Vigne Vin 6:397–404
Moutounet M (1981) Dosages des polyphénols des moûts de raisin. Connaiss Vigne Vin 15:287–301
Moutounet M, Chudzikiewicz J (1980) Frationnement de la matière colorate de vins rouges. Bull Liaison Groupe Polyphenols 9:145–148
Munoz Alcon dH, Marine Font A (1978) Determinación de polifenoles en vinos. An Bromatol 30:107–122
Myers TE, Singleton VL (1979) The nonflavonoid phenolic fraction of wine and its analysis. Am J Enol Vitic 30:98–102
Nagel CW (1985) Application of high-performance liquid chromatography to analysis of flavonoids and phenyl propanoids. Cereal Chem 62:144–147
Nagel CW, Wulf LW (1979) Changes in the anthocyanins, flavonoids and hydroxycinnamic esters during fermentation and aging of Merlot and Cabernet Sauvignon. Am J Enol Vitic 30:111–116
Niketić-Aleksić GK, Hrazdina G (1972) Quantitative analysis of the anthocyanin content in grape juices and wines. Lebensm Wiss Technol 5:163–165
Nozaki K, Yokotsuka K (1985) Chemical studies on coloring and flavoring substances in Japanese grapes and wines. 25. Paper chromatography of phenolic compounds from seeds and skins of 32 grape varieties. J Inst Enol Vitic, Yamanashi Univ 20:1–15
Nykänen L, Suomalainen H (1983) Aroma of beer, wine and distilled alcoholic beverages. Akademie-Verlag, Berlin
Office International de la Vigne et du Vin (1978) Recueil des méthodes internationales d'Analyse des vins. OIV, Paris
Ohta H, Tonohara K, Naitoh T, Kohono K, Osajima Y (1983) Effect of temperature on color of Concord grape juice. Nippon Shokuhin Kogyo Gakkaishi 30:290–295
Okamura S, Watanabe M (1981) Determination of phenolic cinnamates in white wine and their effect on wine quality. Agric Biol Chem 45:2063–2070
Ong BY, Nagel CW (1978) High-pressure liquid chromatographic analysis of hydroxycinnamic acid-tartaric acid esters and their glucose esters in *Vitis vinifera*. J Chromatogr 157:345–355
Otsuka K (1967) Ripe rot disease of grapevine. III. Waxy substance (oleanolic acid) on grape pericarp, and its decomposition by ripe rot fungus. Nippon Jozo Kyokai Zasshi 62:323–325
Pakhomova EG, Georgiyan SA, Klyachko YuA (1980) Determination of phenolic substances of wine by amperometric titration. Izv Vyssh Uchebn Zaved Pishch Tekhnol (5):97–99; Chem Abst 94:45519h

Papaconstantinou E, Pope MT (1970) Heteropoly blues. V. Electronic spectra of one- to six-electron blues of 18-metallodiphosphate anions. Inorg Chem 9:667–669
Paronetto L (1977) Polifenoli e technica enologica. Edagricole, Bologna
Peri C, Pompei C (1971 a) Estimation of different phenolic groups in vegetable extracts. Phytochemistry 10:2187–2189
Peri C, Pompei C (1971 b) An assay of different phenolic fractions in wines. Am J Enol Vitic 22:55–58
Pompei C, Peri C (1971) Determination of catechins in wines. Vitis 9:312–316
Pompei C, Peri C, Montedoro G (1971) Le dosage des leucoanthocyanes dans les vins blancs. Ann Technol Agric 20:21–34
Preston NW, Timberlake CF (1981) Separation of anthocyanin chalcones by high performance liquid chromatography. J Chromatogr 214:222–228
Pridham JB (1964) Methods in polyphenol chemistry. Macmillan, New York
Puisais J, Guiller A, Lacoste J, Huteau P (1968) Dosage spectrophotométrique des tanins. Ann Technol Agric 17:277–285
Quastel JH (1931) A colour test for *o*-dihydroxy-phenols. Analyst 56:311
Rebelein H (1965 a) Beitrag zur Bestimmung des Catechingehaltes in Wein. Dtsch Lebensm Rundsch 61:182–183
Rebelein H (1965 b) Beitrag zum Catechin- und Methanolgehalt von Weinen. Dtsch Lebensm Rundsch 61:239–240
Revilla E, Carpena O, Mataix JJ (1985) Composiciòn polifenolica del limbo foliar de los cultivares de vid Airen y Cencibel durante la fructificaciòn. An Edafol Agrobiol 44:787–797
Revilla E, Alonzo E, Estrella MI (1986 a) Analysis of flavonol aglycones in wine extracts by high-performance liquid chromatography. Chromatographia 22:157–159
Revilla E, Mataix JJ, Carpena O (1986 b) C_6–C_1 and C_6–C_3 phenolic substances in grapevine leaves, cultivars Airen and Cencibel during fruit development. Rev Agroquim Tecnol Aliment 26:234–238
Ribéreau-Gayon P (1959) Recherches sur les anthocyanes des végétaux. Application au genre *Vitis*. Librairie Générale de l'Enseignement, Paris
Ribéreau-Gayon P (1972) Plant Phenolics. Oliver & Boyd, Edinburgh
Ribéreau-Gayon P (1982) The anthocyanins of grapes and wines. In: Markakis P (ed) Anthocyanins as food colors. Academic Press, New York
Ribéreau-Gayon P, Nedeltchev N (1965) Discussion et application des méthode modernes de dosage des anthocyanes et des tanins dans les vins. Ann Technol Agric 14:321–330
Ribéreau-Gayon P, Sartore F (1970) Le dosage des composés phénoliques totaux dans les vins rouges. Chim Anal 52:627–631
Ribéreau-Gayon P, Stonestreet E (1965) Le dosage des anthocyanes dans le vin rouge. Bull Soc Chim France 2649–2652
Ribéreau-Gayon P, Stonestreet E (1966) Dosage des tanins du vin rouge et determination de leur structure. Chim Anal 48:188–196
Ribéreau-Gayon J, Peynaud E, Sudraud S, Ribéreau-Gayon P (1976) Traité d'oenologie: Sciences et techniques du vin. Vol 1 Analyse et contrôle des vins. Dunod, Paris
Roberts EAH (1956) Paper chromatography as an aid to the elucidation of the structure of polyphenols occurring in tea. In: The chemistry of vegetable tannins. Soc Leather Trades Chemists, London, pp 87–97
Roggero JP, Ragonnet B, Coen S (1984) Analyse fine des anthocyanes des vins et des pellicules de raisin par la technique H.P.L.C. Etude de quelques cépages méridionaux. Vigne Vins (327):38–42
Roggero JP, Coen S, Ragonnet B (1986) High performance liquid chromatography survey on changes in pigment content in ripening grapes of Syrah. An approach to anthocyanin metabolism. Am J Enol Vitic 37:77–83
Romeyer FM, Macheix JJ, Goiffon JJ, Reminiac CC, Sapis JC (1983) Browning capacity of grapes. 3. Changes and importance of hydroxycinnamic acid-tartaric acid esters during development and maturation of the fruit. J Agric Food Chem 31:346–349
Roson JP, Jouret C, Goutel AM (1978) Dosage des anthocyanes totals et des anthocyanes ionisées dans les vins rouges. Rev Fr Oenol (69):15–21

Roston DA, Kissinger PT (1981) Identification of phenolic constituents in commercial beverages by liquid chromatography with electrochemical detection. Anal Chem 53:1695–1699

Salagoity-Auguste MH, Bertrand A (1984) Wine phenolics – analysis of low molecular weight components by high performance liquid chromatography. J Sci Food Agric 35:1241–1247

Salem FB (1985) Colorimetric determination of certain sympathomimetic amines. Anal Lett 18(B9):1063–1075

Sankar SK, Howarth RE (1976) Specificity of the vanillin test for flavanols. J Agric Food Chem 24:317–320

Schanderl SH (1970) Tannins and related phenolics. In: Joslyn MA (ed) Methods in food analysis. Physical, chemical and instrumental methods. 2nd edn. Academic Press, New York, pp 701–725

Schreier P, Drawert F (1977) Gaschromatographisch-massenspektrometrische Identifizierung flüchtiger Säuren und Phenole im Trauben- und Weinaroma. In: Appl Spectrom Masse (SM) Reson Magn Nucl (RMN) Ind Aliment (Symp Int Comm Int Ind Agric Aliment) 15:141–150

Seikel MK (1964) Isolation and identification of phenolic compounds in biological materials. In: Harborne JB (ed) Biochemistry of phenolic compounds. Academic Press, New York

Siewek F, Galensa R, Herrmann K (1985) Nachweis eines Zusatzes von Feigensaft zu Traubensaft und daraus hergestellten alkoholischen Erzeugnissen über die HPLC-Bestimmung von Flavon-C-glykosiden. Z Lebensm Unters Forsch 181:391–394

Singleton VL (1974) Analytical fractionation of the phenolic substances of grapes and wine and some practical uses of such analyses. In: Webb AD (ed) Chemistry of winemaking. American Chemical Society, Washington DC. Adv Chem 137:184–211

Singleton VL (1982) Grape and wine phenolics: background and prospects. In: Webb AD (ed) University of Calif Davis Grape and Wine Centennial Symp Proc 1880–1980, pp 219–227. Dept Vitic Enol, Davis, California

Singleton VL, Esau P (1969) Phenolic substances in grapes and wine, and their significance. Adv Food Res Suppl 1 Academic Press, New York

Singleton VL, Noble AC (1976) Wine flavor and phenolic substances. In: Charalambous E, Katz I (eds) Phenolic, sulfur and nitrogen compounds in food flavors. American Chemical Society, Washington DC. Am Chem Soc Symp Ser 26:47–70

Singleton VL, Rossi JA Jr (1965) Colorimetry of total phenolics with phosphomolybdic-phosphotungstic acid reagents. Am J Enol Vitic 16:144–158

Singleton VL, Trousdale E (1983) White wine phenolics: varietal and processing differences as shown by HPLC. Am J Enol Vitic 34:27–34

Singleton VL, Draper DE, Rossi JA Jr (1966) Paper chromatography of phenolic compounds from grapes, particularly seeds, and some variety-ripeness relationships. Am J Enol Vitic 17:206–217

Singleton VL, Sullivan AR, Kramer C (1971) An analysis of wine to indicate aging in wood or treatment with wood chips or tannic acid. Am J Enol Vitic 22:161–166

Singleton VL, Timberlake CF, Whiting GC (1977) Chromatography of natural phenolic cinnamate derivatives on Sephadex LH-20 and G-25. J Chromatogr 140:120–124

Singleton VL, Timberlake CF, Lea AGH (1978) The phenolic cinnamates of white grapes and wine. J Sci Food Agr 29:403–410

Singleton VL, Zaya J, Trousdale E, Salgues M (1984) Caftaric acid in grapes and conversion to a reaction product during processing. Vitis 23:113–120

Singleton VL, Salgues M, Zaya J, Trousdale E (1985) Caftaric acid disappearance and conversion to products of enzymic oxidation in grape must and wine. Am J Enol Vitic 36:50–56

Slinkard K, Singleton VL (1977) Total phenol analysis: automation and comparison with manual methods. Am J Enol Vitic 28:49–55

Somers TC (1968) Pigment profiles of grapes and of wines. Vitis 7:303–320

Somers TC, Evans ME (1977) Spectral evaluation of young red wines: Anthocyanin equilibria, total phenolics, free and molecular SO_2, "chemical age." J Sci Food Agric 28:279–287

Somers TC, Ziemelis G (1980) Gross interference by sulphur dioxide in standard determinations of wine phenolics. J Sci Food Agric 31:600–610
Somers TC, Ziemelis G (1985a) Flavonol haze in white wines. Vitis 24:43–50
Somers TC, Ziemelis G (1985b) Spectral evaluation of total phenolic components in *Vitis vinifera:* grapes and wines. J Sci Food Agric 36:1275–1284
Swain T, Hillis WE (1959) The phenolic constituents of *Prunus domestica.* I. The quantitative analysis of phenolic constituents. J Sci Food Agric 10:63–68
Symonds P (1978) Application de la chromatographie liquid haut performance du dosage de quelques acids organiques. Ann Nutr Aliment 32:957–968
Timberlake CF, Bridle P (1976) Interactions between anthocyanins, phenolic compounds, and acetaldehyde and their significance in red wines. Am J Enol Vitic 27:97–105
Timberlake CF, Bridle P (1977) Anthocyanins: color augmentation with catechin and acetaldehyde. J Sci Food Agric 28:539–544
Timberlake CF, Bridle P (1980) Anthocyanins. In: Walford J (ed) Developments in food colours – I. Applied Science Publishers, Barking, Essex
Trousdale EK, Singleton VL (1983) Astilbin and engeletin in grapes and wines. Phytochemistry 22:619–620
Ubigli M, Barbero L (1983) Studio chimico e fisico-chimico del Barolo. Vignevini 10(10):39–44
Valouiko GG, Pavlenko NM, Stankova NV (1980) Sur la determination des acides phénolcarboxyliques par chromatographie en phase gazeuse. Bull Liaison Groupe Polyphenols,, pp 302–305
Van Buren J (1970) Fruit phenolics. In: Hulme AC (ed) The biochemistry of fruits and their products. Vol 1. Academic Press, New York
Villa D (1985) Caratteristiche chimiche, chimico-fisiche ed organolettiche del vino „Pinot bianco“ in Abruzzo. Vignevini 12(4):37–42
Williams AA, Lewis MJ, May HV (1983) The volatile flavour components of commercial port wines. J Sci Food Agric 34:311–319
Wulf LW, Nagel CW (1976) Analysis of phenolic acids and flavonoids by high-pressure liquid chromatography. J Chromatogr 116:271–279
Wulf LW, Nagel CW (1978) High-pressure liquid chromatographic separation of anthocyanins of *Vitis vinifera.* Am J Enol Vitic 29:42–49
Wulf LW, Nagel CW (1980) Identification and changes of flavonoids in Merlot and Cabernet Sauvignon wines. J Food Sci 45:479–484
Yokotsuka K, Shinkai S, Kushida T (1980) Fraction of wine tannin and nontannin phenolics by Bio-Gel P2 chromatography and characterization of the fraction. J Ferment Technol 58:107–113
Zloch Z (1985) Dukoz a stanoveni antokyanu v barevnych napojich. Kvasny Prum 31(3):58–60

Phenolic Composition of Natural Wine Types

T. C. Somers and E. Vérette

1 Introduction

The diversity of wine types and styles is largely due to extreme variability in concentration and composition of the phenolic constituents, which are also intrinsic to the maturation and ageing of red wines. From vineyard to cellar, the elaboration of fine wines can be broadly interpreted in terms of phenolics management in relation to required sensory characteristics for each particular wine type; for the phenolic composition is affected, qualitatively and quantitatively, by widely variable aspects of vinification and conservation practice.

Research advances have been generally associated with emergence of new analytical methods and improved instrumentation. First indications of the complexity of total phenolic extractives from *Vitis vinifera* and other *Vitis* species came with resolution and identification, by paper chromatography, of the anthocyanin pigments and other monomeric phenolics of wine grapes. This definitive work was initiated and largely conducted at the University of Bordeaux in the period 1953–1964. A significant early discovery was that red wines of *V. vinifera,* the European wine grape species, can be differentiated from those of other species and of inter-specific hybrids with *V. vinifera* by chromatographic inspection of residual anthocyanins (P. Ribéreau-Gayon 1959, 1964, 1974).

First isolations of wine tannins were achieved by gel filtration, enabling recognition of the fact that whereas the colour of new wine is largely due to the monomeric grape anthocyanins, there is dynamic and continuous change towards more stable oligomeric or polymeric pigments during bulk storage and bottle ageing, with progressive decline in anthocyanin levels (Somers 1966, 1968). Of diverse molecular size and heterogeneous stereochemical structures (mol. wt. range 1000–4000), components of this major phenolic fraction have not been amenable to any kind of chromatographic resolution. As chemical complexity increases with time, and there is evidence for involvement of the total phenolics in polymeric pigments and related tannin structures of aged red wines (Somers 1971), the analytical problems are formidable, if not quite impossible, in specific terms.

Since the early 1960's, however, recording spectrophotometry has offered a truly comprehensive approach to investigations of wine phenolics. For red wines, physicochemical phenomena may have extremely large influences on the wine coloration, and have been recognized as central to interpretations of colour density and tint as seen in the glass; detailed analysis of the total phenolic composition, were it ever possible, would always be inadequate in this regard. Spectrophotometric analysis of the intact wine is therefore the more realistic way towards making useful connection between objective method and subjective appreciation of red wine.

The two analytical approaches, chromatography and spectrophotometry, are of course complementary. The advent of high performance liquid chromatography (HPLC), first applied to investigations of grape phenolics by Wulf and Nagel in 1978 and to wine phenolics in 1979 (Nagel and Wulf 1979), has revealed much finer analytical detail than was previously possible.

Longstanding problems in assessment of phenolics in white wines are being resolved by this means, enabling introduction of direct spectrometric analysis, which should find application in commercial practice. Before HPLC, quantitation of individual phenolic components from grapes or wines had been laborious and uncertain; from a few μl wine sample, HPLC enables resolution and quantitative analysis of numerous phenolic monomers in 30 min! There are now many more discrete components to be identified, if there is good reason to do so. Precise interpretation of dynamic change in phenolic composition during white vinification is already possible, and there has been similar advance in knowledge of decreasing concentrations of monomeric phenolics during red vinification and conservation. For the latter, however, the increasing oligomeric and polymeric pigments have continued to defy detailed analysis.

In this chapter, we present our assessment of the present state of the art. Emphases have been given to emerging concepts for dealing with the total phenolics of natural wines, to evaluation of their component categories, to interpretations of physicochemical phenomena operative during the making of red wines, and to relevant methods chosen for their general utility in oenological research and development.

2 Phenolic Components of Wine Grapes

Structural types present in grapes of *V. vinifera* are illustrated in Fig. 1. All types are present in red grapes, with quantitative variabilities, notably in amounts of anthocyanins, for individual grape cultivars and in response to environmental and seasonal influences (Somers 1968; P. Ribéreau-Gayon 1972; Bakker et al. 1986b). There are several hundred cultivars of *V. vinifera* in cultivation around the world, though relatively few are prominent in wine grape production. Genetic factors account for the presence or absence of anthocyanin pigments, and for the 3-monoglucoside character of anthocyanins in red cultivars of *V. vinifera*. Other *Vitis* species, which are generally unsuitable for production of fine wines, are characterized by the presence of 3,5-diglucoside derivatives of the same anthocyanidins (P. Ribéreau-Gayon 1964, 1974).

The distribution of phenolics in the berry structure is most significant for the technology of table wines, and for appreciation of the diversity of wine styles. Thus the flavonoid components, all based on the same C_{15} skeletal structure, are localized in skins, seeds and vascular tissue, whereas the smaller non-flavonoid phenolics are also present in the juice vacuoles. Wine type, determined by the quality and quantity of total phenolics, is then a function of grape maceration during the vinification process.

Astringency in young red wines is due to procyanidins (also termed proanthocyanidins, "condensed tannins", flavans or flavolans) which, along with anthocyanins, catechins and flavonols (low levels) are readily leached from grape skins during fermentation. Stereochemical configurations abound in these reactive oligomers, in which the C_{15} structural units are flavan-3-ols (Fig. 1). Grape seeds are also rich in procyanidins, containing up to 60% of the total grape phenolics;

their contribution may be large where maceration is prolonged, but as skin extraction occurs more rapidly than that from seeds, much depends on the stage at which the ferment or wine is pressed. Grape stalks, normally excluded from the ferment, are a further source of these astringent phenolics. Siegrist (1985) has reported on what appear to be characteristic differences in uptake of tannins from these three sources during fermentation; he remarked on variability of extract and the impossibility of its prediction. Estimates of total phenolics in numerous red wines from many regions have been from 800–4000 mg l^{-1} (Singleton and Esau 1969); most of the phenolic extract is invariably procyanidin in type.

It is, however, important to note that, even with prolonged skin contact and maceration, phenolic extraction is always much less than 50% of the amount

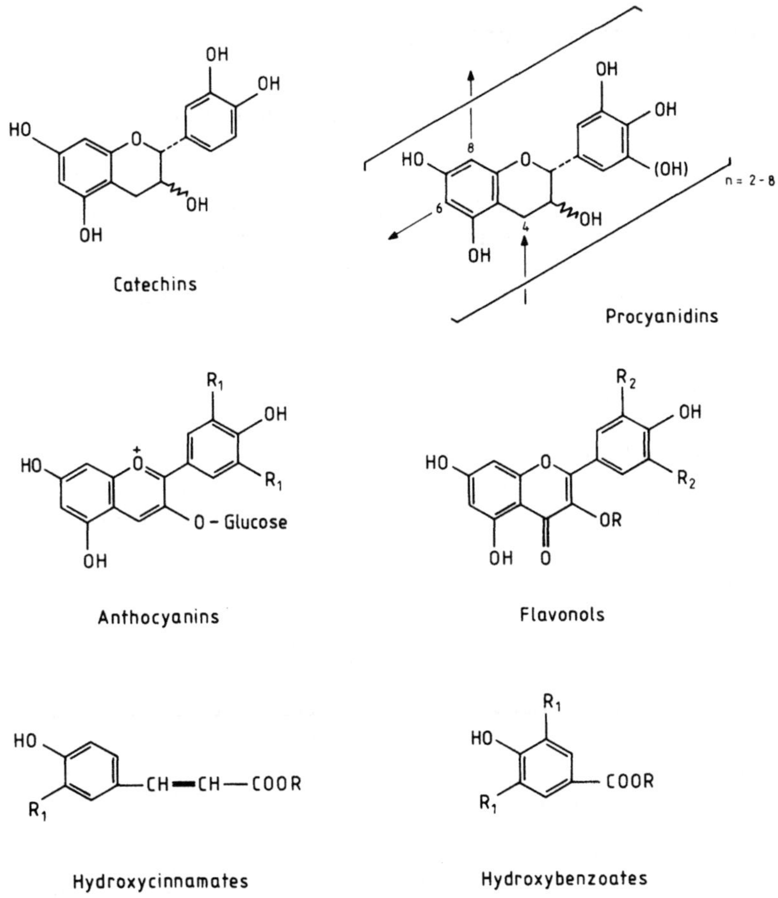

Fig. 1. Phenolic constituents in *V. vinifera*. R_1 = H, OH or OCH_3; R_2 = H or OH. Anthocyanins, in red grapes, are also present as acylated derivatives involving acetic, *p*-coumaric and caffeic acids. Hydroxycinnamates are present principally as trans-esters of tartaric acid, hydroxy benzoates as free acids and unidentified esters, flavonols as 3-glucosides and 3-glucuronides. Procyanidins consist essentially of catechin oligomers and polymers with covalent bonding as indicated; trihydroxy substitution of the B-ring also occurs as a minor structural variation

available and varies greatly in commercial practice; limiting factors may be both viticultural and oenological in origin (Somers and Pocock 1986). The conduct of red vinification is necessarily empirical, and seems likely to remain so in this critical aspect.

For white wines, the problems are less complex, as minimal skin contact after crushing generally ensures the relative absence of C_{15} phenolics. Major phenolic components in white wines are then the juice soluble hydroxycinnamates; hydroxybenzoates are present at much lower concentrations (Fig. 1).

Thus vinification has been aptly described as the "art of fractional extraction of the grape" (E. Peynaud, Bordeaux).

3 Measurement of Total Wine Phenolics

Before consideration of dynamic change in phenolic composition of wine, it is useful to examine the problem of their total measurement at any stage of the wine development. The subject is important because of the recognized relationship between phenolic concentrations and wine type. Some measure of the total phenolics has long been regarded as necessary to adequate description of wine, and much of the technical literature has been presented in these terms (Singleton and Esau 1969; J. Ribéreau-Gayon et al. 1972). Furthermore, it should be noted that modern analytical methods, e.g. HPLC, have no practical application in this regard.

Because of their chemical diversity, the total phenolics must be assessed in arbitrary units, or by reference to a phenolic standard; their quantitation as mg l^{-1} is necessarily indirect and approximate. Earliest measures involved removal of ethanol and titration with permanganate before and after treatment with active carbon (Neubauer 1872). The method was widely used for at least a century, giving "permanganate indices" of total phenolics over a sixfold range for rosé wines to red wine pressings (J. Ribéreau-Gayon et al. 1972). It was superceded by the more reproducible Folin-Denis colorimetric procedure, and this in turn by the related Folin-Ciocalteu assay. This method depends on specific oxidation of phenolics in alkaline solution by complex phosphotungstic-phosphomolybdic reagents, with formation of blue pigments; results are expressed as mg l^{-1} "gallic acid equivalents" (Singleton and Rossi 1965; Singleton and Esau 1969).

For red wines, ultra-violet (UV) absorbance at 280 nm (E_{280}) has also been used as an index of the total phenolic components (Flanzy and Poux 1958). Although good correlations were obtained with data from the Folin-Ciocalteu assay, the latter was considered to be more specific for the phenolic function (J. Ribéreau-Gayon et al. 1972). For white wines, however, there was poor relationship between UV absorbance and the Folin-Ciocalteu assay, which then became established as the preferred procedure for measurement of total phenolics in all table wines (J. Ribéreau-Gayon et al. 1972; Singleton 1974).

Evidence of inherent error in this colorimetric method came with application of gel filtration to interpretation of UV absorbance in white wines, when much lower phenolic concentrations were indicated in UV elution profiles (Somers and

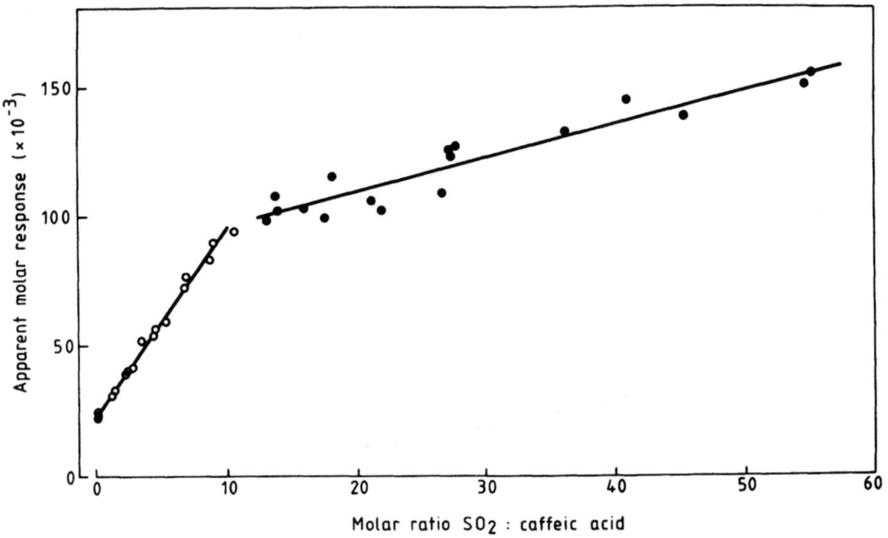

Fig. 2. Influence of relative concentrations on apparent molar response of caffeic acid to the Folin-Ciocalteu reagent in presence of SO_2. For molar ratios to 10:1, $y=23.8+7.09$ x ($r^2=0.99$), and for the range 10:1 to 50:1, $y=84.6+1.3$ x ($r^2=0.89$). (Somers and Ziemelis 1980)

Ziemelis 1972). It was eventually proved that, further to significant response from non-phenolic wine components, there is strong synergistic interaction between sulphur dioxide (SO_2) and *o*-dihydroxy phenols in response to the Folin-Ciocalteu reagent. The actual extent of such interference depends on the molar ratio of SO_2: *o*-dihydroxy phenols (Fig. 2). SO_2 is invariably used in commercial vinification, and this molar ratio may be as high as 10:1 in white wines. Thus the method was shown to be generally invalid in application to white wines; measures of "total phenolics" are frequently magnified more than twofold by this previously unsuspected SO_2 artefact (Somers and Ziemelis 1980). As red wines have much higher phenolic extract than whites, the molar ratio SO_2: *o*-dihydroxy phenols is very low, and error from the SO_2 artefact is then consequently small (Fig. 2).

3.1 Spectral Evaluation of White Wines

Whereas all red wines have UV absorption maxima at about 280 nm, the UV spectra of white wines vary greatly and there is no characteristic λ_{max}. Because of the dominance of hydroxycinnamates in the spectra of grape juices and white wines from free-run juices, principal absorbance is found at about 320 nm. In contrast, grape "pressings" and white wines made via appreciable "skin contact" show progressively higher absorbance in the region 250–350 nm, with λ_{max} at 280 nm (as in red wines). Correction for non-phenolic UV-absorbing compounds would therefore permit direct spectral measurement of total phenolics in white wines (and their more accurate assessment in red wines).

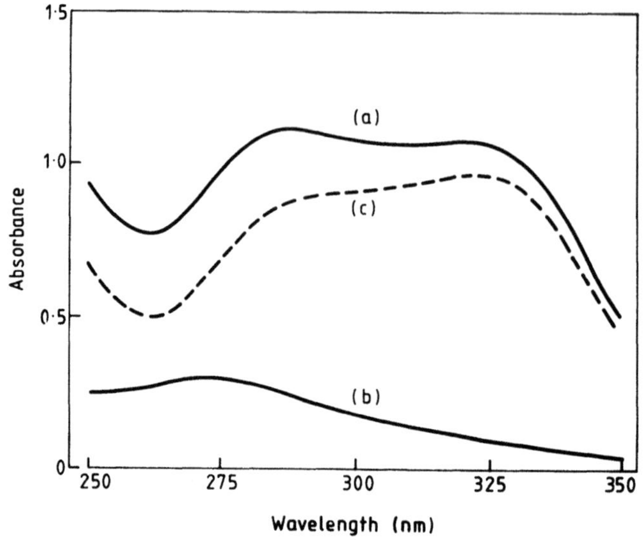

Fig. 3. Effect of heavy fining with PVPP (10% w/v) on the spectrum of a Riesling wine. The intact wine against water as reference (*a*): the wine after PVPP treatment against product from similar treatment of 12% aqueous ethanol saturated with potassium hydrogen tartrate (*b*). Light pathlength 1 mm. Phenolic absorbance is shown by the difference spectrum (*c*)

Gel column analysis enabled introduction of a correction factor (4.0 in 10 mm pathlength) for UV contributions of total non-phenolics in wines and juices at 280 nm (Somers and Ziemelis 1972; Somers and Evans 1977). This was supported by later study of the effects of massive fining (10% w/v) with polyvinyl polypyrrolidone (PVPP) on UV spectra of many white wines and juices (e.g. Fig. 3). Residual absorbance was found to be remarkably uniform, showing mean values (10 mm path) for non-phenolics as 4.0 at 280 nm and 1.4 at 320 nm (Somers and Ziemelis 1985 a).

3.1.1 Total Hydroxycinnamates

As there is no significant contribution from flavonoid phenolics at 320 nm, use of (E_{320}–1.4) as a routine measure of total hydroxycinnamates appears to be well justified. For wines from free-run juice, the difference spectrum obtained after PVPP treatment (Fig. 3) is in fact closely similar to that of caffeic acid, the major phenolic moiety of hydroxycinnamate esters in *V. vinifera*. As other hydroxycinnamates (Sect. 4.1) have similar spectral properties, all contributing to absorbance at 320 nm, the total hydroxycinnamates can be quantitated as "caffeic acid equivalents" (CAE mg l^{-1}). On the basis that caffeic acid has a molar extinction value (ε) of 18,000 in wine at 320 nm.

Total hydroxycinnamates = (E_{320} − 1.4) a.u.
Caffeic acid equivalents, CAE = 10 (E_{320} − 1.4) mg l^{-1}

In surveys of 230 commercial white wines, E_{320} ranged from 2.9–13.1 a.u., mean 7.2, corresponding to CAE values of 15–117 mg l^{-1}, mean 58 (Somers and Ziemelis 1985a). For juices and new wines, in which the hydroxycinnamic acids are present as esters of tartaric acid, the factor 7/4 enables conversion of CAE values to estimates of the total hydroxycinnamoyl tartaric acids. For fresh grape juices and young white wines, there are no known interferences to these spectral estimates, which compare well with HPLC measures of the total individual components (Somers et al. 1987).

3.1.2 Total Flavonoid Extractives

As excessive extraction from grape solids is associated with coarseness and astringency, as well as susceptibility to oxidative browning reactions; this aspect of phenolic composition has negative implications for quality in white wines.

Approximate assessment of flavonoid extractives (λ_{max} 280 nm) can be made from the wine spectrum after correction for absorbance of non-phenolics and of the hydroxycinnamates at 280 nm; E_{280} of the total hydroxycinnamates is about $^2/_3$ that of E_{320} (Somers and Ziemelis 1972). Thus the derivation:

Flavonoid content $= (E_{280} - 4) - {}^2/_3 (E_{320} - 1.4)$ a.u. has been proposed as being applicable to white wines or juices, and also to light rosé wines. For free-run juices and wines in which skin contact has been minimal, use of this equation gives flavonoid values close to zero, usually in the range -1 to $+1$. Spectral characterstics are then defined, as the ratio $(E_{320} - 1.4)$ over $(E_{280} - 4)$ is close to 1.5 (Somers and Ziemelis 1985a).

In development of this analytical concept, approximations were made for the sake of simplicity in application to quality control and to routine analysis of juices and wines. Standard deviations in determination of the absorbance correction factors 4.0 (at 280 nm) and 1.4 (at 320 nm) were 0.95 and 0.31 respectively, illustrating the variability of materials. Experience has shown, however, that the spectral method does permit easy recognition of excessive skin contact in both juices and wines, where analytical interpretations have been most uncertain. Spectral measures of commercial white wines have shown flavonoid levels up to about 3 a.u., but levels as high as 8 a.u. have been observed in analysis of commercial grape pressings, with progressive increase from zero in free-run juice. Conversion of such absorbance measures to "catechin equivalents" is justified on the basis that the flavonoid extract consists of flavan-3-ol units. Thus one a.u. corresponds to approximately 70 mg l^{-1} "catechin equivalents".

Sorbic acid (ε 25000 at λ_{max} 255 nm), which may be used as a preservative in white and rosé wines containing residual sugar, can be responsible for major interference in interpretation of E_{280} values; normal absorbance is restored after extraction with iso-octane (Somers and Ziemelis 1985a).

3.2 Total Phenolics in Red Wines

The only appropriate yet comprehensive measure of total phenolics in red wines is in terms of the maximal UV absorbance at 280 nm. As already indicated, the concept of (E_{280} −4) as a phenolic index in red wines of *V. vinifera* is based on the fact that, whereas the phenolic composition of grape juices is quantitatively similar in all cultivars (with a common background of non-phenolic UV absorbance), the pigment and tannin extractives from grape skins and seeds are responsible for the high E_{280} values of red wines. In analytical surveys of over 400 young commercial red wines from many Australian regions, (E_{280} −4) was found to range three-to fourfold in each of three vintages. The overall range was 23–100 a.u. in normal distribution about a mean of 50 a.u.; values greater than 75 a.u. were rather few (Somers et al. 1983). There is, of course, unknown contribution from non-flavonoid phenolics to E_{280}, but it has been considered unrealistic to attempt corrections for hydroxycinnamates, as was done in spectral assessment of white wine extractives (Sect. 3.1.2).

The nominal interface between rosé and light red wines occurs at about 20 a.u., and it appears that young wines having (E_{280} −4) values less than about 30 a.u. have little capacity to improve during conservation. Robust red wines, with obvious astringency and sufficient phenolic concentrations to permit development during conservation, have (E_{280} −4) values greater than about 40 a.u. High values, in the range 45–65 a.u., have been associated with young red wines from premium areas. Such wines require prolonged bulk maturation, and some bottle ageing, before optimal organoleptic properties are attained.

Despite extensive change in the composition of wine colour during the first year after vintage, and progressively through further conservation (Sect. 5), absorbance at 280 nm is a fairly stable wine feature; normal decline in (E_{280} −4) is 10–20%, notably during the first year when there is invariably marginal loss of phenolics during fining and filtration procedures. In recent studies of young port wines, Bakker et al. (1986 b) have shown a correlation coefficient of 0.96 between (E_{280} −4) and total phenolic measures by the Folin-Ciocalteu reagent. Thus, an indication of phenolic concentration in normal terms is obtained from the UV absorbance by use of the linear regression:

$y = 29.5\,x + 210$, where
y = total phenolics (as gallic acid, mg l^{-1})
and x = total phenolics as (E_{280} −4) a.u.

The mean value of 50 a.u. found for total phenolics as (E_{280} −4) in young Australian wines then corresponds to about 1700 mg l^{-1} "gallic acid equivalents." It is noted that the one unit is as arbitrary as the other. Molar extinction values of components at 280 nm vary widely, but so too does colorimetric response to the reagent (Singleton 1974). Measurement of UV absorbance is, however, incomparably more simple and direct than via the Folin methods.

4 Phenolic Composition of White Wines

4.1 Hydroxycinnamates

Although limited "skin contact" before fermentation may be sometimes considered advantageous in the making of particular wine types, quality has been traditionally associated with fermentation from free-run juice (Terrier and Blouin 1975; Singleton and Noble 1976). Research interest is therefore focussed on the juice-soluble hydroxycinnamates as inevitable (and usually major) phenolic components of the vintage.

Their occurrence in grapes as esters of L-(+)-tartaric acid (Fig. 4) was first reported by P. Ribéreau-Gayon in 1965. Paper and thin layer chromatographic techniques are, however, hardly adequate for resolution, much less accurate measurements, of the individual components and derivatives thereof. During the past 10 years, many new analytical data about this phenolic fraction have come from HPLC studies in various research centres.

Caffeoyl tartaric acid is invariably the dominant phenolic component in fresh juice, together with lesser quantities of *p*-coumaroyl tartaric acid and with feruloyl tartaric acid as a minor component (Fig. 4; J. D. Baranowski and Nagel

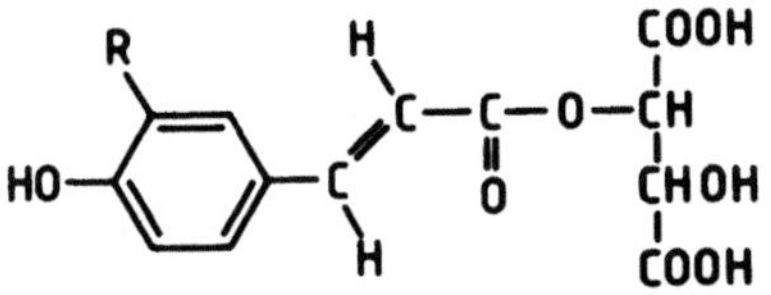

Fig. 4. Hydroxycinnamate components of grapes; R = OH, trans-caffeoyl tartaric acid; R = H, trans-p-coumaroyl tartaric acid; R = OCH_3, trans-feruloyl tartaric acid. A glucoside of trans-p-coumaroyl tartaric acid is also present. Other derivatives also appear in wines (Table 1)

Table 1. Identity and spectral characteristics of hydroxycinnamate fractions[a]

Peak no. in HPLC	Component	Features (nm)		
		λ_{max}	λ_{min}	Shoulder[b]
1	trans-2-S-glutathionyl caffeoyl tartaric acid	329	280	254 (s)
2	trans-caffeoyl tartaric acid	329	263	301 (m)
3	trans-p-coumaroyl tartaric acid glucoside	311	248	–
4	trans-p-coumaroyl tartaric acid	315	250	–
5	trans-feruloyl tartaric acid	330	258	301 (m)
6	trans-caffeic acid	326	260	301 (s)
7	trans-p-coumaric acid	311	248	299 (w)
8	trans-diethyl caffeoyl tartrate	332	264	303 (m)
9	trans-ethyl caffeate	327	263	299 (s)
10	trans-ethyl p-coumarate	312	262	295 (w)

[a] In 10% aq. ethanol saturated with potassium hydrogen tartrate.
[b] s, strong; m, medium; w, weak.

1981). A glucoside of *p*-coumaroyl tartaric acid is also present in grape juice at low relative levels (Ong and Nagel 1978; Somers et al. 1987). Cis- and trans-isomers of each component are separable by HPLC, but the more stable trans-configurations are predominant, with insignificant contributions from cis-forms (Singleton et al. 1986a; Somers et al. 1987). Equilibria between the isomers had been earlier shown to be influenced by UV irradiation (Singleton et al. 1978).

From comparative HPLC studies of the major hydroxycinnamates in 37 grape cultivars of *V. vinifera* (both white and red), it was evident that concentrations in juices are genetically controlled (Singleton et al. 1986a), and that they remain fairly static during ripening of the fruit (Singleton et al. 1986b). For the popular wine grape cultivars examined, concentrations varied about fivefold to over 300 mg l^{-1}, i.e. from approximately 2×10^{-4}M to 10^{-3}M, with no particular difference between white and red cultivars. Large cultivar variation in the proportion of *p*-coumaroyl to caffeoyl tartaric acids (3–33%) was also observed in the main comparative study (Singleton et al. 1986a).

Currently, there is more information about the hydroxycinnamates in grape juices than about their fate during vinification and their uncertain role in relation to wine quality. Spectral measures of total hydroxycinnamates in commercial white wines (Sect. 3.1.1) have shown an eightfold range to about 200 mg l^{-1} (Somers and Ziemelis 1985a). Caffeoyl tartaric acid has been found present at much higher concentrations in Riesling wines of California and Washington than in those of Alsace, posing the question whether the differences are of a regional nature or due to significant variation in wine-making practice (Herrick and Nagel 1985).

Research in this area has been sustained with the expectation of some practical advance in viticulture or oenology. As the major phenolics in expressed juice, the hydroxycinnamates have been considered as likely substrates for browning reactions during vinification and storage; bitterness in white wines has also been associated with high concentrations of these components (Herrick and Nagel 1985). Neither of these propositions has been proved, so that any role the hydroxycinnamates may have in relation to wine quality, type and condition remains undefined.

Other factors are now seen to be important in this regard. Firstly, the hydroxycinnamates are not confined to vacuolar cells of the grape flesh, but are present at much higher concentrations (2×10^{-2}M) in the skin vacuoles (Moskowitz and Hrazdina 1981). In commercial practice, there is therefore progressively higher concentration of total hydroxycinnamates with increasing skin contact, and this is accompanied by progressive extraction of flavonoid components (Somers and Ziemelis 1985a).

Another significant factor is that of continuous change in hydroxycinnamate composition of white wines. Though there is little alteration in the UV spectrum, HPLC studies have amply demonstrated the dynamic nature of phenolic composition during vinification and conservation (Nagel et al. 1979; Somers et al. 1987).

The most rapid and also unexpected effect, which may prove to be of major importance in oenology, is formation of a glutathionyl derivative of the principal phenolic component (caffeoyl tartaric acid) in response even to slight aeration of

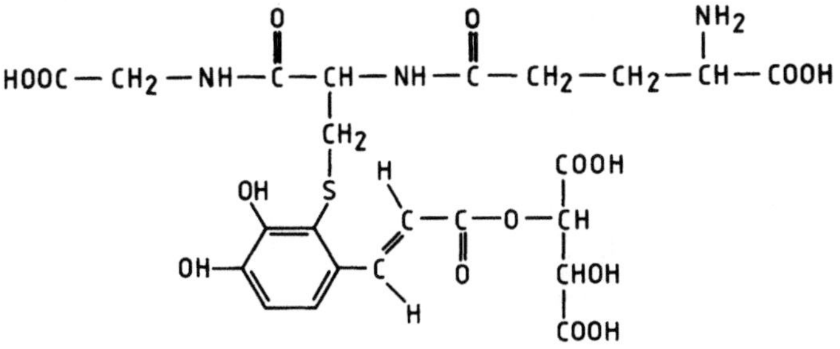

Fig. 5. Trans-2-S-glutathionyl caffeoyl tartaric acid, an enzymic oxidation product of grape juice. (Singleton et al. 1985)

the must, such as inevitably occurs during normal juice preparation. Oxidase activity has been identified as the initiator of reactions leading to appearance of this new hydroxycinnamate component (Fig. 5), for which both caffeoyl and *p*-coumaroyl tartaric acids are substrates (Singleton et al. 1984, 1985; Cheynier et al. 1986). As strict anaerobic procedure is necessary for avoidance of this enzymatic effect on hydroxycinnamate composition, the glutathionyl derivative must be regarded as a normal phenolic component of musts and wines. Its concentration in new wines, relative to that of the main precursor phenolic, is therefore seen as an indicator of aerobic treatment of the juice before fermentation.

Thus the must or clarified juice normally contains at least five hydroxycinnamate components, including the glutathionyl derivative, before fermentation begins. The latter, generally present in commercial wines at low concentrations in comparison with the major components, has proved to be the most stable. Progressive change in hydroxycinnamate composition begins early in the fermentation, and up to ten such components have been identified in new white wines (Somers et al. 1987). HPLC analysis of a young Riesling wine is illustrated (Fig. 6); peak identities and spectral characteristics are shown in Table 1.

The initial but partial fate of the original tartaric acid esters is hydrolysis to free hydroxycinnamic acids, which is accompanied by formation of ethyl esters. Early appearance of these products in the ferment suggests enzymatic influences; rate of change is much decreased in the latter stages of fermentation and during wine conservation. It is significant that, whereas the concentration of total tartaric acid esters decreased by more than 50% during the first 6 months of conservation, there was about 80% retention of total hydroxycinnamates in the various forms found in wine (Somers et al. 1987). In this work, gradient elution was found indispensable to the task of optimizing HPLC conditions for resolution of the total hydroxycinnamate components (Fig. 6).

Availability of a specific enzyme [hydroxycinnamic acid ester hydrolase (HCEH), Kikkoman Corporation; Okamura and Watanabe 1982] appears to offer interesting prospects for advance in this aspect of oenology. Complete hydrolysis of the natural hydroxycinnamoyl esters in wine occurs within a few days in the presence of 1–2 mg l^{-1} HCEH at about 20° C (Fig. 7). The glutathionyl derivative (fraction 1) and the ethyl esters (fractions 9, 10) are unaffected by the hydrolase, although the latter are rapidly hydrolyzed in aqueous solution (Somers et al. 1987).

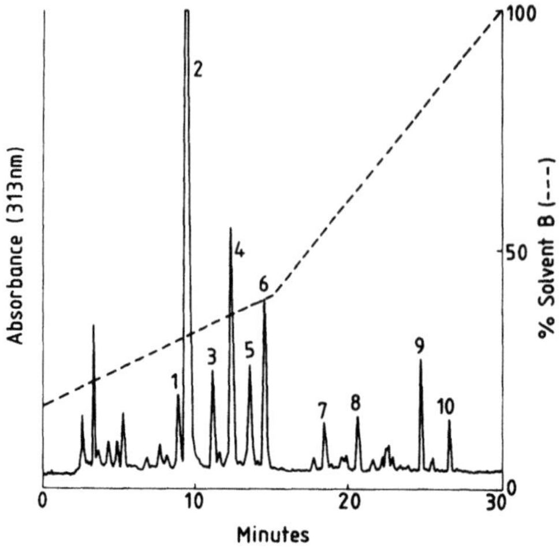

Fig. 6. Analytical HPLC of hydroxycinnamate composition in a young Riesling wine; for details see Table 3. Hydroxycinnamate components are numbered *1–10*. (Somers et al. 1987)

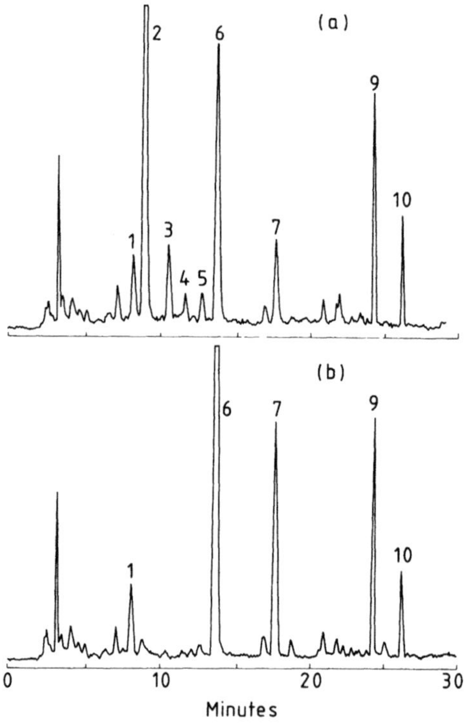

Fig. 7. Effect of enzyme treatment on the hydroxycinnamate composition of a Chardonnay wine. The intact wine (**a**); analysis 6 days after addition of 1.25 mg l^{-1} HCEH, 20° C storage (**b**). (Somers et al. 1987)

Thus the phenolic composition of a white wine can be significantly altered by batch treatment with HCEH, the free hydroxycinnamic acids then appearing as major components. This development has posed the question as to whether sensory properties may be favourably influenced by such intervention. This enzyme should certainly be useful in further appraisal of hydroxycinnamates in wine, and may also find commercial application.

4.2 Flavonoids

As mentioned, the C_{15}-based flavonoid components of grapes are localised in skins, seeds and vascular tissue. Principal components in white grapes are (+)-catechin, (−)-epicatechin and related oligomers of flavan-3-ols (procyanidins, Fig. 1). Only a few (dimeric) procyanidins have been fully determined in stereochemical terms (Lea et al. 1979).

Although the presence of flavonoid extractives is indicated by spectral characteristics of the juice or wine, their assessment by this direct means is merely approximate, with application only to quality control of materials (Sect. 3.1.2). More detailed information about this most variable aspect of white wine composition has been obtained by gel column analysis. This technique, although lacking resolution in comparison with HPLC, has the advantage of separating component categories in a UV elution profile; the flavonoids are eluted last in several fractions from Sephadex G-25 or LH-20 gels in aqueous systems. In a study of 80 commercial white wines, Simpson (1982) showed concentrations of flavonoids to range widely from zero to about 200 mg l^{-1}, with (+)-catechin as a major component (0–80 mg l^{-1}). Strong correlations were found between susceptibility to oxidative browning and concentrations of the catechins and dimeric procyanidins. HPLC analysis for (+)-catechin in particular should therefore serve as a rapid method for comparative assessment of flavonoid extraction and of browning susceptibility in white wines.

As part of a recent investigation of catechins and procyanidins in different parts of the grape and in wines, Bourzeix et al. (1986) have reported on the effect of increasing contact with grape solids during white vinification. Catechins and procyanidin oligomers up to tetramers were resolved and measured by HPLC after prior isolation of the flavonoid fraction on a polyamide column. The qualitative distribution of components was found to vary with different parts of the harvest, and quantitative distribution was as follows: seeds 59%, stems 21%, skins 19% and pulp <1% (Bourzeix et al. 1986).

Coarseness, bitterness and astringency in white wines are principal properties associated with excessive skin contact. Thus the technology is generally designed to minimize the presence of flavonoids. Temperature control at around 10° C is effective in limiting flavonoid extraction during commercial juice preparation (Ramey et al. 1986).

The flavonols (Fig. 1) may be mentioned as providing a further instance of the "nuisance value" of flavonoids in white wines. Although they are merely trace components (a few mg l^{-1}, being detected e.g. as tiny peaks between fractions 8,9 in HPLC of total hydroxycinnamates, Fig. 6) the flavonols have been shown to be responsible for a rare form of phenolic instability. Formation of a haze or yellow sediment in previously clarified and stabilised wine was due to hydrolysis of flavonol glycosides with precipitation of the much less soluble phenolic aglycones (Somers and Ziemelis 1985 b).

4.3 Pigments

The juice of fresh grapes is almost colorless, there being only slight extension of light absorbance into the visible region of the spectrum. White wines, however, normally develop pale yellow to golden hues during vinification and ageing. The chemical nature of these pigments, which are responsible for low absorbance at 380 nm, decreasing to zero at 450 nm, is unknown. They do appear to be quite distinct from the brown polymers arising from oxidation of flavonoid components. Thus the latter may be removed by fining with PVPP and are strongly retained in gel column analysis, whereas the yellow pigments are not adsorbed by PVPP and show polar characteristics in early elution from Sephadex gels.

The formation of yellow pigments by natural wine redox influences on the principal non-flavonoid phenolic, trans-caffeoyl tartaric acid (Fig. 4) and on trans-caffeic acid, is suggested by normal yellowing of their aqueous solutions during storage. Thus both pigment categories (yellow and brown) may be derived oxidatively from the two major categories of phenolics, i.e. from the hydroxycinnamates and from the flavonoids respectively. Whereas the yellow to golden hues are normal attributes, browning is a negative feature in white wines. It is usually accompanied by adverse sensory effect on flavour and aroma, indicative of excessive oxygen contact during the wine conservation.

The greater complexity of actual pigment structures has, however, been suggested by association of wine colour with the nitrogenous and carbohydrate fractions as well as with the phenolic composition (Voyatzis 1984).

5 The Colour of Red Wine

Concentrations of anthocyanins and of total phenolics widely vary in new red wines (Sect. 3.2), which have therefore different requirement for maturation before bottling. Arbitrary division into at least two major categories, such as are generally recognized in commercial production, can be made by use of the phenolic index ($E_{280} - 4$):

a) Light red wines, suitable for marketing within a few months of vintage, have phenolic indices in the range 20–30 a.u. New wines of this type may have bright colour, but have generally low astringency, and are susceptible to oxidative deterioration during long-term storage.
b) Robust red wines, having deeper colour and indices above about 40 a.u., require more prolonged maturation for development of full flavours and acceptable tannin finish. Astringent properties decline as a consequence of changing phenolic composition, and the wine becomes smooth and mellow to the palate, with potential for further development during bottle ageing.

Wines in the first category come generally from regions in which yields per hectare are high, often from irrigation areas. Conversely, wines of high phenolic content, which may become premium aged wines, are associated with relatively low viticultural yields from selected cultivars (Branas 1977; Champagnol 1977).

5.1 Residual Anthocyanins

The colour of red wine is initially and fundamentally due to extraction of anthocyanin pigments from grape skins during vinification, but pigment composition then becomes progressively more complex during the wine conservation.

Following pioneering studies by P. Ribéreau-Gayon (1959), and other works based on use of paper and thin-layer chromatography, the varying composition of total anthocyanins in wine grape cultivars has been examined in subsequent HPLC studies (Wulf and Nagel 1978; Bakker and Timberlake 1985a). These studies have confirmed that malvidin 3-glucoside ("oenin") and its acylated derivatives generally account for 60–80% of total anthocyanins in *V. vinifera*. The pigments are all 3-(β-D-glucosides) of five common anthocyanidins and acylated derivatives thereof; the latter are esters involving mainly acetic acid and *p*-coumaric acid, with minor involvement of caffeic acid (Fig. 8). Thus the use of HPLC has enabled resolution of 21 discrete anthocyanin components in the skins of Cabernet Sauvignon grapes (Wulf and Nagel 1978). The cyanidin pigments are present usually in trace amounts only.

All have similar visible spectra in aqueous acid solutions, with λ_{max} in the region 525–535 nm and low λ_{min} at about 420 nm (E_{420}/E_{520} about 0.2), but there are characteristic tonal differences with the various phenolic substitution patterns (Fig. 8), e.g. delphinidin pigments have bluish tints, peonidin and malvidin pigments are more purely red. There is, however, no similarity between the colours of anthocyanin solutions (or of mixed pigment solutions) and that of new red wine. The latter typically has λ_{max} 520–525 nm, λ_{min} about 420 nm and hue index E_{420}/E_{520} of 0.4 or greater. The involvement of other components in wine coloration has been well indicated by its richness and depth of hue, which contrast with the clear, bright spectral properties of pure anthocyanin solutions.

Extreme limits for concentrations of total anthocyanins in new red wines are from about 100–1000 mg l^{-1}; a level of around 500 mg l^{-1} is fairly typical for a fullbodied wine, in which the total phenolics are also high (Somers et al. 1983). Progressive decline in anthocyanins, and maintenance of wine colour by formation of new, more stable oligomeric to polymeric pigments, is, however, a central feature of wine maturation (Somers 1966, 1968, 1971). HPLC studies of residual anthocyanins in red wines have confirmed their decreasing contributions to wine colour during ageing (Nagel and Wulf 1979; Bakker et al. 1986c).

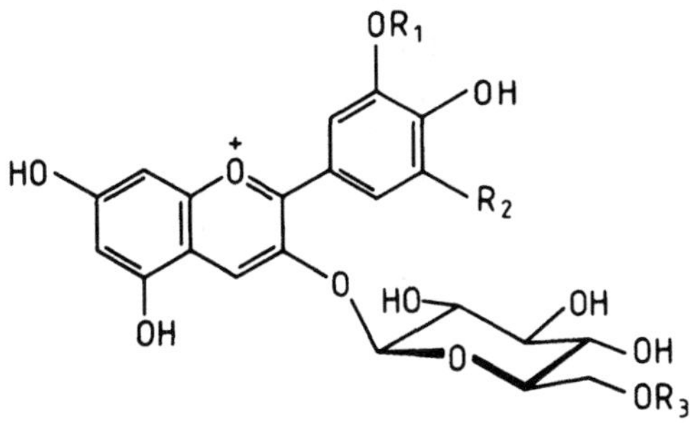

Fig. 8. The anthocyanins of *V. vinifera*. Aglycones of the five anthocyanidin 3-glucosides (R_3 = H) are cyanidin R_1 = H, R_2 = H; peonidin R_1 = CH_3, R_2 = H; delphididin R_1 = H, R_2 = OH; petunidin R_1 = CH_3, R_2 = OH; malvidin R_1 = CH_3, R_2 = OCH_3. Acylated anthocyanins where R_3 = acetyl, p-coumaryl have similar distribution at lower concentrations. The caffeate derivative of malvidin 3-glucoside also occurs at trace levels

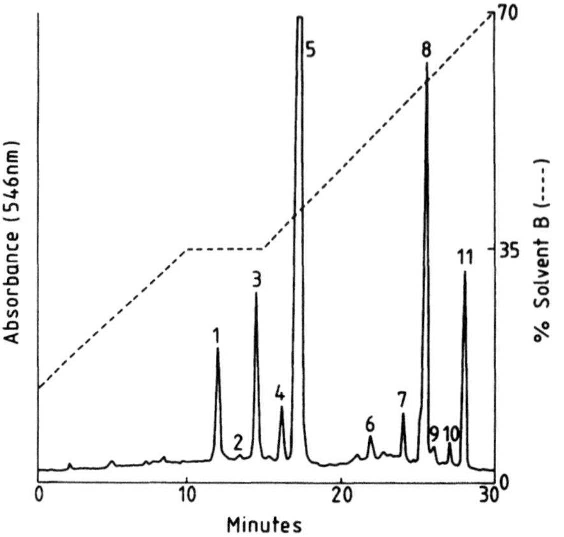

Fig. 9. Analytical HPLC of residual anthocyanins in a young Cabernet Sauvignon wine. Fractions *1–5* are 3-glucosides of delphinidin, cyanidin, petunidin, peonidin and malvidin; fractions *6, 7, 8* are acetate derivatives of 1, 3 and 5; fractions *9, 10, 11* are p-coumarate derivatives of 1, 3 and 5. Column I.C.I/O.D. S2, 5 μm (250 mm × 4.6 mm i.d.), 35° C. Solvents, water (0.6% $HClO_4$, pH 1.3–1.4) and methanol

The rate of decline in anthocyanin concentration is influenced by such factors as temperature, oxygen access, pH, free SO_2 (Somers and Evans 1986); decrease has been reported to be logarithmic with time during ageing of port wines (Bakker 1986; Bakker et al. 1986c). Rapid compositional change in residual anthocyanins, with evident hydrolysis of acylated components, disappearance of minor pigments and appearance of other pigment components, has also been generally indicated in HPLC studies, particularly in work by Bakker and Timberlake (1985b) and by Bakker et al. (1986c). HPLC of residual anthocyanins in a young red wine is illustrated in Fig. 9. Similar analysis of older wines has shown progressively decreased amounts of detectable anthocyanins, which are entirely absent from most red wines aged more than 10 years. HPLC analyses of aged wines typically show an increasingly elevated base-line at later elution. This prolonged "hump", on which anthocyanin peaks can appear, is attributed to oligomeric and polymeric pigments accumulating during wine storage (Bakker and Timberlake 1985b).

HPLC has therefore been of limited application in analysis of wine pigments. Although it does enable resolution and accurate measurement of residual anthocyanins, the latter are of decreasing significance as the wine ages.

There are further considerations which also apply to other chromatographic methods for analysis of anthocyanins in wine or fruit pigments. The pigment coloration is not only pH-dependent, but also much influenced by the presence of free SO_2 and by operation of other physicochemical phenomena (Sect. 5.2). Thus HPLC analysis of residual anthocyanins in red wines requires the use of highly polar solvent systems having pH <2.0; these are media conditions from which all of the natural influences on actual pigment composition have been necessarily eliminated.

5.2 Pigment Equilibria

The young wine is a mass of physicochemical equilibria having direct influences on colour density and hue. Useful interpretations of wine colour then depend primarily upon appreciation of the various equilibrium and kinetic effects on monomeric pigment components. Because of extensive change in the pigment composition during wine conservation, different considerations apply to aged red wines (Sect. 5.3).

5.2.1 pH and SO_2 Factors

Representative anthocyanin structures in pH- and SO_2-dependent equilibria are shown in Fig. 10. The anthocyanins are bright red in acid solutions, the flavyliun cation being the only form at pH <0.5. Colour density decreases with rise in pH because of formation of the colorless carbinol pseudo-base. From about pH 2 in aqueous solution, there is also slight change in hue towards violet from the blue quinonoidal base, and there is formation of a nearly colorless chalcone (Fig. 10). Freshly made neutral or alkaline solutions are blue to violet, but the colour is transient. The distribution with pH of these four structures (of malvidin 3-glucoside) in dilute aqueous solutions – equilibria which involve proton transfer, hydration and tautomeric reactions – has been well defined (Fig. 11); but other factors affecting equilibria and colour stability are also operative in natural systems such as fruit juices and red wines (Brouillard 1982; Timberlake and Bridle 1983; Somers 1987). Before consideration of these factors (Sect. 5.2.2), it is interesting to note that there are as many as ten molecular species in equilibrium for each anthocyanin component in aqueous solution at wine pH 3.3–4.3 (cf. Fig. 10 caption).

Sulphur dioxide (SO_2), which forms a colorless bisulphite addition compound with the flavylium cation (Fig. 10), is a most important and widely variable factor in red wines. The bleacting effect of free SO_2 on anthocyanins is moderated in wine by the preserve of other SO_2-binding compounds, viz. acetaldehyde, pyruvic acid, α-keto glutaric acid, sugars, with all of which free SO_2 is also in equilibrium (Burroughs and Sparks 1973).

Strong correlations have been shown between colour density and quality ranking in comparative studies of young Australian wines. Absence of expected relationship between colour density and anthocyanin content in those wines focussed attention on the pigment equilibria. It was found that the degree of coloration of anthocyanins (α), (i.e. the percentage of total anthocyanins present in coloured forms) varied from 6–25%; α value was also correlated with relative wine quality (Somers and Evans 1974). Relationships between colour composition and overall wine quality were confirmed in other investigations (Tromp and van Wyk 1977; Jackson et al. 1978; Somers et al. 1983). Even larger variation in the degree of coloration (up to tenfold) was observed in the latter study, which involved over 400 young red wines from three vintages.

Spectral methods for such comparative assessment of colour composition – regarded as an integration of contributions from monomeric and oligomeric to polymeric pigment forms – were based on deliberate displacements of pH- and

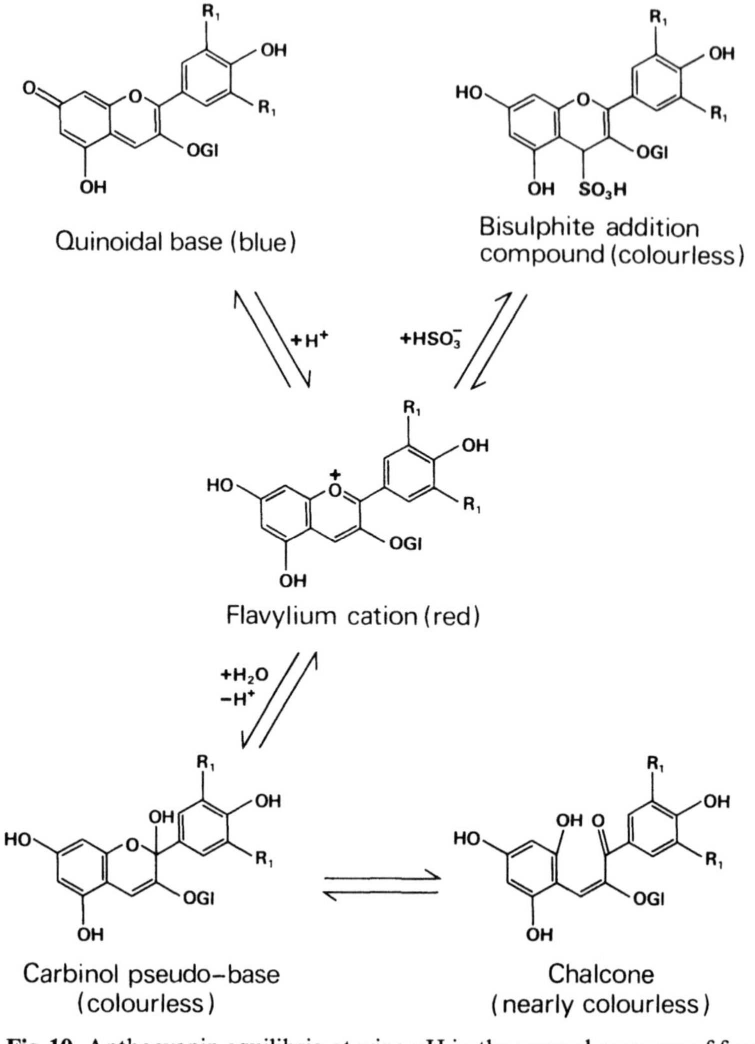

Fig. 10. Anthocyanin equilibria at wine pH in the normal presence of free SO_2. Other molecular species are also present. Seven different structures of malvidin 3-glucoside have been shown by PMR spectrometry. These include three epimers of the carbinol base and two chalcone isomers. Three tautomers of the quinonoidal base can exist, but equilibrium is too fast for their discrete detection. (Cheminat and Brouillard 1986)

SO_2-dependent equilibria (Fig. 10). The relative insensitivities of non-anthocyanin pigments to pH change and to bleaching by SO_2 enabled their approximate estimation; interpretations of the anthocyanin equilibria could then be made (Somers and Evans 1977). Typical variability in parameters of colour composition is shown in Table 2. In an analytical survey of wine industry practice, the level of free SO_2 in vinification and conservation, much more than the wine pH, was shown to be the major influence on pigment equilibria and degree of coloration in young red wines (Somers and Wescombe 1982).

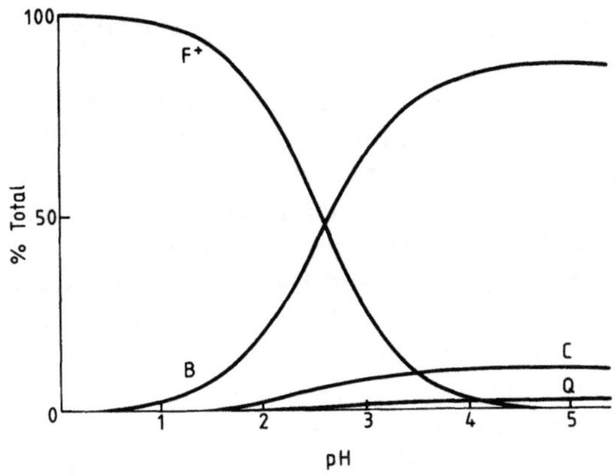

Fig. 11. Equilibrium distribution of malvidin 3-glucoside as a function of pH in dilute aqueous solution at 25° C. Red flavylium cation F^+; carbinol pseudo-base B; chalcone C; blue quinoidal base Q. (Brouillard and Delaporte 1978)

Table 2. Variability in wine colour composition, 1976 vintage (Somers et al. 1983)

Analytical measure	Shiraz (103 wines)			Cabernet Sauvignon (79 wines)		
	Min.	Mean	Max.	Min.	Mean	Max.
Colour density	2.7	7.6	16.5	2.5	7.6	15.7
Colour hue	0.45	0.70	1.0	0.48	0.72	0.99
Actual degree of coloration (α)	3	15	33	4	14	32
Natural degree of coloration (α')	10	21	36	11	20	36
Free SO_2 (mg l^{-1})	0	2.4	17.5	0	2.5	11.5
Total anthocyanins (mg l^1)	125	357	733	107	366	620
Total phenolics (a.u.)	26	50	96	23	50	67
"Chemical age" index (i)	0.16	0.35	0.58	0.19	0.36	0.58
"Chemical age" index (ii)	0.03	0.10	0.19	0.04	0.10	0.18
pH	3.38	3.79	4.34	3.37	3.81	4.40
Quality rating[a]	10.0	14.5	18.5	9.6	14.3	18.5

[a] Statistical study of these and other wine groups showed quality to be positively correlated with colour density, the parameter α, and total phenolics. Negative relationships involved colour hue and free SO_2. There was no correlation with wine pH. Wines were 7–8 months old at time of inspection.

5.2.2 Self-Association and Co-Pigmentation Phenomena

Anthocyanin concentrations in young red wines (2×10^{-4} to 2×10^{-3}M) are much higher than those employed in model equilibria studies (about 4×10^{-5}M), and there are, additionally, relatively larger quantities of other flavonoid phenolics. Pigment composition and equilibria are then far more complex, with numerous coloration effects in dynamic operation (Brouillard 1982; Osawa 1982).

Thus concentration factors, termed self-association and co-pigmentation, which have long been known as vital influences in flower coloration (Robinson and Robinson 1931; Scheffeldt and Hrazdina 1978; Osawa 1982), play fundamental roles in determining "la robe" of a young red wine, and the subtle changes in colour density and hue during early maturation.

The intensely bluish red colour of berry juices is largely due to these phenomena, by which the degree of coloration of anthocyanins may be as high as 50%. Much of the steep decline in colour density during fermentation of red juices (prepared via heat treatment of wine grapes) or during fermentation on skins, is due to the instability of deeply coloured molecular aggregates to ethanol; and these effects are reversible by removal of ethanol (Somers and Evans 1979). Residual coloration phenomena, attributed to self-association of the red cationic forms and to stabilization of the blue quinonoid chromophores (Fig. 10) by co-pigmentation with non-coloured phenolics, are now known to persist in red wines, with widely varying influences relating both to wine type and wine age.

Asen et al. (1972) observed gross positive deviations from Beer's Law in solutions of cyanidin 3,5-diglucoside (10^{-4} to 10^{-2}M) at pH 3.2, and examined co-pigmentation effects with various phenolics and amino acids. Maximum effects with anthocyanidin 3-glucosides in the presence of rutin occur at about pH 4.2 (Williams and Hrazdina 1979); effects are greater at higher molar ratios of co-pigment to pigment (Scheffeldt and Hrazdina 1978).

Thus there are positive deviations from Beer's Law in the wine pH range due to both phenomena. Self-association is the dominant effect at lower pH, where concentration of flavylium cations is high. Co-pigmentation effects are greater at higher pH, giving a bathochromic shift to higher λ_{max} (Timberlake and Bridle 1983). Following much debate about structures, recent studies based on circular dichroism and proton NMR have indicated that both types of coloration phenomena are due to formation of vertically stacked molecular aggregates; the driving forces are hydrophobic interactions between aromatic nuclei stacked parallel to each other and surrounded by hydrophilic glucose moieties (Hoshino et al. 1981; Iacobucci and Sweeny 1983; Goto et al. 1986).

It is evident that the integrated effect of these colour synergisms in a particular wine is determined by many compositional factors, viz. pH and SO_2 as primary influences on the distribution of anthocyanin species (Fig. 10, 11), the composition and concentrations of anthocyanins and of non-pigment phenolics, and the ratio of pigments to total phenolics. As indicated by the data of Asen et al. (1972), amino acid composition may also be a contributing factor in co-pigmentation effects.

As the coloration phenomena are concentration-dependent, there is a simple way of detecting their presence, while also measuring the extent of their influences on the degree of coloration of anthocyanins and on colour density of red wines. Maximum effect is observed when the wine is young, and in the absence of any free SO_2, as with new wines fresh from primary fermentation. Positive deviation from Beer's Law, i.e. "colour synergism" at wine pH, is then indicated by decline in corrected absorbance at 520 nm after serial dilution with 10% aqueous alcohol saturated with potassium hydrogen tartrate (Fig. 12). For the young wine illustrated, colour absorbance at 520 nm was more than doubled by the concentration effects. These phenomena are hardly operative in wines having residual anthocyanins less than about 50 mg l^{-1} (10^{-4}M), as in young red wines of low phenolic extract and in aged red wines (Somers 1987).

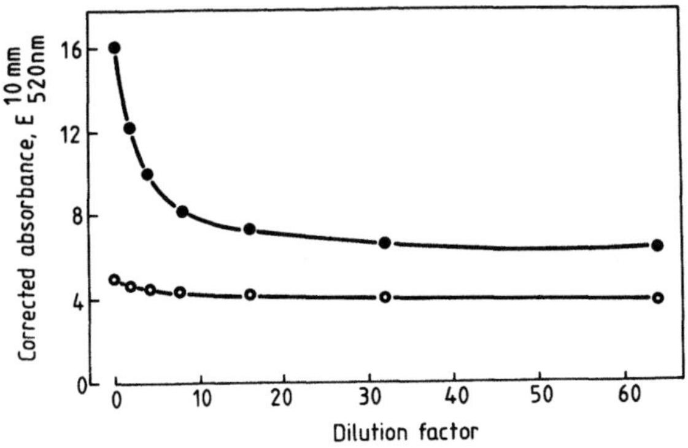

Fig. 12. Effects of dilution on pigment concentration in a young red wine containing 430 mg l^{-1} total anthocyanins (●), and in an aged wine containing only 30 mg l^{-1} total anthocyanins (○). Wine media conditions were maintained by dilution with 12% aqueous ethanol saturated with potassium hydrogen tartrate (pH 3.7). (Somers 1987)

The finely poised nature of the colour equilibria is also indicated by the fact that temperature has appreciable influence on colour density in new wines of high anthocyanin content, e.g. 20% decrease in E_{520} during warming from 5° to 20° C. The natural decline of self-association and co-pigmentation effects, consequent upon progressive decrease in anthocyanin concentrations, is largely responsible for the dynamic colour characteristics of young wines during early maturation, i.e. subtle change from deep purple-red hues towards crimson and red (Somers 1987).

5.3 Maturation and Ageing

Continuous change in the composition of red wine colour is a direct consequence of the chemical reactivities and instabilities of the phenolic grape extractives. Most rapid change occurs during the first year or two after vinification, when the wine is normally in bulk storage. This "maturation phase" should be considered as being quite distinct from the later "ageing phase", when the wine is in bottle and well protected from any further contact with air (P. Ribéreau-Gayon et al. 1983).

The rate and course of phenolic change in young wines, with associated effects on colour, astringency and flavour, are however, extremely variable. Temperature and oxygen access have been shown to be major influences on the rate of decline in anthocyanin concentration, and on the formation of polymeric wine pigments (Somers and Evans 1986). Whereas premium quality red wines are traditionally matured in oak cooperage for up to 3 years before bottling, similar phenolic change occurs in stainless steel or other storage. Compositional differences may, however, arise from greater oxygen access in wood storage (P. Ribéreau-Gayon et al. 1983). Such wines may then improve in bottle for up to 40 years under favourable cellar conditions; longevity and long-term resistance to oxidative deterioration are generally associated with deep colour and high total phenolics in the young wine (Somers 1983).

Whereas the brightness and purple tints of young red wines are fundamentally due to monomeric anthocyanin pigments derived intact from grape skin tissue, there is continuous displacement of the monomers by more stable, darker pig-

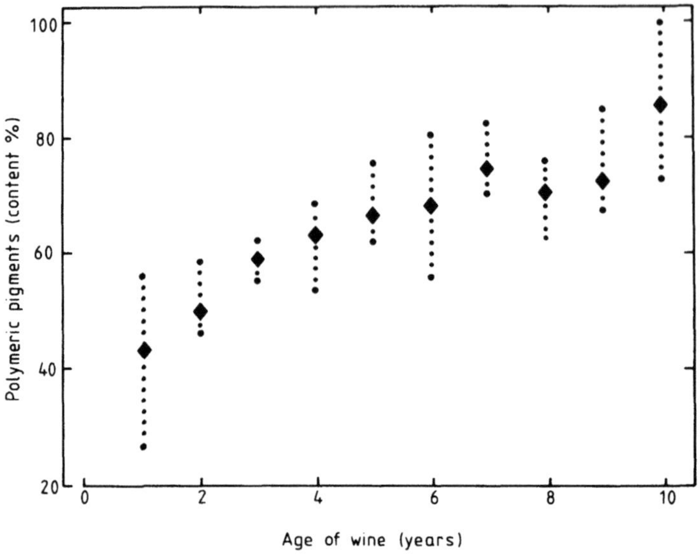

Fig. 13. Increasing contribution of polymeric pigments to red wine colour during ageing. Average values (▲) in each year are plotted, and the extremes of (●) of each age group are shown. The first value, at 1 year, is the mean of 40 different analyses, and the others are means of four or more different wines from each year. % polymeric pigment is that remaining ($E_{420}+E_{520}$) after exhaustive extraction with iso-amyl alcohol. (Somers 1971)

ments which are inevitable artefacts of the vinification process (Fig. 13). Mean molecular weights of these heterogeneous tannin pigments (isolated by gel filtration) have been reported as being about 1000 in young wines, increasing to 2000 at 5 years and to about 4000 in wines aged 20 years (Ribéreau-Gayon and Glories 1971).

Thus there are characteristic changes in the wine spectrum during maturation and ageing. The wine hue or tint, routinely measured as the ratio of absorbance at 420 nm to that at 520 nm, increases from 0.4–0.5 in new wines (pH 3.5–3.7) to around 0.8–0.9 in mature red wines. pH adjustment has been shown to give linear increase in tint values, and linear decrease in colour density with increasing pH (Bakker et al. 1986 b). With inevitable oxidative influences (involving formation of *o*-quinonoid structures from *o*-dihydroxy phenolic moieties), the hue index increases to above 1.0 in aged wines (Fig. 14); the latter have distinctly brown tints.

Although spectral methods have demonstrated wide variability in the colour composition of young red wines, with significant implications in relation to wine quality (Sect. 5.2.1, Tab 2), Bakker et al. (1986 c) have shown that early decline in total anthocyanins, measured by HPLC, is greater than indicated by the spectral measures. Their work confirms evidence from polyamide chromatography of wine pigments (Glories 1984 b) that there is a gradual transition from monomeric anthocyanins through oligomers to polymeric pigments, with progressive decrease in response to pH change and progressive increase in resistance to bleach-

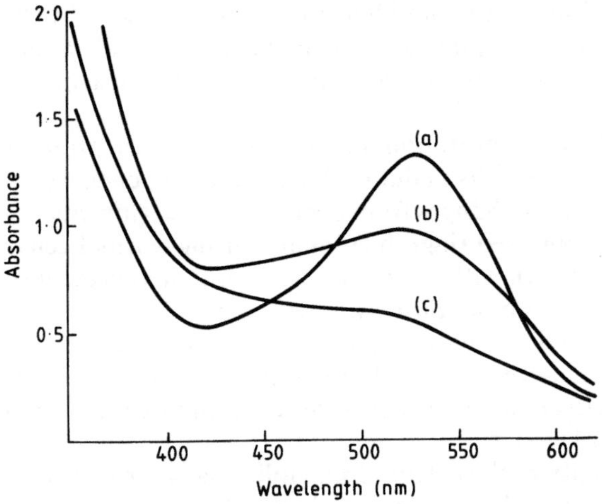

Fig. 14. Characteristic change in the visible spectrum of red wine during ageing. A wine at 6 months after vintage (*a*); a wine aged 5 years (*b*); a wine aged 12 years (*c*). Light path-length 1 mm

ing by SO_2 – whereas an abrupt change in such responses is implicit in application of the spectral methods. Thus HPLC studies of an ageing port wine have shown the rate of decrease in anthocyanin concentration to be logarithmic with time at constant temperature, suggesting a similar approach to future work with red wine (Bakker 1986; Bakker and Timberlake 1986 c). It seems clear that dynamic change in phenolic composition of red wines during early maturation is better described with reference to the rate of decrease in anthocyanins (measured by HPLC) than by spectral measures of increase in percentage levels of polymeric pigments.

Interpretations of wine colour composition are likely to remain rather speculative. Whereas anthocyanins can be accurately measured by HPLC only at pH <2.0 (Sect. 5.1), the colour composition of these monomers is influenced by many factors affecting the pigment equilibria (Sect. 5.2). Gel column analysis (Somers 1966, 1968; Ribéreau-Gayon and Glories 1971) and PVPP column analysis (Glories 1984 b) have demonstrated great complexity in the polymeric pigments, which are increasingly dominant during maturation and ageing; but these methods do not enable assessment of their contribution to wine colour. Spectral measures and interpretations, with attention to coloration phenomena (Sect. 5.2), continue to provide the best practical means for comparative assessment of wine colour composition (Sect. 5.3.1).

It should be noted, however, that tristimulus colorimetry, now facilitated by computer calculation of tristimulus values, enables better description of wine colour than do simple absorbance measures. Comparison of such data with conventional measures of port wines has been recently reported (Bakker et al. 1986 a). The use of tristimulus values has been recommended by the Office International de la Vigne et du Vin (1978) for commercial description of wine colour; it has limited application in oenological research.

5.3.1 Chemical Age

Progressive change in the wine spectrum (Fig. 14) is the most obvious feature of ageing reactions in red wines. The pigments of young wines are evidently in a highly ordered condition, i.e. colour density and hue are much influenced by the various parameters affecting pigment equilibria and the aggregation of monomeric pigment structures, rather than by anthocyanin concentrations per se (Sect. 5.2).

Decrease in anthocyanins during maturation and ageing, with formation of oligomeric and polymeric pigments, results in diminishing colour dependence on pH and SO_2 regimes. The wine colour changes from deep purple-red (low hue index E_{420}/E_{520}) towards crimson and red (high hue index) during normal cellar and bottle storage conditions (Fig. 14), with increasing structural involvement of the total flavonoid phenolics in colour of maturing wines (Somers 1971). Thus the improving relationship between colour density and the concentration of total phenolics during ageing – wherein anthocyanins decrease to zero – demonstrates the importance of having a relatively high level of total flavonoid extract in red vinification (P. Ribéreau-Gayon 1982).

Observations of wide variability in the extent of phenolic change in commercial wines of equal age (Table 2) led to the concept of "chemical age" as a measure of wine maturation (Somers and Evans 1977). Chemical age indices, which refer to increasing dominance of the wine colour by oligomeric and polymeric pigments having progressively decreased susceptibilities to pH change and to bleaching by SO_2, are based on the following spectral measures:

i) E_{520}, wine absorbance at 520 nm.

ii) $E_{520}^{CH_3CHO}$, wine absorbance at 520 nm after restoration of SO_2-bleached pigments by addition of acetaldehyde. For young wines in bulk storage, $E_{520}^{CH_3CHO}$ may be considerably greater than E_{520}. For aged wines in bottle, there is generally no difference, as pigments are then less susceptible to SO_2; free SO_2 is, in any case, much decreased, if not zero.

iii) $E_{520}^{SO_2}$, SO_2-resistant pigments, measured after addition of SO_2 (2000 mg l^{-1}); this value increases during maturation and ageing, approaching that of E_{520}.

iv) E_{520}^{HCl}, total wine pigments, measured after high dilution in 1M-HCl to eliminate equilibria and concentration effects; this value decreases during maturation and ageing.

v) E_{280}^{HCl}, absorbance of the above solution at 280 nm. This is the most stable of the spectral parameters after the first year of conservation (during which there is 10–20% decline).

Use of spectral ratios enables comparative assessment of change in the composition of wine pigments, e.g. the ratios $E_{520}^{SO_2}/E_{520}^{CH_3CHO}$ and $E_{520}^{SO_2}/E_{520}^{HCl}$ both increase towards 1.0 during ageing, in consequence of increasing colour stability against bleaching by SO_2 and against change in pH (Fig. 15).

The ratio $E_{520}^{HCl}/E_{280}^{HCl}$ is perhaps the most objective index of change in phenolic composition, as it emphasizes the decreasing presence of pH-responsive pigments during wine ageing. Initial values, which depend on the proportion of total anthocyanins to total phenolics, may be as high as 1.2 (Bakker et al. 1986c), but are generally 0.8–0.9 in new red wines. The rate of decline in this ratio depends largely on the conditions of wine conservation. In our experience of Australian wines, the index falls to about 0.4 during the first year of vintage, to about 0.2 in mature

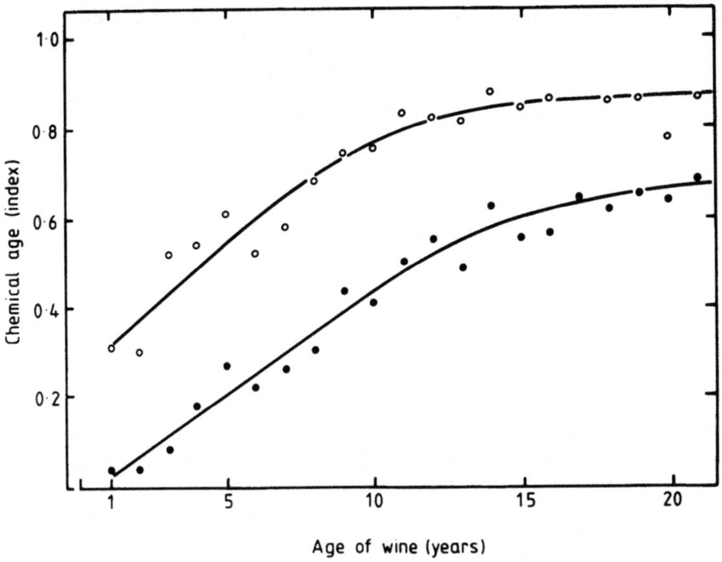

Fig. 15. Progression of chemical age indices against age of vintage red wines from the same vineyard. $E_{520}^{SO_2}/E_{520}^{CH_3CHO}$ (○); $E_{520}^{SO_2}/E_{520}^{HCl}$ (●). (Somers 1983)

wines (3–5 years), and to below 0.1 in wines aged more than 10 years. Very low values (<0.1) have, however, been noted in fairly young wines (<5 years); these circumstances have been generally associated with poor wine quality and with low concentrations of total phenolics.

5.3.2 Phenolic Condensation Mechanisms

Natural stabilization of wine colour by formation of polymeric pigments as artefacts of the vinification process is perhaps the most notable chemical phenomenon in oenology. Precise mechanisms and structures have, however, remained speculative. The following observations demand interpretation in chemical and physicochemical terms:

i) Bright colour may be maintained for many years despite decline in anthocyanins towards zero concentrations in aged wines. Subtle change in hue is accompanied by decrease in astringency and mellowing of wine taste and flavour.
ii) Initial (non-enzymatic) phenolic interactions are extremely fast, beginning with crushing of the berry and mixing of plant components (Somers 1971). Polymeric pigments are present from the earliest stage of primary fermentation (Bakker et al. 1986c).
iii) The polymers are resistant to pH change and to decolorization by SO_2 (Somers 1971).
iv) Their heterogeneous nature has been established by chromatography, including HPLC.
v) The anthocyanins alone are unstable in model wine media; red colour is gradually lost. Colour stability by formation of polymeric pigments occurs only in the (normal) presence of other flavonoid phenolics (P. Ribéreau-Gayon 1982; di Stefano and Ciolfi 1983), and high relative concentration of the non-pigment phenolic fraction is a significant positive factor in stabilization of wine colour (P. Ribéreau-Gayon 1982; Glories 1984a).

Interpretations of phenolic condensation reactions have been based on consideration of the ferment, and of new red wine, as acidic solutions of many electrophilic and nucleophilic reagents, with numerous and diverse reaction products. The red anthocyanin chromophores, though represented as positively charged oxonium ions (Fig. 8), behave as benzylic cations with the (+) charge at the 4- or 2-positions of the heterocyclic ring (I, II).

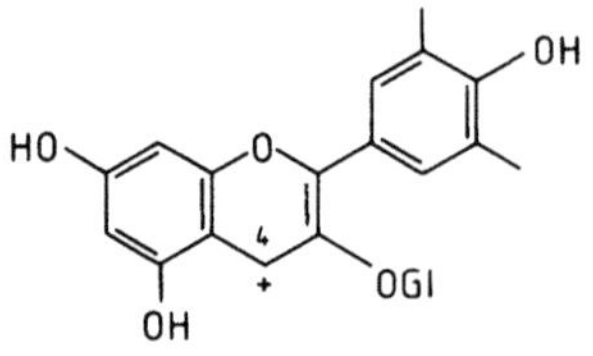

I

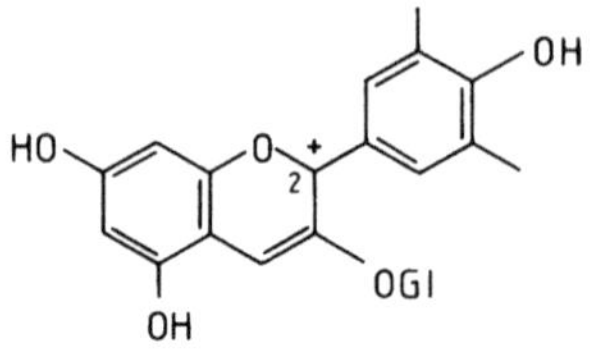

II

As carbocations are reactive species, the anthocyanins in wine are essentially unstable, being susceptible to attack by nucleophilic reagents, e.g. by water and by SO_2, to give the colourless carbinol base and the bisulphite-addition compound respectively (Fig. 10). Carbocation interactions with the nucleophilic 6- and 8-positions of flavan-3-ol units in procyanidins and of catechins (Fig. 1) are of most importance for wine colour stability. Reaction of a synthetic flavylium salt with various phenolics in aqueous solution at pH 3–5 has been demonstrated; all products involved substitution at the 4-position of the heterocyclic ring (Jurd and Waiss 1965; Jurd 1967). Substitution at the 4-position of simple flavylium pigments has been shown to confer stability against pH change and against decolorization by SO_2 (Timberlake and Bridle 1968; Somers 1971).

Following mechanistic concepts for formation of wine tannins (Jurd 1969), similar reactions were proposed for direct condensation of anthocyanins with procyanidins (Somers 1971; P. Ribéreau-Gayon 1973). Kinetics of malvidin 3-glucoside reaction with d-catechin in model wine solution have been examined (E. S. Baranowski and Nagel 1983), but no condensed pigment structures involving anthocyanins have yet been characterized.

Whereas earlier speculation about phenolic mechanisms (Jurd 1969, P. Ribéreau-Gayon 1973) was based on the inferred presence of flavan-3,4-diols ("leucocyanidins"), grape tannins, extracted during fermentation from skins and seeds are now recognized as oligomeric and polymeric procyanidins having up to eight flavan-3-ol units (Weinges and Piretti 1971; Haslam 1980; Romeyer et al. 1986). Haslam (1980) has suggested that the acid-catalyzed C-C bond-breaking and bond-making reactions – biomimetic mechanisms which characterize procyanidin chemistry – could be principal reaction processes during wine

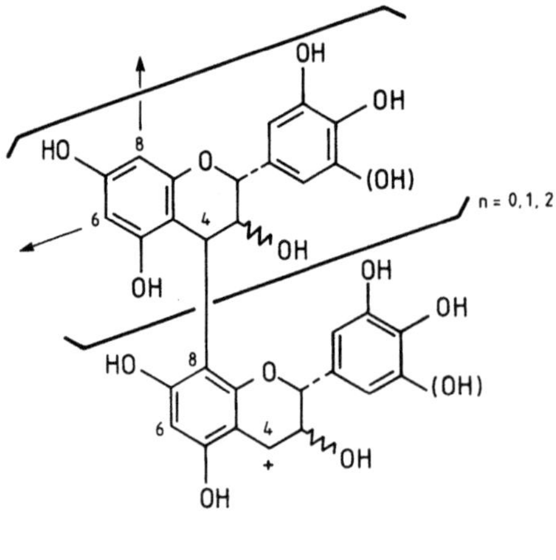

III

maturation and ageing. His proposal, which is supported by earlier observations of Timberlake and Bridle (1976), is that initial reactions involve degradations of polymeric procyanidins, e.g. to catechins, dimeric or trimeric procyanidins (Fig. 1, n = 2,3) and to the carbocations III. As there is essentially random fission of inter-flavan bonds, oligomeric procyanidins and carbocations of varying size are produced (Haslam 1980).

There may then be numerous carbocation species (I, II, III) available for attack by a variety of phenolic and other nucleophilic reagents. Phenolic nucleophilic reaction sites are principally the vacant 6- and 8-positions of procyanidins and of catechins, but it should be noted that only the anthocyanin carbocations can conceivably contribute red colour, whether as monomers or in condensed structures. From anthocyanins, the initial reaction product with a nucleophile R would be the labile flavene IV, which is readily oxidized to the condensed anthocyanin species V; quinonoid structures VI may then be formed by deprotonation (Somers 1971). Various mechanisms for phenolic condensation reactions in red wines have also been discussed by P. Ribéreau-Gayon (1973, 1974, 1982), Timberlake and Bridle (1976), Glories (1984a) and Vogel (1985).

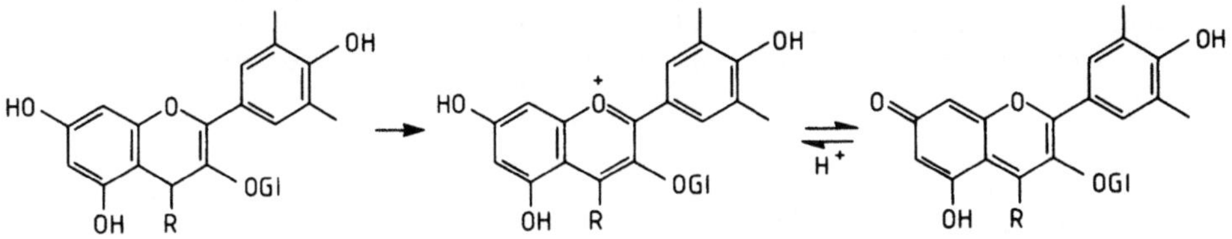

IV V VI

Facile as such reactions may seem, they have not been confirmed by experiment, and are therefore dubious. Reflection on molecular aggretates, now known to be responsible for coloration phenomena in young red wines (Sect. 5.2.2), and on the dynamic characteristics of procyanidin structures, leads us to resurrect the original hypothesis for phenolic change during natural stabilization of wine colour (Somers 1966).

Thus an ordered system of phenolic development, rather than random and varied condensation reactions, is suggested by the need to account for the facts that wine colour density and tint may remain scarcely altered during the first year after vintage, despite gross change in pigment composition and decrease in concentration of free anthocyanins (Somers and Evans 1986). The hypothesis is that the anthocyanin chromophores, which are already loosely associated (in the ferment and new wine) with the generally much larger concentrations of reactive procyanidins, are physicochemically rather than covalently incorporated into a procyanidin matrix during wine ageing. This concept, according to which phenolic aggregates responsible for coloration phenomena in new wines are structural precursors to more stable tannin pigments of mature wines, fits well with general oenological observations, and with particular physicochemical properties of anthocyanins.

Progressive loss of anthocyanin colour in aqueous solution occurs via nucleophilic hydration of the red carbocation; the equilibrium is displaced towards the colorless carbinol base and subsequent chalcone formation (Fig. 10; Brouillard and Delaporte 1977, 1978; Brouillard 1982). Significantly, colour loss at wine pH is decreased by the presence of ethanol, and prevented by the additional presence of tannins at high concentrations relative to anthocyanins (P. Ribéreau-Gayon 1982, Glories 1984a). Protection of anthocyanin colour in red wine from nucleophilic attack by water (and by SO_2) is therefore attributed to the nucleophilic characteristics of other wine constituents. These are principally the procyanidins and catechins, now known to be intimately associated in macromolecular complexes with the anthocyanins in new red wines (Sect. 5.2.2).

These phenolic aggregates, the instability of which has been demonstrated (Fig. 12), are logical precursors to the more stable polymeric pigment structures of mature red wines. Covalent bonding is probably not operative in the initial stabilization phenomena, which are envisaged as being purely physicochemical; cohesive forces are likely to be ionic, hydrogen bonding and stereochemical in nature.

The existance of solvation shells surrounding these macromolecules in wine has been suggested by the high aqueous affinity of the polymeric pigments in relation to that of the free anthocyanins (Somers 1971). Structures, which are more easily imagined than illustrated, may then consist of an anthocyanin carbocation core enclosed within a nucleophilic procyanidin matrix and an outer shell of hydronium ions H_3O^+.

Degradative change in the protected anthocyanins must inevitably occur, however, with retention of red colour during ageing dependent on there being high concentrations of procyanidins ("tannin extract") in relation to original anthocyanin extract. Thus the tannin spectrum, initially colourless but tending to yellow and brown with slow oxidation of o-dihydroxy moieties, is superimposed on those of residual pigments (both free and combined within the three-dimensional complex of flavan-3-ol units). Recovery of intact anthocyanins from polymeric pigment structures, which would support this general hypothesis, is a subject of current work.

Whatever the fundamental condensation reactions may be, it is certain that acetaldehyde can be a major influence on phenolic composition (Trillat 1908; Timberlake and Bridle 1976). Acetaldehyde may be generated (mg l^{-1}) by coupled oxidation of ethanol with phenolics in wine (Wildenradt and Singleton 1974), but actual levels attained depend on particular wine-making and storage conditions (Trillat 1908; Somers and Wescombe 1987). Slow interaction with phenolics, probably by nucleophilic attack on the carbonyl group (Baeyer reaction), leads to $-CH(CH_3)$-bridging between centres 6 and 8 of different procyanidin and other flavonoid structures (Timberlake and Bridle 1976). This type of linkage results in precipitation of insoluble phenolic polymers when free acetaldehyde is allowed to accumulate continually. Maintenance of free SO_2 levels provides control of such reactions during bulk storage of wine, but does not prevent long-term influence on phenolic composition in aged wines (Somers and Wescombe 1987).

It has indeed been suggested that phenolic ageing reactions in red wine are actually initiated by the presence of free acetaldehyde (P. Ribéreau-Gayon et al. 1983). Studies of small-scale fermentations and wine storage under controlled conditions have, however, confirmed the occurrence of two general mechanisms, viz. one involving direct interactions between anthocyanins and other phenolics, the second involving acetaldehyde-induced reactions (Somers and Evans 1986; Bakker 1986; Bakker and Timberlake 1986c; Somers and Wescombe 1987). The rate and extent of the latter reactions obviously depend on the amount of acetaldehyde available during conservation. Wide variation (from zero upwards) in acetaldehyde formation and consumption – as a direct consequence of varied wine-making practice and different cellar conditions – with inevitable interactions and modification of total phenolic composition, contributes to the range of red wine style and quality (Somers and Wescombe 1987).

6 Analytical Methods

The practical aim is for routine assessment of dynamic change in total phenolic composition during vinification and maturation of the natural wine types. Inherent limitations of pure analytical method, as by HPLC, are then most evident in relation to coloration phenomena in young red wines, which are dependent on actual phenolic concentrations in wines, and are influenced by other chemical features of the wine media (Sect. 5.2). There is, furthermore, the problem that polymeric pigment forms are not amenable to analysis by chromatography (Sect. 5.3).

A balanced approach to the task of evaluating phenolic composition of the intact wine requires use of HPLC in conjunction with techniques based on recording uv/vis. spectrophotometry. The latter uniquely provides an integrated measure of all phenolic components, and permits recognition of coloration phenomena. Whereas up to 80% of UV absorbance (250–350 nm) in white wines is due to phenolics, the phenolic contribution in red wines, with much higher UV absorbance, is generally more than 95%; and it is 100% in the visible region.

Since wine colour is a primary attribute of wine quality, and is associated with other sensory properties, there is obvious need for closer study of the physicochemical phenomena involved in wine pigmentation, and of the mechanisms by which red wine colour is stabilized during maturation and ageing. For white wines, where phenolic extract has negative implications, HPLC is invaluable to further exploration of spectral parameters relating to wine quality and stability.

6.1 Spectral

It is important to appreciate the oenological applications and limitations of the Beer and Lambert Laws which are fundamental to spectral analysis in the UV and visible regions.

As in spectrophotometry generally, there are no exceptions to Lambert's Law concerning light pathlength and absorbance. For white juices and wines, Beer's Law, by which absorbance is a linear function of concentration, is also obeyed. Thus analyses may be made directly from UV spectra in short light pathlength cells, or the sample may be suitably diluted at similar pH for measurement in 10 mm cells.

As indicated, however, in discussion of pigment equilibria and coloration phenomena (Sect. 5.2), there are several reasons why Beer's Law is not obeyed in red wines. The extent of deviation from Beer's Law is in fact used as a measure of coloration phenomena in red wines (Fig. 12). Wine colour density must therefore be measured directly on an undiluted sample. Beer's Law is obeyed at high dilution, however, enabling spectral measurement of total pigments and of total phenolics.

6.1.1 White Juices, Wines

The UV spectra of free-run grape juices and wines from free-run juice can generally be recorded directly by use of 1- or 2-mm quartz cells, with water as reference. Optical clarity is essential, and use of a 5-ml syringe with 0.4-µm membrane filter is convenient for sample preparation, though usually unnecessary with finished wines. In dealing with raw grape juice (must), dilution (1.0 to 10.0 ml) with 3% aqueous acetic acid (pH 3.5) before membrane filtration and measurement in a 10-mm quartz cell is often an easier procedure than direct filtration of the sample. Such dilution may be necessary in dealing with grape pressings or wines from pressings, which have high UV absorbance. In any event, absorbance readings at 280 nm and 320 nm are converted to values for undiluted juice or wine in 10-mm pathlength, i.e. E^{10}_{280}, E^{10}_{320} a.u.

Spectral estimates of total flavonoid extract and of total hydroxycinnamates in juices and wines have been based on use of correction factors 4 and 1.4 for non-phenolic UV absorbance at 280 nm and 320 nm respectively, with correction also for absorbance of total hydroxycinnamates at 280 nm (Sect. 3.1).

As mentioned, for any particular juice or wine sample, the residual absorbance factors can be determined after massive fining with PVPP. The sample (5 ml) is thoroughly mixed with PVPP (0.5 g) in a 10-ml screw-capped centrifuge tube. The tube is left in horizontal position for about 30 min, then centrifuged for 10 min at 4000 g. After membrane filtration of supernatant fluid, absorbances at 280 nm and 320 nm are measured against a reference sample from similar treatment of aqueous potassium hydrogen tartrate containing 12% ethanol (for wines) or 20% glucose (for juices); convert to E_{280}^{10}, E_{320}^{10} a.u. for use as correction factors in that sample.

The spectrum from the above treatment can be used to enable display of the UV spectrum of total phenolics by difference from that of the intact sample (Fig. 3). The difference spectrum cannot be obtained directly (i.e. by recording the sample spectrum against PVPP treated sample as reference) because of interference from soluble PVPP contaminants at wavelengths below 275 nm (Somers and Ziemelis 1985 a).

6.1.2 Red Wines

The conventional analytical description of wine colour density and tint, as first proposed by Sudraud (1958), depends on direct absorbance measures at 420 and 520 nm. Spectral readings are taken in cells of 1- or 2-mm pathlength against water reference, and converted to values for 10-mm lightpath.

Wine colour density $= E_{420}^{10} + E_{520}^{10}$
Wine hue or tint $= E_{420}/E_{520}$.

This simple concept, which refers specifically to yellow and brown hues (420 nm) and to red (520 nm), has served well in dealing with wine colour as a function of ageing (cf. Fig. 14). It has, however, been considered less satisfactory for young wines, which may have colour hues ranging from bright to dull; the latter typically at higher pH, when absorbance at 620 nm (blue region) is a significant factor. Thus Glories (1984 b) has proposed use of $(E_{420} + E_{520} + E_{620})$ as a more comprehensive measure of colour density in young red wines.

For the latter, displacements of pigment equilibria at wine pH (with excess SO_2, CH_3CHO) can show wide variability in colour composition (Table 2, Sect. 5.2). From recent comparative measures of residual anthocyanins by HPLC and spectral analysis, it is, however, evident that oligomeric pigments are also bleached by SO_2, so that broader interpretations must be made of these effects (Bakker et al. 1986 c). The spectral measures, which remain useful in relation to pigment equilibria (Sect. 5.2) and to indices of chemical age (Sect. 5.3.1) are conducted as follows:

$E_{520}^{SO_2}$. After recording the wine spectrum in a 1- or 2-mm cell, add 20% sodium metabisulphite (5 or 10 μl by capillary pipette of freshly prepared aqueous solution) and mix by inversion. Note absorbance at 520 nm after 1 min.

$E_{520}^{CH_3CHO}$. Add 10% aqueous acetaldehyde (20 μl) to a fresh wine sample (2.0 ml). After about 45 min, measure absorbance at 520 nm in a 1 or 2 mm cell.

Total wine pigments E_{520}^{HCl}, and E_{280}^{HCl} as a measure of UV absorbance, are read from solutions made by addition of wine (100 μl, or 200 μl for light reds) to

1 M HCl (10.0 ml). After about 1 h, measure absorbances at 520 and 280 nm in a 10-mm cell against water as reference. Correct for the dilution used, i.e. $\times 101$ or $\times 51$.

The earlier concept of degree of ionization (α) of anthocyanins (Somers and Evans 1974, 1977) has been modified in the light of more recent information (Timberlake 1982; Bakker et al. 1986c). The symbol α is retained for reference to degree of coloration of total wine pigments at a selected pH, 3.5 or 3.7, depending on which value is more suited to the wine group being examined; adjustment of pH to a uniform value eliminates the variable effect of pH on coloration phenomena. After pH adjustment, measure E_{520} and correct to 10 mm pathlength:

$$\text{Degree of pigment coloration, } \alpha_{3.5,\,3.7} = E_{520}/E_{520}^{\mathrm{HCl}} \times 100\%$$

Deviation from Beer's Law and the extent of colour synergism in a particular wine can be determined by diluting the wine 50-fold with 12% aq. ethanol saturated with potassium hydrogen tartrate (pH 3.7). The ratio of original E_{520} (after adjustment to the same pH 3.7) to that of the corrected E_{520} value for diluted wine (i.e. $\times 50$) is a measure of self-association and co-pigmentation effects operative in that wine (Fig. 12, Sect. 5.2.2):

$$\text{Colour synergism factor, } s_{3.7} = \frac{E_{520}\ \text{(intact wine)}}{E_{520}\ \text{(after treatment)}} .$$

It should be noted, however, that both colour parameters (α, s) are immediately susceptible to the presence of free SO_2. Maximal values of the synergism factor (s), which refer to high initial concentrations of anthocyanins, are seen in new red wines; the factor decreases to unity in aged wines, in which Beer's Law is obeyed.

In contrast, the degree of coloration (α), as defined above, is found to vary widely in new red wines, but then increases towards higher percentage values during ageing. This is because of increasing contribution of polymeric pigments to wine colour (E_{520}) and consequent decrease in response of total pigments (E_{520}^{HCl}) to acid pH.

6.2 High Performance Liquid Chromatography (HPLC)

Increasing application of HPLC during the past decade has led to much new information about the phenolic constituents of grapes and wines. The technique is rapid and relatively simple to perform, providing both qualitative and quantitative data not previously accessible by earlier chromatographic methods. There is the further advantage that no preliminary sample preparation may be required.

Reference texts on HPLC theory (e.g. Snyder and Kirkland 1979) and on applications to food analysis (Charalambous 1979; Macrae 1982) are a sufficient general background to the subject. Isocratic systems, as used in the first HPLC investigations of grape and wine phenolics (Wulf and Nagel 1976), have been largely displaced by more efficient gradient elution systems. A selection of analytical systems which have been applied to study of different phenolic categories is given in Table 3. The following features are noted:

i) Reverse phase chromatography, in which components are separated by partition between a non-polar stationary phase (e.g. C_{18} alkyl groupings bonded to a silica support) and a polar mobile phase, is the general rule. The various commercially available columns differ in dimensions, type of packing and particle size. Highest efficiencies are obtained with 5-μm rather than the 10-μm particle size packings in earlier use.

ii) Maintenance of column temperature at a constant value, e.g. 35° C, is recommended practice. Solvent viscosity is reduced, efficiency is improved and HPLC profiles are more reproducible, particularly when gradient systems are in use.

iii) Eluting solvents are acidified mixtures of water usually with methanol or acetonitrile. The latter has a lower viscosity and solvent strength than methanol, and gives somewhat different elution profiles. Whereas isocratic systems are useful for separation of components having closely similar polarities, gradient systems are more suitable for the range of phenolics and other detector-responsive components in grapes and wines. The duration of analysis is then shortened, but adequate regeneration of the column is necessary before its further use.

iv) Acidification depresses ionization of the colourless phenolics, and greatly improves their resolution in reverse-phase conditions. Preferred acids have been H_3PO_4, $HClO_4$, formic or acetic acids. In our experience, formic acid has proved most satisfactory; there have been no corrosion problems during its prolonged use at low concentrations, and its volatility has been advantageous in recovery of components from preparative HPLC (Somers et al. 1987).

v) Acidity requirements for analysis of anthocyanins are rather special. Satisfactory resolution occurs only with use of strongly acidic solvents, pH close to 1.5 (Preston and Timberlake 1981). Such low pH imposes stress on the column packing material, as it may induce hydrolysis of the bonded alkyl groups from the silica support. To extend column use, it is therefore essential that the column be thoroughly flushed with methanol (or acetonitrile) after each analysis. Use of less acidic mobile phase, e.g. pH 2.5, leads to peak broadening as a consequence of the presence of other equilibrium forms of the anthocyanins (Wulf and Nagel 1978; Fig. 11).

vi) HPLC detection of phenolics is by spectrophotometry in the UV (or visible region for anthocyanins). Early detectors were photometers fixed at one available wavelength, 254 nm or 280 nm, later followed by variable wavelength detectors covering the range 190–700 nm. Simultaneous detection at two or more wavelengths offers an analytical advantage because of the characteristically different spectral properties of major phenolic categories, viz. hydroxycinnamates, catechins and procyanidins, flavonols, acylated vs. nonacylated anthocyanins. Thus absorbance ratios in a single analysis can help to identify compounds and to verify purity.

vii) The ultimate in detector capability appears to be the photodiode array detector. Its use enables acquisition and recall of the complete spectrum of each

Table 3. HPLC systems for analysis of phenolic components in grapes and wines

Category	Column	Mobile phase	Detection (nm)	Reference
1. Hydroxy-cinnamates	Zorbax ODS (Du Pont Inst.)	Isocratic and gradient A: H_2O, pH 2.6 (H_3PO_4) B: CH_3CN	320	Ong and Nagel (1978)
	RP-C_{18}, 5 µm (Brownlee Labs)	Isocratic H_2O, pH 2.6 ($H_2NH_4PO_4$, H_3PO_4) and CH_3OH (84:16)	320	Singleton et al. (1984)
	Lichrosorb RP 18, 5 µm (Merck)	Gradient A: $HCOOH-H_2O$ (0.5:99.5) B: $HCOOH-CH_3OH$ (2:98)	280, 313	Somers et al. (1987)
2. Anthocyanins	Lichrosorb ODS	Gradient A: $HCOOH-H_2O$ (10:90) B: $HCOOH-CH_3OH-H_2O$ (10:50:40)	520	Wulf and Nagel (1978)
	Spherisorb Hexyl, 5 µm (Knauer)	Gradient A: H_2O, pH 1.5 ($HClO_4$) B: CH_3OH	280, 525	Preston and Timberlake (1981)
	Bondapak C_{18} Microbore (Waters Assoc.)	Gradient A: $HCOOH-H_2O$ (10:90) B: $HCOOH-(CH_3)_2CO-H_2O$ (10:25:65)	546	Roggero et al. (1986)
3. Catechins and procyanidins	Spherisorb Hexyl, 5 µm	Gradient, with pH shift A: H_2O, pH 2.5 ($HClO_4$) B: CH_3OH; C (H_2O, pH 7.0)	280	Lea (1982)[a]
	Hypersil SAS, 5 µm Micropak MCH, 10 µm (Varian)	Gradient A: H_2O, pH 2.5 ($HClO_4$) B: CH_3OH	280	Salagoïty-Auguste and Bertrand (1984)
	Lichrosorb RP 18, 5 µm	Gradient A: H_2O, pH 2.6 (H_3PO_4) B: CH_3CN		Romeyer et al. (1986)
	Sil C_{18}, 10 µm (Alltech)	Gradient A: H_2O B: CH_3COOH	280	Bourzeix et al. (1986)
4. Flavonols	Spherisorb ODS 2, 5 µm (Knauer)	Gradient A: $CH_3COOH-H_2O$ (4:96) B: CH_3CN	280, 340 or 365	Cheynier and Rigaud (1986)

[a] Used apple juice, see text.

peak detected at a chosen wavelength, along with "purity parameters" which provide qualitative assessment of the resolution obtained for any component.

Direct injection of the sample, juice or wine, gives satisfactory results for both hydroxycinnamates and anthocyanins, which are principal monomeric phenolics of white and red wines respectively. Detailed analysis of flavonols, which are minor phenolic constituents of grape skins, has required their prior separation from other components (Cheynier and Rigaud 1986). Similarly, the best results with HPLC of procyanidins, in which oligomers from catechins through to procyanidin tetramers have been resolved, also involves preliminary sample preparation by use of a polyamide column (Bourzeix et al. 1986). For grape juices and white wines, a better prospect could be that of using a pH shift technique (7.0 to 2.5), which has enabled early elution of interfering phenolic acids and direct HPLC of procyanidins in apple juices (Lea 1982).

It is noted, however, that HPLC recovery of procyanidins from red wines, along with estimates of residual anthocyanins and other phenolics, falls far short of total phenolic concentrations. As discussed earlier, polymeric pigments are increasingly the major phenolic fraction during wine maturation, and these are not resolved by any current method. In HPLC the polymeric pigments are eluted as a broad hump on which peaks for accompanying monomers and procyanidin oligomers are superimposed. Their isolation, as a heterogeneous mixture, has been possible only by gel filtration.

7 Summary

The major phenolic components of *Vitis vinifera* are the procyanidins ("tannins"), catechins and hydroxycinnamates, with anthocyanin pigments additionally present in the skins of red wine grapes. Widely variable concentrations of these component categories, consequent upon particular grape cultivar, environmental influences on grape composition and variation in the vinification method, are directly associated with the entire range of natural wine types, from delicate whites through rosés to astringent young red wines.

Spectral methods for evaluation of total hydroxycinnamates and of total flavonoid extractives in white wines are presented, along with HPLC description of phenolic composition. Physicochemical phenomena affecting the coloration of young red wines are discussed, viz. the influences of pH, free SO_2, self-association and co-pigmentation of anthocyanins on pigment equilibria, and on wine colour density and hue.

Progressive formation of oligomeric and polymeric pigments during vinification, maturation and ageing of red wines is described in terms of chemical age indices based on spectral analysis. Phenolic condensation mechanisms are critically discussed, and an earlier hypothesis for general interpretation of the phenomenon by which wine colour is stabilized during conservation is reconsidered in light of more recent information on molecular pigment aggregates in new red wines.

Analytical methods for spectral and HPLC assessments of phenolic composition in grape juices, white wines and red wines are comprehensively described.

References

Asen S, Stewart RN, Norris KH (1972) Co-pigmentation of anthocyanins in plant tissues, and its effect on colour. Phytochemistry 11:1139–1144

Bakker J (1986) HPLC of anthocyanins in port wines: determination of ageing rates. Vitis 25:203–214

Bakker J, Timberlake CF (1985a) The distribution of anthocyanins in grape skin extracts of port wine cultivars as determined by high performance liquid chromatography. J Sci Food Agric 36:1315–1324

Bakker J, Timberlake CF (1985b) The distribution and content of anthocyanins in young port wines as determined by high performance liquid chromatography. J Sci Food Agric 36:1325–1333

Bakker J, Timberlake CF (1986c) The mechanism of colour changes in aging port wine. Am J Enol Vitic 37:288–292

Bakker J, Bridle P, Timberlake CF (1986a) Tristimulus measurements (CIELAB 76) of port wine colour. Vitis 25:67–78

Bakker J, Bridle P, Timberlake CF, Arnold GM (1986b) The colours, pigment and phenol contents of young port wines: effects of cultivar season and site. Vitis 25:40–52

Bakker J, Preston NW, Timberlake CF (1986c) The determination of anthocyanins in aging red wines: comparison of HPLC and spectral methods. Am J Enol Vitic 37:121–126

Baranowski ES, Nagel CW (1983) Kinetics of malvidin-3-glucoside condensation in wine model systems. J Food Sci 48:419–422

Baranowski JD, Nagel CW (1981) Isolation and identification of the hydroxycinnamic acid derivatives in White Riesling wine. Am J Enol Vitic 32:5–13

Bourzeix M, Weyland D, Heredia N, Desfeux C (1986) Étude des catéchines et des procyanidols de la grappe de raisin, du vin et d'autres dérivés de la vigne. Bull OIV 59:1171–1254

Branas J (1977) Introduction to the work of the symposium. OIV symposium on the quality of the vintage 14–21 February 1977, Cape Town. Oenological and Viticultural Res Inst, Stellenbosch, pp 13–28

Brouillard R (1982) Chemical structure of anthocyanins. In: Markakis P (ed) Anthocyanins as food colours. Academic Press, New York, pp 1–40

Brouillard R, Delaporte B (1977) Chemistry of anthocyanin pigments. 2. Kinetic and thermodynamic study of proton transfer, hydratation and tautomeric reactions of malvidin 3-glucoside. J Am Chem Soc 99:8461–8468

Brouillard R, Delaporte B (1978) Étude par relaxation chimiques des équilibres de transfert de proton, d'hydratation et de tautoméric prototropique de l'anthocyane majoritaire de *Vitis vinifera,* malvidine. In: Laszlo P (ed) Protons and ions involved in fast dynamic phenomena, Elsevier, Amsterdam, pp 403–412

Burroughs LF, Sparks AH (1973) Sulphite-binding powers of wines and ciders II. Theoretical consideration and calculation of sulphite binding equilibria. J Sci Food Agric 24:199–206

Champagnol F (1977) Physiological state of the vine and the quality of the harvest. OIV symposium on the quality of the vintage 14–21 February 1977, Cape Town. Oenological and Viticultural Res Inst, Stellenbosch, pp 107–116

Charalambous G (ed) (1979) Liquid chromatographic analysis of food and beverages, vol 1. Academic Press, London, p 236

Cheminat A, Brouillard R (1986) PMR investigation of 3-0-(β-D-glucosyl) malvidin. Structural transformations in aqueous solutions. Tetrahedron Lett 27:4457–4460

Cheynier V, Rigaud J (1986) HPLC separation and characterization of flavonols in the skins of *Vitis vinifera* var. Cinsault. Am J Enol Vitic 37:248–252
Cheynier VF, Trousdale EK, Singleton VL, Salgues MJ, Wylde R (1986) Characterization of 2-S-glutathionyl-caftaric acid, and its hydrolysis in relation to grape wines. J Agric Food Chem 46:297–299
Di Stefano R, Ciolfi G (1983) Formazione di antociani polimeri in presenza di flavani ed evoluzione degli antociani monomeri durante la fermentazione. Riv Vitic Enol 36:325–338
Flanzy M, Poux C (1958) Les possibilités de la microvinification, application à l'étude de la macération. Ann Technol Agric 7:377–401
Glories Y (1984a) La couleur des vins rouges 1. Les équilibres des anthocyanes et des tanins. Connaiss Vigne Vin 18:195–217
Glories Y (1984b) La couleur des vins rouges 2. Mesure, origine et interprétation. Connaiss Vigne Vin 18:253–271
Goto T, Tamura H, Kawai T, Hoshino T, Harada N, Kondo T (1986) Chemistry of metalloanthocyanins. In: Breslow R (ed) Symp Bioorg Chem 6–8 May 1985, NY, Ann NY Acad Sci 471:155–173
Haslam E (1980) In vino veritas: oligomeric procyanidins and the ageing of red wines. Phytochemistry 19:2577–2582
Herrick IW, Nagel CW (1985) The caffeoyl tartrate content of White Riesling wines from California, Washington and Alsace. Am J Enol Vitic 36:95–97
Hoshino T, Matsumoto U, Goto T (1981) Self-association of some anthocyanins in neutral aqueous solutions. Phytochemistry 20:1971–1976
Iacobucci GA, Sweeny JG (1983) The chemistry of anthocyanins, anthocyanidins and related flavylium salts. Tetrahedron 39:3005–3038
Jackson MG, Timberlake CF, Bridle P, Vallis L (1978) Red wine quality: Correlations between colour, aroma and flavour and pigment, and other parameters of young Beaujolais. J Sci Food Agric 29:715–727
Jurd L (1967) Catechin-flavylium salt condensation reactions. Tetrahedron 23:1057–1064
Jurd L (1969) Review of polyphenol condensation reactions and their possible occurrence in the aging of wines. Am J Enol Vitic 20:191–195
Jurd L, Waiss AC (1965) Anthocyanidins and related compounds VI. Flavylium salt-phloroglucinol condensation products. Tetrahedron 21:1471–1483
Lea AGH (1982) Reversed-phase high-performance liquid chromatography of procyanidins and other phenolics in fresh and oxidising apple juices using a pH shift technique. J Chromatogr 238:253–257
Lea AGH, Bridle P, Timberlake CF, Singleton VL (1979) The procyanidins of white grapes and wines. Am J Enol Vitic 30:289–300
Macrae R (ed) (1982) HPLC in food analysis. Food Science and Technology Series. Academic Press, London, p 340
Moskowitz AH, Hrazdina G (1981) Vacuolar contents of fruit subepidermal cells from *Vitis* species. Plant Physiol 68:686–692
Nagel CW, Wulf LW (1979) Changes in the anthocyanins, flavonoids and hydroxycinnamic acid esters during fermentation and aging of Merlot and Cabernet Sauvignon. Am J Enol Vitic 30:111–116
Nagel CW, Baranowski JD, Wulf LW, Powers JR (1979) The hydroxycinnamic acid-tartaric ester content of musts and grape varietes grown in the Pacific Northwest. Am J Enol Vitic 30:198–201
Neubauer C (1872) Studien über den Rotwein. Ann Oenol 2:1–41
Office International de la Vigne et du vin (ed) (1978) Recueil des méthodes internationales d'analyse des vins caractérisques, chromatiques, 16–30 Paris
Ong BY, Nagel CW (1978) High pressure liquid chromatographic analysis of hydroxycinnamic acid tartaric acid esters, and their glucose esters in *Vitis vinifera*. J Chromatogr 157:345–355
Okamura S, Watanabe M (1982) Purification and properties of hydroxycinnamic acid ester hydrolase from *Aspergillus japonicus*. Agric Biol Chem 46:1839–1848
Osawa Y (1982) Co-pigmentation of anthocyanins. In: Markakis P (ed) Anthocyanins as food colours. Academic Press, New York, pp 41–68

Preston NW,Timberlake CF (1981) Separation of anthocyanin chalcones by high performance liquid chromatography. J Chromatogr 214:222–228
Ramey D, Bertrand A, Ough CS, Singleton VL, Sanders E (1986) Effects of skin contact temperature on Chardonnay must and wine composition. Am J Enol Vitic 37:99–106
Ribéreau-Gayon J, Peynaud E, Sudraud P, Ribéreau-Gayon P (1972) Sciences et techniques du vin, vol 1. Dunod, Paris, Ch 13, pp 471–514
Ribéreau-Gayon P (1959) Recherches sur les anthocyanes des végétaux. Application au genre *Vitis*. Libr Gen Enseignement, Paris
Ribéreau-Gayon P (1964) Les composés phénoliques du raisin et du vin. Inst Natl Rech Agron Paris
Ribéreau-Gayon P (1965) Identification d'esters des acides cinnamiques et de l'acide tartrique dans les limbes et les baies de *V. vinifera*. CR Acad Sci Paris 260:341–343
Ribéreau-Gayon P (1972) Evolution des composés phénoliques au cours de la maturation du raisin. II Discussion des résultats obtenus en 1969, 1970 et 1971. Connaiss Vigne Vin 6:161–175
Ribéreau-Gayon P (1973) Interprétation chimique de la couleur des vins rouges. Vitis 12:119–142
Ribéreau-Gayon P (1974) The chemistry of red wine colour. In: Webb AD (ed) Chemistry of winemaking. Am Chem Soc, Washington, pp 50–87
Ribéreau-Gayon P (1982) The anthocyanins of grapes and wines. In: Markakis P (ed) Anthocyanins as food colours. Academic Press New York, pp 209–244
Ribéreau-Gayon P, Glories Y (1971) Detérmination de l'état de condensation des tanins du vin rouge. CR Acad Sci Paris 273:2369–2371
Ribéreau-Gayon P, Pontallier P, Glories Y (1983) Some interpretations of colour changes in young red wines during their conservation. J Sci Food Agric 34:505–516
Robinson GM, Robinson R (1931) A survey of anthocyanins I. Biochem J 25:1687–1705
Roggero JP, Coen S, Ragonnet B (1986) High performance liquid chromatography survey on changes in pigment content in ripening grapes of Syrah. An approach to anthocyanin metabolism. Am J Enol Vitic 37:77–83
Romeyer FM, Macheix JJ, Sapis JC (1986) Changes and importance of oligomeric procyanidins during maturation of grape seeds. Phytochemistry 25:219–221
Salagoïty-Auguste MH, Bertrand A (1984) Wine phenolics. Analysis of low molecular weight components by high performance liquid chromatography. J Sci Food Agric 35:1241–1247
Scheffeldt P, Hrazdina G (1978) Co-pigmentation of anthocyanins under physiological conditions. J Food Sci 43:517–520
Siegrist J (1985) Les tanins et les anthocyanes du Pinot et les phénomènes de macération. Rev Oenol 11:11–13
Simpson RF (1982) Factors affecting oxidative browning of white wine. Vitis 21:233–239
Singleton VL (1974) Analytical fractionation of the phenolic substances of grapes and wine, and some practical uses of such analyses. In: Webb AD (ed) Chemistry of winemaking. ACS Symp ser 137 Ch 9
Singleton VL, Esau P (1969) Phenolic substances in grapes and wine, and their significance. Academic Press, New York
Singleton VL, Noble AC (1976) Wine flavor and phenolic substances. In: Charalambous G, Katz I (eds) Phenolic sulphur and nitrogen compounds in food flavors. ACS Symp ser 26 Ch 3, pp 47–70
Singleton VL, Rossi JA (1965) Colorimetry of total phenolics with phosphomolybdic-phosphotungstic acid reagents. Am J Enol Vitic 16:144–158
Singleton VL, Timberlake CF, Lea AGH (1978) The phenolic cinnamates of white grapes and wine. J Sci Food Agric 29:403–410
Singleton VL, Zaya J, Trousdale EK, Salgues MJ (1984) Caftaric acid in grapes, and conversion to a reaction product during processing. Vitis 23:113–120
Singleton VL, Salgues M, Zaya J, Trousdale EK (1985) Caftaric acid disappearance, and conversion to products of enzymic oxidation in grape must and wine. Am J Enol Vitic 36:50–56
Singleton VL, Zaya J, Trousdale EK (1986a) Caftaric and coutaric acids in fruit of Vitis. Phytochemistry 23:2127–2133

Singleton VL, Zaya J, Trousdale EK (1986b) Compositional changes in ripening grapes: caftaric and coutaric acids. Vitis 25:107–117
Somers TC (1966) Wine tannins – isolation of condensed flavonoid pigments by gel filtration. Nature 209:368–370
Somers TC (1968) Pigment profiles of grapes and of wines. Vitis 7:303–320
Somers TC (1971) The polymeric nature of wine pigments. Phytochemistry 10:2175–2186
Somers TC (1983) Influence du facteur temps de conservation sur les caracteristiques physico-chimiques et organoleptiques des vins. Bull OIV 56:172–188
Somers TC (1987) Assessment of phenolics in viticulture and oenology. In: Lee TH (ed) Proc Sixth Aust Wine Ind Tech Conf, 14–17 July 1986, Adelaide SA. Australian Wine Research Institute and Australian Society of Viticulture and Oenology
Somers TC, Evans ME (1974) Wine quality: correlations with colour density and anthocyanin equilibria in a group of young red wines. J Sci Food Agric 25:1369–1379
Somers TC, Evans ME (1977) Spectral evaluation of young red wines: anthocyanin equilibria, total phenolics, free and molecular SO_2, chemical age. J Sci Food Agric 28:279–287
Somers TC, Evans ME (1979) Grape pigment phenomena: interpretation of major colour losses during vinification. J Sci Food Agric 30:623–633
Somers TC, Evans ME (1986) Evolution of red wines. I. Ambient influences on colour composition during early maturation. Vitis 25:31–39
Somers TC, Pocock KF (1986) Phenolic harvest criteria for red vinification. Aust Grapegrower Winemaker No 268:24–30
Somers TC, Wescombe LG (1982) Red wine quality. The critical role of SO_2 during vinification and conservation. Aust Grapegrower Winemaker (220):68–74
Somers TC, Wescombe LG (1987) Evolution of red wines II. An assessment of the role of acetaldehyde. Vitis 26:27–36
Somers TC, Ziemelis G (1972) Interpretations of ultraviolet absorption in white wines. J Sci Food Agric 23:441–453
Somers TC, Ziemelis G (1980) Gross interference by sulphur dioxide in standard determinations of wine phenolics. J Sci Food Agric 31:600–610
Somers TC, Ziemelis G (1985a) Spectral evaluation of total phenolic components in *Vitis vinifera:* grapes and wines. J Sci Food Agric 36:1275–1284
Somers TC, Ziemelis G (1985b) Flavonol haze in white wines. Vitis 24:43–50
Somers TC, Evans ME, Cellier KM (1983) Red wine quality and style: diversities of composition and adverse influences from free SO_2. Vitis 22:348–353
Somers TC, Vérette E, Pocock KF (1987) Hydroxycinnamate esters of *Vitis vinifera:* compositional changes during white vinification and effects of exogenous enzymatic hydrolysis. J Sci Food Agric 40:67–78
Snyder LR, Kirkland JJ (1979) Introduction to modern liquid chromatography, 2nd edn. John Wiley, New York
Sudraud P (1958) Interprétation des courbes d'absorption des vins rouges. Ann Technol Agric 7:203–208
Terrier A, Blouin J (1975) Observations sur l'extraction des jus des raisins blancs. Connaiss Vigne Vin 9:273–303
Timberlake CF (1982) Factors affecting red wine colour: The use of a "coloration" constant in evaluating red wine color. Symposium proceedings, Grape and wine Centennial, 18–21 June 1980, Davis CA. Davis CA, University of California 1982, pp 240–244
Timberlake CF, Bridle P (1968) Flavylium salts resistant to sulphur dioxide. Chem Ind 1489
Timberlake CF, Bridle P (1976) Interactions between anthocyanins, phenolic compounds, and acetaldehyde and their significance in red wines. Am J Enol Vitic 27:97–105
Timberlake CF, Bridle P (1983) Colour in beverages. In: Williams AA, Atkin RK (eds) Sensory quality in foods and beverages. Chichester Ellis Horwood, pp 140–154
Timberlake CF, Bridle P, Jackson MG, Vallis L (1978) Correlations between quality and pigment parameters in young Beaujolais red wines. Ann Nutr Aliment 32:1095–1101

Trillat A (1908) L'aldéhyde acétique dans le vin: son origine et ses effects. III. Rôle de l'aldéhyde acétique dans quelques modifications du vin: vieillissement, jaunissement et amertume. Ann Inst Pasteur 22:876–895

Tromp A, van Wyk CJ (1977) The influence of colour on the assessment of red wine quality. Proc S Afr Soc Enol Vitic, pp 107–118

Vogel P (1985) Carbocation chemistry. Ch. 1. In: Studies in Organic Chemistry Series, vol 21. Elsevier, Amsterdam

Voyatzis I (1984) Recherches sur les composés phénoliques des vins blancs. Interprétation de la couleur. Thesis University of Bordeaux II, Bordeaux, No 116

Weinges K, Piretti MV (1971) Isolierung des $C_{30}H_{26}O_{12}$-Procyanidins Bl aus Weintrauben. Liebigs Ann Chem 748:218–220

Wildenradt HL, Singleton VL (1974) The production of aldehydes as a result of oxidation of polyphenolic compounds and its relation to wine aging. Am J Enol Vitic 25:119–126

Williams M, Hrazdina G (1979) Anthocyanins as food colorants: Effect of pH on the formation of anthocyanin-rutin complexes. J Food Sci 44:66–68

Wulf LW, Nagel CW (1976) Analysis of phenolic acids and flavonoids by high-pressure liquid chromatography. J Chromatogr 116:271–276

Wulf LW, Nagel CW (1978) HPLC separation of anthocyanins of *Vitis vinifera*. Am J Enol Vitic 29:42–49

The Site-Specific Natural Isotope Fractionation-NMR Method Applied to the Study of Wines

G. J. Martin and M. L. Martin

1 Stable Isotopes and Site-Specific Natural Isotope Fractionation

Stable and radioactive isotopes are important probes for mechanistic investigations in plant physiology and numerous experiments have been designed for following the fate of a given precursor, artificially labeled with ^{2}H or ^{3}H, ^{13}C or ^{14}C, ^{15}N etc. in the course of biosynthetic reactions (Schmidt et al. 1982; Stothers 1974; Simpson 1975; Garson and Staunton 1979; Hutchinson 1982; Simon et al. 1968). Interestingly, as regards the study of products synthetized in natural conditions, the stable isotopes present at natural abundance show themselves to be important sources of information about the history of each chemical species.

1.1 Natural Isotope Distribution

The main atoms which make up organic matter frequently exist under several isotopic forms, one of them being largely predominant. Such is the case for carbon, characterized by a natural ratio $^{13}C/^{12}C$ of about 1.1×10^{-2}, for oxygen in which $^{18}O/^{16}O$ is about 2×10^{-3} and for hydrogen, which contains a still lower proportion of the deuterium isotope ($D/H \simeq 1.5 \times 10^{-4}$) etc.

About 50 years ago, it was recognized that the isotope content of a given molecular species may differ according to its origin. A striking example is that of water, which exhibits a deuterium content of 150×10^{-6} for Nantes tap water, whereas the proportion of deuterium drops to 89×10^{-6} in the case of water from the Antarctic region (Confiantini 1978).

$^{13}C/^{12}C$ isotope ratio is usually accessible as a whole by mass spectrometry measurements performed on carbon dioxide issued from the combustion of the sample. These ratios provide extensive and fruitful information about the different biochemical and physiological pathways which govern the photosynthesis of plants (Wickman 1952; Park and Epstein 1960; Krueger and Reesman 1982). Mass spectrometry determinations of the hydrogen and oxygen isotope ratios of water have also been helpful in improving our knowledge of the water cycle in plants of different vegetal species or grown in different climatic conditions (Dunbar 1982a, c; Sudraud and Koziet 1978; Ziegler 1976; Bricout 1981). More recently, the behavior of the total deuterium content of dry matter or of more elaborated extracts of plants such as lipids, starch, cellulose, or proteins was investigated through measurements involving the water issued from appropriate burning of the sample (Estep and Hoering 1980; Ziegler et al. 1976; Sternberg et al.

1984). Due to experimental difficulties the $^{18}O/^{16}O$ ratios of organic matter have not received much attention and very few results are concerned with plant physiology (Fehri and Letolle 1977, 1979; Sternberg et al. 1984).

As far as the nature of the plant is concerned, most isotope studies resort to $^{13}C/^{12}C$ measurements and compare Calvin C_3, Hatch-Slack C_4 or CAM metabolisms (Bender 1971; Lerman 1974; Bricout and Fontes 1974, Ziegler et al. 1976; Smith 1975). In particular, much effort has been devoted to explaining the origin of the stronger discriminating effect against ^{13}C exhibited in the carboxylation step of C_3 plants as compared to C_4 species.

For the purpose of distinguishing the relative roles of metabolic and environmental factors, the D/H ratios are expected to be more sensitive probes than the $^{13}C/^{12}C$ ratios since the water cycle of a plant greatly depends on the meteorological conditions, whereas the carbon isotope content of atmospheric CO_2 is almost constant at the earth's surface. Moreover, although valuable results concerning plant physiology and related disciplines such as food science can be obtained from the study of overall isotope contents, deeper insight into the biochemical mechanisms can be expected from determination of isotope contents at specific molecular sites. In this respect the technical requirements of mass spectrometry, which usually proceeds by burning the sample before analysis, are not suited to the measurement of an "intramolecular" isotope distribution. This severe limitation can sometimes be overcome by tedious and time-consuming chemical transformations which, in addition, carry the risk of indesirable isotope effects. Nevertheless a number of interesting types of behavior concerning carbon isotope ratios at strategic molecular positions have been observed in this way (Monson and Hayes 1980).

Exploiting the joint structural and quantitative dimensions of NMR spectroscopy, we showed in 1981 (G.J. Martin and M.L. Martin 1981) that deuterium is far from being randomly distributed in organic molecules. The large variations detected in the deuterium contents at the different molecular sites are related to specific kinetic (enzymatic or chemical) or thermodynamic isotope effects and the determination of site-specific natural isotope fractionation by nuclear magnetic resonance (SNIF-NMR) was proposed as a method of labeling without enrichment (G.J. Martin et al. 1982b).

This method has proved to be particularly powerful in the field of wines which also remains one of its privileged applications since the initial challenge was to find a way of detecting the chaptalization of wines (G. J. Martin and M. L. Martin 1983, G. J. Martin et al. 1986b)

1.2 Isotopic Filiation of Wines

Besides its considerable economic importance, the vine is also, if considered scientifically, an interesting example of a C_3 plant, since it grows in a number of countries with different climates and has undergone drastic phenotypic selection. This makes possible a comparative investigation of the influence of varieties and environmental factors. Vine products have been the subject of a number of studies based on mass spectrometry determinations of the total isotope contents of different components and especially of water. Bricout (1973, 1978), Dunbar (1982a), Förstel and Hützen (1984) and Förstel (1985), analyzing the $^{18}O/^{16}O$ ratio of water from grape and wine, concluded that fermentation has no noticeable influence on the isotope content, which depends, in contrast, on climatological parameters such as evapo-transpiration. Thus an enrichment in the ^{18}O and ^{2}H contents of water from grape must is observed when the plant is grown in hot countries, whereas the wine variety shows no significant effect (Bricout et al. 1975; Förstel and Hützen 1984). Determination of the isotope content of water extracted from leaves and fruits also enables the effect of temperature, rain, maturation, parasitism, physical treatments or additives to be characterized (Bricout 1973, 1982; Dunbar 1982b). In the case of sparkling wines (Dunbar 1982d), the method of production can be identified using the ^{13}C content of carbon dioxide originating – from C_3 plants such as grapes – from sugarcane, a species belonging to the C_4 family and used in Champagne fermentation – or from industrial gas used in the carbonation procedure.

Although mass spectrometry determination of the total isotope ratios of more elaborated molecules such as alcohols or sugars presents some experimental difficulties, the carbon isotope ratio of ethanol has been successfully used for identifying the C_3 or C_4 origin of the ethanol precursors (Förstel 1986; Bricout 1982; Dunbar et al. 1983). Similarly, the average D/H ratios of ethanols from grapes differ significantly from those of ethanols from C_4 plants such as sugarcane or maize (Rauschenbach et al. 1979; Bricout et al. 1975). More recently, measurement of the isotope ratio of the nonexchangeable hydrogens of residual sugars has been shown to provide a method for detecting added unfermented sugars in wines (Dunbar et al. 1983). However, on the basis of mass spectrometry determination of the isotope ratios of water and of ethanol as a whole, no unambiguous method could be proposed for identifying the variety of the vine, the geographical origin of the wine, or the addition before fermentation of a C_3 sugar such as beet sugar.

The site-specific dimension of the SNIF-NMR method may be the source of selective and independent information about the mechanistic steps of plant photosynthesis. Thus in the case of an alcoholic mixture fermented from biomass, and namely from grape must, the natural labeling of ethanol and water is a key to determining the origin, botanic or geographic, as well as the year of production of the mother plant, and to detecting several kinds of adulteration.

2 Theoretical Bases of Isotopic Characterization

A multi-isotopic strategy of analysis of a wine is expected to be generally applicable to the extent that the isotopic parameters of the wine constituents faithfully represent the isotopic properties of its precursors. The various components, successively involved in the course of wine production and analysis, and labeled according to the scheme shown below, will be described in terms of their various isotopomers:

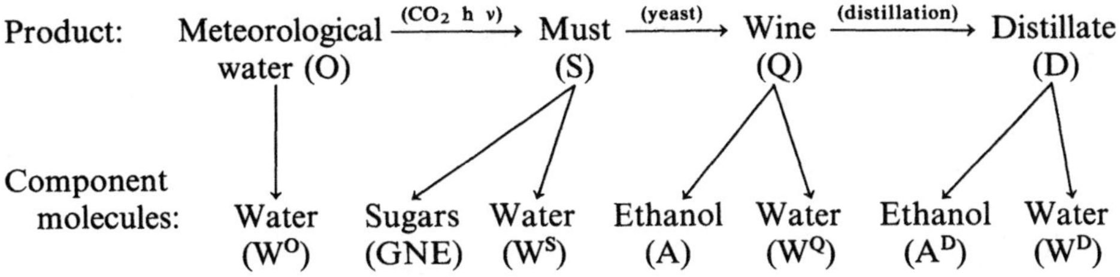

For our purpose, wine can be reduced to a mixture of water and ethanol, and due to the very low natural abundance of deuterium, only the 5 monodeuterated isotopomers will have to be considered in the SNIF-NMR method (Fig. 1).

$CH_2D{-}CH_2{-}OH$	$CH_3{-}CHD{-}OH$	$CH_3{-}CH_2{-}OD$	HDO
I	II R and II S	III	W^Q .

In fact, in achiral media, the two enantiomeric methylene sites of ethanol are not distinguished in the NMR spectrum and only one mean isotope ratio is available for sites II. In order to investigate the enantiomeric balance, a diastereotopic relationship must be introduced in the ethyl moiety by appropriate reaction or interaction with a chiral substrate (ML Martin et al. 1983).

If N_i and N_H denote the numbers of monodeuterated and fully protonated molecules and P_i the stoechiometric number of hydrogens in site i, the statistical,

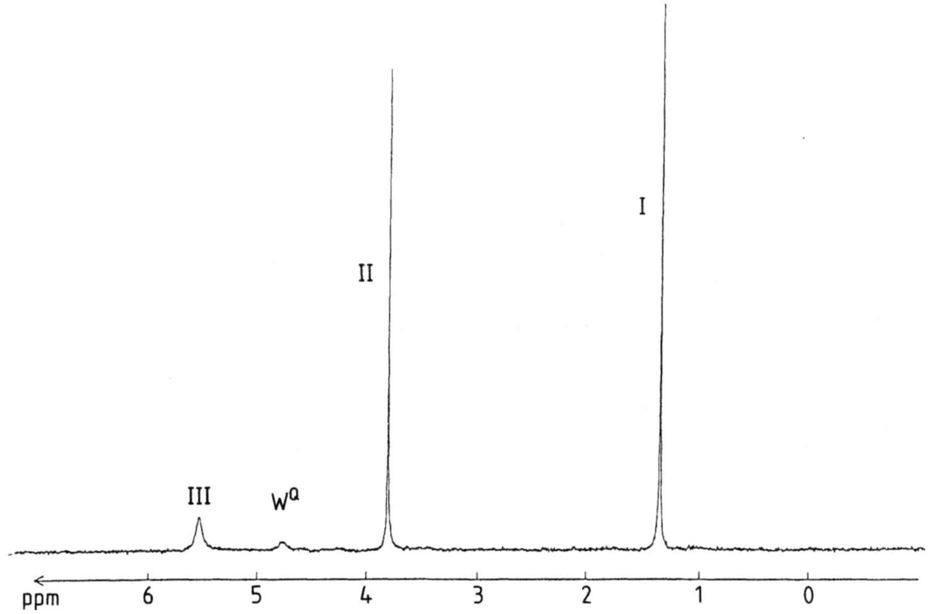

Fig. 1. Proton decoupled natural abundance ^{2}H NMR spectrum of sample of 96% ethanol. The nominal resonance frequency of the spectrometer is 61.4 MH_z. 128 scans have been accumulated

F_i and effective, f_i, mole fractions of isotopomer i are defined by Eqs. (1) and (2):

$$F_i = \frac{P_i}{\Sigma_i P_i} \tag{1}$$

$$f_i = \frac{N_i}{\Sigma_i N_i}, \tag{2}$$

where the summations hold over the different monodeuterated isotopomers of the considered molecule.

At the natural abundance level of deuterium the specific isotope ratio of site i is approximately given by

$$(D/H)_i = \frac{N_i}{P_i N_H}. \tag{3}$$

The average or molecular isotope ratio, $(\overline{D/H})$, is related to $(D/H)_i$ by Eqs. (4) and (5)

$$(\overline{D/H}) = \Sigma_i F_i (D/H)_i \tag{4}$$

$$(D/H)_i = \frac{f_i}{F_i} (\overline{D/H}). \tag{5}$$

For ethanol, Eq. (4) becomes

$$(\overline{D/H}) = \tfrac{1}{2}(D/H)_I + \tfrac{1}{3}(D/H)_{II} + \tfrac{1}{6}(D/H)_{III}, \tag{6}$$

and it is convenient to introduce the parameter R

$$R = 2\frac{(D/H)_{II}}{(D/H)_I}, \tag{7}$$

which expresses the relative enrichment or depletion of the methylene site, the methyl site being arbitrarily given the statistical weight 3. A random distribution of deuterium within the ethyl fragment would therefore be characterized by a value of the ratio R equal to 2.

Since, in appropriate technical conditions, the area of an NMR signal is proportional to the number of nuclei resonating at the considered frequency, the isotope parameters are directly accessible from surface or, in strictly defined conditions, from signal height measurements. Thus if S_I and S_{II} denote the intensities of the methyl and methylene signals in the deuterium spectrum of a wine distillate D:

$$R = 3\frac{S_{II}}{S_I}, \tag{8}$$

and

$$f_i = \frac{S_i}{\Sigma_i S_i} \quad \text{with } (i = I \text{ to } III). \tag{9}$$

Determination of the $(D/H)_i$ ratios requires the use of an external or internal reference (G. J. Martin et al. 1985). In the first procedure the working standard WS, is situated in a sealed tube coaxial with the cell containing the distillate, whereas in the second procedure the working standard is mixed with the distillate. This working standard must be independently scaled with respect to the international standard V. SMOW (Craig 1961; Hagemann et al. 1970; Gonfiantini 1978).

In the internal referencing technique for example the $(D/H)_j^Q$ parameter associated with isotopomer j of ethanol in the distillate is given by (G. J. Martin et al. 1985):

$$(D/H)_j^D = \frac{P^{WS}}{P_j^A} \frac{M^A}{M^{WS}} \frac{m^{WS}}{m^D} \frac{T_j^D}{t_m^D} (D/H)^{WS} . \tag{10}$$

T_j^D is the ratio of signal intensities S_j^D/S^{WS}, m^D and m^{WS} are the masses of the distillate and of the working standard, M^A, M^{WS} the molecular weights of ethanol and WS, and the t_m^D % is the alcoholic grade in weight.

In order to convey botanical, geographical and climatological information, the isotope parameters of wine must reflect properties of the grape sugars and must water. In this respect a knowledge of the relative probabilities that deuterium atoms in a given site of the products, Q_j, arise from different sites, S_i, of the reactants may enable the isotope distribution in inaccessible glucose-fructose and water starting media to be inferred from measurements performed on the alcoholic solution. On a quantitative basis it may be considered that the deuterium content in n sites of the end products is expressible in terms of the deuterium contents in the m different sites of the starting materials by a set of linear equations. Since at the natural level, the fractional abundance of deuterium reduces approximatively to the D/H ratio, the redistribution occurring in the bio-conversion may be tentatively described by a set of n equations of type (11)

$$(D/H)_j^Q = a_{j1} (D/H)_1^S + \ldots + a_{ji} (D/H)_i^S + \ldots + a_{jm} (D/H)_m^S, \tag{11}$$

or in matrix notation

$$[A] \vec{D^S} = \vec{D^Q}, \tag{12}$$

where $\vec{D^Q}$ and $\vec{D^S}$ are the column vectors of the deuterium contents in the end products and starting materials respectively and [A] is the redistribution matrix which reflects the technical and mechanistic aspects of the bio-conversion. Thus, providing the matrix [A] is previously determined by series of calibration experiments, it becomes possible to produce, from the vector $\vec{D^Q}$ deduced from SNIF-NMR measurements performed on the wine, a fingerprint of the deuterium contents of the disappeared starting materials. Access to complete sets of reciprocal redistribution coefficients is unfortunately restricted by the limited discriminating power of ^{2}H NMR, even when a high field spectrometer is used. Thus only three sets of sites, (1), (5,5,6′) and (2,3,4) can be clearly distinguished in the 61.4 MH_z spectrum of the nonexchangeable sites of glucose (GNE). Due to this limitation, some of the average coefficients of the reduced redistribution matrix have therefore no simple mechanistic significance and should only be considered as empirical factors. Nevertheless, by exploiting the described treatment, redistribution parameters conveying invaluable filiation information can be obtained, as will be discussed in Section 4.

Finally it should be emphasized that the multi-isotopic method of wine characterization is generally applicable to commercial products only to the extent that the [A] matrix is mainly governed by reproducible biochemical effects and is not significantly affected by uncontrolled technological effects.

3 Experimental Techniques of Isotopic Characterization

In principle the higher the number of independent isotopic parameters and the better the accuracy of their determination, the more efficient will the istotope fingerprint method be for providing an authentification and quality label for wines. The SNIF-NMR method is a powerful tool for determining site-specific isotope parameters, and average $\overline{D/H}$ values can also be obtained. However, the average $\overline{D/H}$ parameter of the nonexchangeable sites of glucose, that of ethanol, and eventually the $^{13}C/^{12}C$ ratios, may be alternatively or more conveniently measured by mass spectrometry. This technique is also more appropriate to a fast and precise determination of the D/H ratio of water and it is the only method for measuring the $^{18}O/^{16}O$ ratio, since the ^{18}O isotope has no nuclear spin and its quadrupolar ^{17}O partner is characterized by a low natural abundance and broad NMR signals.

A description of mass spectrometry methods can be found in Gonfiantini (1981) and in Volume 4 of this Series. We shall restrict ourselves to a brief presentation of the SNIF-NMR methodology (M. L. Martin et al. 1980, G. J. Martin 1985).

3.1 Methodology of the SNIF-NMR Method

Due to sensitivity and dynamic range problems, the NMR measurements are performed on ethanolic distillates prepared in strictly standardized conditions. The alcoholic grade in weight of the distillate which intervenes in Eq. (10), is determined by the Karl-Fischer method and eventually converted to volume content with the aid of alcoholic tables. The value of the isotope ratio of water remaining at the end of the distillation $(D/H)_W^R$, can be determined by mass spectrometry in the usual way. However, it should be noted that this value may differ from that of the wine, $(D/H)_W^Q$, since the fraction of water which passes during the distillation is greatly impoverished with respect of that which remains in the flask. A quantitative treatment appropriate to the selected experimental conditions shows that the error admitted in assimilating $(D/H)_W^Q$ to $(D/H)_W^R$ is not higher than 0.15 ppm.

An analogous situation occurs in principle in the case of the SNIF-NMR measurements of the ethanol distillate since the yield of the distillation is not strictly quantitative. However, it is shown that for the alcohol grades in volume $t_V^D \simeq 0.95$ and $t_V^R \simeq 0.0015$ currently adopted, the ethanol site-specific ratios measured on the distillate differ only negligigly from those of the wine.

It may also be desirable to determine the average ratio $(\overline{D/H})^A$ of the ethanol molecule. This parameter, which is accessible by mass spectrometry, was the only one previously considered before the use of NMR (Bricout et al. 1975). In fact this overall ratio is not very specific to the investigated wine, since it averages different contributions of the initial sugars and of the fermentation medium. Moreover, when measuring $(\overline{D/H})^A$ by mass spectrometry of an anhydrous ethanol, one must be aware of the fact that the contribution of the exchangeable site may differ from that which it ought to be in the wine, since noticeable fractionation is likely to occur in the coursse of the physical treatments. The value of the total deuterium content of ethanol or of its ethyl fragment may therefore be more conveniently accessible from the NMR experiments.

As regards the hydroxyl sites of the ethanolic solution, it should be noted that either one average NMR signal or two separate signals are observed according to whether the chemical exchange between ethanol and water is respectively fast or slow on the NMR time-scale. But whatever situation occurs, the deuterium contents in the hydroxyl site, III, and in water, W, are systematically related through the equilibrium constant $Ke \simeq 1.03$ of the protium-deuterium exchange. In order to illustrate the range of wine isotope parameters, a typical set of hydrogen ratios can be listed: $(D/H)_I \simeq 100$ ppm, $(D/H)_W^Q \simeq 160$ ppm, $(D/H)_{III} \simeq 165$ ppm, $(\overline{D/H})^A \simeq 121$ ppm.

As mentioned previously, determination of absolute values of the D/H ratios requires the use of external or internal referencing. The first procedure is fast and easily implemented, but more accurate results have been obtained by resorting to a chemically inert and nonvolatile internal standard. N,N, tetramethylurea (TMU), which provides a signal of suitable frequency and intensity, has been conveniently used. In some cases an intramolecular reference can be introduced by reacting the investigated molecules with a common pool of an appropriate substance of known isotopic ratio (G. J. Martin et al. 1985). In this respect, acetylation of glucose is advantageous both as a means for obtaining better resolved spectra and as a source of intramolecular referencing.

Since the characterization of wines exploits tiny differences in deuterium contents, a great deal of effort has been devoted to the implementation of accurate quantitative procedures and to statistical treatments of repetitive experiments. A dedicated software ISOLOG (G. G. Martin et al. 1987) written in Pascal conveys the data from the spectrometer computer towards an IBM-AT or Digital computer, which administrates an isotopic data base. This program also governs an automatic treatment of the NMR measurements and after appropriate statistical calculations presents the data in the form of $(D/H)_I$, $(D/H)_{II}$, R and $(\overline{D/H})$ values. The corresponding $^{13}C/^{12}C$, $^{18}O/^{16}O$ and $(D/H)_W$ ratios obtained from mass spectrometry measurements may be jointly stored. Appendage data bases contain the meteorological (precipitation, temperature, relative humidity, insulation) and geographical (longitude, latitude, distance from the sea) characteristics of the region and year of production, as well as data on the meteoric water of the region. ISOLOG manages various routines including I/O operations, data retrieval, and statistical packages for analyzing the raw data and helping identification of an unknown sample by use of discriminant factor analysis.

3.2 Experimental Conditions

3.2.1 Preparation of the Sample

A test sample of 300 ml (measured exactly) of wine of known alcohol content, t_V^O, is distilled using a Cadiot-type rotating strip column. The rate of reflux is maintained constant at 1/10 and the azeotrope which boils between 78.2 and 78.5° C is collected. For a wine containing approximatively 10% alcohol, about 31 ml of ethanol are collected with a purity of about 95%; the remaining liquid thus contains no more than 0.15% ethanol.

The alcohol level of the distillate is determined by the Karl-Fischer method. The use of an automatic titrator associated with an electronic balance fitted with a small printer is an advantage, as the ethanol solution containing the internal reference can be prepared in situ.

Ten ml of distillate and working standard (pure analytical N,N-tetramethylurea) are measured with an automatic pipet. These volumes of reagents are weighed to an accuracy of 0.1 mg using the balance of the titrator and both the ratio m^{WS}/m^D and the alcohol content, t_m^D, are recorded on the printed list with the registry number of the sample. C_6F_6 is introduced into the mixture for the NMR measurement.

3.2.2 SNIF-NMR Measurements

A 9.4 T (400 MHz) spectrometer fitted with a specific probe tuned on the deuterium frequency, accepting 15-mm-diameter tubes and equipped with fluorine field-frequency locking and an automatic sample-changer, gives quite satisfactory results (an optimized dedicated version is developed with Bruker). In order to deal with a large number of spectra over the week-end, the computer of the spectrometer should be interfaced with a hard disk ($\simeq$96 Mo).

The preliminary adjustments of the spectrometer must be carried out correctly (M. L. Martin et al. 1980) and sealed test tubes will be produced in collaboration with the Community Bureau of Reference from the Commission of the European Communities in Brussels.

The repeatability of NMR measurements performed on the same sample with the same spectrometer is better than 0.2% in relative value. The $(D/H)_i$ and R ratios are obtained with an accuracy of $\pm$0.15 ppm and $\pm$0.005 units respectively. The reproducibility of measurements carried out in long term or in different laboratories is of the order of 0.5% in relative value, on condition that test samples are used.

A signal-to-noise ratio of at least 175 is necessary. This is obtained with 128 accumulations characterized by an acquisition time of 6.8 s over a range of 1200 Hz (active memory 16 K words). Ten spectra of each sample are successively recorded for statistical treatment and eight samples can be automatically studied in 24 h with the aid of the changer robot.

4 Basic Mechanisms of Isotope Fractionation from Sugar Solutions to Fermented Alcohol-Water Mixtures

It has been observed, from mass spectrometry measurements, that the overall deuterium content of natural ethanols is lower than that of the sugars (Bricout et al. 1975). Fractionation effects are therefore likely to occur in the course of the fermentation process and it might be suspected (Simon and Medina 1968; Rauschenbach et al. 1979) that ethanol does not faithfully reflect the properties of the juice materials, which are themselves representative of the physiological, geographical and climatological context of the photosynthetic reactions.

A better understanding of the fractionation mechanisms which lead from the isotope repartition in the sugar medium to that in the alcoholic medium can be attained by analyzing series of test experiments in which glucoses from different starches and sucroses are fermented in water media with different isotope ratios. Thus SNIF-NMR measurements performed on glucose acetates enable isotope ratios of several sets of sites to be estimated, and, by transforming glucose into nitrate, the mean $\overline{D/H}$ ratio of the nonexchangeable sites can be determined by mass spectrometry.

By applying a theoretical analysis of the data as described in Section 2, it is demonstrated that in the conventional conditions of wine production, constant redistribution parameters relate the isotope contents in molecular species of the grape juice and wine media (Martin et al. 1983b–1986b). These parameters, which characterize the sensitivity of the isotope ratios of the products towards the starting components, show that the deuterium content of the methyl site of ethanol is about five times more sensitive towards the nonexchangeable sites of glucose than towards the starting water. In terms of variations with respect to a given wine taken as a standard, Eq. (13) is approximately satisfied:

$$\Delta(D/H)_I = 1.1\ \Delta(D/H)_{GNE} + 0.23\ \Delta(D/H)^S_W. \quad (13)$$

In contrast, the methylene site of ethanol has no direct connection with glucose and responds with a fractionation factor of about 0.7 to variations in the isotope ratio of the juice water:

$$\Delta/D/H)_{II} = 0.03\ \Delta\ (D/H)_{GNE} + 0.74\ \Delta\ (D/H)^S_W. \quad (14)$$

This behavior substantiates the occurrence of strong isotope effects in the enzymatic incorporation of atoms from water into organic molecules.

As regards the hydroxyl sites of glucose and ethanol, it should be recalled that they are involved in a chemical exchange with the water hydrogens. Although thermodynamic effects may introduce differences in the isotope ratios of these exchangeable sites, no independent mechanistic information is provided by the individual parameters. In contrast, it should be emphasized that no oxygen exchange is detected between water and ethanol in normal conditions (Dunbar 1982e).

For fermentation media containing a concentration of sugar of approximately 100 g l^{-1}, the redistribution coefficients which relate variations in the isotope content of wine water to variations in the D/H ratios of must water and of the nonexchangeable sites of glucose, are given in Eq. (15):

$$\Delta(D/H)^{O}_{W} = 0.055\ \Delta(D/H)^{S}_{GNE} + 0.97\ \Delta(D/H)^{S}_{W}. \tag{15}$$

This behavior corroborates the participation of the non-exchangeable sites of glucose in the water medium.

More generally the existence of a basic matrix of coefficients nearly independent of the nature of the yeast strain, of the temperature and of the concentration, within the limits usually admitted in wine production, makes the water and ethanol isotopomers of wine faithful and powerful probes for characterizing the physiological and biological effects which have governed the photosynthesis of grapes in natural conditions.

5 Isotope Fractionation as a Source of Information on Environmental Factors

As expected from the above considerations, the site-specific isotope ratios of ethanol and water in natural wines may provide valuable criteria for characterizing the geographical origin of the wine, the nature of the vine variety and the year of production. A systematic investigation of the influence of these factors has been performed on several hundreds of genuine wines and typical behaviors could be characterized by appropriate discriminant analyses of the data.

Thus concerning the recognition of the production area, the different countries are primarily differentiated by their latitude, a situation nearer to the equator being associated with higher values of the isotope ratios. This behavior is partly due to the enrichment in heavy isotopes in the rains of tropical countries (Dangsgaard 1964). In addition, the temperature-precipitation duality of a given country contributes to slightly increasing or decreasing the site-specific $(D/H)_j$ ratios. A dry and hot climate produces a deuterium enrichment of vine leaves and leads to higher values of the isotope ratios of the wine. Consequently, even for countries situated in the same range (about 500 km) of latitudes, a clear distinction can usually be made between oceanic, mediterranean, and continental production area (Fig. 2).

The vine variety in itself exerts only a very small effect – if any – on the isotope ratios of ethanol and water. In fact, when an influence of the variety is identified, it can actually be related to differences in the climate undergone during the growth period of the grapes. For example, it is observed that, for a given country and a given year, the vine variety Cabernet-Sauvignon is significantly enriched in deuterium as compared to the Carignan species. This behavior may be related to the fact that both the photosynthetic and maturation cycles of Cabernet-Sauvignon take place during hotter periods than in the case of Carignan. Such phenomena of physiological nature probably explain the distinction which can eventually be made, for a given production area, between early and late varieties. More generally multiple correlations can be defined between the various site-specific isotope ratios and the decadal temperatures and precipitation parameters associated with the vineyard region.

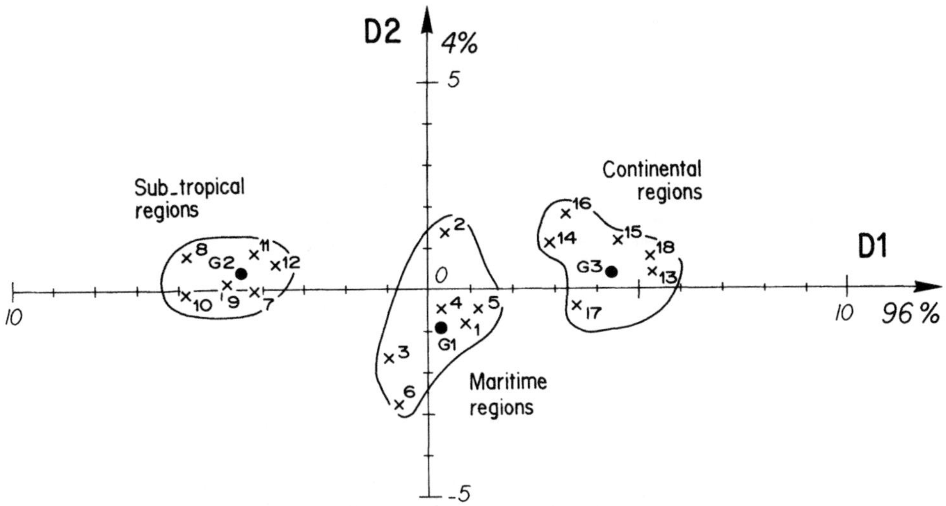

Fig. 2. Discriminant factor analysis of natural wines from different production area arranged in three groups of countries. *Group 1:* Maritime regions (wet and temperate) *1* France (Atlantic Coast); *2* Spain (North West); *3* Brazil (P. Alegre); *4* New Zealand; *5* France (Mediterranean Coast); *6* Greece. *Group 2:* Subtropical regions (dry and hot) *7* Morocco; *8* South Africa; *9* Australia (Adelaıde); *10* Tunisia; *11* Algeria; *12* Israel. *Group 3:* Continental regions (wet and cold) *13* Germany: *14* France (Continent); *15* Switzerland; *16* Bulgaria; *17* Hungary; *18* Yugoslavia. The discriminant analysis was performed on six variables and the individuals are fully assigned

These results have encouraged us to investigate the possibility of using the stable isotope fingerprint as a new instrument for dating wines and spirits. In this respect, highly significative correlations between the isotope ratios and appropriate combinations of the climatological parameters associated with every year of production could be derived for a series of cognacs. In the case of wines, purely statistical considerations involving a two-factor analysis of the variance for three groups of wines of different varieties (Ugni Blanc, Carignan, Cabernet-Sauvignon) produced in the course of 10 different years (1976 to 1981) demonstrate that a significant effect of the year of production can be defined, whereas the variety dependence remains negligible (Fig. 3).

Multi-variate analyses of appropriate sets of isotopic and meteorological data are presently under study. These treatments are the source of helpful criteria for predicting the year of production of a fermented beverage. However, on the basis of the phenomenological considerations previously discussed, it is expected that the stable isotope approach may not always afford unequivocal conclusions, since two different years may suffer from comparable climatological conditions. Nevertheless, the search for simulations of the isotopic behaviors on the basis of detailed meteorological parameters should provide invaluable information on the environmental dependency of the biosynthesis of sugars. Moreover, the stable isotope approach appears as a useful complement to the method of dating by radio-isotope tracers which becomes les and less efficient – in the absence of nuclear accidents.

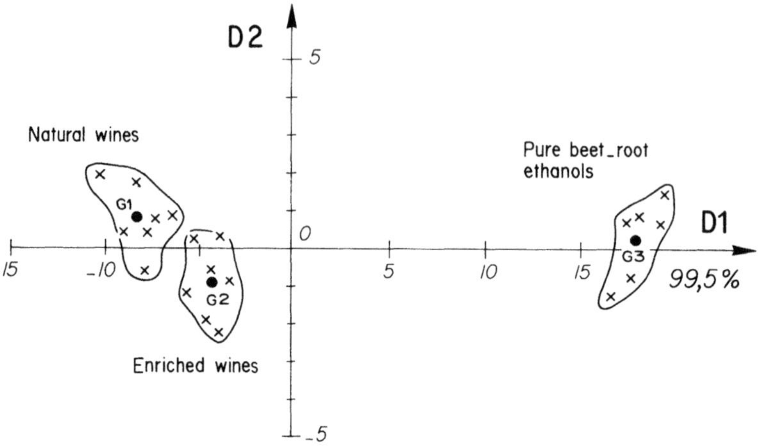

Fig. 3. Discriminant factor analysis of natural and enriched wines from different districts, different years of production and different varieties: *Group 1 and Group 2:* Mediterranean Coast; 1982 Pyrenées Orientales, Aude, Gard, Bouches du Rhóne, Hérault; 1983 Bouches du Rhône, Hérault. *Group 3:* 96% ethanol samples from 11 years (1972 to 1982) and from five districts of production (Seine Saint-Denis, Loiret, Côte d'Or, Oise, and Nord). Each entry in the group is the mean of the values for several samples

6 The Control of Chaptalization and Watering of Wines

In increasing the ethanol consumption, the excessive and unlawful enrichment of wines with sucrose or glucose has pernicious effects on health. Moreover, it unduly affects the finances of the agricultural policy in the EEC, since the ethanol resulting from added commercial sucroses (beet, cane) or glucoses (maize, wheat) is significantly cheaper than wine ethanol. The direct and indirect additional charge induced by this practice is certainly not far from one billion Ecus. In other circumstances, profits are sought by increasing the available volume of wines or in extreme cases by making synthetic wines from water, sucroses, molasses, industrial ethanols, etc.

Although the addition of sugar from a C_4 plant origin can be revealed by mass spectrometry measurment of the overall $^{13}C/^{12}C$ ratio, no efficient method for detecting the enrichment and watering of wines was available before the development of the procedure based on SNIF-NMR determinations.

Since the isotope fingerprint of wine retains the mark of the botanic nature of the precursor sugar and of the must water, the detection of exogeneous sugars and water is based on weighting equations of type (16) and (17), in which variations in the isotope ratios of glucose and water are expressed in terms of contributions of several origins:

$$\Delta(D/H)^S_G = c\ \Delta(D/H)^S_{GB} + (1\text{-}c)\ \Delta(D/H)^S_{GG} \qquad (16)$$

$$\Delta(D/H)^S_W = m\ \Delta(D/H)^S_{WT} + (1\text{-}m)\ \Delta(D/H)^S_{WG}, \qquad (17)$$

c and m denote the mole fractions of exogeneous sugar and water respectively and the subscripts GB (GG) and WT (WG) refer to glucose from beetroot (or grape) and to tap water (or grape water).

By comparing an adulterated wine to a natural wine of the same region and year, the change in the isotope parameters of ethanol and water can be predicted on the basis of Eqs. (13) to (15) and Eqs. (16) to (17). By adopting mean values $\Delta(D/H)^S_{GB} = -8$ ppm and $\Delta(D/H)^S_{WT} = -5$ ppm for the depletions of beet sugar with respect to grape sugar and of tap water with respect to must water, the following relationships between the variations in the isotope parameters of wine and the concentrations of added sugar and water are calculated:

$$\Delta(D/H)_I \simeq -8.8\,c - 1.1\,m \tag{18}$$

$$\Delta(D/H)_{II} \simeq -0.2\,c - 3.7\,m \tag{19}$$

$$\Delta(D/H)^Q_W \simeq -0.45\,c - 4.85\,m \tag{20}$$

In the absence of watering down, Eq. (18) predicts a decrease of 0.88 ppm for an enrichment of 10% (for a wine which contains 10% v/v ethanol an enrichment of 10% corresponds to 1% v/v ethanol from beetroot). A systematic investigation performed on 80 different pairs of natural and 10% enriched wines from France leads to an average value of −0.85 ppm per percent v/v with a standard deviation of 0.2 (G. J. Martin et al. 1986c). The experimental results are therefore fully consistent with the theoretical predictions (Fig. 3).

In the same way, a watering down of 10% is expected to introduce a decrease of −0.48 ppm, whereas an experimental value of −0.5 ppm per percent v/v is determined.

Different kinds of wine-making procedures have been thoroughly investigated in connection with the problems of enrichment and watering, and for a wine of known origin very small deviations in the site-specific isotope contents with respect to the parameters of the corresponding natural test wine can now be detected and explained.

7 SNIF-NMR Applied to the Characterization of Natural Products and to the Study of Metabolism

As illustrated in the case of wine, site-specific natural isotope fractionation factors, as determined by NMR, offer the basis of a very powerful methodology suited for characterizing the origin of natural products, for describing the fate of hydrogen atoms in the course of bioconversions such as those which lead from vine products to wine, for measuring isotope effects which may be helpful in the understanding of chemical or biochemical mechanisms, for appreciating the influence of physical parameters such as temperature and humidity on the biosynthesis of raw materials, in relation with plant physiology, and for detecting and eventually quantifying different types of adulteration or substitution of natural products.

A wide field of applications is therefore opened to the method which has already proved its efficiency in origin recognition of anetholes of vanilla, of alcohols derived from fruits, cereals, or tubers, of amino acids, camphors, pinenes, and of a number of aromas (Martin GJ et al. 1982h, 1983a, 1986; Toulemonde et al. 1983; Grant et al. 1982). Interestingly, the method is capable of detecting a natural chirality of compounds, such as ethanol, usually considered as achiral (Martin et al. 1983). Indeed, since the methylenic hydrogens are introduced in the course of different stereospecific steps of sugar fermentation, an enantiomeric unbalance may occur. The determination of optical purity requires a diastereotopy of the methylenic sites which can be obtained by reacting the ethanol sample with an optically active substrate. In this way it is shown, for example, that an enantiomeric enrichment in the R configuration of monodeuterated ethanol occurs when water with an isotope ratio higher than that of the must is added in the presence of yeast at the end of the fermentation process. This phenomenon is the result of a stereospecific exchange reaction, between R methylenic hydrogens and water hydrogens, promoted by a flavine enzyme. More generally, since chemical reaction usually lead to racemic products, the SNIF-NMR detection of a methylenic chirality becomes a good indication for a natural origin of the product or at least for its production by enzymatic reactions.

From a related point of view, simple measurements of site-specific isotope ratios may offer in favorable cases, such as that of α-pinenes, an easy way for determining optical purity without resorting to the tedious use of optically pure reagents (G.J. Martin et al. 1986a). An interesting use of the method also lies in the possibility of studying reaction mechanisms and determining kinetic and thermodynamic isotope effects of chemical and biochemical reactions without the need for isotope labeling (G.J. Martin et al. 1982b, 1986c; Pascal et al. 1984, Zhang et al. 1986).

As regards plant physiology, the pathways followed by the hydrogen atoms in their fate from water to the various molecular sites of plant components can be investigated and the role of the physical parameters which govern the climate can be specified in the natural conditions of plant growth.

References

Bender NM (1971) Variations in the $^{13}C/^{12}C$ ratios of plants in relation to the pathway of photosynthetic CO_2 fixation. Phytochemistry 10:1239–1244

Bricout J (1973) Fruit and fruit products – control of authenticity of fruit juices by isotopic analysis. J Assoc Anal Chem 56:739–742

Bricout J (1978) Recherches sur le fractionnement des isotopes stables de l'hydrogène et de l'oxygène dans quelques végétaux. Rev Cytol Biol Vég Bot 1:133–209

Bricout J (1982) Possibilities of stable isotope analysis in the control of food products. In: Schmidt HL, Förstel H, Keinzinger K (eds) Stable isotopes. Elsevier, Amsterdam

Bricout J, Fontes JC (1974) Distinction analytique entre sucre de canne et sucre de betterave. Ann Falsif Expert Chim 211–215

Bricout J, Fontes JC, Merlivat L (1975) Sur la composition en isotopes stables de l'éthanol. Ind Aliment Agric 92:375–378

Bricout J, Koziet J, Derbessy M, Beccat B (1981) Nouvelles possibilités de l'analyse des isotopes stables du carbone dans le contrôle de la qualité des vanilla. Ann Falsit Expert Chim 74:691–696
Craig H (1961) Standard for reporting concentrations of deuterium and oxygen 18 in natural waters. Science 133:1833–1834
Dansgaard W (1964) Stable isotopes in precipitations. Tellus 16:435–468
Dunbar J (1982a) A study of the factors affecting the $^{18}O/^{16}O$ ratio of the water of wine. Z Lebensm Unters Forsch 174:355–359
Dunbar J (1982b) Detection of added water and sugar in New Zealand commercial wines. In: Schmidt HL, Förstel H, Heinzinger K (eds) Stable isotopes. Elsevier, Amsterdam, p 495
Dunbar J (1982c) Oxygen isotope studies on some New Zealand grape juices. Z Lebensm Unters Forsch 175:253–257
Dunbar J (1982d) Use of $^{13}C/^{12}C$ ratios for studying the origin of CO_2 in sparkling wines. Fresenius Z Anal Chem 311:578–580
Dunbar J (1982e) The non-exchangeability of oxygen. Z Phys Chem 130:247–250
Dunbar J, Schmidt HL (1984) Measurement of the $^2H/^1H$ ratios of the carbon-bound hydrogen atoms in sugars. Fresenius Z Anal Chem 317:853–857
Dunbar J, Schmidt HL, Woller R (1983) Möglichkeiten des Nachweises der Zuckerung von Wein über die Bestimmung von Wasserstoff-Isotopenverhältnissen. Vitis 22:375–386
Estep MF, Hoering TC (1980) Biochemistry of the stable hydrogen isotopes. Geochim Cosmochim Acta 44:1197–1206
Fehri A, Letolle R (1977) Transpiration and evaporation as the principal factors in oxygen isotope variations of organic matter in land plants. Physiol Vég 15:363–370
Fehri A, Letolle R (1979) Relation entre le milieu climatique et les teneurs en oxygène 18 de la cellulose des plantes terrestres. Physiol. Vég 17:107–117
Förstel H (1985) Die natürliche Fraktionierung der stabilen Sauerstoff-Isotope als Indikator für Reinheit und Herkunft von Wein. Naturwissenschaften 72:449–455
Förstel H (1986) Neue Möglichkeiten der Weinanalyse. Die natürliche Variation der stabilen Isotope 122:202–208
Förstel H, Hützen H (1983) Oxygen isotope ratios in German groundwater. Nature 304:614–616
Förstel H, Hützen H (1984) Stabile Sauerstoff-Isotope als natürliche Markierung von Weinen. Weinwirtschaft-Technik 120:71–76
Garson MJ, Staunton J (1979) Some new NMR methods for tracing the fate of hydrogen in biosynthesis. Chem Soc Rev 8:539–560
Gonfiantini R (1978) Standards for stable isotope measurements in natural compounds. Nature 271:534–536
Gonfiantini R (1981) The S notation and the mass-spectrometric measurement techniques. In: Techn Rep Ser N° 210 Stable Isotope Hydrology Int Atomic Energy Agency Vienna, pp 35–84
Grant DM, Curtis J, Croasmun WR, Dalling DK, Wehrli FW and Wehrli S (1982) NMR determination of site-specific deuterium isotope effects. J Am Chem Soc 104:4492–4494
Hagemann R, Nief G, Roth E (1970) Absolute isotopic scale for deuterium analysis of natural waters – absolute D/H ratio for SMOW. Tellus 22:712–715
Hutchinson CR (1982) The use of isotopic hydrogen and NMR spectroscopic techniques for the analysis of biosynthetic pathways. J Nat Prod 45:27–37
Krueger HW, Reesman RH (1982) Carbon isotope analyses in food technology. Mass Spectrom Rev 1:205–236
Lerman JC, Deleens E, Nato A, Moyse A (1974) Variation in the carbon isotope composition of a plant with crassulacean acid metabolism. Plant Physiol 53:581–584
Martin GG, Pelissolo FJC, Martin GJ (1986) Isolog: a diagnosis system for origin recognition of natural products through isotope analysis. Comput Enhanced Spectrosc 3:147–152
Martin GJ, Martin ML (1981) Deuterium labelling at the natural abundance level as studied by high field quantitative ^2H NMR. Tetrahedron Lett 22:3525–3528

Martin GJ, Martin ML (1983) Détermination par résonance magnétique nucléaire du deutérium du fractionnement isotopique spécifique naturel. Application à la détection de la chaptalisation des vins. J Chim Phys 80:294–297

Martin GJ, Martin ML, Mabon F (1982a) A new method for the identification of the origin of natural products. Quantitative ^{2}H NMR at the natural abundance level applied to the characterization of anetholes. J Am Chem Soc 104:2656–2659

Martin GJ, Martin ML, Mabon F, Michon MJ (1982b) Natural selective ^{2}H labelling applied to the study of chemical mechanisms: labelling without enrichment. J Chem Soc Chem Commun 616–617

Martin GJ, Martin ML, Mabon F, Michon MJ (1983a) A new method for the identification of the origin of the ethanols in grain and fruit spirits: high field quantitative deuterium NMR at the natural abundance level. J Agric Food. Chem 31:311–315

Martin GJ, Zhang BL, Martin ML, Dupuy P (1983b) Application of quantitative deuterium NMR to the study of isotope fractionation in the conversion of saccharides to ethanols. Biophys Res Commun 111:890–896

Martin GJ, Sun XY, Guillou C, Martin ML (1985) NMR determination of absolute site-specific natural isotope ratios of hydrogen in organic molecules – analytical and mechanistic applications. Tetrahedron 41:3285–3296

Martin GJ, Janvier P, Akoka S, Mabon F, Jurczak J (1986a) A relation between the site-specific natural deuterium contents in α-pinemes and their optical activity. Tetrahedron Lett 27:2855–2858

Martin GJ, Guillou C, Naulet N, Brun S, Tep Y, Cabanis JC, Cabanis MT, Sudraud P (1986b) Contrôle de l'origine et de l'enrichissement des vins par analyse isotopique spécifique – étude des différentes techniques d'enrichissement des vins. Sci Aliment 6:385–405

Martin GJ, Zhang BL, Naulet N, Martin ML (1986c) Deuterium transfer in the bioconversion of glucose to ethanol studied by specific isotope labelling at the natural abundance level. J Am Chem Soc 108:5116–5122

Martin ML, Delpuech JJ, Martin GJ (1980) Practical NMR spectroscopy. Heyden-Wiley, London, Ch 9

Martin Ml, Zhang BL, Martin GJ (1983) Natural chirality of methylene sites applied to the recognition of origin and to the study of biochemical mechanisms. FEBS Lett 158:131–133

Monson KD, Hayes JM (1980) Biosynthetic control of the natural abundance of carbon-13 at specific positions within fatty acids in *Escherichia coli*. J Biol Chem 225:11435–11441

Park R, Epstein S (1960) Carbon isotope fractionation during photosynthesis. Geochim Cosmochim Acta 21:110–126

Pascal RA, Baum MW, Wagner CK, Rodgers LR (1984) Measurement of deuterium kinetic isotope effects in organic reactions by natural abundance deuterium NMR spectroscopy. J Am Chem Soc 106:5377–5378

Rauschenbach P, Simon H, Stichler W, Moser H (1979) Vergleich der Deuterium und Kohlenstoff-13-Gehalte in Fermentations- und Syntheseethanol. Z Naturforsch 34c:1–4

Schmidt HL, Förstel H, Heinzinger Eds (1982) Stable isotopes. Proceedings of the fourth International Conference. Elsevier Amsterdam

Simon H, Medina R (1968) Messung der T-Fixierung in Äthanol nach Hefe-Gärung in H_2O/HOT oder mit verschiedenen T-markierten Zuckern. Z Naturforsch 23b:326–329

Simon H, Medina R, Müllhofer G (1968) Messung der T-Fixierung bei einigen Teilschritten der Glykolyse. Z Naturforsch 23b:59–64

Simpson TJ (1975) Carbon-13 nuclear magnetic resonance in biosynthetic studies. Chem Soc Rev 497–522

Smith BN (1975) Carbon and hydrogen isotopes of sucrose from various sources. Naturwissenschaften 62:390–391

Sternberg LR, Deniro MJ, Ajie H. (1984) Stable hydrogen isotope ratios of saponifiable lipids and cellulose nitrate from CAM, C_3 and C_4 plants. Phytochemistry 23:2475–2477

Sternberg LR, Deniro MJ, Ting IP (1984) Carbon, hydrogen and oxygen isotope ratios of cellulose from plants having intermediary photosynthetic modes. Plant Physiol 74:104–107

Stothers JB (1974) ^{13}C NMR-studies of reaction mechanisms and reactive intermediates. In: Levy GC (ed) Topics in carbon-13 NMR spectroscopy. Wiley-Interscience, New York 1:231–244

Sudraud P, Koziet J (1978) Recherche de nouveaux critères analytique de caractérisation des vins. Ann Nutr Aliment 32:1063–1072

Toulemonde B, Horman I, Egli H, Derbesy M (1983) Food related applications of high resolution NMR differentiation between natural and synthetic vanillin samples using ^{2}H NMR. Helv Chim Acta 66:2342–2345

Wickman FE (1952) Variations in the relative abundance of the carbon isotopes in plants. Geochim Cosmochim Acta 2:243–254

Winkler FJ, Schmidt HL (1980) Einsatzmöglichkeiten der ^{13}C-Isotopen-Massenspektrometrie in der Lebensmitteluntersuchung. Z Lebensm Unters Forsch 171:85–94

Zhang BL, Wu WX, Gao ZH, Sun XY (1986) Acta chim Sin 44:437–441

Ziegler H, Osmond CB, Stichler W, Trimborn P (1976) Hydrogen Isotope discrimination in higher plants: correlations with photosynthetic pathway and environment. Planta 128:85–92

Yeast and Bacterial Control in Winemaking

Th. Henick-Kling

1 Introduction

Wine is the product of microbial fermentation of grape juice. If left unattended, juice, released from the grape berry, is spontaneously fermented by yeast and bacteria naturally present mainly in the winery environment but also on grape skin, stems, leaves, in soil and air, on human hands, or carried by insects (Ribereau-Gayon et al. 1975; Dittrich 1977).

Wine fermentation differs very much from other industrial fermentations in that it does not start with a sterile fermentation substrate. Depending on the quality of the grapes at harvest and the quality of sanitation performed on the equipment used in the juice preparation, the juice contains a varying number of microorganisms. Only some of the microorganisms carried into the juice are able to grow in it and very few are able to compete with highly specialized fermentative yeasts which quickly dominate the alcoholic fermentation.

The task of the winemaker is to ensure that the desired fermentative yeasts predominate in the juice and carry out the fermentation. Uncontrolled growth of other yeasts and bacteria negatively affects the final wine quality. To ensure the development of the desired yeast, the winemaker can modify the fermentation conditions. Today, most major producers of wine inoculate the juice with selected strains of yeast to control yeast growth and fermentation rate.

In spontaneous fermentations, two principal fermentations occur successively or sometimes simultaneously. During the primary or alcoholic fermentation, yeasts convert grape sugar into ethanol and CO_2; then lactic acid bacteria grow and convert malic acid into lactic acid and CO_2. This secondary or malolactic fermentation is common in red wines and in some white wine styles.

Under aerobic conditions, acetic acid bacteria and some lactic acid bacteria can produce large amounts of acetic acid, causing vinegary spoilage of wine. Only one other group of bacteria has been found to grow in juice and wine, *Streptomyces* sp. (Fleet et al. 1984). Under conditions favorable to their growth, they can produce aroma-active metabolites which might contribute earthy flavors to the wine.

In wine fermentation, the perhaps earliest use of biotechnology, the final quality of the wine can be modified in two major ways, (1) by the quality of the grapes and juice preparation, and (2) by the selection of most suitable yeasts and bacteria and control of fermentation conditions.

During the course of vinification, the changing composition of the medium selects for various types of yeast and bacteria (Ribereau-Gayon et al. 1975 and 1977; Dittrich 1977; Lafon-Lafourcade et al. 1983a; Fleet et al. 1984; Wibowo et al. 1985; Heard and Fleet 1986). Growth and metabolism of yeast and bacteria during vinification can be controlled by modification of the fermentation substrate and fermentation conditions and by the use of pure cultures of yeast and bacteria (Lafon-Lafourcade 1986; Wibowo et al. 1985).

Primary and secondary fermentation in wine may also be modified by the growth and metabolic products of other microorganisms. Some metabolites of molds growing on grapes can inhibit yeast growth and fermentation (Sponholz 1988). Various types of yeasts and bacteria compete against each other by competing for energy sources and nutrients and by producing inhibitory substances active against other yeast and bacteria. Finally, bacterial viruses interfere in the fermentation process by infecting and destroying bacterial cells.

2 Yeasts

Yeasts are a group of fungi in which the unicellular form is predominating and in which the most common form of vegetative propagation is budding (Stewart and Russel 1985). Compared to other groups of microorganisms, the yeasts comprise relatively few genera and species which are apparently closely related. Due to strong special selection, strain differences among yeasts are often more important than species differences. Yeasts are commonly used in food preservation because of their ability to rapidly convert sugar into ethanol.

2.1 Types, Selection, and Succession

Yeasts are fermentative and can utilize a wide variety of sugars. Various types of yeast are found in sugar solutions excreted by trees, in blossoms, and on fruits (Dittrich 1977).

On grapes, yeasts are especially concentrated in areas of the berry where the skin has cracked and the juice is exposed, near stomata, and on stems (Ribereau-Gayon et al. 1975; Dittrich 1977). The yeasts on the surface of the fruit are carried through grape harvest, crushing, and pressing into the juice. During harvest, winery equipment frequently in contact with grapes and juice becomes the major source of yeast and bacteria entering the juice.

The largest number of different types of yeast is found at the start of alcoholic fermentation. The various yeasts found in grape juice are for practical reasons grouped into useful and non-useful yeasts. The useful yeasts are the strongly fermentative yeasts which are commonly called "wine yeasts"; the non-useful yeasts are all other yeasts and are commonly called "wild yeasts". Grape juice at the beginning of alcoholic fermentation contains strains of the genera *Saccharomyces, Kloeckera, Metschnikowia, Pichia, Schizosaccharomyces, Hanseniaspora, Candida,* and others (Kunkee and Amerine 1970; Kunkee and Goswell 1977; Dittrich 1977; Heard and Fleet 1986).

Ethanol, SO_2, and O_2 are the major factors which limit the growth of the various yeast types initially present in the juice (Dittrich 1977). Weakly fermentative yeast are inhibited by the increasing ethanol concentration during wine fermentation. Fermenting juice with 2–5% (v/v) ethanol contains almost solely *Saccharomyces* sp. During the early stage of fermentation, the oxygen available in the juice is consumed and the carbon dioxide produced blankets the fermenting juice and prohibits the uptake of oxygen from the air, thus inhibiting growth of aerobic yeasts and aerobic spoilage bacteria. The addition of sulfur dioxide to the juice before fermentation also strongly changes the composition of the yeast flora, again favoring the development of *Saccharomyces* sp. All strains of winemaking yeasts belong to one species, *Saccharomyces cerevisiae* (Barnett et al. 1983; Kreger-van Rij 1984). In some wines, yeasts other than *S. cerevisiae* may be present for longer periods and may contribute significantly to the flavor of the final wine (Heard and Fleet 1986).

2.2 Physiology of *Saccharomyces cerevisiae*

Growth of *S. cerevisiae* depends on the availability of carbohydrates, oxygen, nitrogenous compounds, and vitamins. The content of these substrates in grape juice is highly variable, depending on grape variety, growth conditions, and fermentation control.

2.2.1 Morphology and Growth

The cells of *S. cerevisiae* typically are round to oval and can form short chains of two to eight cells (Dittrich 1977) (Fig. 1). The cell size and shape vary strongly among the different strains (2.5–10.5) × (5.0–21.0) μm. The cells replicate vegetatively by budding and after mating of two haploid cells form one to four asco spores (Stewart and Russell 1985).

Depending on the availability of oxygen, energy for cell maintenace and growth is provided from aerobic or anaerobic metabolism of sugars. For the fermentation of grape juice to wine, air is excluded from the juice, the yeast grows anaerobically, converting the grape sugars into ethanol and CO_2.

The growth cycle of yeast in grape juice during fermentation follows the classical phases of microbial growth, (a) lag phase, (b) exponential phase, (c) linear growth, (d) stationary phase, (e) decline or death phase (Ribereau-Gayon et al. 1975; Lafon-Lafourcade 1986; Monk 1986).

The lag phase is short when the juice receives a large inoculum of yeast from grape skins and machinery surfaces or from a prepared starter culture. Low temperature, high concentration of sugar, addition of large amounts of SO_2, and lack of oxygen increase the lag phase and decrease the extent and rate of growth during the exponential phase of growth.

At the beginning of the exponential phase, the cells begin to bud. Sugar degradation at this stage is directed towards the production of cell biomass, only a small amount of ethanol is formed. During exponential growth, the yeast population doubles at regular intervals. After a period of exponential growth, yeast growth changes to a linear rate. The greater portion of sugar is converted into ethanol during this period. Towards the end of the growth phase, the number of budding yeast cells decreases rapidly and growth of the yeast population stops.

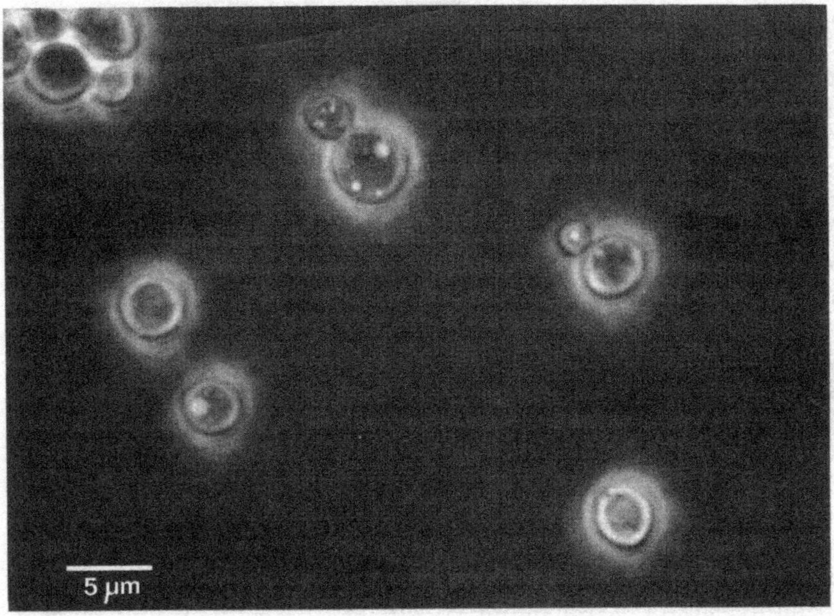

Fig. 1. *Saccharomyces cerevisiae,* a pure culture wine-making yeast, with budding cells, grown in a synthetic medium, seen under phase contrast.· 765.5; 20 µm ≅ 15.5 mm

At this point the juice still contains approximately 50% of its initial sugar concentration.

The number of viable yeasts starts to decrease during the stationary phase and decreases rapidly during the decline phase, during which the yeast cell density also decreases (Lafon-Lafourcade 1986; Sablayrolles and Barre 1986). With the declining number of viable yeasts, the metabolic activity of the yeast culture decreases. In addition, the metabolic activity of the yeast cell decreases with age and due to inhibition by the increasing concentration of ethanol (Dittrich 1977, Strehaiano and Goma 1983). At stationary phase, half of the initial amount of sugar may still be present (Lafon-Lafourcade 1986; Monk 1986). For the fermentation to go to completion at this stage, it is important that the wine contains adequate amounts of amino acids and ammonia ion and that the temperature is maintained near the physiological optimum for the yeast. Strain selection and culture preparation also influence the survival and metabolic activity of the yeast at this stage.

Yeast and bacterial growth can be measured optically, by microscopic counting, by plating and by special staining methods as number of viable yeast cells (Dittrich 1977; Gerhardt et al. 1981). It is important to realize the different principles of these methods; optical methods measure the amount of light deflected or adsorped by the total cell mass, and in the microscopic method all yeast cells, dead or alive, are counted. Plating and staining techniques determine the number of indivdual live yeast cells. During exponential growth, when essentially all yeast cells in the population are viable, both methods give the same answer. Yet when an optical method is used for determining the status of yeast growth during the stationary and decline phases, it gives the total number of viable and dead yeast present. Thus it cannot be used for an assessment of the fermentative activity of the cell population at this stage. To estimate the fermentative capacity of the yeast population, especially during stationary and decline phase, the number of viable yeast cells must be determined by plate count or staining method.

There are several reasons why yeast growth stops before all sugar is utilized. The main reason is likely to be lack of energy (Dittrich 1977). The anaerobic metabolism of one mole of glucose provides only 2 mol of ATP, compared to 38 mol under aerobic conditions. Lack of energy is likely to stop synthesis of new cell components, thus reducing the rate of cell growth and eventually cell viability. Sterols and fatty acids which are formed from acetyl-CoA, a product of the metabolism of sugars, are not produced under anaerobic conditions (see Sect. 2.2.4). Other causes for inhibition of growth are ethanol, temperature, high concentrations of ethanol, and inhibitory compounds originating from the grape, spray residue, and microbial metabolism (Lafon-Lafourcade 1986). Growth rate and metabolic activity vary with yeast strain and starter culture preparation (Müller-Späth and Loescher 1985; Lafon-Lafourcade (1986). Cessation of yeast growth can be due to exhaustion of intracellular reserve materials or an essential nutrient, inhibition by high concentration of ethanol, CO_2, or SO_2, inhibition by high or low temperature, lack of oxygen during growth phase or starter culture preparation, high concentration of sugar, clarification of must before fermentation, residual pesticides, microbial antagonism (Kirsop and Brown 1972; Ribereau-Gayon et al. 1975; Sapis-Domercq et al. 1976; Dittrich 1977; Traverso-Rueda 1980; Lafon-Lafourcade et al. 1980; Monk 1982 and 1986; Monk and Castello 1984; Gnaegi et al. 1983; Strehaiano and Goma 1983; Lafon-Lafourcade 1983 and 1986; Houtman and du Plessis 1986; Sablayrolles and Barre 1986).

2.2.2 Carbohydrate Metabolism

Saccharomyces cerevisiae ferments a wide range of sugars. Before the yeast can utilize any sugar, the sugar must enter the cell. Polysaccharides might be hydrolyzed outside the cell before all or some of their hydrolysis products enter the cell.

Important for yeast growth and alcoholic fermentation are the main grape sugars, glucose and fructose, and when added, sucrose (Dittrich 1977). Other sugars are usually present in concentrations of less than 1 g l^{-1} and not metabolized by *S. cerevisiae* during fermentation. Glucose and fructose are usually present in a 1 : 1 ratio and at a concentration of 140–240 g l^{-1}. In juice from raisined and *Botrytis*-affected grapes, the content of glucose plus fructose may reach 500 g l^{-1} and the ratio of glucose to fructose decreases as glucose is preferentially metabolized by the mold (Amerine and Ough 1980; Sponholz 1988). Other hexoses are present in juice and wine in very small concentration. Mannose and galactose were found at a concentration of less than 50 and 200 mg l^{-1}, respectively (Sponholz 1988). The other sugar important in winemaking is sucrose. The sucrose concentration in the grape berry can reach 10 g l^{-1}, but soon after extraction of the juice, the sucrose is converted by grape and yeast enzymes into glucose and fructose (Sponholz 1988). In some wine-growing regions sucrose is added to juices which do not contain a sufficient concentration of glucose and fructose. *Saccharomyces cerevisiae* produces the extracellular enzyme invertase (beta-D-fructofuranoside fructosehydrolase) which hydrolyzes sucrose into its constituents, glucose and fructose (Dittrich 1977; Stewart and Russell 1985).

Saccharomycea cerevisiae has a constitutive transport system which is common for glucose, fructose, and mannose; it is equilibrative and does not require energy (Stewart and Russell 1985). The rate of transport varies with the type of sugar and with aerobic and anaerobic conditions (Dittrich 1977). In most strains of *S. cerevisiae,* the rate of uptake for glucose is approximately five times higher than that of fructose and mannose. As a result, the rate of glucose utilization is higher than that of fructose and in incomplete fermentations a larger proportion of fructose will remain in the wine than if the fermentation goes to completion. After completed fermentation, the residual concentration of fructose and glucose is 1–3 $g\,l^{-1}$, total (Dittrich 1977; Wibowo et al. 1985). Upon entry into the cell, glucose and fructose are phosphorylated, then degraded via the Embden-Meyerhof-Parnas pathway (Dittrich 1977; Ayres et al. 1980). Under anaerobic conditions, the pyruvic acid formed in this pathway is decarboxylated to acetaldehyde, which in turn is reduced to ethanol. The enzyme switching the yeast metabolism to the production of ethanol is pyruvate decarboxylase. Under aerobic conditions, the activity of the pyruvate decarboxylase increases when the medium contains an excess amount of sugar, its activity may be regulated by the intracellular concentration of pyruvate (Van Dijken et al. 1984). At low activity of the pyruvate decarboxylate, a small amount of lactic acid may be formed. Under aerobic conditions the pyrvate is completely degraded in the citric acid cycle and the respiratory chain to CO_2 and H_2O. Alcoholic fermentation of one mole glucose or fructose yields 2 mol ATP. Under aerobic conditions, when O_2 is available and glucose is oxydized to CO_2 and H_2O, 1 mol of glucose or fructose yields 38 mol of ATP.

The low yield of ATP under anaerobic conditions may stop yeast growth during fermentation because not sufficient energy is available for biosynthesis (Dittrich 1977).

For the preparation of yeast starter cultures and the control of alcoholic fermentation, it is most important to consider the role of oxygen in controlling yeast metabolism and growth. When oxygen is available the yeast metabolism will switch to respiration yielding 38 mol of ATP for each mole of glucose, thus allowing for faster growth rate, production of a larger biomass, and the synthesis of reserve materials, sterols, and fatty acids. Therefore aeration is used in the preparation of yeast starter cultures where a large biomass is needed. Conversely, in juice inoculated for alcoholic fermentation with a large population of yeast, oxygen is excluded to optimize the production of ethanol. If a yeast starter culture is not used for induction of fermentation, the juice might be aerated during the early stages of yeast growth to encourage rapid growth of indigenous yeast and the formation of a sufficiently large yeast population to complete alcoholic fermentation. Sablayrolles and Barre (1986) defined the oxygen need of a wine yeast in a synthetic grape juice medium, free of fatty acids. The yeast required a minimum of 10 $mg\,l^{-1}$ oxygen during its growth to completely ferment 230 $g\,l^{-1}$ of glucose. The amount of oxygen required for growth in grape juice will depend on the availability of other oxidizing compounds, the concentration of fatty acids and sterols in the juice, and on the provision of fatty acids and sterols in the cells of the starter culture.

Care must be taken not to over-oxidize the juice; beyond what is needed for sufficient yeast growth overoxidation will reduce the final wine quality. Aeration of fermenting juice stimulates yeast growth by providing oxygen for aerobic utilization of sugar, by mixing the yeast and the substrate, and by removing inhibitory substances such as carbon dioxide, ethanol, acetaldehyde, and acetate (Dittrich 1977). The provision of oxygen to yeast during growth in grape juice does not result in a complete switch from anaerobic to aerobic metabolism, since high sugar concentration does inhibit respiration. Therefore it is advantageous to prepare a sufficiently large yeast biomass by aerobic preculture of the yeast in a medium containing less carbohydrates.

Limited aerobic growth of yeast is possible in the juice based on oxygen which entered the juice during pressing and by using electron acceptors other than oxygen (Dittrich 1977).

During fermentation, the yeast incorporates only 0.8 to 1.3% of the glucose and fructose carbon in its biomass (Dittrich 1977). When yeast growth stops, nearly all the sugar is converted to ethanol and carbon dioxide. The nongrowing yeast cells require only a small amount of energy and nutrients for maintaining their viability.

At sugar concentrations above 250 g l^{-1}, yeast fermentation is significantly delayed and fermentation incomplete. This can result in wines of lower alcohol content and in high amounts of residual sugar (Ribereau-Gayon et al. 1975; Dittrich 1977). Such wines, produced by the partial fermentation of naturally highly concentrated juices (produced by the infection of the grapes by the mold *Botrytis cinerea*) are the famous Sauternes and Beerenauslese wines made in France and Germany, respectively.

The high concentration of sugar, especially in juices prepared from botrytized grapes, is inhibitory to yeast growth because of its osmotic pressure and because of accumulation of larger concentrations of ethanol inside the cell (Stewart and Russel 1985). In such juices, the yeast cell loses water, it shrinks and loses viability (Dittrich 1977).

Fermentation of these botrytized juices is difficult. So-called osmotolerant or xerotolerant yeast strains are required to ferment these highly concentrated juices, and because growth is strongly inhibited juice must be inoculated with a large number of viable yeasts (Peynaud and Domercq 1955; Dittrich 1977; Stewart and Russel 1985). In some cases the juice may also require the addition of vitamins and amino acids because it has been depleted of these nutrients of *Botrytis cinerea* to such an extent that yeast growth is inhibited. Occasionally, strongly fermentative, osmotolerant yeast may produce up to 16% (v/v ethanol in these late harvest juices. But this does not give the desired balance of alcohol, acidity, and sweetness in such wines. Excessive fermentative activity under this condition can be best controlled by lowering the fermentation temperature.

Osmotolerant yeasts are also more tolerant against inhibition by ethanol, sulfur dioxide, acetate, and other preservatives (Dittrich 1977). The higher tolerance to large concentrations of sugar apparently is due to differences in the composition of the cytoplasmic membrane (Stewart and Russel 1985). Under water stress yeasts accumulate polyols. the polyols act as osmoregulators protecting the intracellular enzymes. Both, osmotolerant and non-osmotolerant strains *Saccharomyces cerevisiae* produce glycerol. Osmotolerant and non-osmotolerant strains vary in their ability to accumulate glycerol in their cells. While osmotolerant strains are able to accumulate large amounts of glycerol, non-osmotolerant strains leak increasing amounts of glycerol into the growth medium, thus their metabolic response becomes futile. The amount of glycerol formed during fermentation varies strongly with the yeast strains carrying out the fermentation and the juice composition (Mayer and Pause 1970; Sponholz et al. 1986). Dittrich (1964) found between 4.3 and 9.5 g/l glycerol were produced during fermentation of various wines. The juice itself contains between 0 and 9.2 g/l glycerol depending on the degree of *Botrytis* infection of the grapes.

Beta-glucosides like cellobiose and gentobiose present in grape juice are not metabolized by wine yeast (Dittrich 1977). But heteroglucosides common in grape juice are enzymatically cleaved and the glucose part is utilized. This is one reason for loss of color during fermentation. The red pigment, anthocyan, is a beta-glucoside (anthocyanidin plus glucose). Depending on the yeast strain, pH, and rate of fermentation, varying amounts of anthocyan are hydrolyzed during fermentation, resulting in a loss of color of up to 27% (Dittrich 1977).

2.2.3 Ammonia, Amino Acids, and Vitamins

Yeast generally can synthesize all amino acids and nitrogenous bases required for their cellular growth from ammonium ion. Yeast growth is accelerated when ready-made building blocks, amino acids, for the cells' enzymes and structural components are available in the growth medium.

Lack of ammonia ion, amino acids, and vitamins is not usually apparent in the fermentation of juices in the European wine regions. Yet it has been suspected that competition between yeasts and between yeast and lactic acid bacteria for limited nutrients may inhibit yeast growth and alcoholic fermentation (Lafon-Lafourcade 1986). Also growth of the mold *Botrytis cinerea* on the grapes can strongly reduce the amount of amino acids in the grape (Sponholz 1986), and treatment of juice with ion exchange and addition of large amounts of sulfur dioxide can remove vitamins (Amerine et al. 1980; Monk and Costello 1984).

The amount of nitrogen in grape juice varies, depending on the method of juice preparation, degree of fruit ripeness, grape variety, and viticultural practice (Sponholz 1986). It is typically between 200 and 1400 mg l^{-1} and the ammonium content is between 0 and 300 mg l^{-1}.

Most of the nitrogen assimilated by the yeast during growth in grape juice is taken up in the form of amino acids and ammonia, nitrate is usually not metabolized (Dittrich 1977). During growth in grape juice, *S. cerevisiae* utilizes 1–2 g l^{-1} of amino acids (Dittrich 1977). During the decline phase of yeast growth, the concentration of amino acids in wine increases again due to release of amino acids and proteins from decaying yeast cells (autolysis) and due to the action of proteolytic enzymes on grape protein.

In European juices the content of amino acids varies between 0.4 and 6.5 g l^{-1}; they typically contain 2–4 g l^{-1} (Sponholz 1986). In Australian juices the concentration of amino acids is lower, typically between 0.6 and 2.5 g l^{-1} (The Australian Wine Research Institute, unpublished). The amount of available nitrogen in the form of amino acids and free ammonia ion has been found limiting to yeast growth and rate of fermentation in Australian juices (Monk 1982; Monk and Costello 1984). Infection of the grape berries with *Botrytis cinerea* can reduce the amount of amino acids by 33 to 80%, it can reach as little as 0.4 g l^{-1} (Sponholz 1986). Such low concentrations of amino acids can be inhibitory to yeast growth and result in stuck fermentations. Particularly growth during the linear phase is influenced by the availability of amino acids and ammonia ion and fermentation temperature (Monk 1986).

In grape juice, 75–85% of the total amount of amino acids consists of glutamine, arginine, proline, glutamate, serine, threonine, and alanine (Sponholz 1988). Jones and Pierce (1964) studied the utilization of amino acids by *S. cerevisiae* from wort during beer fermentation. They classified the wort amino acids into four groups, depending on their rate of removal. The important amino acids arginine, glutamate, glutamine, serine, and threonine, constituting a majority of the amino acids in grape juice, were removed fastest, alanine was removed at a slower rate, and proline, although the most plentiful amino acid in wort, was scarcely utilized. The differences in the rate of utilization of various amino acids are based on different activity of transport systems for various groups of amino acids and for individual amino acids (Stewart and Russell 1985). *Saccharomyces cerevisiae* has two classes of mechanisms for the transport of amino acids across the cytoplasmic membrane, (1) a general amino acid permease (GAP) which can transport all basic and neutral amino acids, except proline, and (2) a range of at least 11 transport systems specific for one or a small number of amino acids. It appears that during the initial growth of *S. cerevisiae* in wort, amino acids are predominantly taken up by specific transport systems. As long as ammonia ions are available, the synthesis of the GAP is repressed and deamination of specifically transported amino acids may also suppress synthesis of GAP. In grape juice the rate of amino acid uptake may be influenced by the pH, yeast strain, and preculture conditions.

Certain juice treatments can reduce the amount of amino acids and vitamins. The addition of bentonite to the juice and removal of the precipitate depletes the juice of amino acids (Dittrich 1977). Also, juices which have been stabilized for long-term storage by addition of large amounts of sulfur dioxide and by lowering the pH by ion exchange have been found difficult to ferment because of amino acid and vitamin deficiencies (Monk and Costello 1984). High concentrations of sulfur dioxide inactivate thiamine and ion exchange treatment removes some thiamine, pantothenic acid, pyrodoxin, and nicotinic acid (Amerine et al. 1980); the concentration of vitamins can be expected to decrease further during long periods of storage.

Monk (1982) and Monk and Costello (1984) have shown that nutritional deficiencies of grape juice can be overcome by additions of ammonia, amino acids, and vitamins. In fresh Riesling grape juice, the addition of amino acids and ammonia increased the linear growth rate by 40 to 300%; the increased growth resulted in increased sugar utilization during this period. In vitamin-deficient juices, resulting from ion exchange treatment and long-term storage with high concentration of sulfur dioxide, the addition of thiamine and diammonium phosphate improved yeast growth.

Yeast utilize small amounts of proteins and they can release up to 100 mg l^{-1} during autolysis (Dittrich 1977). The proteolytic activity is not significant during fermentation as long as free ammonia ion and amino acids are available in the substrate. Because of the differences in the uptake of ammonia ion, amino acids, and proteins, *S. cerevisiae* first utilizes the ammonia present, then the various amino acids, and the proteins last. Ammonia ion pass freely through the cell membrane and amino acids are taken up by general and specialized amino acid transferases. Proteins generally do not pass freely through the cell membrane and are not transported. To utilize the protein-bound nitrogen, the yeast must produce extracellular proteolytic enzymes which cut the protein into its constituent amino acids, which are then taken up by the yeast cell. The synthesis of new enzymes requiring large amounts of energy is highly uneconomical for the yeast cell during anaerobic growth, where the energy yield per mole of stubstrate is very small.

Yeasts require sulfate or organic sulfur (cysteine, ethionine), potassium, magnesium, sodium, and calcium, and in trace amounts boron, copper, zinc, manganese, iron, iodine, and molybdenum. Sufficient amounts of these elements are provided in grape juice (Dittrich 1977).

2.2.4 Sterols and Fatty Acids

Sterols and fatty acids are components of the cytoplasmic membrane essential in the functioning of the cell membrane (Demel and De Kruyff 1976). The sterol and fatty acid composition of the cell membrane is affected by the growth conditions. Changes in the lipid composition of the membrane enable it to maintain its fluidity and function under adverse external conditions. When growing under aerobic conditions, *S. cerevisiae* can synthesize sterols and fatty acids from acetyl-CoA, it cannot synthesize them anaerobically (Kirsop 1974; Dittrich 1977). During anaerobic growth, the concentration of sterols in the yeast population decreases rapidly as the cells divide and no new sterols are synthesized; as the cellular sterol concentration is depleted, growth rate slows, and eventually stops, slowing down fermentation and possibly causing an incomplete fermentation (Andreasen and Stier 1954; Kirsop 1974; Rose 1977; Lafon-Lafourcade 1986). When sterols are added to the juice, viability of the yeast increases and the fermentation rate increases (Larue et al. 1980). The principal sterol in yeast cells is ergosterol. Thus, the addition of ergosterol and unsaturated fatty acids to grape juice stimulates the rate of growth and fermentation (Brechot et al. 1971; Lafon-Lafourcade 1986). To be able to grow anaerobically, the yeast added in a starter culture must contain sufficient amounts of ergosterol and unsaturated fatty acids. When a yeast starter culture is prepared correctly under aerobic conditions, the yeast will incorporate sufficient amounts of sterols and fatty acids to grow for about five generations under anaerobic conditions which is sufficient for complete fermentation of the sugar initially present (Monk 1986). Fatty acids which can serve as precursors for the synthesis of ergosterol are found in the waxy skin on the grape berry; Brechot et al. (1971) found that oleanolic acid isolated present in this wax coating is stimulatory to yeast growth both alone and in combination with oleic acid, which is also found on the grape.

2.2.5 Temperature and pH

Temperature strongly influences yeast growth. *Saccharomyces cerevisiae* can grow over the range of 0–45° C; the maximum rate of growth is between 20 and

30° C. In most winemaking countries the fermentation temperature is below this temperature, either because of the low cellar temperature due to a cool climate or because the juice temperature is controlled by refrigeration. Amerine et al. (1980) recommended for the production of white wines a fermentation temperature of 10–15° C and for red wines a temperature of 20–30° C. The low fermentation temperature of 10–15° C preserves the fresh, fruity aroma of the grapes, while warmer temperatures produce a wine less dominated by fruit flavors and the warmer temperature enhances the extraction of color from the skin of red grapes.

For control of fermentation rate and the final wine quality, the heat production by the yeast must be considered. During fermentation of 1 mol of glucose (180 g l^{-1}), approximately 23.5 kcal are generated (Troost 1972). The rate of heat production depends on the rate of fermentation and the amount of heat loss depends on size, shape, and material of which the fermentation vessel is built. For example, at a room temperature of 13° C, the juice temperature in a 600-, 1200-, 7200-l wooden barrel increased to 19–22°, 21–25°, 30°, and 33° C, respectively (Dittrich 1977). Boulton (1980) developed a kinetic model which allows the prediction of heat production during fermentation. If the temperature in the juice during fermentation exceeds 40° C, the yeast cells are killed and the fermentation sticks. In order to avoid overheating of the fermenting juice in large vessels and at high cellar temperature, the fermentation tanks must be refrigerated, and in fact in most larger wineries the fermentation temperature is controlled within narrow limits. Controlling the juice temperature between 20 and 22° C ensures rapid and complete fermentation, prolonged fermentation at low temperature increases the risk of incomplete fermentation. Rapid fermentation at 20–22° C also results in the production of only small amounts of sulfur dioxide binding compounds (Dittrich 1977). Thus the finished wine will require smaller additions of sulfur dioxide to protect it from oxydation. Cold fermentations at temperatures below 10° C are preferred by some wine-makers in order to minimize loss of fruit flavor due to higher temperature and rapid evolution of carbon dioxide. The low temperature also inhibits growth of acetic and lactic acid bacteria and of most spoilage yeast. The constant low temperature causes long, slow fermentations with little yeast growth. Since yeast growth is strongly inhibited at the preferred low temperatures, fermentation under these conditions requires that the juice is inoculated with a yeast starter culture. The yeast starter culture is added at 5–10% (v/v) to juice to be fermented at very low temperature.

The fermentation temperature also influences the production of aroma-active compounds (Dittrich 1977; Cottrell and McLellan 1986). Growth of *S. cerevisiae* is not inhibited at the pH of wine, which can range from 2.8 to 3.8, also the length of the lage phase after inoculation of a yeast culture into the juice is not affected by the pH in this range (Benda 1982).

2.2.6 Inhibition by Ethanol and Carbon Dioxide

Ethanol. Yeast growth is partially inhibited in grape juice at 2% (v/v) ethanol; some strains are completely inhibited at 6–8% (v/v) ethanol, other strains only at concentrations above 13% (v/v); maximal concentrations of 18% (v/v) ethanol may be achieved (Dittrich 1977). Inhibition of growth by ethanol is stronger under anaerobic conditions than under aerobic conditions. Ethanol affects the metabolism of the yeast cell at two levels, (1) at the cytoplasmic membrane and (2) at its cell-internal enzymes (Stewart and Russell 1985). Ethanol can accumulate in the cell membrane and inhibit transport processes and ethanol inhibits the glycolytic enzymes to varying degrees (Nagodawithana et al. 1977). In studies with two wine yeasts, Strehaiano and Goma (1983) found that sensitivity to inhibition by ethanol is related to the degree of ethanol accumulation inside the cell. The amount of ethanol retained in the cell depends on the osmotic pressure and the concentration of ethanol in the medium (Stewart and Russell 1985; Guijarro and Lagunas 1984).

Carbon Dioxide. Only a small part of the carbon dioxide produced during fermentation remains in the wine. The carbon dioxide retained in the wine reacts with water to form carbonic acid. Even though carbonic acid is a weak acid it can inhibit yeast growth (Dittrich 1977). Fermentation under carbon dioxide pressure reduces the fermentation rate and can result in incomplete fermentation. Also, carbon dioxide accumulated during prolonged, slow fermentation at low temperature can inhibit the fermentation. Yeast growth is completely inhibited at 15 g l^{-1} carbon dioxide which is equal to a pressure of 440 kPa at 15° C. Although yeast growth is inhibited at 440 kPa, the formation of ethanol is only inhibited at a pressure of 850 kPa, and the yeast cell finally dies at a pressure of 1830 kPa. The carbon dioxide pressure becomes growth-limiting during bottle fermentation in the production of champagne-style wines. The final pressure of carbon dioxide in these wines is 500–600 kPa pressure at 15° C (Wilkinson 1986).

2.2.7 Inhibition by Sulfur Dioxide, Acetic Acid, Metals, Phenolics, and Tannins

Sulfur Dioxide is added to juice and wine to control oxidation and the number of contaminating microorganisms (Troost 1972; Amerine et al. 1980). The sulfur dioxide added enters a complex equilibrium of bound sulfur dioxide, free (molecular) sulfur dioxide, and bisulfite and sulfite ions (Burroughs and Sparks 1973a, 1973b; Somers and Wescombe 1982; Hood 1983). The active form of sulfur dioxide is the free sulfur dioxide, but it also has a limited inhibitory effect bound (Lafon-Lafourcade and Peynaud 1974; Macris and Markakis 1974; Carr et al. 1976; Hood 1983). pH affects the dissociation of sulfur dioxide, lower pH values giving higher proportions of free sulfur dioxide. Inhibition by sulfur dioxide is stronger in the presence of ethanol.

Yeast strains vary strongly in their sensitivity to inhibition by sulfur dioxide (Dittrich 1977). Additions of 100 mg l^{-1} SO_2 can delay the start of spontaneous yeast fermentation, but the rate of fermentation is not affected by 100 and 300 mg l^{-1} SO_2. Pure culture wine yeasts are usually not inhibited by concentrations below 150 mg l^{-1} SO_2.

Acetic Acid. Large amounts of acetic acid may be formed on *Botrytis*-infected grapes. The hyphe of the mold penetrating the skin of the grape berry causes some juice to leak out onto the berry surface where it can be metabolized aerobically by acetic acid bacteria present on the berry (Dittrich 1977). Already 1 g l^{-1} of acetic acid noticeably inhibits yeast fermentation; it is strongly inhibited at concentrations above 2 g l^{-1}. Such large concentrations of acetic acid also are not acceptable for table wines; the maximum permitted concentration for table wines is between 1 and 2.5 g l^{-1} (Amerine and Ough 1980).

Metals. Due to its low pH, grape juice is corrosive to exposed metal surfaces. High concentrations of iron are known to inhibit yeast growth. White grape juice on average contains 0.58 mg l^{-1} iron, red juice contains 1.24 mg l^{-1} (Sponholz 1988). In wine, the content of iron varies widely, concentrations of less than 1 mg l^{-1} and up to 25 mg l^{-1} have been found (Amerine and Ough 1980). High concentrations of iron, as found in some wines, can inhibit refermentation of base wine for champagne production (Dittrich 1977). Large concentration of copper can be inhibitory in the same way as iron. Excess amounts of iron and copper can be removed with potassium ferrocyanide (blue fining) (Troost 1972; Amerine et al. 1980). Iron and copper concentrations in excess of 7–10 mg l^{-1} and 0.2–0.4 mg l^{-1}, respectively cause cloudiness in wines (Amerine and Ough 1980).

Phenolic Compounds and Tannins are not inhibitory to yeast growth and fermentation in the concentrations normally found in juices (Dittrich 1977).

2.2.8 Effect of Suspended Solids

Nonclarified juice ferments more rapidly and is less likely to stick than highly clarified juice (Dittrich 1977). Suspended solids in the form of bentonite, cotton, wheat flour, activated charcoal, yeast hulls, etc. will increase the rate of fermentation when added to highly clarified juice. This effect has several reasons, (1) suspended particles act as nuclei for the formation of bubbles of carbon dioxide which is then released from the wine, thus the inhibitory effect of carbon dioxide on the yeast is reduced and the movement generated by the escaping gas bubbles mixes the yeast and the medium, increasing the supply of fermentation substrate and nutrients in the immediate environment of the yeast cell (Dittrich 1977), and (2) grape solids, activated charcoal, and yeast hulls can bind and thus inactivate possibly inhibitory substances (Lafon-Lafourcade 1986). Some preparations of yeast hulls may also add sterols and fatty acids to the juice which stimulate yeast growth and fermentation activity.

2.3 Use of Yeast Starter Cultures

2.3.1. Spontaneous Versus Pure Culture Fermentation

Large amounts of wine are still made by traditional spontaneous yeast fermentation, especially in the old wine-producing countries and by small wineries.

Arguments in favor of spontaneous fermentation are (1) the traditional spontaneous fermentation has been working well for hundreds of years, (2) wine fermented by the indigenous population of yeast may have a different quality than wine fermented with a pure culture yeast, (3) preparation of yeast starter cultures is an additional cost.

Arguments in favor of the use of yeast starter cultures are (1) traditional methods do not always work, (2) inoculation with selected yeast strains give control over the type of yeast which conducts the fermentation and can improve the wine quality, and (3) it is more economical because it allows better control of the time required for completion of fermentation and it allows control over the style and quality of wine produced.

The use of yeast starter cultures enables the wine-maker to select the type of yeast most suitable to the fermentation conditions he chooses and the style of wine he wants to produce, such as wines produced at low temperatures or from juices with high sugar concentrations. With the use of yeast starter cultures, delays in the onset of fermentation can be avoided, which could result in a lower wine quality caused by excessive oxidation of the juice or by spoilage of the juice due to uncontrolled growth of other yeast and bacteria. By inoculation with a large mass of active yeast, problems of stuck fermentation can be overcome. Finally, the timely production is economically very important.

2.3.2 Preparation of Starter Cultures

There are basically three types of yeast starter cultures, (1) the traditional method of inoculating a freshly prepared juice with an amount of actively fermenting juice, (2) commercially prepared active dry yeast, and (3) liquid starter cultures.

Traditional Starter Cultures. The traditional use of starter cultures is the inoculation of a freshly prepared juice with an actively fermenting juice. In this use of starter cultures, the winemaker relies on the fact that in an actively fermenting juice the desirable wine yeast will predominate. This yeast is added with the fermenting wine to subsequent lots of juice. The problem of this starter culture system is that it can be difficult to have the starter culture ready, and in sufficient amount, when needed, the need being determined by the ripening and harvest conditions. Also, yeast cultivated over several subcultures under semi-anaerobic conditions loses its viability due to lack of ergosterol and certain fatty acids in the cell membrane (Andreasen and Stier 1954; Kirsop 1974; Rose 1977).

Active Dry Yeast. The availability of commercially prepared dry active wine yeast eliminates the need of preparing large volumes of starter culture in the winery and makes any amount of starter culture available when it is needed. The yeast strains used in the active dry yeast preparations have been selected for special fermentation characteristics, such as fermentation at low temperature, at high concentrations of sugar, rapid of slow fermentation, flocculation, non-foaming, production of desirable aromatic compounds, and resistance to concentrations of sulfur dioxide commonly used in winemaking.

The manufacturer of the active dry yeast grows the yeast with adequate supply of oxygen and nutrients to produce a yeast which contains optimal amounts of protein, ergosterol, unsaturated fatty acids, and reserve materials (Monk 1986). The yeast is then dried to preserve it for transport and storage. In the drying process, the water content of the cell is reduced from 70% to 8%. To revitalize the yeast, it must be rehydrated. Rehydration of the dry yeast is very critical and must be done carefully in accordance with the instructions provided by the manufacturer. If the yeast is rehydrated improperly it can leak large amounts of cellular components and lose viability. The rehydrated yeast is then added to the juice.

Liquid Starter Cultures may be prepared from selected wine yeast, in the winery or by a commercial supplier. The main difference of these cultures to the active dry yeast is that they are not subjected to drying and rehydration. This does result in a culture with a higher percentage of viable yeast; but this disadvantage of the dry yeast preparation is outweighed by a higher concentration of yeast cells in the dried product and the rehydrated cell suspension. Liquid yeast cultures are usually prepared in diluted grape juice, which might have added vitamine, ammonia, and amino acids. The cultures should be aerated with an air flow providing per minute 0.5–1.0 l air l^{-1} juice (approximately 2 mg l^{-1} dissolved oxygen) (Monk 1986). When the culture reaches maximal cell density, the liquid starter culture is added directly to the juice. When the culture is allowed to age after it has reached maximal cell density (stationary phase) it will lose viability.

When using liquid and dry starter cultures, care must be taken that the yeast is not exposed to sudden large differences in temperature, since temperature shifts of 10° C or greater can cause mutations and loss of viability. A suffiently large amount of starter culture should be used to ensure that the initial cell density in the juice is equal to or greater than 5 Mio cells ml^{-1}. The starter culture should not be added concurrently with the addition of sulfur dioxide, since at the time of the addition, most of the sulfur dioxide is present as molecular sulfur dioxide, its inhibitory form.

Starter Culture for Champagne Production. To adapt the yeast to growth in wine, starter cultures for the production of champagne style wines have traditionally been prepared in the champagne base wine supplemented with 5 g l^{-1} of sugar, and 0.5–1.0 g l^{-1} ammonium sulfate, and oxygenation at the time of inoculation to simulate yeast growth (Amerine et al. 1980). A variation of the traditional method of starter culture preparation is the use of a mixture of equal parts of base wine, water, and tirage liqueur (500 g l^{-1} sugar dissolved in wine) (Wilkinson 1986). The culture is continuously aerated at 0.1 l (air) l^{-1} min^{-1}. A final cell density of 10^8 cells ml^{-1} can be achieved. Although this preparation improves the tolerance of the yeast to SO_2, tolerance to ethanol is not improved. Tolerance to ethanol it depends on the composition of lipids in the cell membrane, the osmotic pressure and concentration of ethanol in the medium, and varies among yeast strains (Thomas et al. 1978; Jones et al. 1981; Stewart and Russell 1985; Guijarro and Lagunas 1984.

Monk and Storer (1986) compared the traditional method (Amerine et al. 1980) with the method of preparing the starter culture aerobically in grape juice. The starter culture prepared aerobically in grape juice reached maximum cell density after 36 h, the culture prepared anaerobically or aerobically in wine required 76 h to reach maximum cell density. The grape juice culture also contained a larger amount of yeast with a higher proportion of viable yeast. When the cultures were inoculated at a cell density of more than 4 Mio cells ml^{-1} into the champagne base wine, they did not differ in their performance. Thus either method may be used, but it appears that the method of preparing the starter culture aerobically in grape juice has the advantage of producing a starter culture with a higher cell density and viable cell numbers after short incubation periods. Because smaller volumes of starter culture are needed and preparation time is shorter, the aerobic starter culture preparation in juice is more economical for the preparation of starter cultures in champagne production.

To avoid temperature shock on inoculation of the starter culture into the base wine, the starter culture must be prepared at a temperature similar to that of the base wine or it must be slowly accustomed to the lower temperature.

2.4 Stuck Fermentations

Overheating and very low temperature, insufficient energy for synthesis of new cell components, lack of nutrients, sterols and reserve materials, high concentrations of sugar and ethanol, biological inhibitors, spray residues, and highly clarified must can all cause stuck fermentations (Dittrich 1977; Lafon-Lafourcade 1986).

Stuck ferments are difficult to referment, because they may lack nutrients after having sustained aborted yeast growth. The old yeast should be removed, the juice clarified using bentonite and if the presence of inhibitory substance is suspected it may be treated with yeast hulls or activated charcoal (Dittrich 1977, Lafon-Lafourcade 1986). The partially fermented juice should then be inoculated with a large number of active yeasts (2–5%, osmotolerant yeasts for juice with high concentration of sugar), nutrients may be added. Fermentation temperature should be adjusted for optimal growth to approximately 18–22° C; if the alcohol content of the partially fermented wine is above 6–8%, a lower temperature should be used because of the increasing inhibitory effect of alcohol with increasing temperature.

2.5 Sterilization

To eliminate contaminating microorganisms, juice is sometimes heated before incubation with selected yeast cultures. The juice is either flash-heated to 80–90° C for 15–30 s or pasteurized at 65° C for 10–15 min. These treatments destroy most of the microorganisms present but do not sterilize the juice. The treatment is more effective in juices of low pH and with small amounts of suspended solids. Care must be taken not to prolong the heat treatment unnecessarily, since it will result in cooked flavors in the final wine. To prevent cooked flavors it is especially important to heat and cool the juice quickly to minimize the holding time at elevated temperature.

Sulfur dioxide is commonly added to the juice at crushing to minimize oxidation and to reduce the number of contaminating microorganisms. To sound grapes with unbroken skin and with little or no mold infection, no or no more than 50 mg l^{-1} SO_2 are added. Juice from *Botrytis*-infected grapes also carries more acetic acid bacteria. To prevent further growth of acetic acid bacteria and aerobic yeast, the juice must be processed quickly. It is cooled, clarified well, and 50–80 mg l^{-1} SO_2 are added, or it can be cooled, clarified, and pasteurized (Dittrich 1977). All processing is designed to reduce excessively by high numbers of potential spoilage organisms and to create conditions unfavorable to their growth. The juice is then fermented with a pure culture winemaking yeast which is tolerant to these concentrations of SO_2. In response to the presence of SO_2 the yeasts produce more acetaldehyde, the main SO_2-binding substance (Somers and Wescombe 1982). The increased amount of acetaldehyde requires larger additions of SO_2 after fermentation in order to establish the desired concentration of free SO_2 in the wine at bottling.

Alternatively, juice can be filter-sterilized, but this is expensive and usually not done unless it is intended to store the juice for a longer period before fermentation.

Because wine always contains residual amounts of fermentable carbohydrates, yeast and bacteria might grow in the wine after it is bottled. It is therefore generally filter-sterilized when bottled. To reduce the number of contaminating yeasts, wine may also be filled hot into the bottle. Heating of wine containing 12% (v/v ethanol to 50° C for 2 min, generally reduces the number of contaminating yeasts and bacteria sufficiently to stabilize the bottled wine (Troost 1972; Dittrich 1977). Additions of sulfur dioxide to result in 20–30 mg l^{-1} of free SO_2 are sufficient to suppress further growth of bacteria, but will not suppress yeast growth.

2.6 Genetics

To molecular biologists, *Saccharomyces* is an important tool in assaying DNA sequences introduced into cells by various methods. Strathern et al. (1981 and 1982) describe the developments in the molecular biology of *Saccharomyces*. Many genes and gene mutations have been characterized and yeasts are used for the manufacture of foreign proteins. Yet most of this work has been done with brewing yeast strains and comparatively little is known about the molecular biology of wine yeast strains.

Saccharomyces cerevisiae can grow stably as haploid, diploid, and polyploid cells (Dittrich 1977; Thornton 1983 and 1986; Stewart and Russell 1985). Heterothallic strains have two mating types, alpha and a. In a typical life-cycle, the haploid cells formed from the ascus spores soon mate and produce diploid cells. Unter favorable growth conditions, the diploid cells reproduce vegetatively by budding. Some strains form ascus spores when growth (budding) stops. When the conditions are again favorable, the spore germinates and the resting cell starts budding again. Polyploid yeast strains have been selected for in the classical fermentation industries, brewing and bread-making, where a form of starter cultures has been used for many years. The advantage of polyploid yeast strains is that they are genetically more stable because they possess multiple genes. Their disadvantage is that it is more difficult to genetically modify polyploid yeast.

Most of the commercial wine yeast strains are diploid, some are haploid, and some polyploid (Dittrich 1977; Thornton 1983). In some strains of wine yeast, sporulation has not been observed. The fermentation activity is genetically dominant; it might be influenced by polymeric genes, the effect of which can be accumulative. The most strongly fermentative yeasts are usually polyploid. Thornton (1986) characterized 25 strains of Australian and New Zealand wine yeast. Twenty one strains were homothallic, one heterothallic, and one possibly triploid. Of 19 potentially valuable wine yeasts, the majority formed asci with four viable spores. Fifteen strains of 19 wine yeasts possibly were octaploid and homothallic. No mating was observed with 18 of the 25 strains, suggesting that the spores formed were not haploid.

2.6.1 Genetic Manipulation

Various systems of transfer of nucleic, mitochondrial, and plasmid DNA and RNA have been developed for *Saccharomyces* (Spencer and Spencer 1977; Barney et al. 1980; Hara et al. 1980; Skatrud et al. 1980; Ouchi et al. 1983; Thornton 1983; Williams et al. 1984; Henderson et al. 1985; Russel and Stewart 1985; Stewart and Russell 1985). A valuable technique for genetic manipulation of polyploid yeast is the rare-mating or cytoduction technique. It allows the introduction of (1) mitochondrial DNA which allows alteration in respiratory functions, (2) double-stranded RNA encoding for killer toxins and the resistance to the toxin, and (3) DNA plasmids carrying new characteristics constructed with gene-cloning and transformation techniques. Another important technique is spheroplast and protoplast fusion, during which DNA exchange between two yeast cells occurs. Yet this exchange is rather unspecific and does not allow the exchange of a specific gene(s). The resulting yeasts are always different from the original yeasts. A very potent tool for the introduction of specific characters is transformation using specifically constructed plasmid carriers. In this way, foreign DNA can be introduced, a single yeast gene can be isolated and modified.

Genetic manipulation of fermentation characteristics of wine yeast has just started. Using hybridization techniques, Thornton (1978) produced a nonfoaming wine yeast from a foaming wine yeast strain, introduced the property of flocculation (Thornton 1985), and improved the fermentation efficiency of a wine yeast strain from 84 to 93% (Thornton 1980 and 1982). Hara et al. (1980) produced a wine yeast strain which produced a killer toxin active against a range of other *Saccharomyces* strains.

Some important physiological characteristics are coded for outside the nucleus, in the mitochondria and on plasmids. One such characteristic is the coding for a cytochrome which is part of the electron transport chain in the respiratory pathway (Dittrich 1977). In the so-called petite mutants, the respiratory pathway is incomplete and the mutants are impaired in their ability to generate energy from the sugars metabolized. Thus cell growth and fermentation rate are reduced. The spontaneous formation of petite mutants is of concern, especially in the production of champagne wines. The rate of mutation under the conditions of secondary alcoholic fermentation appears to be higher than during primary alcoholic fermentation for table wine production (Monk, pers. commun.).

Another important genetic difference between wine yeasts is that between killer and non-killer yeast. Killer yeast were discovered by Bevan and Makower (1963). The so-called killer yeast strains produce a toxin which kills sensitive yeasts (Young and Philliskirk 1975). The toxin has been identified as a protein (Palfree and Bussey 1979; Radler 1986). Some killer proteins might be glycoproteins, but the carbohydrates of these killer toxins have not been identified (Bussey 1972; Pfeiffer and Radler 1983; Radler 1986). The killer toxin is coded for by a plasmid which also carries the information for the resistance against the killer toxin (Wood and Bevan 1968). Killer yeast have been found in laboratory, brewing, and wine strains of *S. cerevisiae* and in other yeasts such as *Candida, Debaromyces, Hansenula, Kluyveromyces, Pichia,* and *Torulopsis* (Imamura et al. 1974; Maule and Thomas 1973; Philliskirk and Young 1975).

Of 19 Australian and New Zealand strains of wine yeast examined, eight were killer yeasts and were resistant to one or two killer strains (Thornton 1986). The same frequency of occurrence of killer yeast is found in other wine-growing regions (Naumova and Naumov 1973; Naumov et al. 1973).

2.6.2 Use of Killer Yeast in Winemaking

A wine yeast strain capable of producing a killer toxin which inhibits the growth of contaminating yeast would be of benefit in that it could ensure that growth of contaminating yeast is prevented and the wine is produced with the characteristics of the desired wine yeast. To be of benefit, the killer toxin(s) must be active against a wide range of contaminating yeast and must be active under wine-making conditions of low pH and in the presence of ethanol. Killer toxins of different yeast strains vary in their sensitivity to low pH. The optimum pH of their activity is usually in the range of pH 4 to 5, and a number of these toxins are not active at the low pH of grape juice and wine (pH 3.0–4.0) (Pfeiffer and Radler 1983, 1984; Radler 1986; Gockowiak 1987). Pfeiffer and Radler (1983) identified a killer toxin which was produced by *S. cerevisiae* during growth in grape juice of pH 3.5, but the activity of the toxin is strongly inhibited at the pH of wine. Hara et al. (1980) produced wine yeast which produced a killer toxin active against a range of strains of *Saccharomyces;* the toxin was not active against contaminating yeast of other genera. The introduced toxin was active under winemaking conditions (pH 3.4, 12–13% ethanol, 22–26° C). It is likely that other killer toxins will be found which are not inhibited at the low pH of wine. The killer toxins also vary in their range of effectiveness against different yeast strains (Gockowiak 1987). Some killer toxins produced by *S. cerevisiae* are effective against only a few other strains of *S. cerevisiae,* but other killer toxins have been found which are effective against yeast of other genera as well. The value of the killer toxins known thus far is limited for use in winemaking because of their generally low activity at low pH and of their limitation in the range of yeasts against which they are effective. Yet the incorporation of killer toxins into wine yeast strains by genetic manipulation has another advantage. Yeasts producing a killer toxin are also resistant to the toxin because the killer plasmid coding for the killer toxin also carries the immunity against the toxin. By genetic manipulation, plasmids coding for different killer toxins can be introduced into a wine yeast strain, thus making it resistant to a range of killer toxins. Such novel yeast strains would be better protected against killer toxins which might be produced by wild yeast present in the grape must, thus reducing the risk of starter culture failure.

3 Bacteria

Bacteria are present in wine at all stages of processing and storage. Because of strong growth-restricting conditions, only a few groups of specialized bacteria are able to grow in grape juice and wine (Ribereau-Gayon et al. 1975; Dittrich 1977; Benda 1982).

Lactic acid bacteria predominate under the anaerobic conditions of vinification and wine storage (Maret and Sozzi 1977, 1979; Maret et al. 1979; Lafon-Lafourcade and Joyeux 1981; Costello et al. 1983; Lafon-Lafourcade et al. 1983a; Fleet and Lafon-Lafourcade 1984; Lafon-Lafourcade and Ribereau-Gayon 1984; Wilbowo et al. 1985; Davis et al. 1986). They are important in the winemaking process for their role in malolactic fermentation.

In the presence of oxygen, acetic acid bacteria can grow in juice and wine (Ribereau-Gayon et al. 1975; Dittrich 1977; Lafon-Lafourcade and Joyeux 1981; Lafon-Lafourcade and Ribereau-Gayon 1984). Acetic acid bacteria are spoilage bacteria in wine.

Species of *Streptomyces, Actionomyces,* and *Bacillus* can be present at low cell density on grapes and in wine (Fleet and Lafon-Lafourcade 1984; Fleet et al. 1984). *Streptomyces* and *Actinomyces* bacteria can be the cause of earthy, musty, and woody off-flavors in wine when wine bottles are sealed with corks which have been infected with these bacteria and the flavor-active substances are leaked from the cork into the wine (Lefebvre et al. 1983). Although *Streptomyces, Actinomyces,* and *Bacillus* have been isolated from wine, it is not known to what extent they are able to grow in wine and what contribution their limited growth in wine has to wine flavor. Generally, due to higher demand for oxygen and lower tolerance to low pH, *Streptomyces, Actinomyces,* and *Bacillus* species usually grow only in oxidized, spoiled wines.

3.1 Acetic Acid Bacteria

Acetic acid bacteria are strictly aerobic, producing acetate from sugars and ethanol (Ribereau-Gayon et al. 1975; Dittrich 1977). During growth they produce large amounts of acetate and small amounts of other fatty acids. The acetic acid produced cannot be removed from the wine. Wines containing more than 1.1–2.5 $g\,l^{-1}$ acetic acid are considered spoiled (Amerine and Ough 1980), but concentrations of less than 1 $g\,l^{-1}$ can already reduce the wine quality. Usual concentrations of acetic acid in wine are 0.3–0.5 $g\,l^{-1}$ (Dittrich 1977).

Growth of acetic acid bacteria is strongly inhibited by low temperatures; below 10° C their growth is almost completely inhibited (Dittrich 1977). They are present during all stages of vinification (Lafon-Lafourcade and Joyeux 1981). Yet in juice prepared from undamaged grapes which are free of growth of mold and carry only few bacteria, the concentration of acetic acid bacteria is very low (less than 10^2 cells ml^{-1}). In contrast, in juice prepared from damaged fruit, the number of acetic acid bacteria can reach 10^7 cells ml^{-1}. During alcoholic fermentation, their number decreases to less than 10^2 cells ml^{-1} as the yeasts consume the available oxygen. A small number of acetic acid bacteria remain in the wine during subsequent storage even in the presence of 25 $mg\,l^{-1}$ of free SO_2. The presence of small numbers of acetic acid bacteria over longer periods of storage can contribute to increased concentration of acetate. When monitoring the number of viable acetic acid bacteria in the wine in order to estimate their potential for producing acetate, it must be considered that acetate can also be produced by lactic acid bacteria when oxygen or an alternative proton acceptor is available (Sect. 3.2).

To protect juice prepared from grapes heavily infected with acetic acid bacteria against spoilage, it is necessary (1) to reduce the number of acetic acid bacteria in the juice by rapid clarification and by addition of 50–80 mg l^{-1} of SO_2, the juice may be heat- or filter-sterilized, and (2) to reduce the chance of reinfection from equipment surfaces by properly sanitizing the equipment. In wine, growth of acetic acid bacteria is controlled by exclusion of air, by fermenting and storing at low temperature (below 15° C), and the use of SO_2. During storage of juice and wine, air should be excluded by keeping storage containers completely filled and by blanketing the surface of exposed wine with carbon dioxide or nitrogen gas. A concentration of 15–30 mg l^{-1} free SO_2 should be maintained in the wine during storage.

3.2 Lactic Acid Bacteria

Lactic acid bacteria (LAB) enter the juice and wine from grapes, leaves, soil, and equipment surface (Webb and Ingraham 1960; Peynaud and Domercq 1961; Gini and Vaughn 1962; Schanderl 1969; Weiller and Radler 1970; Donnelly 1977; Sapis-Domercq et al. 1978; Lafon-Lafourcade et al. 1983 a). The LAB of wine belong to three genera, *Leuconostoc, Lactobacillus,* and *Pediococcus;* they are microaerophilic and require carbohydrates and some preformed amino acids and vitamins for growth (Radler 1966; Kunkee 1967; Weiller and Radler 1972, 1976; Ribereau-Gayon et al. 1975; Dittrich 1977; Benda 1982; Sedewitz et al. 1984; Murphy et al. 1985; Wibowo et al. 1985). The only species of *Leuconostoc* able to grow in wine is *Lc. oenos* (Garvie 1967, 1981; Davis 1985). In older literature isolates of *Leuconostoc* from wine are called *Lc. mesenteroides* or *Lc. citrovorum,* these isolates should be called *Lc. oenos* based on the classification by Garvie (1967, 1981; Garvie and Farrow 1980). Three species of *Pediococcus* and seven species of *Lactobacillus* are commonly found in wine (Davis 1985). The pH of wine is highly selective. Thus wines of pH below 3.5 generally contain only strains of *Lc. oenos,* while in wines with pH values above 3.5 various species of *Pediococcus* and *Lactobacillus* predominate (Fornachon 1964; Mayer 1974; Ribereau-Gayon et al. 1975; Costello et al. 1983; Davis 1985; Wibowo et al. 1985).

3.2.1 Nutritional Requirements

Oxygen and Carbon Dioxide. Growth of LAB is stimulated by CO_2 (Charpentier 1954; Dittrich 1977; Stamer 1979). *Leuconostoc oenos* requires small amounts of CO_2 for growth (Mayer 1974). Oxygen can be used as a terminal electron acceptor by *Lc. mesenteroides* and *L. plantarum* (Lucey and Condon 1984, 1986; Sedewitz et al. 1984; Murphy et al. 1985).

Carbohydrates. The LAB of wine produce energy from the fermentation of carbohydrates, producing either exclusively lactate (homofermentative) or a mixture of lactate, CO_2, acetate, and ethanol (heterofermentative) (Kandler 1983). *Leuconostoc* sp. are heterofermentative, the genus *Lactobacillus* contains hetero- and homofermentative species, and *Pediococcus* sp. are homofermentative. The homofermentative bacteria follow the glycolysis pathway which yields 2 mol of lactate and 2 mol of ATP from 1 mol of glucose. The heterofermentative pathway (6P-gluconase pathway or pentose-P pathway) yields from 1 mol of glucose 1 mol of lactate, 1 mol of CO_2, and 1 mol of acetate and ethanol with the gain of 1 ATP. The ratio of acetate and ethanol depends on the presence of alternative proton acceptors in the medium; if an alternative proton acceptor is present, an additional ATP can be produced from acetyl-P. Oxygen can be reduced to H_2O_2 and H_2O and fructose to mannitol. Under aerobic conditions, the homofermentative *L. plantarum* can oxidize lactate to acetate with the gain of 1 ATP (Sedewitz et al. 1984; Murphy et al. 1985).

Growth of LAB depends on the presence of fermentable carbohydrates (Radler 1966; Kandler et al. 1973). During malolactic fermentation (MLF), 0.4 to 0.8 $g\,l^{-1}$ sugar is degraded (Melamed 1962). Radler (1958a) calculated that for the degradation of 1 $g\,l^{-1}$ of malate within 10 d, approximately 0.01 g of bacteria is required; 0.1 g of glucose provides sufficient energy to produce 0.01 g of bacterial cell mass. Most dry wines contain 1 to 3 $g\,l^{-1}$ glucose and fructose which is more than what is required to sustain sufficient bacterial growth to ensure MLF. But MLF may be inhibited in wines containing less than 0.2 $g\,l^{-1}$ glucose and fructose (Vaughn 1955; Dittrich 1977). Studies with washed cell suspensions of *Lc. oenos* and *L. plantarum* show that the rate of glucose uptake decreases strongly at a concentration of approximately 0.2 $g\,l^{-1}$ glucose (Henick-Kling 1986a).

Glucose and fructose are the most important source of energy for the LAB during growth in wine, since nearly all wines contain sufficient amounts of these sugars; also, in washed cell suspension, *Lc. oenos* and *L. plantarum* utilize glucose preferentially over other sugars (Henick-Kling unpubl.). Wine does contain a number of other sugars which might be utilized if the concentration of glucose and fructose is insufficient. Of the pentoses, arabinose and xylose are apparently degraded by LAB during growth in wine (Melamed 1962; Ribereau-Gayon et al. 1975). Other hexoses in wine are present only at very low concentration; mannose and galactose have been found at concentrations of less than 0.2 $g\,l^{-1}$ Sponholz 1988) and are therefore not important to the growth of LAB. Oligosaccharides, polysaccharides, and glycosidic compounds can be substrate for growth of LAB. Most strains of *Lc. oenos* produce beta-glucosidases (Garvie 1967); also the wine can contain residual activity of glucosidases from the grape which slowly release glucose (Costello et al. 1985).

Organic Acids. Tartrate and malate are the major organic acids in wine; together they account for approximately 90% of the total acidity of the wine. In addition to these two acids, many other organic acids are present in low-concentration (typically 0–0.5 $g\,l^{-1}$) (Radler 1975, Amerine and Ough 1980).

Tartrate. Small decreases in the concentration of tartrate (3–30%) are sometimes observed during MLF (Bousbouras and Kunkee 1971; Pilone et al. 1966; Rice and Mattick 1970; Shimazu and Watanabe 1979). The changes in the concentration of tartrate are most likely due to changes in the solubility of the tartrate during fermentation rather than to microbial degradation (Kunkee 1967; Rice and Mattic 1970). Only some strains of *L. plantarum, L. brevis,* and L. arabinosus have been found to degrade tartrate (Krumperman and Vaughn 1966; Flesch 1968a; Peynaud 1968; Barre 1969; Radler and Yannissis 1972). Radler and Yannissis (1972) examined 78 strains of LAB belonging to the genera *Lactobacillus, Leuco-*

nostoc, and *Pediococcus*, of which only four strains of *L. plantarum* and one strain of *L. brevis* did degrade tartrate. Complete degradation of tartrate spoils the wine because tartrate is most important for taste and structure of the wine (Radler 1975; Dittrich 1977; Lafon-Lafourcade 1986). *Lactobacillus plantarum* and *L. brevis* anaerobically degrade tartrate via two different pathways resulting in different end products, *L. plantarum* produces CO_2, acetate, and lactate, *L. brevis* produces CO_2, acetate, and pyruvate (Radler and Yannissis 1972). Neither bacterium can grow on tartrate as sole carbon source. In *L. plantarum*, metabolism of tartrate is inhibited in the presence of glucose, glucose does not affect tartrate metabolism in *L. brevis*. Tartrate metabolism by lactobacilli apparently is completely inhibited at pH values below 3.5 (Ribereau-Gayon et al. 1975), the optimum pH for tartrate degradation by *L. plantarum* is 5.0; no measurable amounts of tartrate are degraded at pH 3.0 (Radler and Yannissis 1972).

Wine can be protected against bacterial spoilage due to degradation of tartrate by maintaining the wine pH below 3.5, which strongly inhibits growth of potentially harmful lactobacilli, and by monitoring the types and total number of LAB present in the wine during storage. Inoculation of wine with a pure culture of *Lc. oenos* which does not degrade tartrate also can protect wine against growth of lactobacilli; but inoculation must be done under conditions which favor growth of *Lc. oenos* and before indigenous lactobacilli are able to grow to high cell density and predominate during MLF.

Malate. All malate formed in the grape is of the L(−) form (Amerine and Ough 1980; Sponholz 1988). D-malate is not naturally present in grape juice and is not metabolized by LAB of wine. Most LAB isolated from wine and apparently all strains of *Lc. oenos* metabolize L-malate (Weiller and Radler 1970; Ribereau-Gayon et al. 1975; Radler and Broehl 1984; Izuagbe et al. 1985; Radler 1986). The conversion of L-malate to L-lactate and $\cdot CO_2$ (malolactic fermentation) is an important step in vinification. Its enological significance has recently been reviewed by Davis et al. (1985 a). Three pathways of malate degradation have been found in LAB (Radler 1975, 1986).

Most LAB contain the malolactic enzyme with converts L-malate stoichiometrically into L-lactate and CO_2, without free intermediates (Schuetz and Radler 1973, 1974; Lonvaud-Funel and Strasser de Saad 1982; Caspritz and Radler 1983; Spettoli et al. 1984). No ATP is produced in this reaction and the LAB containing this enzyme are not able to grow on malate as sole carbon source (Kandler et al. 1973).

A second pathway involves the malic enzyme which has been found only in strains of *L. casei* and *Streptococcus faecalis* (London and Meyer 1969; London et al. 1970, 1971; Schuetz and Radler 1974). The malic enzyme decarboxylates L-malate to pyruvate which is then either reduced by lactate dehydrogenase (LDH) to lactate or oxydized to acetyl-P and acetate. Due to the presence of L-LDH and D-LDH both, L- and D-lactate, are produced from malate in this reaction. The aerobic pathway of oxidation to acetate yields one ATP, which allows strains containing the malic enzyme to grow on malate as sole energy source (Schuetz and Radler 1974; Kandler 1983). The malic enzyme in *L. casei* is induced by growth on malate as sole carbon source, the malolactic enzyme is induced by growth on malate and glucose (Schuetz and Radler 1974).

A third pathway exists in some strains of *L. fermentum* which contain neither the malolactic nor the malic enzyme (Radler 1986). These strains oxidize L-malate to oxaloacetate via a malate dehydrogenase, oxaloacetate is decarboxylated to pyruvate, L- and D-lactate, acetate, succinate, and CO_2 are produced from pyruvate. This pathway is little active at pH values below 4.0.

Citrate. The concentration of citrate in grape juice is usually less than 0.2 g l^{-1} but can reach more than 0.6 g l^{-1} in juice from *Botrytis*-infected grapes (Sponholz 1988). The metabolism of citrate yields extra pyruvate, from which diacetyl, acetaldehyde, acetoin, and butyleneglycol, and ethanol are produced (Kandler 1983). Alternatively, some LAB can split pyruvate (pyruvate lyase) to acetyl-CoA and formate, acetyl-CoA is further metabolized to acetyl-P and acetate producing 1 ATP (Kandler 1983). Under glucose limitation almost all pyruvate is metabolized to acetate and formate. Leuconostocs and *L. casei* contain the pyruvate lyase. Apparently strains of *P. cerevisiae* and some other lactobacilli are unable to degrade citrate (Du Plessis 1964; Whiting and Coggins 1964; Wibowo et al. 1985).

Pyruvate and other organic acids are only found in grape juice and wine in concentrations of less than 0.1 g l^{-1}; their concentration is increased in grapes from *Botrytis*-infected fruit and in wine due to production by the yeast (Radler 1975, 1986; Sponholz 1988).

Pyruvate, 2-oxoglutarate, gluconate, and succinate are metabolized by LAB but it is unknown to what extent these acids are metabolized at the low concentration that they are present in wine (Radler and Broehl 1984; Radler 1986).

Sorbate and Fumarate. Although fumarate is metabolized by yeast, it inhibits growth of LAB by restricting pyrimidine biosynthesis (Cofran and Meyer 1970; Pilone et al. 1974; Silver and Leighton 1982). Fumarate may be added to wine after completion of alcoholic fermentation to inhibit growth of LAB. The amount of fumarate necessary to inhibit growth of LAB is usually 0.7–1.5 g l^{-1}, but varies with wine pH and the presence of other inhibitory substances (Silver and Leighton 1982; Pilone et al. 1974). Inhibition by fumarate in wine can be overcome by LAB which possess fumarase or by the presence of residual yeasts which metabolize the fumarate. Few leuconostocs contain fumarase. Radler and Broehl (1984) found that of the LAB studied 77 of 83 lactobacilli, 8 of 19 leuconostocs, and 10 of 28 pediococci degraded fumarate; Silver and Leighton (1982) could not find fumarase activity in the strains of *Leuconostoc* (Fig. 2 A), *Lactobacillus*, and *Pediocossus* which they examined.

Sorbate is also used as preservative in wine. It inhibits yeast growth but it is not effective against LAB (Radler 1986). Some LAB can reduce sorbate to its corresponding alcohol (2,4-hexadien-1-ol), which at the low pH of wine esterifies with ethanol and rearranges to 2-ethoxy-hexa-3,5-diene (Crowell and Guymon 1975). This ester produces the so-called geranium off-flavor which spoils the wine (Wuerdig et al. 1974).

Amino Acids. All LAB isolated from wine require some preformed amino acids for growth (Radler 1966; Weiller and Radler 1972, 1976; Ribereau-Gayon et al. 1975; Dittrich 1977; Benda 1982; Wibowo et al. 1985). The requirement for specific amino acids varies between strains; thus strains of *L. plantarum* and *Lc. mesenteroides* isolated from vine leaves required three to four amino acids, whereas those of *Lc. oenos* and *P. pentosaceus* isolated from wine required up to 16 amino acids (Weiller and Radler 1972). Generally, arginine, cystine, cysteine, glutamate, histidine, leucine, phenylalanine, serine, tryptophan, tyrosine, and valine are essential for growth (Benda 1982). During growth in wine, the concentration of several individual amino acids may decrease strongly, while that of others increases (Wibowo et al. 1985; Henick-Kling 1986a). Yet the availability of amino acids in wine has not been found to be growth-limiting (Radler 1966). Proteins and peptides in wine can also be used by LAB as a source of amino acids (Feuillat et al. 1977). Lactic acid bacteria have been shown to produce extra- and intracellular proteases (Akuzawa et al. 1983; Hickey et al. 1983; Law and Kolstad 1983; Exterkate 1984; Kolstad and Law 1985). Furthermore, yeasts produce extracellular proteases during alcoholic fermentation (Feuillat et al. 1980). In wine, the concentration of free amino acids has been observed to increase during growth of LAB (Ribereau-Gayon et al. 1975; Henick-Kling 1986a).

Alternatively, some amino acids can serve as energy source; for example, arginine is degraded completely by several strains of *Lc. oenos* and *L. brevis*, and histidine, glutamate, and tyrosine are degraded by some strains of *P. cerevisiae* and *L. brevis* (Weiller and Radler 1976). Yet during growth in wine, arginine appears to be less important as a source of energy than are glucose and fructose. Brechot et al. (1984) speculated that arginine is the major substrate for growth of *Lc. oenos* in wine. They found that arginine was incorporated into the biomass in preference to glucose. Yet, the calculated amount of ATP gained from the arginine degraded would yield only one tenth of the amount of biomass actually formed in the wine. End products of arginine degradation are CO_2 and NH_3, with a gain of 1 mol of ATP. During growth of the bacteria, D-lactate was also produced, presumably the end product of glucose metabolism (Garvie 1967). The amount of D-lactate produced indicades that sufficient glucose was degraded to yield a biomass of cells of *Lc. oenos* which is even larger than that determined in the wine.

Vitamins, Minerals, and Other Growth Factors. Lactic acid bacteria of wine require several B-group vitamins, with all strains apparently requiring folic and nicotinic acid (Radler 1966). The requirement for biotin, riboflavin, panthothenic acid, and pyridoxin is strain-dependent.

Leucovorin has been described as a specific growth factor for pediococci (Weiller and Radler 1976). The so-called tomato-juice factor, a derivative of pantothenic acid, is stimulatory for growth of *Lc. oenos* (Amachi and Yoshizumi 1969; Amachi 1975).

Purine and pyrimidine derivatives are also required or are stimulatory for the growth of LAB (Radler 1966).

Lactic acid bacteria of wine require K^+, Na^+, and relatively large amounts of Mn^{2+}; their concentration in wine is generally sufficient for growth and metabolism of LAB (Radler 1966; Dittrich 1977).

3.2.2 Inhibition of Bacterial Growth in Wine

Composition of the wine, the method of vinification, and interrelationships between LAB and other microorganisms affect the survival and growth of LAB in wine (Radler 1966; Ribereau-Gayon et al. 1975; Wibowo et al. 1985; Lafon-Lafourcade 1986).

Composition of Wine. In order to be able to maintain their viability and grow in wine, the bacteria must be able to produce and store energy under adverse conditions of low pH, low temperature, presence of SO_2 and ethanol. Energy necessary for cell growth and cell maintenance is conserved in two ways: (1) substrate level phosphorylation in the cytoplasm, and (2) via the electrochemical proton gradient (protonmotive force, Δp) across cell membranes (Mitchell 1966 and 1981). The two forms of energy conservation are closely interrelated and both are affected by the composition of the medium.

pH. The pH of wine is one of the most important factors which limits growth and MLF in wine (Radler 1966; Castino et al. 1975; Dittrich 1977; Wibowo et al. 1985). The pH of juice and wine is between 2.8 and 4; ideally for table wines it should be between 3.1 and 3.6 (Amerine and Ough 1980). Growth of LAB is optimal at near-neutral pH values and is inhibited at lower pH values (Radler 1966; Henick-Kling 1983). The minimum pH for growth and metabolism of sugar is between 2.8 and 4, and varies between species (Fornachon 1964; Peynaud and Domercq 1968; Peynaud and Sapis-Domercq 1970). In general, strains of *Lc. oenos* are most tolerant to low pH and they predominate in wines below pH 3.5 (Wibowo et al. 1985); furthermore, strains of *Lc. oenos* have the lowest pH optimum for growth. *Leuconostoc oenos* can grow optimally at pH values ranging from 4.5 to 6.5 (Radler 1966; Henick-Kling 1983).

At the low pH of wine, the ability of the bacteria to obtain energy from the metabolism of glucose is inhibited. The optimum pH for catabolism of glucose by *Lc. oenos* and *L. plantarum* is between 4.5 and 6.0 (Henick-Kling 1986 a).

At low pH, glucose metabolism is inhibited likely by a lower intracellular pH (Henick-Kling 1986 a). At an extracellular pH of 5.5, which is optimal for growth, the intracellular pH (pH_i) of *Lc. oenos* and *L. plantarum* is approximately 6.5; at low extracellular pH (3.5), pH_i of *Lc. oenos* and *L. plantarum* is approximately 5.9. The pH_i is also affected by the metabolism of malate. In cells catabolizing L-malate, the pH is increased by up to 0.5 pH units (Henick-Kling 1986 a).

Cells of *Lc. oenos* can be adapted to growth at low pH by subculture in media of low pH (Flesch 1968 b; Henick-Kling 1986 a). Cells adapted to low pH have higher rates of malate catabolism, higher pH_i, and higher Δp (Henick-Kling 1986 a).

Temperature. Lactic acid bacteria of wine are mesophilic bacteria with optima for growth between 25 and 30° C (Peynaud 1967; Benda 1982; Dohman 1982). The optimal temperature for malate degradation by nongrowing cells is 30° C (Lafon-Lafourcade 1970). The rates of growth and MLF are strongly inhibited by low temperature (Lafon-Lafourcade 1970; Dittrich 1977; Henick-Kling 1983). In practice, MLF is rarely observed at temperatures below 10° C (Dittrich 1977).

Ethanol. Growth and MLF by LAB are increasingly inhibited at ethanol concentrations above 5% (v/v) (Radler 1966; Peynaud and Domercq 1968; Henick-Kling 1986 a). Sensitivity to ethanol varies strongly between strains. Most LAB of wine are unapable to grow at ethanol concentrations above 15% (v/v), but some LAB are able to grow in wine of 20% (v/v) ethanol (Wibowo et al. 1985). Yet, ethanol apparently is not a major growth inhibiting factor (Dittrich 1977).

MLF can be more rapid in high alcohol wine (14–15% v/v), given that other growth-limiting factors are not prohibiting growth (Dittrich 1977; Wibowo et al. 1985).

Sulfur Dioxide strongly inhibits growth of LAB (Wibowo et al. 1985). The sensitivity of LAB to SO_2 varies; generally, concentrations of 50 to 100 mg l^{-1} bound SO_2 and 1–10 mg l^{-1} free SO_2 can inhibit growth of LAB and malolactic fermentation. Sulfur dioxide is more inhibitory at low pH.

Winery Practices. Clarification of juice and wine by sedimentation, centrifugation, or filtration can remove a large portion of LAB and so reduce the incidence of bacterial growth and its effect on wine quality. During clarification also some nutrients are removed which are stimulatory to bacterial growth (Fornachon 1957; van Wyk 1976). Bacterial growth in wine can also be inhibited by the removal of suspended particles and tannin-protein complexes which aid flocculation and protect bacterial cultures against phage infection (Atkinson and Daoud 1976; Ito 1967; Barnet et al. 1984). Yeast and bacteria can be completely removed from wine by filtration through 0.45 μm membranes (Radler and Schoenig 1977).

Malolactic fermentation occurs more frequently and more consistently in wine fermented with grape skins than in wine fermented in their absence (Kunkee 1967; Beelman and Gallander 1970). The grape skins presumably act as a support and provide a protective micro-environment, analogous to flocculation by some microorganisms (Atkinson and Daoud 1976). Moreover, materials stimulatory to the growth of LAB might be leached from the skins (Amachi and Yoshizumi 1969). Yeasts attached to the skins might also release growth-stimulatory nutrients (Lafon-Lafourcade and Peynaud 1961; Amerine and Kunkee 1968).

It is practice in some countries to heat the juice (thermovinification) in order to improve color extraction, to reduce enzymatic browning, and to sterilize it before fermentation with pure culture yeast (Troost 1972; Ribereau-Gayon et al. 1975; Amerine et al. 1980). Wines produced in this way have been found to be less suitable for growth of LAB than those produced without heat treatment (Beelman and Gallander 1970; Martiniere et al. 1974; Beelman et al. 1980).

3.2.3 Malolactic Fermentation

During malolactic fermentation (MLF), the LAB degrade malate forming lactacte and CO_2. This conversion of malate into lactate reduces the acidity of the wine (Radler 1966; Kunkee 1967; Ribereau-Gayon et al. 1975; Dittrich 1977; Davis et al. 1985a; Wibowo et al. 1985). Other metabolites produced during growth further modify the wine flavor. Growth of LAB is encouraged in wines in which MLF is desired to reduce the acidity of the wine and to modify flavor. The reduction of the acidity of wine is beneficial to the quality of wine made in cool winegrowing regions, since the grapes grown in cool areas often contain excessive amounts of organic acids. The flavor change associated with the growth of LAB is desirable in certain styles of wines. The growth of LAB in wine must be tightly controlled to ensure the growth of desirable LAB which produce no off-flavors and that MLF is completed rapidly to save processing time of the wine.

The malolactic conversion is catalyzed by the malolactic enzyme (Schuetz and Radler 1973; Lonvaud-Funel and Strasser de Saad 1982; Caspritz and Radler 1983; Spettoli et al. 1984). Two other pathways for the degradation of malate are found in LAB (see above) but most LAB and all *Lc. oenos* contain the malolactic enzyme.

The malolactic enzyme converts L-malate stoichiometrically into L-lactate and CO_2, without free intermediates. The conversion requires NAD and Mn^{2+}; reduction of NAD is not detected (Schuetz and Radler 1973; Lonvaud-Funel and Strasser de Saad 1982; Caspritz and Radler 1983; Spettoli et al. 1984). The malolactic enzyme from *L. delbrueckii* has recently been cloned (Williams et al. 1984).

Physiological Role. During growth in wine, LAB rapidly degrade malate. Yet LAB are able to grow in wine which is free of malate (Wibowo et al. 1985). It has been shown that the conversion of malate to lactate does not yield utilizable energy (Radler 1958 b; Kandler et al. 1973; Schuetz and Radler 1974). The metabolism of malate accelerates and extends the growth of *Lc. oenos* at low pH (Radler 1958 b; Kandler et al. 1973; Kunkee 1974); but it does not influence the molar growth yield from glucose (Kandler et al. 1973). Due to MLF, wine acidity is reduced by 0.1–0.3% and pH increases by between 0.1 and 0.3 pH units (Davis et al. 1985 a).

Because ATP is not produced in the malolactic conversion catalyzed by the malolactic enzyme, it has been difficult to explain the biological function of this reaction (Davis et al. 1985 a). Kandler et al. (1973) reason that the physiological advantage from the degradation of malate comes from the pH increase which allows extended growth in the acidic natural environment of the malolactic bacteria. Kunkee (1974 and 1975), Morenzoni (1974), and Pukrushpan and Kunkee (1977) propose that the malolactic enzyme has a secondary activity. In this secondary activity, 0.2% of the malate metabolized is converted to pyruvate, which in turn serves as substrate for growth; but this appeares unlikely (Henick-Kling 1986 a). Based on measurements of pH_i and Δp in *Lc. oenos* and *L. plantarum*, another explanation has been proposed for the biological role of MLF (Henick-Kling 1986 a). Measurements associated with the catabolism of L-malate have shown that pH_i and Δp are increased in cells metabolizing malate. This gives the cell an energetic advantage without yielding metabolizable intermediates.

Effect of pH on Rate of Malolactic Fermentation. During growth in wine, *Lc. oenos* concurrently degrades L-malate (Henick-Kling 1983). The rate of malate degradation depends on the number of viable cells present in the wine, on the rate of growth, and on the pH (Henick-Kling 1986 a). Low pH, low temperature, and the presence of ethanol and SO_2 inhibit the rate of growth of the malate-converting bacteria and reduce the viability of cells inoculated into the wine to induce MLF. Because of the limiting effect of pH on growth of LAB, MLF occurs more frequently and is completed more rapidly at higher pH values (Bousbouras and Kunkee 1971; Castino et al. 1975; Costello et al. 1983; Mayer and Vetsch 1973; Henick-Kling 1983). In contrast, degradation of malate by nongrowing cells of *Lc. oenos* is most rapid at low pH values (Lafon-Lafourcade 1970; de Menezes et al. 1972; Henick-Kling 1985 a).

The malolactic enzyme is constitutive in *Lc. oenos*, thus its activity does not change during growth of the cell population (Henick-Kling 1986a). In *L. plantarum*, the malolactic enzyme is induced during growth in medium containing malate; thus, the onset of MLF by *L. plantarum* may be delayed during its growth in wine. The rate of malate metabolism by nongrowing cells with a constitutive or fully induced malolactic enzyme depends on the rate of transport of malate into the cell and on the intracellular pH (Henick-Kling 1986a). The rate of malate degradation is highest at low pH values between 3.0 and 4.5. The pH of the growth medium affects the intracellular pH. In cells of *Lc. oenos* grown in media of low pH (3.5), the intracellular pH is increased by approximately 0.5 pH units over that of cells grown in media of high pH (5.5). In those cells grown in media of low pH the intracellular pH was 6.3, which is the optimum for the activity of the malolactic enzyme (Caspritz and Radler 1983; Henick-Kling 1986a).

Prevention of MLF. If MLF is not desired for the wine style, growth of LAB in must and wine must be suppressed by removing or inactivating the bacteria present. An addition of 50–100 mg l^{-1} SO_2 to the must, depending on pH and concentration of SO_2-binding compounds, destroys more than 90% of the viable bacteria in the must; generally, a concentration of 10–30 mg l^{-1} free SO_2 completely inhibits growth of LAB in wine. Juice clarification by filtration or centrifugation removes a large number of bacteria. Rapid alcoholic fermentation with high yeast cell density and an ethanol concentration of more than 8% (v/v) strongly inhibits growth of LAB. After alcoholic fermentation, the wine should be clarified quickly; by removing yeast cells, wine is made more inhospitable for bacteria, because a source of nutrients and a protection against physical, biological, and chemical inhibitors is removed. After alcoholic fermentation and clarification and during storage; a concentration of 15–30 mg l^{-1} free SO_2 should be maintained in the wine. The wine should be controlled at regular intervals during storage for bacterial growth. If significant growth occurs (i.e., $>10^2$ cells ml^{-1}), the wine should be filtered to remove the bacteria present. Storage at temperatures below 15° C also strongly inhibits growth of LAB.

Induction of MLF. It MLF and bacterial production of flavor compounds are desired, the winemaker can encourage the growth of indigenous LAB or inoculate the wine with a pure culture of *Lc. oenos* (Henick-Kling 1986b). The preferred bacteria for carrying out MLF are strains of *Lc. oenos*, because they can grow in wine at low pH and degrade malate without producing undesirable flavor compounds or degrading wine components which are necessary for wine quality and stability, i.e., tartrate, ethanol, and glycerol.

To encourage the growth of *Lc. oenos*, the pH of the wine should be between 3.1 and 3.5. At pH values above 3.5, potentially harmful lactobacilli and pediococci rapidly outgrow *Lc. oenos* (Mayer and Vetsch 1973). Addition of SO_2 should be avoided, since it is very inhibitory to *Lc. oenos* (Wibowo et al. 1985). Also, the presence of SO_2 favors growth of pediococci over that of leuconostocs (Mayer and Vetsch 1978). Therefore, the wine should contain no free SO_2 and no more than 50 mg l^{-1} bound SO_2. The wine should not be heavily clarified, i.e., at most the wine should be racked off gross lees. Wine temperature should be no lower than 18° C at the time of inoculation.

Growth of indigenous *Lc. oenos* will be slow at pH values below 3.1 and, depending on the initial cell density and other growth inhibitory factors, onset and completion of fermentation can be delayed considerably.

Inoculation with a bacterial starter culture has three advantages: (1) less time is required for the bacteria to grow up to a cell density high enough to rapidly degrade the malate present in the wine; (2) it is possible to select bacteria with desirable characteristics to carry out the fermentation; (3) inoculation with a mixed strain starter culture gives protection against phage attack.

Preparation of Starter Culture. Bacterial starter cultures for the induction of MLF are available commercially as freeze-dried preparations and frozen concentrate. The commercial preparations can be used for direct inoculation into wine. Yet subculture in a grape juice medium for activation and adaptation is recommended (Lafon-Lafourcade et al. 1983 b; Henick-Kling 1986 b). Because of the high cell numbers in the commercial preparations, only two subcultures are needed before an adapted and suitably large amount of starter culture is available for inoculation into wine. For successful inoculation into wine, the bacteria must be adapted to wine conditions, i.e., low pH, low temperature, presence of ethanol. *Leuconostoc oenos* can be adapted to growth in wine by preparing the starter culture in a medium containing 40–80% wine (Hayman and Monk 1982). For the preparation of starter cultures, the use of dry wine, diluted 50% with water or grape juice has been recommended (Henick-Kling 1986 b). The medium is adjusted to pH 4.0–4.5 and yeast extract is added at 0.2 g l^{-1}. The wine used in the starter culture medium should contain no free SO_2 and more than 50 mg l^{-1} total SO_2. For optimal growth, the cultures should be prepared at 20–25° C. The wine temperature at the time of inoculation should be no less than 18° C. If the wine temperature is lower, the starter culture must be gradually adapted to the lower temperature to avoid temperature shock.

3.2.4 Bacteriophage of *Leuconostoc oenos*

Bacteriophages (phages) can inhibit malolactic fermentation in wine (Cazelles and Gnaegi 1982; Sozzi et al. 1982; Gnaegi et al. 1984; Henick-Kling et al. 1986 a). An increasing number of wineries use bacterial starter cultures to induce MLF. To avoid failure of the starter culture, it is important to protect the bacteria against phage-induced lysis (Sozzi and Gnaegi 1984; Henick-Kling 1986 a, b).

Phages of *Lc. oenos* belong to the morphological group B of Bradleys' classification (Bradley 1971; Sozzi et al. 1982; Gnaegi et al. 1984; Davis et al. 1985 b; Henick-Kling et al. 1985 a) (Fig. b). Wine apparently contains various strains phages of *Lc. oenos* which differ in host range and sensitivity to low pH, SO_2, and phenolic compounds (Henick-Kling et al. 1986 b). The phage attach with their base plate to the cell wall of the bacterial host and inject their nucleic acid; the nucleic acid of the phage directs the bacterial cell to build new phage particles which are released with the lyses of the host cell (Sozzi and Gnaegi 1984; Henick-Kling 1985) (Fig. 2 c).

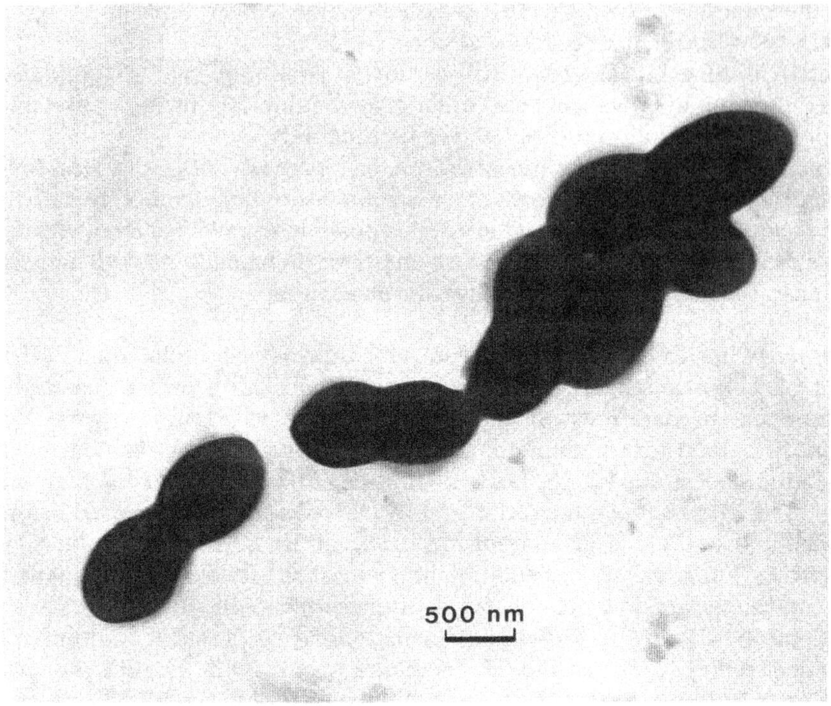

A

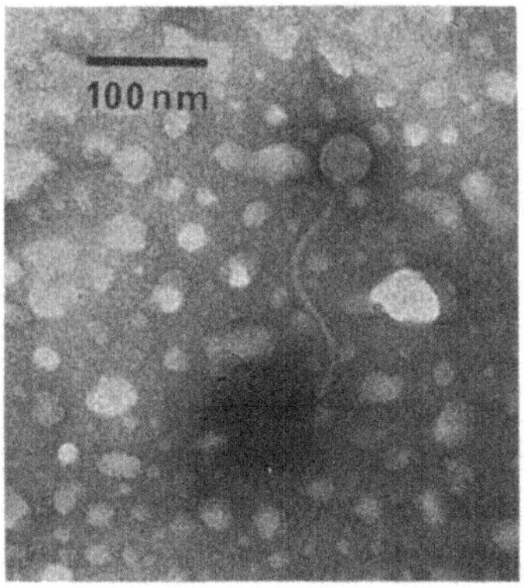

B

Fig. 2A–C. Electronmicrograph of *Leuconostoc oenos*. Negative staining. 10 mm = 500 nm. **B** Bacterophage of *Leuconostoc oenos*. **C** Bacteriophage of *Leuconostoc oenos*, infection and lysis of host cells

To avoid starter culture failure due to phage attack, the starter culture should be composed of strains which are resistant to lysis by bacteriophage. Strains resistant to a number of known phages can be combined in a multiple-strain starter culture to give a high degree of protection against phage interference (Sozzi and Gnaegi 1984; Henick-Kling 1986b). Phage inhibition is strongest when phage infection occurs at the beginning of the bacterial growth phase, at low temperature, and at low pH (Henick-Kling et al. 1986b). *Leuconostoc oenos* can overcome phage inhibition under conditions which favor its growth, i.e., temperature above

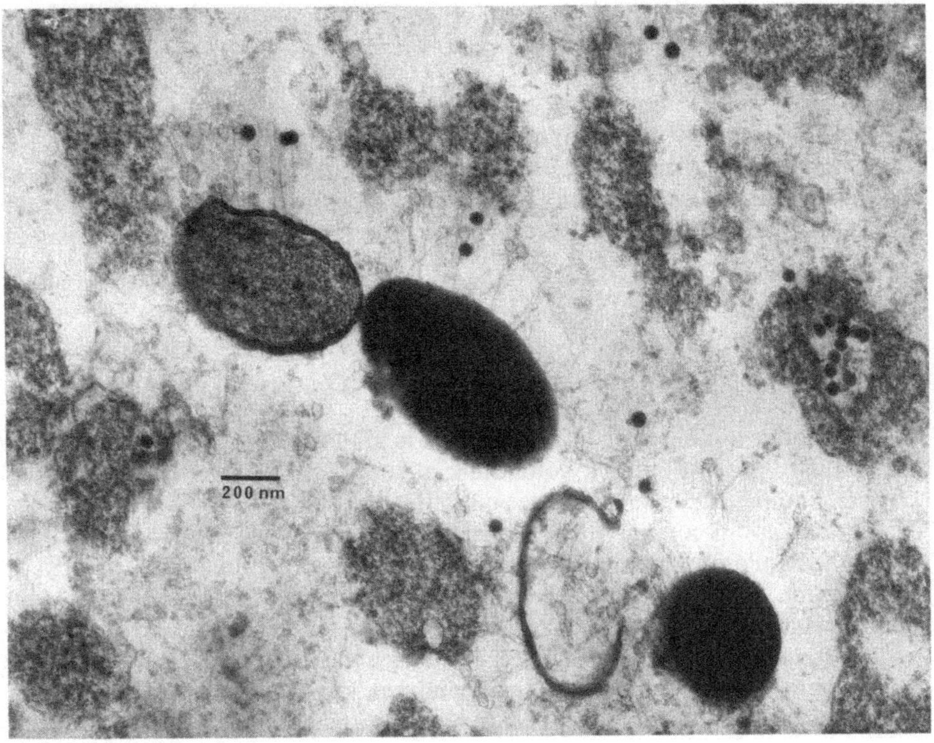

Fig. 2. C

15° C, pH 3.3–3.5, and absence of SO_2. Thus the risk of phage interference in MLF can be reduced by creating favorable conditions for growth of *Lc. oenos* and by inoculating the wine with an active starter culture at high cell density avoiding long growth phases during which phage present in the wine could multiply and inhibit MLF. The risk of phage infection is also reduced by good sanitization of equipment which comes in contact with the wine.

3.2.5 Spoilage by Lactic Acid Bacteria

Production of Acetic Acid. Under aerobic conditions homo- and heterofermentative LAB can produce increased amounts of acetic acid (Kandler 1983; Lucey and Condon 1984, 1986; Murphy et al. 1985; Sedewitz et al. 1984). Growth of the types of LAB and their metabolism must be controlled by limiting the amount of oxygen available and by controlling juice and wine pH.

Formation of Haze and CO_2. Haze and excessive amounts of CO_2 reduce the quality of wine. Growth of LAB in the wine after bottling forms a haze and sediment in the bottle. In addition, heterolactic LAB produce CO_2 from the metabolism of carbohydrates. Sterile filtration and bottling under aseptic conditions avoid the contamination of the bottled wine with LAB and their secondary growth.

Also, additions of SO_2, sorbate, and fumarate suppress growth of LAB (Troost 1972; Amerine et al. 1980; Davis et al. 1985a; Wibowo et al. 1985).

Ropyness in Wine. Polysaccharides produced by pediococci and *Leuconostoc* can increase the viscosity of the wine (Dittrich 1977). Strains of *Leuconostoc* vary in the amount of polysaccharide produced; not all leuconostocs produce polysaccharides from sucrose (Fornachon 1964). It is not known under conditions strains of *Lc. oenos* used in starter cultures produce polysaccharides in wine.

4 Conclusion

Growth and metabolism of yeast and bacteria in juice and wine are controlled by the strongly growth-limiting conditions of low pH, low temperature, lack of oxygen, and the presence of various inhibitory substances.

For successful control of the winemaking process, it is important to understand the regulating mechanisms governing the growth and metabolism of yeast and bacteria. By manipulating the conditions in juice or wine, the winemaker can control what type of yeast and bacteria conduct the fermentation and the fermentation rate. Another very powerful tool for fermentation control is the use of yeast and bacterial starter cultures. In starter cultures, most desirable microorganisms can be selected which dominate the fermentation and complete it within a predictable time period.

References

Akuzawa R, Ito O, Yokoyama K (1983) Two types of intracellular proteinase from dairy lactic acid bacteria. Jpn J Zootechn Sci 54:685–689

Amachi T (1975) Chemical structure of a growth factor (TJF) and its physical significance for malo-lactic bacteria. In: Carr JG, Cutting CV, Whiting GC (eds) Lactic acid bacteria in beverages and food. Academic Press, London, pp 103–118

Amachi T, Yoshizumi H (1969) Studies on the bacteria isolated from wine. Part V. Isolation and properties of the growth factor from tomato juice for a bacterium inducing malo-lactic fermentation. Agric Biol Chem 33:139–146

Amerine MA, Kunkee RE (1968) Microbiology of winemaking. Annu Rev Microbiol 22:323–358

Amerine MA, Ough CS (1980) Methods for analysis of musts and wines. John Wiley, New York

Amerine MA, Berg HW, Kunkee RE, Ough CS, Singelton VL, Webb AD (1980) The technology of winemaking. 4th edn. AVI Publishing, Westport

Andreasen AA, Stier TJB (1954) Anaerobic nutrition of *Saccharomyces cerevisiae*. I. Ergosterol requirement for growth in a defined medium. J Cell Comp Physiol 43:271–284

Atkinson B, Daoud JS (1976) Microbial flocs and flocculation in fermentation process engineering. In: Ghose TK, Fiechter A, Blakeborough N (eds) Advances in biochemical engineering, vol 4. Springer, Berlin Heidelberg New York, pp 41–124

Ayres JC, Mundt JO, Sandine WE (1980) Microbiology of foods. Freeman, San Fancisco

Barnet YM, Daft MJ, Stewart WDP (1984) The effect of suspended particulate material on cyanobacteria-cyanophage interactions in liquid culture. J Appl Bacteriol 56:109–115

Barnett JA, Payne RW, Yarrow D (1983) Yeasts: Characteristics and identification. Cambridge Univ Press, Cambridge

Barney MC, Jansen GP, Helbert JR (1980) Use of spheroblast fusion and genetic transformation to introduce dextrin utilization into *Saccharomyces uvarum*. ASBC 38:1–5

Barre P (1969) Taxonomie numérique de lactobacilles isolés du vin. Arch Mikrobiol 68:74–86

Beelman RB Gallander JF (1970) The effect of grape skin treatments on induced malo-lactic fermentation in Ohio wines. Am J Enol Vitic 21:193–200

Beelman RB, Mc Ardle FJ, Duke GR (1980) Comparison of *Leuconostoc oenos* strains ML34 and PSU-1 to induce malo-lactic fermentation in Pennsylvania red table wines. Am J Enol Vitic 31:269–276

Benda I (1982) Wine and brandy. In: Reed G (ed) Prescott and Dunn's industrial microbiology. AVI Publishing Westport, pp 293–402

Bevan EA, Makower M (1963) The physical basis of the killer character in yeast. Proc 11th Int 1963 Congr Gent 1:202–203

Boulton R (1980) The prediction of fermentation behavior by a kinetic model. Am J Enol Vitic 31:40–45

Bousbouras GE, Kunkee RE (1971) Effect of pH on malo-lactic fermentation in wine. Am J Enol Vitic 22:121–126

Bradley DE (1971) A comparative study of the structure and biological properties of bacteriophages. In: Maramorosch K, Kurstax K (eds) Comparative virology. Academic Press, New York, pp 208–250

Brechot P, Chauvet J, Dupuy P, Croson M, Rabatu A (1971) Acide oléanolique facteur de croissance anaérobique de la lavure du vin. Ann Technol Agric 20:103–110

Brechot P, Chauvet J, Dubois C, Dupuy P (1984) Substrates used by the malo-lactic bacteria for their growth in wine. Proceedings of the 34th Annual Meeting, San Diego, CA, 22 June 1984. Am Soc Enol Vitic, Davis, CA

Burroughs LF, Sparks AH (1973 a) Sulphite-binding power of wines and ciders. I. Equilibrium constants for the dissociation of carbonyl bisulphite compounds. J Sci Food Agric 24:187–198

Burroughs LF, Sparks AH (1973 b) Sulphite-binding power of wines and ciders. II. Theoretical consideration and calculation of sulphite-binding equilibria. J Sci Food Agric 24:199–206

Bussey H (1972) Effect of yeast killer factor on sensitive cells. Nature, New Biol 235:73–75

Carr JG, Davis PA, Sparks AH (1976) The toxicity of sulphur dioxide towards certain lactic acid bacteria from fermented apple juice. J Appl Bacteriol 40:201–212

Caspritz G, Radler F (1983) Malolactic enzyme of *Lactobacillus plantarum*, purification, properties, and distribution among bacteria. J Biol Chem 258:4907–4910

Castino M, Usseglio-Tomasset L, Gandini A (1975) Factors which affect the spontaneous initiation of the malo-lactic fermentation in wines. The possibility of transmission of inoculation and its effect on organoleptic properties. In. Carr JG, Whiting CV, Cutting GC (eds) Lactic acid bacteria in beverages and food. Academic Press, London, pp 139–148

Cazelles O, Gnaegi F (1982) Enquête sur l'importance pratique du problème des bactériophages dans le vin. Rev Suisse Vitic Arboric Hortic 14:267–270

Charpentier Y (1954) Contribution a l'étude biochimique des facteurs de l'acidité des vins. Ann Technol Agric 3:89–167

Cofran DR, Meyer J (1970) The effect of fumaric acid on malo-lactic fermentation. Am J Enol Vitic 21:189–192

Costello PJ, Morrison GJ, Lee TH, Fleet GH (1983) Numbers and species of lactic acid bacteria in wines during vinification. Food Technol Aust 35:14–18

Costello PJ, Monk PR, Lee TH (1985) An evaluation of two commercial *Leuconostoc oenos* strains for induction of malolactic fermentation under winery conditions. Food Technol Aust 37:21–23, 30

Cottrell THE, McLellan MR (1986) The effect of fermentation temperatuve on chemical and sensory characteristics of wines from seven wine grape cultivars grown in New York State. Am J Enol Vitic 37:190–194
Crowell EA, Guymon JF (1975) Wine constituents arising from sorbic acid addition, and identification of 2-ethoxyhexa-3,5-diene as source of the geranium-like off-odor. Am J Enol Vitic 26:97–102
Davis CR (1985) Taxonomy and ecology of lactic acid bacteria. In: Lee TH (ed) Malolactic fermentation. Proc Seminar, 16 August 1984, Melbourne. The Australian Wine Research Institute, Glen Osmond, South Australia, pp 3–17
Davis CR, Wibowo D, Eschenbruch R, Lee TH, Fleet GH (1985 a) Practical implications of malolactic fermentation: A review. Am J Enol Vitic 36:290–301
Davis CR, Silveira NFA, Fleet GH (1985 b) Occurrence and properties of bacteriophages of *Leuconostoc oenos* in Australian wines. Appl Environ Microbiol 50:872–876
Davis CR, Wibowo D, Lee TH, Fleet GH (1986) Growth and metabolism of lactic acid bacteria during fermentation and conservation of some Australian wines. Food Technol Aust 38:35–40
Demel RA, De Kruyff B (1976) The function of sterols in membranes. Biochim Biophys Acta 457:109–132
de Menezes BTJ, Splittstoesser DF, Stamer JR (1972) Induced malo-lactic fermentations of New York State wines. NY Food Life Sci 1:24–26
Dittrich HH (1964) Über die Glycerinbildung von *Botrytis cinerea* auf Traubenbeeren und Traubenmosten sowie über den Glyceringehalt von Beeren- und Trockenbeerenausleseweinen. Wein Wiss 19:12–20
Dittrich HH (1977) Mikrobiologie des Weines. Handbuch der Getränketechnologie. Ulmer, Stuttgart
Dohman TP (1982) Characterization of Oregon-derived malo-lactic bacteria, fermentation properties and storage stability. MS thesis. Oregon State University, Corvallis, OR
Donnelly DM (1977) Airborne microbial contamination in a winery bottling room. Am J Enol Vitic 28:176–181
Du Plessis L de W (1964) Degradation of citric acid and L-malic acid by lactic acid bacteria from dry wines. S Afr J Agric Sci 7:31–42
Exterkate FA (1984) Location of peptidases outside and inside the membrane of *Treptococcus cremoris*. Appl Environ Microbiol 47:177–183
Feuillat M, Bidan P, Rosier Y (1977) Croissance des bactéries lactiques à partir des pricipaux constituants azotes du vin. Ann Technol Agric 26:435–447
Feuillat M, Brillant G, Rochard J (1980) Mise en évidence d'une production de protéases exocellulaires par les levures au cours de la fermentation alcoolique du moût de raisin. Connaiss Vigne Vin 14:37–52
Fleet GH, Lafon-Lafourcade S (1984) The ecology of yeasts and lactic acid bacteria accociated with Bordeaux wines. In: Lee TH, Somers CS (eds) Advances in viticulture and oenology for economic gain. Proc 5th Aust. Wine Ind Tech Conf, 29 Nov–1 Dec 83. Perth. The Australian Wine Research Institute, Glen Osmond, South Australia
Fleet GH, Lafon-Lafourcade S, Ribereau-Gayon P (1984) Evolution of yeasts and lactic acid bacteria during fermentation and storage of Bordeaux wines. Appl Environ Microbiol 48:1034–1038
Flesch P (1968 a) Morphologie, Stoffwechselphysiologie und Charakterisierung der Malic-Enzym-Aktivität L-Äpfelsäure abbauender Bakterien. Arch Mikrobiol 60:285–302
Flesch P (1968 b) Malic-Enzym-Synthese and pH-Adaptation bei L-Äpfelsäure abbauenden Bakterien. Arch Mikrobiol 64:9–22
Fornachon JCM (1957) The occurrence of malo-lactic fermentation in Australian wines. Aust J Appl Sci 8:120–129
Fornachon JCM (1964) A leuconostoc causing malo-lactic fermentation in Australian wines. Am J Enol Vitic
Garvie EI (1967) *Leuconostoc oenos* sp. nov. J Gen Microbiol 48:431–438
Garvie EI (1981) Sub-divisions within the genus *Leuconostoc* as shown by RNA/DNA hybridization. J Gen Microbiol 127:209–212

Garvie EI, Farrow JAE (1980) The differentiation of *Leuconostoc oenos* from non-acidophilic species of leuconostoc, and the identification of five strains from the American Type Culture Collection. Am J Enol Vitic 31:154–157

Gerhardt PH, Murray RGE, Costilow RN, Nester EW, Wood WA, Krieg NR, Phillips GB (1981) Manual of methods for general bacteriology. American Society for Microbiology, Washington

Gini B, Vaughn RH (1962) Characteristics of some bacteria associated with the spoilage of California dessert wines. Am J Enol Vitic 13:20–31

Gnaegi F, Aerny J, Bolay A, Crettenand J (1983) Influence des traitements viticoles antifongiques sur la vinification et la qualité du vin. Rev Suisse Vitic Arboric Hortic 15:243–250

Gnaegi F, Cazelles O, Sozzi T, d'Amico N (1984) Connaissance sur les bactériophages de *Leuconostoc oenos* et progrès dans la maitrîse da la fermentation malolactique des vins. Rev Suisse Vitic Arboric Hortic 16:59–65

Gockowiak H (1987) Killer yeast in wine-making. In: Lee TH (ed) Proceedings of the 6th Australian Wine Technical Conference. The Australian Wine Research Institute, Glen Osmond, South Australia

Guijarro JM, Lagunas R (1984) *Saccharomyces cerevisiae* does not accumulate ethanol against a concentration gradient. J Bacteriol 160:874–878

Hara S, Iimura Y, Otsuka K (1980) Breeding of useful killer wine yeasts. Am J Enol Vitic 31:28–33

Hayman DC, Monk PR (1982) Starter culture preparation for the induction of malolactic fermentation in wine. Food Technol Aust 34:14–18

Heard GM, Fleet GH (1986) Occurrence and growth of yeast species during the fermentation of some Australian wines. Food Technol Aust 38:22–25

Henderson RCA, Cox BS, Tubb R (1985) The transformation of brewing yeasts with a plasmid containing the gene for copper resistance. Curr Genet 9:133–138

Henick-Kling T (1983) Comparison of Oregon-derived malolactic bacteria in pilot scale wine production. MS thesis. Oregon State University, Corvallis, Or

Henick-Kling T (1985) Phage infection of malolactic fermentation. In: Lee YH (ed) Malolactic fermentation. Proc. Seminar, 16 August 1984, Melbourne. The Australian Wine Research Institute, Glen Osmond, South Australia, pp 128–143

Henick-Kling T (1986a) Growth and metabolism of *Leuconostoc oenos* and *Actobacillus plantarum* in wine. PhD thesis. University of Adelaide, South Australia

Henick-Kling T (1986b) Control of malolactic fermentation. Technical Review No 41. The Australian Wine Research Institute, Glen Osmond, South Australia, pp 3–6

Henick-Kling T, Lee TH, Nicholas DJD (1986a) Inhibition of bacterial growth and malolactic fermentation in wine by bacteriophage of *Leuconostoc oenos*. J Appl Bacteriol 61:287–293

Henick-Kling T, Lee TH, Nicholas DJD (1986b) Characterization of the lytic activity of bacteriophages of *Leuconostoc oenos* isolated from wine. J Appl Bacteriol (in press)

Hickey MW, Hillier AJ, Jago GR (1983) Peptidase activities in lactobacilli. Aust J Dairy Technol, September, pp 118–123

Hood AV (1983) Inhibition of growth of wine lactic acid bacteria by acetaldehyde-bound sulphur dioxide. Aust Grapegrow Wine-maker 232:34–43

Houtman AC, du Plessis CS (1986) Nutritional deficiencies of clarified white grape juices and their correction in relation to fermentation. S Afr J Enol Vitic 7:39–46

Imamura T, Kawamoto M, Takaoka Y (1974) Characteristics of a main mash infected the killer yeast in sake brewing and nature of its killer factor. J Ferment Technol 52:293–299

Ito U (1967) The mechanisms of flocculation of flocculent yeast during the wort fermentation. Part III. The action of tannin-protein complex in wort to accelerate the yeast flocculation, and the mechanism of yeast flocculation during the wort fermentation. Mem College Sci Univ Kyoto Ser A 31(2):127–135

Izuagbe YS, Dohman TP, Sandine WE, Heatherbell DA (1985) Characterization of *Leuconostoc oenos* isolated from Oregon wines. Appl Environ Microbiol 50:680–684

Jones M, Pierce JS (1964) Adsorption of amino acids from wort by yeasts. J Inst Brew 70:307–315
Jones RP, Pamment N, Greenfield PF (1981) Alcohol fermentation by yeasts – the effect of environment and other variables. Proc Biochem April/May, pp 42–49
Kandler O (1983) Carbohydrate metabolism in lactic acid bacteria. Antonie Leeuwenhoek 49:209–224
Kandler O, Winter J, Stetter KO (1973) Zur Frage der Beeinflussung der Glucosevergärung durch L-Malat bei *Leuconostoc mesenteroides*. Arch Mikrobiol 90:65–75
Kirsop BH (1974) Oxygen in brewery fermentation. J Inst Brew 80:252–259
Kirsop BH, Brown ML (1972) Some effects of wort composition on the rate and extent of fermentation by brewing yeasts. J Inst Brew 78:51–57
Kolstad J, Law BA (1985) Comparative peptide specificity of cell wall, membrane and intracellular peptidases of group N streptococci. J Appl Bacteriol 58:449–456
Kreger-van Rij NJW (1984) The yeasts, a taxonomic study. 3rd Ed, Elsevier Science Publ, Amsterdam
Krumperman PH, Vaughn RH (1966) Some lactobacilli associated with decomposition of tartaric acid in wine. Am J Enol Vitic 17:185–190
Kunkee RE (1967) Malo-lactic fermentation. Adv Appl Microbiol 9:235–279
Kunkee RE (1974) Malo-lactic fermentation in winemaking. In: Webb AD (ed) Chemistry of winemaking. Adv Chem Ser 137:151–170
Kunkee RE (1975) A second enzymatic activity for the decomposition of malic acid by malo-lactic bacteria. In: Carr JG, Cutting CV, Whiting GC (eds) Lactic acid bacteria in beverages and food. Academic Press, London, pp 29–42
Kunkee RE, Amerine MA (1970) Yeasts in winemaking. In: Rose AH, Harrison JS (eds) The yeasts 3. Academic Press, London, pp 5–72
Kunkee RE, Goswell RW (1977) Table wines. In: Rose AH (ed) Alcoholic beverages. Economic microbiology, vol 1. Academic Press, London, pp 315–385
Lafon-Lafourcade S (1970) Étude de la dégradation de l'acide malique par les bactéries lactiques non-proliferantes isolées des vins. Ann Technol Agric 19:141–154
Lafon-Lafourcade S (1983) Wine and Brandy. In: Rehm HJ, Reed G (eds) Biotechnology. Verlag Chemie, Weinheim, pp 81–163
Lafon-Lafourcade S (1986) Applied microbiology. Experientia 42:904–914
Lafon-Lafourcade S, Joyeux A (1981) Les bactéries acétiques du vin. Bull OIV 54:803–829
Lafon-Lafourcade S, Peynaud E (1961) Composition azotée des vins en fonction des conditions de vinification. Ann Technol Agric 10:143–160
Lafon-Lafourcade S, Peynaud E (1974) Sur l'action anti-bactérienne de l'anhydride sulfureux sous forme libre et sous forme combinée. Connaiss Vigne Vin 10:187–203
Lafon-Lafourcade S, Ribereau-Gayon P (1984) Les altérations des vins par les bactéries acétiques et les bactéries lactiques. Connaiss Vigne Vin 18:67–82
Lafon-Lafourcade S, Dubourdieu D, Hadjinicolaou D, Ribereau-Gayon P (1980) Incidence des conditions de travail des vendanges blanches sur la clarification et la fermentation des moûts. Connaiss Vigne Vin 14:97–109
Lafon-Lafourcade S, Carre E, Ribereau-Gayon P (1983 a) Occurrence of lactic acid bacteria during different stages of vinification and conservation of wines. Appl Environ Microbiol 46:874–880
Lafon-Lafourcade S, Carre E, Lonvaud-Funel A, Ribereau-Gayon P (1983 b) Induction de la fermentation malolactique des vins par inoculation d'une biomasse industrielle congelée de *L. oenos* après réactivation. Connaiss Vigne Vin 17:55–71
Larue F, Lafon-Lafourcade S, Ribereau-Gayon P (1980) Relationship between the sterol content of yeast cells and their fermentation activity in must. Appl Environ Microbiol 39:808–811
Law BA, Kolstad J (1983) Pectolytic systems in lactic acid bacteria. Antonie Leeuwenhoek 49:225–245
Lefebvre A, Riboulet JM, Boidron JN, Ribereau-Gayon P (1983) Incidence des microorganisms du liège sur les altérations olfactives du vin. Sci Aliment 3:265–278
London J, Meyer EY (1969) Malate utilization by a group D *Streptococcus*: physiological properties and purification of an inducible malic enzyme. J Bacteriol 98:705–711

London J, Meyer EY, Kulczyk SR (1970) Allosteric control of a *Lactobacillus* malic enzyme by glycolytic intermediate products. Biochim Biophys Acta 212:512–514
London J, Meyer EY, Kulczyk SR (1971) Comparative biochemical and immunological study of malic enzyme from two species of lactic acid bacteria: evolutionary implications. J Bacteriol 106:126–137
Lonvaud-Funel A, Strasser de Saad AM (1982) Purification and properties of a malolactic enzyme from a strain of *Leuconostoc mesenteroides* isolated from grapes. Appl Environ Microbiol 43:357–361
Lucey CA, Condon S (1984) Stimulation by aeration of growth rate and biomass yield of leuconostocs. Ir J Food Sci Technol 8:154–155
Lucey CA, Condon S (1986) Active role of oxygen and NADH oxidase in growth and energy metabolism of *Leuconostoc*. J Gen Microbiol 132:1789–1796
Marcis BJ, Markakis P (1974) Transport and toxicity of sulphur dioxide in *Saccharomyces cerevisiae* var. *ellipsoideus*. J Sci Food Agric 25:21–29
Maret R, Sozzi T (1977) Flore malolactique de moûts et de vins du Canton Du Valais (Suisse). I. Lactobacilles et pédiocoques. Ann Technol Agric 27:255–273
Maret R, Sozzi T (1979) Flore malolactique de moûts et de vins du Canton du Valais (Suisse). II. Evolution des populations de lactobacilles et de pédiocoques au cours de la vinification d'un vin blanc (un Fendant) et d'un vin rouge (une Dole) Ann Technol Agric 28:31–40
Maret R, Sozzi T, Schellenberg D (1979) Flore malolactique de moûts et de vins du Canton du Valais (Suisse). III. Les leuconostocs. Ann Technol Agric 28:41–55
Martiniere P, Sapis JC, Ribereau-Gayon P (1974) Évolution du nombre de bactéries lactiques vivantes au cours de la vinification et de la conservation des vins. CR Seances Acad Agric Fr 60:255–261
Maule AP, Thomas PD (1973) Strains of yeast lethal to brewery yeasts. J Inst Brew 79:137–141
Mayer K (1974) Mikrobiologisch und kellertechnisch wichtige neue Erkenntnisse in bezug auf den biologischen Säureabbau. Schweiz Z Obst Weinbau 110:291–297
Mayer K, Pause G (1970) Überprüfung einiger Weinhefen auf Alkohol- und Glycerinbildung. Schweiz Z Obst- Weinb 106:490–492
Mayer K, Vetsch U (1973) pH und biologischer Säureabbau in Wein. Schweiz Z Obst Weinbau 109:635–639
Mayer K, Vetsch U (1978) Biologischer Säureabbau in Wein: Ungünstige Selektivwirkung der schwefligen Säure. Schweiz Z Obst Weinbau 114:642–647
Melamed N (1962) Determination des sucres résiduels des vins, leur relation avec la fermentation malolactique. Ann Techn Agric 11:5–11, 107–119
Mitchell P (1966) Chemiosmotic coupling in oxidative and photosynthetic phosporylation. Biol Rev 41:445–502
Mitchell P (1981) From black box bioenergetics to molecular mechanisms: vectorial ligand-conduction mechanisms in biochemistry. In: Skulachev VP, Hinkle PD (eds) Chemiosmotic proton circuits in biological membranes. Addison Welsey, Reading, MA, pp 611–633
Monk PR (1982) Effect on nitrogen and vitamin supplements on yeast growth and rate of fermentation of Rhine Riesling grape juice. Food Technol Aust 34:328–332
Monk PR (1986) Rehydration and propagation of active dry wine yeast. Aust Wine Ind J 1:3–5
Monk PR, Costello PJ (1984) Effect of ammonium phosphate and vitamin mixtures on yeast growth in preserved grapejuice. Food Technol Aust 36:25–28
Monk PR, Storer RJ (1986) The kinetics of yeast growth and sugar utilization in tirage: the influence of different methods of starter culture preparation and inoculation levels. Am J Enol Vitic 37:72–76
Morenzoni R (1974) The enzymology of malo-lactic fermentation. In: Webb AD (ed) Chemistry of winemaking. Ser 137. Am Chem Soc, Washington, DC, pp 171–183
Müller-Späth H, Loescher Th (1985) Positive und negative Einflüsse auf den Metabolismus der Hefe bei der Wein- und Sektbereitung. Weinwirtschaft Technik 10:326–335
Murphy MG, O'Connor L, Condon S (1985) Oxygen-depened lactate utilization by *Lactobacillus plantarum*. Arch Microbiol 141:75–79

Nagodawithana TW, With H, Cutaia AY (1977) Study of the feed-back control on selected enzymes of the glycolytic pathway. J Am Soc Brew 35:179–193
Naumova GI, Naumov TI (1973) Comparative genetics of yeast. XIII. Comparative study of killer strains of *Saccharomyces* from different collections. Genetika 9:140
Naumov GI, Tyurina LV, Bur'yan NI, Naumova TI (1973) Wine-making, an ecological niche of type K2 killer *Saccharomyces*. Biol Nauki 16:103–107
Ouchi K, Nishiya T, Akiyama H (1983) UV-killed protoplast fusion as a method for breeding killer yeasts. J Ferment Technol 61:631–635
Palfree GE, Bussey H (1979) Yeast killer toxin: purification and characterization of the protein from *Saccharomyces cerevisiae*. Eur J Biochem 93:487–493
Peynaud E (1968) Etudes récentes sur les bactéries du vin. Fermentation et vinification, vol I, Institut national de la recherche agromonique, Paris, pp 219–256
Peynaud E, Domercq S (1955) Sur les espèces de levures fermentant sélectivement le fructose. Ann Pasteur 89:346–351
Peynaud E, Domercq S (1961) Études sur les bactéries lactiques des vins. Ann Technol Agric 10:43–60
Peynaud E, Domercq S (1968) Études sur les quatre cents souches de coques hétérolactiques isolés de vins. Ann Inst Pasteur Lille 19:159–170
Peynaud E, Sapis-Domercq S (1970) Étude de deuxcentcinquantes souches de bacilles hétérolactiques isolés de vins. Arch Mikrobiol 70:348–360
Pfeiffer P, Radler F (1983) Purification and characterization of extracellular and intracellular killer toxin of *Saccharomyces cervisiae* strain 28. J Gen Microbiol 128:2699–2706
Pfeiffer P, Radler F (1984) Comparison of the killer toxin of several yeasts and the purification of a toxin of type K2. Arch Microbiol 137:357–361
Philliskirk G, Young TW (1975) The occurrence of killer character in yeasts of various genera. Antonie Leeuwenhoek J Microbiol Serol 41:147–151
Pilone GJ, Kunkee RE, Webb AD (1966) Chemical characterization of wines fermented with various malo-lactic bacteria. Appl Microbiol 14:608–615
Pilone GJ, Rankine BC, Pilone DA (1974) Inhibiting malo-lactic fermentation in Australian dry red wines by adding fumaric acid. Am J Enol Vitic 25:99–107
Pukrushpan L, Kunkee RE (1977) Role of malate in the metabolism of malo-lactic bacteria. Abstracts of the Annual Meeting of the ASM 1977. Am Soc Microbiol Washington, DC, No 77
Radler F (1958a) Untersuchung des biologischen Säureabbaus im Wein. III. Die Energiequelle der Äpfelsäure-abbauenden Bakterien. Arch Mikrobiol 31:224–230
Radler F (1958b) Untersuchung des biologischen Säureabbaus im Wein. IV. Über Faktoren, die das Wachstum der Äpfelsäure-abbauenden Bakterien beeinflussen. Vitis 1:288–297
Radler F (1966) Die mikrobiologischen Grundlagen des Säureabbaus im Wein. Zentralbl Bakteriol Parasiten Abt II 120:237–287
Radler F (1975) Die organischen Säuren im Wein und ihr mikrobieller Stoffwechsel. Dtsch Rundsch 71:20–26
Radler F (1986) Microbial biochemistry. Experientia 42:884–893
Radler F, Broehl K (1984) The metabolism of several carboxylic acids by lactic acid bacteria. Z Lebensm Unters Forsch 179:228–231
Radler F, Schoenig I (1977) Entkeimungsfiltration von Milchsäurebakterien aus Wein mit Filterschichten verschiedener Stoffzusammensetzung. Weinwirtschaft 27:752–760
Radler F, Yannissis C (1972) Weinsäureabbau bei Milchsäurebakterien. Arch Mikrobiol 82:219–239
Ribereau-Gayon J, Peynaud E, Ribereau-Gayon P, Sudraud P (1975) Sciences et techniques du vin, vol 2. Dunod, Paris
Rice AC, Mattick LR (1970) Natural malo-lactic fermentation in New York State wines. Am J Enol Vitic 21:145–152
Rose AH (1977) In: Forsander O, Eriksson K, Oura E, Jounela-Eriksson P (eds) In: Alcohol industry and research. Alkon Keskaslavboratorio, Helsinki, p 179
Russell I, Stewart GG (1985) Valuable techniques in the genetic manipulation of industrial yeast strains. ASBC J 43:84–90

Sablayrolles JM, Barre P (1986) Évaluation des besoins en oxygène de fermentations alcooliques en conditions oenologiques simulées. Sci Aliment 6:373–383
Sapis-Domercq S, Bertrand A, Mur F, Sarre C (1976) Influence des produits de traîtement de la vigne sur la microflore levurienne. Connaiss Vigne Vin 10:369–390
Sapis-Domercq S, Bertrand A, Joyeux A, Lucmaret V, Sarre C (1978) Etude de l'influence des produits de traitements de la vigne sur la microflore des raisins et des vins. Experimentation 1977. Comparaison avec les resultats de 1976 et 1975. Connaiss Vigne Vin 12:245–275
Schanderl H (1969) Zur Frage der Herkunft der Weinbakterien. Dtsch Weinbau-Jahrb 20:142–147
Schuetz M, Radler F (1973) Das Malatenzym von *Lactobacillus plantarum* und *Leuconostoc mesenteroides*. Arch Mikrobiol 91:183–202
Schuetz M, Radler F (1974) Das Vorkommen von Malatenzym und Malo-Lactat-Enzym bei verschiedenen Milchsäurebakterien. Arch Mikrobiol 96:329–339
Sedewitz B, Schleifer KH, Goetz F (1984) Physiological role of pyruvate oxidase in the aerobic metabolism of *Lactobacillus plantarum*. J Bacteriol 160:462–465
Shimazu Y, Watanabe M (1979) Malolactic fermentation in sparkling wine. J Ferment Technol 57:512–518
Silver J, Leighton T (1982) Control of malolactic fermentation in wine. 1. Mechanisms of fumaric acid inhibition and considerations concerning malate metabolism by lactic acid bacteria. In: Webb AD (ed) Proc Univ Calif Davis Grape Wine Cent Symp, pp 170–175
Skatrud PL, Jaeck DM, Kot EJ, Helbert JR (1980) Fusion of *Saccharomyces uvarum* with *Saccharomyces cerevisiae:* genetic manipulation and reconstruction of a brewers yeast. ASBC 38:49–53
Somers TC, Wescombe LG (1982) The critical role of SO_2 during vinification and conservation. Aust Grapegrow Winemaker 220:68–74
Sozzi T, Gnaegi F (1984) The problem of bacteriophages in wine. In: Lee TH, Somers TC (ed) Advances in viticulture and oenology for economic gain. Proc 5th Aust Wine Ind Tech Conf, 29 November–1 December 1983, Perth, pp 391–402
Sozzi T, Gnaegi F, d'Amico N, Hose H (1982) Difficultées de fermentation malolactique du vin dues à des bacteriophages de *Leuconostoc oenos*. Rev Suisse Vitic Arboric Hortic 14:17–23
Spencer JFT, Spencer DM (1977) Hybridization of non-sporulating and weakly sporulating strains of brewers and destillers yeasts. J Inst Brew 83:287–289
Spettoli P, Nuti MP, Zamorani A (1984) Properties of malolactic activity purified from *Leuconostoc oenos* ML34 by affinity chromatography. Appl Environ Microbiol 48:900–901
Sponholz WR (1988) Traubenmost. In: Würdig G, Woller R (eds) Chemie des Weines. Ulmer, Stuttgart
Sponholz WR, Lacher M, Dittrich HH (1986) Die Bildung von Alditolen durch die Hefen des Weines. Chem Mikrobiol Technol Lebensm 10:19–24
Stamer JR (1979) The lactic acid bacteria: microbes of diversity. Food Technol 33:60–65
Stewart GG, Russell I (1985) The biology of *Saccharomyces*. In: Demain AL, Solomon NA (eds) Biology of industrial microorganisms. Biotechnology Series. The Benjamin/Cummings, Menlo Park, pp 511–536
Strathern JN, Jones EW, Broach JR (1981) The molecular biology of the yeast *Saccharomyces* – life cycle and inheritance. Cold Spring Harbor Laboratory. Cold Spring Harbor, New York
Strathern JN, Jones EW, Broach JR (1982) the molcular biology of the yeast *Saccharomyces* – metabolism and gene expression. Cold Spring Harbor Laboratories, Cold Spring Harbor, New York
Strehaiano P, Goma G (1983) Effect of initial substrate concentration on two wine yeasts: relation between glucose sensitivity and ethanol inhibition. Am j Enol Vitic 34:1–5
Thomas DS, Hossak JA, Rose AH (1978) Plasmamembrane lipid composition and ethanol tolerance in *Saccharomyces cerevisiae*. Arch Microbiol 117:239–245
Thornton RJ (1978) Investigations of the genetics of foaming in wine yeasts. Eur J Appl Microbiol Biotechnol 5:103–107

Thornton RJ (1980) Genetic investigation and modification of wine yeast characteristics. Grape and wine centennial symposium proceedings, June 1980. University of California Davis, Davis, CA
Thornton RJ (1982) Selective hybridisation of pure culture wine yeasts. II. Improvement of fermentation efficiency and inheritence of SO_2 tolerance. Eur J Appl Microbiol Biotechnology 14:159–164
Thornton RJ (1983) New yeast strains from old – the application of genetics of wine yeasts. Food Technol Aust 35:46–50
Thornton RJ (1985) The introduction of flocculation into a homothallic wine yeast. A practical example of the modification of winemaking properties by the ues of genetic techniques. Am J Enol Vitic 36:47–49
Thornton RJ (1986) Genetic characterization of New Zealand and Australian wine yeasts. Occurrence of killer systems and homothallism. Antonie Leeuwenhoek 52:97–103
Traverso-Rueda S (1980) A quantification of the stimulatory effect of sterols on the growth rate of *Saccharomyces cerevisiae* under vinification conditions. PhD Thesis, University of California, Davis
Troost G (1972) Technologie des Weines. 4th edn. Ulmer, Stuttgart
Van Dijken JP, Jonker R, Bruinenberg PM, Bwee Houweling-Tan G, Scheffers WA (1984) Regulation of fermentation capacity in *Saccharomyces cerevisiae*. Antonie Leeuwenhoek 50:87–88
Van Wyk CJ (1976) Malo-lactic fermentation in South African table wines. Am J Enol Vitic 27:181–185
Vaughn RH (1955) Bacterial spoilage of wines with special reference to California conditions. Adv Food Res 6:67–108
Webb RB, Ingraham JL (1960) Induced malo-lactic fermentations. Am J Enol Vitic 11:59–63
Weiller HG, Radler F (1970) Milchsäurebakterien aus Wein und von Rebenblättern. Zentralbl Bakteriol Parasitenk Infektionskr Hyg Abt 2 124:707–732
Weiller HG, Radler F (1972) Vitamin- und Aminosäurebedarf von Milchsäurebakterien aus Wein und von Rebenblättern. Mitt Klosterneuburg 22:4–18
Weiller HG, Radler F (1976) Über den Aminosäurestoffwechsel von Milchsäurebakterien aus Wein. Z Lebensm Unters Forsch 161:259–266
Whiting GC, Coggins RA (1964) Metabolism of malate and citrate by lactic acid bacteria. Rep Agric Hortic Res Sta Bristol for 1963, pp 157–167
Wibowo D, Eschenbruch R, Davis DR, Fleet GH, Lee TH (1985) Occurrence and growth of lactic acid bacteria in wine: a review. Am J Enol Vitic 36:302–313
Wilkinson J (1986) Secondary fermentation and maturation. In: Lee TH, Lester DC (eds) Production of sparkling wine by the methode champenoise. Proc. Seminar Aust Soc Vitic Oenol 14 Nov 85. The Australian Wine Research Institute, Glen Osmond, South Australia, pp 125–131
Williams SA, Hodges RA, Strike TL, Snow R, Kunkee RE (1984) Cloning the gene for malolactic fermentation of wine from *Lactobacillus delbrueckii* in *Escherichia coli* and yeasts. Appl Environ Microbiol 47:288–293
Wood DR, Bevan EA (1968) Studies on the nature of the killer factor produced by *Saccharomyces cerevisiae*. J Gen Microbiol 51:115–126
Wuerdig G, Schlotter H-A, Klein E (1974) Über die Ursachen des sogenannten Geranientones. Allg Dtsch Weinfachztg 110:578–583
Young TW, Philliskirk G (1975) Killer yeasts and fermentation – the effect of killer yeasts in mixed batch and continuous cultures of wort. Eur Brew Conv Proc Congr, pp 337–347

Detection of Illicit Spirits

W. Simpkins

1 Introduction

Excise on alcoholic products constitutes a major source of revenue to most governments throughout the world. In most countries there exists an analytical requirement to distinguish licit from illicit spirits.

In Australia, excise is paid on all spirituous beverages except wine. Illicit spirits can be divided into four main types, all of which which contravene customs legislation. These originate from:

a) Illicit distillation, mostly of traditional "ethnic" products known as grappa, rakija or komovica, slivovica, vodka and arak from grapes, plums and other fruits. Cane sugar and molasses are also used. These are readily available, easily fermented, and the waste products can be disposed of via the sewerage.
b) Extension of authentic spirits with those of type (a) or with commercial grades of ethanol which are duty-free because of their special end use. Denatured industrial alcohol and grape spirit intended for wine fortification do not attract excise.
c) Smuggling and unauthorized removal from bond, for example by robbery, of authentic spirits.
d) Manufacture of spirits in contravention of the regulations such as by the addition of synthetic flavours or more than the permitted level of dry sherry to brandy, or the inclusion of synthetic alcohol from a petroleum source into spirits for human consumption.

In the broadest sense, the definition also covers those spirits on which excise has not been foregone, but which have been prohibited because they are under the minimum legal strength, or have been contaminated with pesticides, "wood alcohol", nitrosamines, lead, radioactivity etc. Responsibility for these spirits generally lies with Health departments.

The chemist's tasks can therefore be seen as having three separate objectives: to determine the anomalies in type (a) and (b) illicit spirits that set them apart from the great number of licit spirits available to the public; to establish that type (c) spirit can be uniquely matched with other existing stock, and differentiated from all other spirits of that type; and to detect the contaminants that make a spirit unpotable. Of these, the last is perhaps easiest to present to a court of law. The programs undertaken to detect illicit spirits are required to be costed to fit into a budgetary framework that is acceptable to Government. The emphasis is therefore on tests which are quick yet effective, so that as many samples might be screened annually as possible. These samples include seizures, bottles purchased at retail outlets perhaps because of their low price or as a result of information or a complaint by industry. As well, they originate from various operations conducted by the Australian Customs Service (ACS) to compare randomly selected bulk and the corresponding retail products for differences in their composition.

2 Major Volatile Congener Profiles

2.1 Volatile Congeners

The congeneric by-products of fermentation co-distil with ethanol, providing a "fingerprint" that can assist in identifying spirit type. The ultimate concentration of congeners within a spirit depends on a wide range of factors, including starting materials, the way that fermentation is performed, and ultimately the distillation process. Inconsistencies in the profile are an indicator of illicit manufacture.

In our laboratory, preliminary testing is limited to the ten most common congeners (see Fig. 1). Levels are expressed, as is conventional, in mg per 100 ml.

2.2 Methods of Analysis

Volatile congeners are usually determined by gas chromatography (GC). A number of stationary phases are suitable. Packed columns, which are robust, have a higher capacity and are less costly than capillary columns, are still generally popular, perhaps in part because of the court's preference for a methodology which appears in standard texts. The AOAC (1984) lists either Carbowax 1500 (23%) on Chromosorb W or glycerol (2%)+1,2,6-hexanetriol (2%) on Gas-Chrom R. Carbowax 20M in combination with a graphitized carbon black support such as Carbopack (Supelco) or Graphpac (Alltech) is also used. Di Corcia (1980) reported the application of PEG 20 on Carbopack to the analysis of whisky. Porous polymer solid phases (Chromosorb 101, Porapak Q and QS and

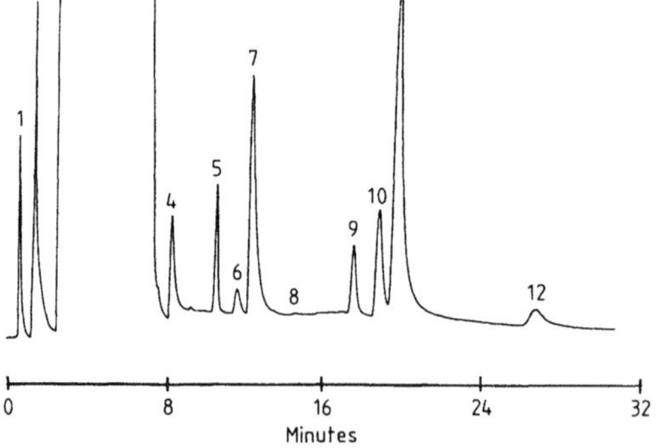

Fig. 1. GC profile of brandy: *1* acetaldehyde; *2* methanol; *3* ethanol; *4* 1-propanol; *5* ethyl acetate; *6* 2-butanol (sec-butanol); *7* 2-methyl-1-propanol (isobutanol); *8* 1-butanol; *9* 3-pentanol (internal standard); *10* 2-methyl-1-butanol (active amyl alcohol); *11* 3-methyl-1-butanol (iso-amyl alcohol); *12* ethyl lactate. Column: 2 m × 2 mm 1.D.Glass, 3% Carbowax 20M on Graphpac 80/100 mesh. Operating conditions: Initial temperature 40° C for 2 min, programmed to 115° C at 5° C min^{-1}, final hold 13 min; carrier gas N_2, 30 ml min^{-1}; Injection volume: 6 µl

Tenax GC) are useful, particularly for determining methanol in marc and fruit brandies where the levels cause overloading of other columns. They will not separate the methylbutanol isomers. As a general rule, retention time, peak shape and peak area are affected by the amount of ethanol in the sample with all packed columns. Calibration error will occur unless standards of the same strength as the sample are used. Bonded phase capillary columns suitable for congener determinations have eliminated many of the shortcomings previously experienced with liquid-coated glass and fused silica, such as susceptibility to oxidation and column bleed due to solvation. MacNamara (1984) has reported the separation of 26 constituents in malt whisky in less than 43 minutes using Chrompack's CP Wax 57 CB (50 m × 0.32 mm) capillary column. Congeners may also be determined by high performance liquid chromatography (HPLC). For example, Bio-Rad's "Fruit quality analysis" cation exchange resin, and Aminex HPX ion exclusion columns will separate many of them, and provide additional useful information on nonvolatile materials such as sugars and wine constituents, if they are present.

Review of Previous Work. Most investigators concerned with the detection of illicit spirits have used GC to establish a reference library of congener levels in authentic spirits. From these, they have evolved classification rules for spirit type, mostly from selected congener ratios. Bober and Haddaway (1963) suggested that Jamaica rums contained twice the 1-propanol found in other rums. In his two classic papers, Singer (1966a, 1966b) reported that the proportions of C_3, C_4 and C_5 alcohols expressed as the ratios of isopentanol/isobutanol and n-propanol/isobutanol fell into distinct ranges typical of the particular spirit type, with the 2-methyl-1-butanol/3-methyl-1-butanol ratio being particularly characteristic. He reported this latter ratio to be 0.19–0.24 for good quality cognac, 0.24–0.26 for inferior cognac, 0.29–0.33 for marc de Bourgogne and 0.36–0.44 for whisky. Cavazza (1975) reported that this ratio would also separate rums from brandies and whiskies. Hall (1976) reported that Scotch, malt, grain, rye and Canadian whisky types could be differentiated by reference to their iso-pentanol/isobutanol ratios, drawing on results obtained from his own analyses and those of Schoeneman and Dyer (1968, 1973) and Shoeneman et al. (1971). His chief concern was to detect the fraudulent substitution of expensive liquors by a cheaper variety. However, in formulating his classification rules, Hall rejected "a few of Schoeneman's individual results... on statistical grounds as 'outliers'". Mesley et al. (1975) examined approximately 600 samples of diverse origins and concluded that rums could be distinguished from all other spirits with a 98% success rate because (1) their methanol content was less than 23 g Hl^{-1}; (b) their n-propanol/2-methylpropanol ratio was less than 0.7, (c) their 2-methylbutanol/3-methylbutanol ratio was less than 0.38, and (d) their ethyl acetate/methanol ratio was greater than 1.8. Hogben and Mular (1976) reported that a minimum amyl/i-butyl alcohol ratio of 1.6 and a maximum i-butyl/n-propyl alcohol ratio of 5.0 appeared to be the most promising criteria in attempting to assess the authenticity of Australian brandies, provided that these limits were used in a complementary manner and not independently. Australia's large grape harvest provides cheap starting materials for illicit "brandy" production. Ng and Woo (1980) applied Hogben's crite-

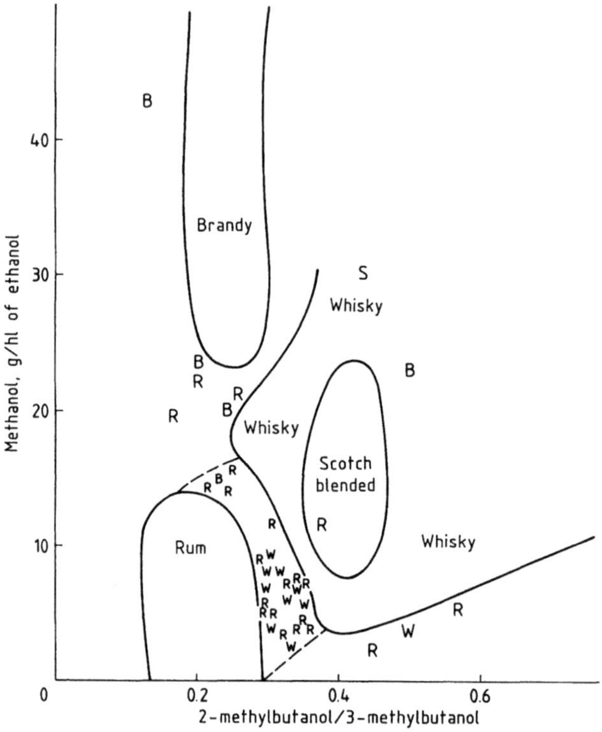

Fig. 2. Spirit types in relation to methanol content and ratio of pentanols. *B* brandy; *R* rum; *S* Scotch whisky; *W* whisky. (Lisle et al. 1978) (Reproduced by permission of the Controller of Her Britannic Majesty's Stationery Office)

ria to Chinese products to detect illicit "samsoo" of rice origin. Lisle et al. (1978) published data for 695 spirits representing the main types available in the United Kingdom. Their conclusions were that a methanol level of more than 50 g Hl^{-1} distinguished most brandies from whisky and rum; a two-dimensional plot of methanol against the 2-methylbutanol/3-methylbutanol ratio separated 90% of the remaining brandies from whisky and rum (see Fig. 2). In particular, they found that neither Hall's nor Cavazza's criteria, both of which had been formulated from relatively small sample populations and restricted to only a few spirit types, would extend to the samples drawn in their own survey. However, if a sample were known to be whisky, consideration of the 1-propanol content in relation to the ratio of the total pentanols/2-methyl-1-propanol was a useful guide to its origin (see Fig. 2). Interestingly, their survey revealed a number of spirits which were rum by name, odour and taste, but had a pattern of higher alcohols typical of whiskies. It has been demonstrated by Suomalainen (1971) that yeast type is a major factor in determining the quantity and proportion of congeners.

Other reports which list congener profiles include those on Scotch (Postel and Adam 1977), Irish Whisky (Postel and Adam 1978), pear brandy (Hildenbrand 1982), plum brandy (Filajdic and Djukovic 1973; Nosko 1974), grappa (Casagrande 1980) and sundry other beverages (Bonte et al. 1978). Nykanen and Suomalainen (1983) list individual congener levels in a number of beverages.

2.3 Identification of Spirits

Recently we surveyed over 1000 authentic spirits sampled by ACS officers in NSW (Simpkins 1985), with the view of applying the criteria listed in previous reports to the interpretation of congener values when screening for illicit spirits. The results were incorporated into a computerized data bank which is summarized in Table 1. We came to the following conclusions:

a) The range of most congeners was significant. For example, methanol, the concentration of which is related to pectin hydrolysis, and which is not easily removed during distillation because it forms an azeotrope with ethanol, was highest in fortifying spirits produced from marc, and in brandies derived from domestic fruits such as plums, peaches, apples, and pears, and lowest in rums. However, six Australian brandies and two fortifying spirits contained less methanol than the maximum found in spirits originating from either cereals or cane sugar. 2-Butanol was present in all fruit brandies, 80% of the grappas and at lower levels, 45% of the imported brandies and 10% of Australian brandies. Some fruit brandies contained very high levels of 1-propanol. Malt whisky and bourbon whisky contained more higher alcohols than any of the blended whiskies. Few blending or fortifying spirits, or the rectified spirits used to make gin and vodka, contained higher alcohols other than 1-propanol. Ethyl lactate was highest in plum brandies, and those products labelled as originating from rice wine. Some differentiations within spirit type were noted. These might have been idiosyncratic of a regional style, or even that of a particular distiller. For example, brandies from Greece and Cyprus differed from both Australian and French brandies because of their paucity of higher alcohols. The vodka produced by one Australian distiller from grain had consistently higher 1-propanol levels than any another brand.

b) Little additional information could be obtained from binary combinations other than those involving methanol, 2-butanol and the methylbutanols. The correlation coefficient between 2-methyl-1-butanol and 2-methyl-1-propanol, and also 2-methyl-1-butanol and 3-methyl-1-butanol, was higher than 0.9 for most types, but a graphical representation showed similar distributions. It is to be expected that a high degree of correlation within spirit types occurs for these congeners because of their close boiling points.
The congener combination involving Hall's criteria separated bourbon from Scotch, as he reported. It also separated most Australian brandies from French. Such a result seems serendipidous and may merely reflect the fact that many of the imported brandies were of the same style. Most other spirit types shared similar distributions (summarized in Table 2).
The methanol to acetaldehyde ratio separated rum (0.1 to 0.3) from most of the blended whiskies but not the malt whiskies.

c) Binary combinations having one congener on the ordinate and a ratio of congeners on the abscissa were effective in two cases only. Most of the whiskies could be typed using Lisle's method (Fig. 3). This more complex approach did not separate whisky from the other spirit types. The only combination which effectively identified spirit type was that originally implied by Singer – a plot

Table 1. Maximum, minimum and median values of major volatile congeners in the main spirit types, expressed as mg 100 ml^{-1} absolute alcohol

Type	No.	MeOH	AcH	1-PrOH	2-MePrOH	2-BuOH	1-BuOH	2-MeBuOH	3-MeBuOH	THA[a]	EtOAc	EtLact
Australian brandy	129	296	50	98	188	6	3	196	616	1031	254	115
		5	1	9	5	–	–	1	4	31	1	–
		68	10	28	39	< 1	< 1	40	160	280	29	7
Imported brandy	103	202	39	88	105	47	7	72	245	453	105	44
		14	1	1	1	–	–	–	–	1	1	–
		73	14	23	52	4	< 1	29	97	227	37	5
Fortifying spirit	147	1200	67	281	469	–	–	140	480	928	660	99
		10	0	–	–	–	–	–	–	–	–	–
		93	6	12	1	–	–	–	–	18	3	< 1
Fruit brandy	54	1076	37	1700	130	218	64	93	351	2462	655	240
		110	1	35	3	1	0	2	7	54	0	8
		560	16	135	47	23	12	29	115	366	152	71
Grappa	44	830	67	104	89	290	13	79	203	654	369	138
		30	2	8	18	–	–	9	24	84	2	–
		214	14	53	64	45	2	44	140	339	46	17
Tequila	30	167	36	98	136	–	–	96	300	560	84	–
		29	2	12	28	–	–	13	51	131	2	–
		57	9	38	66	–	–	35	169	298	18	–
Vodka	39	40	16	95	–	–	–	2	8	–	–	–
		1	1	–	–	–	–	–	–	–	–	–
		6	2	< 1	–	–	–	< 1	< 1	–	–	–
Bourbon	32	28	24	36	125	–	1	173	324	570	135	< 1
		8	2	9	46	–	–	70	166	337	14	–
		16	9	23	79	–	< 1	112	240	463	60	–
Rum (coloured)	62	10	38	134	184	1	5	88	409	775	190	12
		2	4	11	2	–	–	0	0	33	0	–
		5	17	50	28	–	2	17	114	225	59	< 1

Rum	21	10	39	100	48	–	–	25	165	325	158	12
(white)		2	5	4	3	–	–	1	5	13	2	–
		4	12	19	14	–	–	8	34	65	34	< 1
Whisky	184	22	22	246	97	–	–	52	128	322	93	< 1
(Scotch)		3	1	6	26	–	–	6	13	87	1	–
		10	9	36	57	–	–	16	38	154	24	–
Whisky	26	22	15	60	81	–	–	45	104	256	95	< 1
(other)		4	4	4	9	–	–	9	21	43	2	–
		9	8	15	23	–	–	13	31	103	16	–
Whisky	12	11	15	45	126	–	–	70	200	420	91	< 1
(malt)		5	3	32	90	–	–	54	145	336	14	–
		5	10	35	102	–	–	62	160	350	44	–
Gin	53	99	29	98	8	–	–	2	5	92	14	–
		1	3	–	–	–	–	–	–	–	–	–
		4	5	< 1	< 1	–	–	< 1	< 1	< 1	< 1	–

[a] Total higher alcohols.
–, Not detected.

of the methylbutanol isomers, taken collectively, against their ratio. Figure 4a to d shows the extent to which this succeeded. Used in conjunction with methanol and 2-butanol, it enabled classification of most of the main spirit types.

d) Binary combinations involving multiple ratios were ineffective.

Table 2. Distribution ratios of the sum of the methylbutanols to 2-methyl-1-propanol for principal spirit types

	Frequency of ratio (%)							
Range	0–1	1–2	2–3	3–4	4–5	5–6	6–7	>7
Australian brandy	1	2	6	12	18	37	11	13
Imported brandy	20	2	28	32	7	2	6	3
Fortifying spirit	17	9	15	19	7	15	7	4
Fruit brandy	0	2	58	34	4	2	0	0
Grappa	3	5	50	38	2	0	2	0
Tequila	0	0	48	48	4	0	0	0
Bourbon	0	0	13	34	22	22	9	0
Rum	2	0	14	24	33	15	7	5
Blended Scotch	58	39	3	0	0	0	0	0
Whisky (other)	4	39	23	34	0	0	0	0

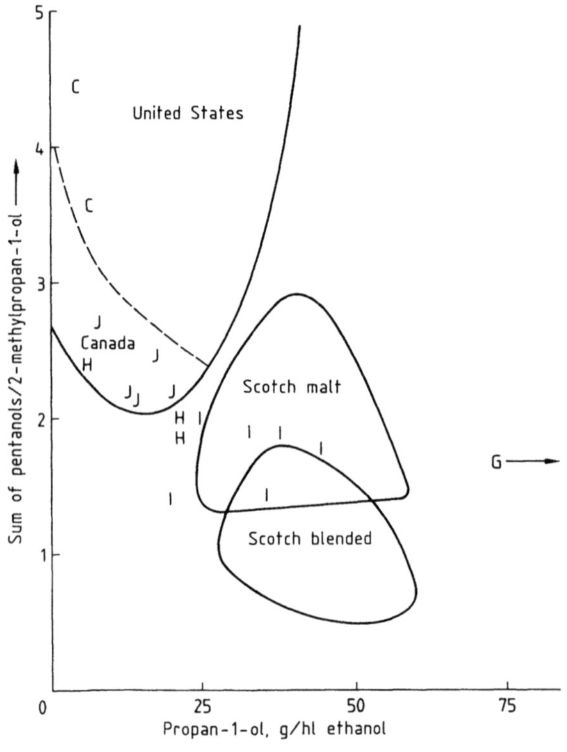

Fig. 3. Distinction between whiskies according to 1-propanol and the ratio of total pentanols to 2-methyl-1-propanol. *C* Canada; *H* Holland; *G* Germany; *I* Ireland; *J* Japan (Lisle et al 1978). Reproduced by permission of the Controller of Her Britannic Majesty's Stationery Office)

2.4 Detection of Illicit Spirits

The diversity of licit spirits which are available is enormous, as any dictionary of alcoholic beverages shows. Italian aqua vita, Turkish raki, Lebanese arak, Peruvian pisco brandy, and Australian grape vodka all originate from grapes; Polish schnapps, Genever gin, Scandinavian acquavit and Chinese mai-tai from grain; German "Inlander" rum from sugarbeet, Sri Lankan arak from coconut, and

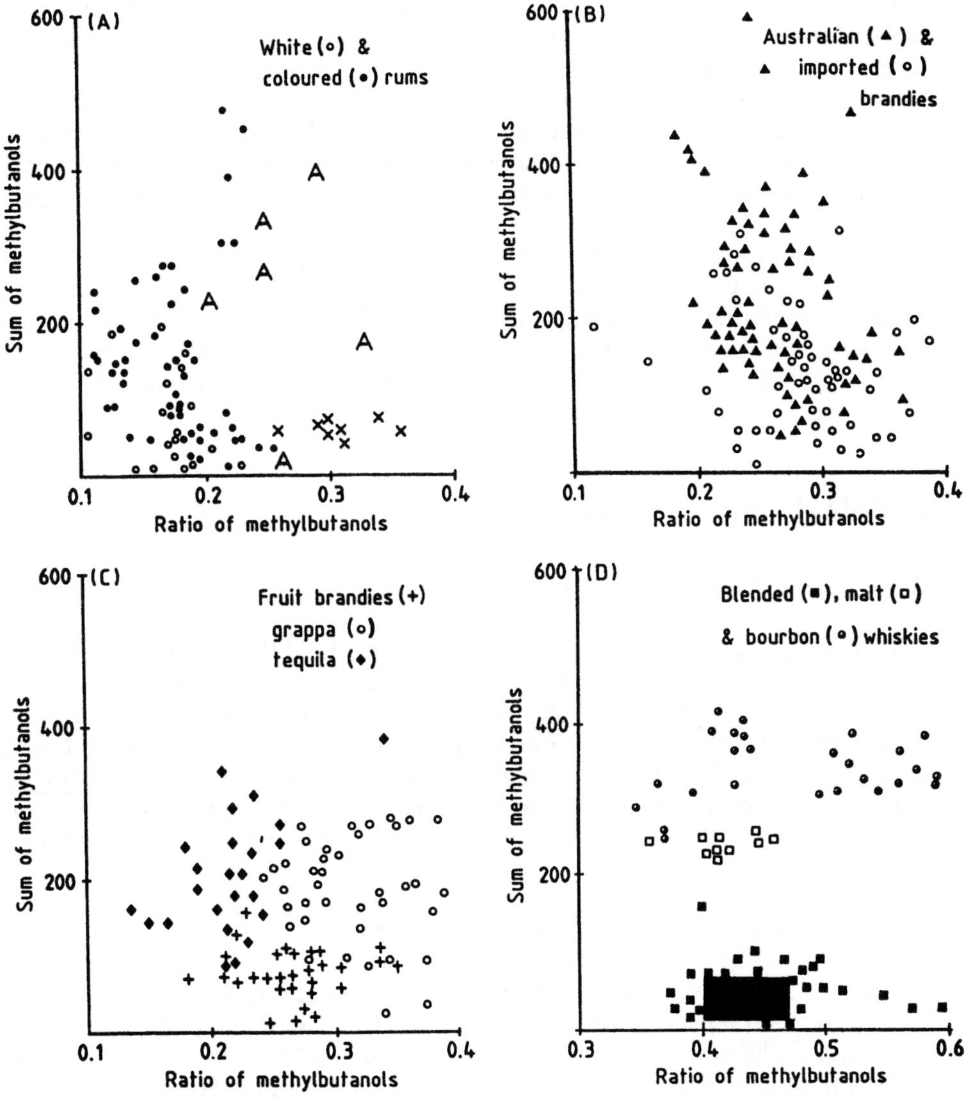

Fig. 4a–d. Identification of principal spirit types from their methylbutanol contents. Note that Australian (*A*) and imported brandies (x) having less methanol than 30 mg per 100 ml have been included in 4a (see text)

most liqueurs contain rectified molasses spirit. Any of these might be imitated. The congener profiles of most of them are consistent with the main spirit types produced in the same way. The task of maintaining an up-to-date library of all types is not feasible, and for routine screening of "unknown" samples, the use of Tables 1 and 2 and Figs. 2–4 have been found to be effective. Continuous updating with authentic samples must, of course, be maintained to monitor for changes in style.

Many illicit type (a) or "backyard" spirits analyzed in our laboratory have had anomalous congener profiles. Those considered to have originated from grapes, for example, contained methanol and higher alcohol concentrations within the same range as brandies, but outlying acetaldehyde and ethyl acetate levels. Poor control of secondary infection of the wash by microorganisms, excessive use of sulphur dioxide, and inefficient distillation are probable causes for these properties. Narayanaswamy and Golani (1977) have reported high levels of acetaldehyde, isobutanol and sometimes ethyl acetate in illicit Indian country liquor. The detection of illicitly extended spirits has proved more difficult, particularly if the fraud is practised with technical expertise and the patience to extend authentic spirit with relatively small amounts of illicit product. It requires a history of the profiles of the brands under scrutiny, from the bulk product at time of manufacture or importation, to the retail product. Sampling can be performed clandestinely by Customs officers, purportedly for fiscal purposes. It is then possible to identify the type of extender by a simple profile subtraction.

3 Stable Isotope Ratio Analysis

3.1 Limitations of Congener Profiling

Stable isotope ratio analysis provides information about the origin of the ethanol component of spirits that may not be readily apparent from congener profiles. Situations sometimes occur where the congener contents and ratios lie on the borderline between the zones for different spirit types, or are almost totally absent, such as in rectified spirits like vodka. As already stated, the addition of foreign spirit may not greatly distort the profile of some authentic spirits. The addition of "nature-identical" essences may also give a false impression of authenticity.

3.2 Stable Carbon Isotope Ratio Analysis (SCIRA)

Isotope ratios are routinely measured using mass spectrometry. It has been known for several decades that carbon isotope fractionation occurs during photosynthesis. The difference in the carbon-13/carbon-12 ratio in plants compared to that in atmospheric carbon dioxide is controlled by the plant's metabolism. Smith and Epstein (1971) reported two categories of higher plants:

a) C_3 plants, that follow the Calvin-Bensen (Calvin and Bassham 1962) pathway to fix carbon dioxide directly into the three-carbon molecule, 3-phosphoglyceric acid. The majority of plants whose sugar or starch is used to make ethanol belong to this group. These include fruits such as grapes, apples, peaches, pears, plums, cherries and berries, the cereals wheat, barley, oats and rice, and vegetables like potato and cassava.
b) C_4 plants, that use the Hatch-Slack (Hatch et al. 1967) photosynthetic pathway wherein bicarbonate is incorporated into the four carbon molecule, oxaloacetic acid. These evolutionarily more modern plants have a lower isotopic depletion than those in the first group. Those used in alcohol production include sugar cane, maize and sorghum.
c) The list of photosynthetic pathways was extended by Bender et al. (1973) with the discovery of the crassulacean acid metabolism (CAM), which involves a combination of the C_3 and C_4 mechanisms. The agave cactus used to produce tequila has this type of metabolism.
Importantly, it was demonstrated by Bricout and Menoret (1975) that the carbon-13/carbon-12 ratio was unaffected by fermentation, remaining the same in the resulting ethanol as in the sugar used to produce it, and that a linear relationship existed between the proportion of C_3 to C_4 carbohydrate in the substrate and the carbon-13/carbon-12 ratio of the final spirit.

An investigation into the application of SCIRA to the detection of illicit spirits (Simpkins and Rigby 1982) established a range of values for 11 spirit types, of known authenticity, available in New South Wales. The method of Kaplan et al. (1970) was used for sample preparation. Spirits (10 μl) were sealed into glass melting point tubes, which were then placed into a nickel pipe small enough to lie in the ceramic boat normally used to hold solid samples. This protected the quartz combustion chamber from flying glass fragments when the melting point tubes exploded after application of heat. (It is noted that Sofer (1980) has since developed a far simpler method to convert organic material to the carbon dioxide form necessary for mass-spectral determination of isotope ratios. It entails sealing the sample in a glass tube together with a measured amount of copper oxide before combusting at 550° C. Since alcohol samples are free from corrosive elements, the resulting gas can be introduced directly into the inlet system of the mass spectrometer.) Ratios were reported as "delta values" relative to the PD Bellemnite International Standard as:

$$\delta^{13}\text{C‰} = \frac{{}^{13}\text{C}:{}^{12}\text{C sample} - {}^{13}\text{C}:{}^{12}\text{C standard}}{{}^{13}\text{C}:{}^{12}\text{C standard}} \times 1000$$

with units being in per mil (parts per thousand). The investigation established a typical range of delta values for commercial spirits (see Table 3).

Delta values for single-source liquors such as grape brandy (C_3) or sugar cane spirit (C_4) studied in this survey were found to be 25.8 ± 1.3 and 11.8 ± 0.6 per mil respectively. Variations in the plant growth rate are generally considered to cause this small variation. Some isotopic fractionation may also have occurred during distillation, although this wold be unlikely in a continuous still. For those liquors

Table 3. $\delta^{13}C$ values of authentic spirits of various origins

Type	Source	No.	$\delta^{13}C$ (% PDB)
Scotch blended whisky	Maize; barley	66	−17.8 to −12.0
Scotch malt whisky	Barley	8	−24.9 to −22.9
Australian whisky	Barley plus maize, sorghum, wheat	3	−24.2 to −14.6
Gin	Maize, wheat, barley, molasses	11	−24.3 to −10.5
Bourbon	Maize, rye, barley	7	−13.8 to −12.3
Tequila	Agave	5	−12.2 to −10.8
Vodka	Wheat, cane sugar	15	−20.3 to −10.4
Rum	Molasses	8	−12.4 to −11.1
Australian brandy	Grapes	44	−27.1 to −24.5
Imported brandy	Grapes	19	−27.4 to −20.2
Plum brandies, slivovitz	Plums	6	−26.4 to −15.8

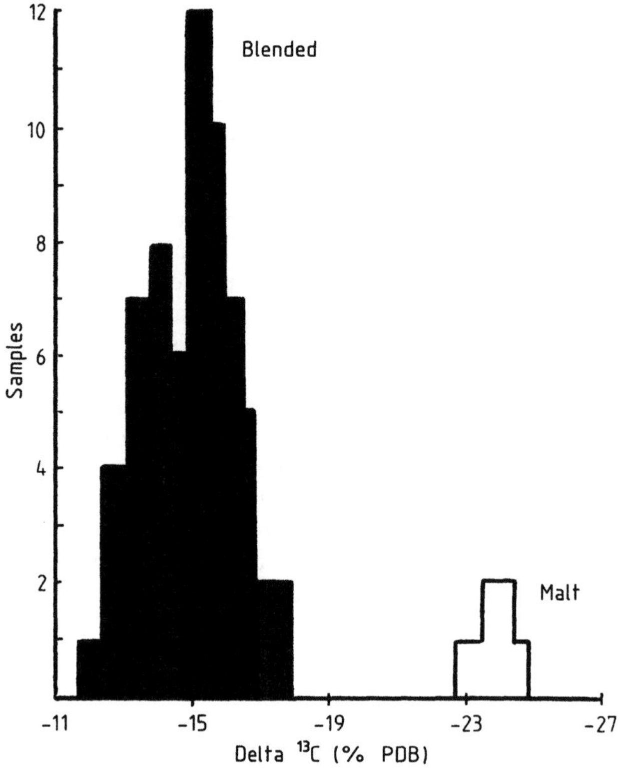

Fig. 5. Delta values for a range of blended Scotch and malt whiskies imported into Australia

that are legally defined as a blend, the spread can be attributed to variations in the composition and relative proportions of the base materials. For example, Scotch whisky is a mixture of malted barley spirit (C_3) which has a mean delta value of approximately -24 per mil, and grain spirit, usually maize (C_4), whose mean delta value lies around -11 per mil. It can be seen from Fig. 5 that the barley component of the cereals used to make the blended whiskies in the survey ranged from less than 10% to more than 50%, with a median value around 29%. Blends containing less than 25% barley spirit contravene the legislation if manufactured in Australia.

SCIRA can therefore effectively detect malpractices where spirits having different delta values are admixed, or the starting materials are not as defined by legislation or standard practice. Its efficacy is governed by the experimental error of the method, generally accepted as ± 0.2 per mil, and the availability of bulk samples to provide a benchmark against which suspect spirits can be compared. Where these control samples are not procurable, a conclusion at the 99% confidence level can only be achieved if the delta value lies at least three standard deviations from the mean for authentic spirits of the same type. It follows that cane

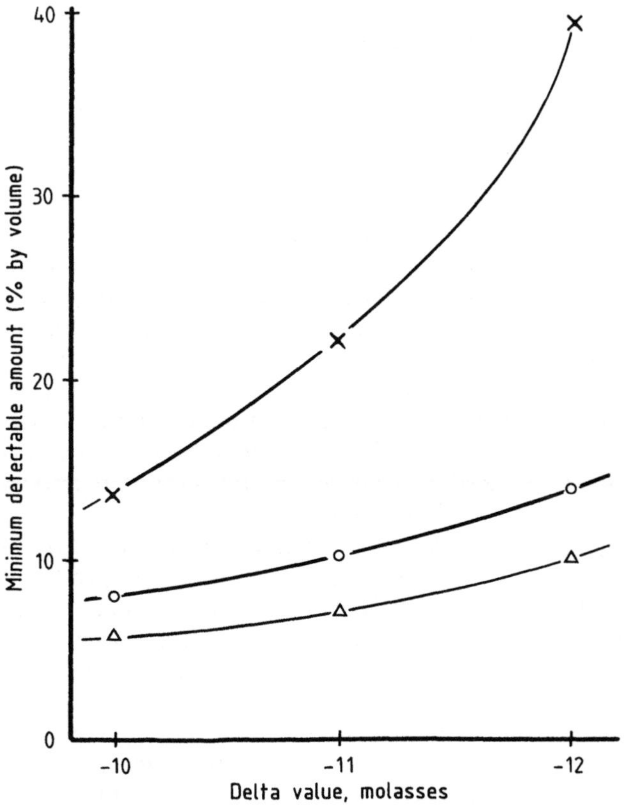

Fig. 6. Minimum detectable amount of cane sugar spirit in extended Scotch having delta ^{13}C values of –×–: –13, –o–: –15, –Δ–: –17

sugar spirit cannot be detected in whisky in any amount without the control, but as little as 25% of it in brandy will produce a significant departure from the normal reading for pure grape spirit.

The method becomes far more sensitive if the bulk spirit is available Figure 6 has been constructed from the relationship

$$\text{detection limit \% (v/v)} = \frac{0.4^{\star}}{\delta^{13}\text{C authentic whisky} - \delta^{13}\text{C “foreign spirit”}} \times 100$$

⋆ Twice the experimental error.

to show that in favourable circumstances, 7% sugar cane spirit can be detected in whisky by comparing the delta values of bulk and bottled samples. Even less can be detected in brandy because of the greater difference in respective delta values.

We have used SCIRA to compile a history of spirit types whose starting products are not legally defined. Vodka, for example, is made from cane sugar, cereals, grapes, and even lactose by different manufacturers. If a vodka is suspected as being type (c) or (d), the delta values will help point to its origin.

3.3 Stable Hydrogen Isotope Ratio Analysis (SHIRA)

Deuterium/hydrogen ratios have also been applied to the characterization of ethanol using mass-spectrometry (Bricout and Menoret 1975). However, sample preparation is somewhat onerous. It again requires combustion of the ethanol, followed by separation of the water produced, and its catalytic conversion to gaseous hydrogen. A review of techniques for sample preparation for stable hydrogen and oxygen ratio analyses has recently been presented by Wong and Klein (1986). Two variables make the interpretation of deuterium/hydrogen ratios complex. Unlike carbon isotope ratios, which are essentially uniform, the distribution of deuterium undergoes a twofold variation from the equator to the poles on the Standard Mean Ocean Water scale (Craig 1961). This variation is reflected in the composition of the plant carbohydrates. As well, some of the deuterium in the aqueous fermentation matrix is bonded during fermentation to the carbon atoms in the ethanol that is eventually produced. In whiskies and rums this is an important consideration, because water is added during mashing.

Martin et al. (1982, 1983) have used natural abundance deuterium-NMR to show that site-specific natural isotope fractionation occurs in alcohols produced either by fermentation or synthetically. For a given sugar fermented under standard conditions by *Saccharomyces cerevisiae,* the deuterium content originating from water amounted to 15% of that in the CH_2DCH_2OH isotopomer and 70% of that in the CH_3CHDOH isotopomer. The remainder comes from the sugar itself. The type of yeast also makes a small contribution. By means of an R-parameter to express the internal distribution of deuterium in the methylenic and methyl groups, Martin was able to distinguish between many of the main spirit types, including those that follow similar biosynthetic path ways. Although the main thrust of his work was to put an end to the unscrupulous practices of chapitalization and of sugaring and then refermenting spoilt wines, thus taking advantage of the French Government's commitment to purchase all alcohol distilled from poor quality wines, its value in detecting illicit spirits is apparent.

3.4 Stable Oxygen Isotope Ratio Analysis

Oxygen-18/oxygen-16 ratios have mainly been applied to detecting the watering of wine. However, they have been used to establish data for multivariate analysis of spirits (Misselhorn and Grafahrend 1984).

4 Trace Elements

Type (a) illicit spirits are usually made on a small scale. The pot still maybe as crude as a cream can, or laundry tub, soldered to a coil of copper tubing which functions as the condenser. Condensers have even been made from car radiators (Kreysa and Buscemi 1977). The "Turk's head" drip still is also utilized (see Fig. 7). Leaching of heavy metals into the final product often occurs, and is facilitated if high volatile acid levels develop in the wash or mash because of bacterial activity. Medical reports of copper, arsenic and lead poisoning through consumption of contaminated "moonshine" are numerous (Gerhardt 1980). As a result of his investigations on Georgian (USA) illicit whisky using X-ray fluorescence spectroscopy, Gerhardt reported potentially toxic levels of copper, lead, zinc and arsenic in some of the samples he tested. Hoffman et al. (1968) included mercury in the list of potentially toxic elements after using neutron activation analysis to determine 22 elements (8 were duplicated using atomic absorption spectroscopy) in a range of illicit and commercial spirits. He showed that the trace metal fingerprint could distinguish these spirits on a statistical basis. Interestingly, some quite high levels of antimony, samarium, europium and even gold were also reported. Reilly (1973) found high levels of zinc in Zambian distilled beers. These were attributed to the practice of storing spirits in galvanized metal drums. The determi-

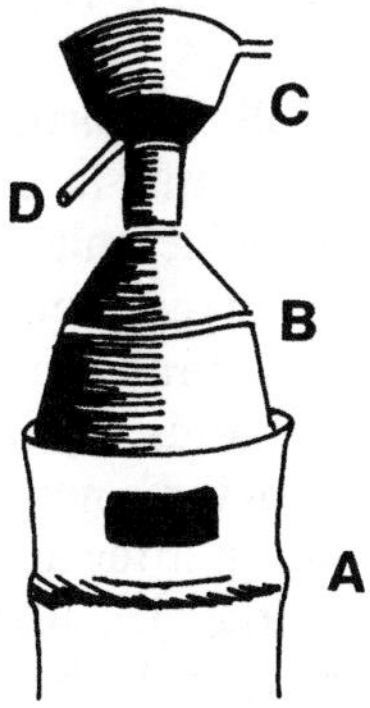

Fig. 7. Turk's head still. Discarded oil drum *A* is used to boil the fermented grape juice in a spot *B*. Vapour is condensed in the "head" *C* which contains an open dish of water, and is collected around the rim of the upper chamber before running out of the overflow *D*

Table 4. Comparison of distribution of some trace elements in licit and illicit spirits

Copper			Lead		
$mg\,l^{-1}$	% Licit	% Illicit	$mg\,l^{-1}$	% Licit	% Illicit
< 2	69	30	<0.01	90	10
2– 4	15	36	0.01–0.05	7	20
4– 6	6	11	0.05–0.10	–	10
6– 8	5	13	0.10–0.20	1	15
8–10	3	3	0.20–0.50	2	15
10–20	3	5	0.50–1.00	–	15
>20	NIL	2	>1.00	–	15

Zinc			Calcium		
$mg\,l^{-1}$	% Licit	% Illicit	$mg\,l^{-1}$	% Licit	% Illicit
<0.01	21	2	<0.01	20	10
0.01–0.05	17	20	0.01–0.05	8	4
0.05–0.10	11	25	0.05–0.10	8	9
0.10–0.20	8	25	0.10–0.20	15	32
0.20–0.50	25	24	0.20–0.50	17	23
0.50–1.00	7	2	0.50–1.00	5	13
>1.00	11	2	>1.00	21	9

nation of trace elements in spirits presents few matrix problems for the most common techniques. Solids are essentially carbohydrates. Sensitivity can be greatly increased simply by evaporative concentration, followed by low-temperature ashing before making up in weak acid. Ethanol will interfere in the determination of many elements by flame AAS (Meranger and Somers 1968) and must be removed.

Most of the type (a) illicit spirits submitted to AGAL by ACS have contained higher copper and lead levels than authentic spirits (see Table 4). However, a few of the smaller Australian distilleries have had trouble in eliminating traces of metallic contamination. Therefore a note of caution should be introduced into the interpretation of results. Water pipes frequently give rise to high concentrations of copper when new, lead is a ubiquitous contaminant because of automobile emissions, and those elements associated with water hardness may have been added during breaking down.

Trace metal profiles also have a role in detecting types (b)–(d) illicit spirits. Their presence in whiskies and brandies is generally to be avoided, because large amounts affect colour and clarity. However, not all manufacturers use deionized water for breaking down spirits. Thus calcium, iron, potassium, magnesium, manganese, sodium and zinc reflect to some extent the nature of the manufacturing process and can be used as a guide to differentiate between some brands. Varja (1972), in a study of 18 brandies, concluded that the content of calcium, magnesium, and sodium depended on the quality of water used for dilution after distillation. Potassium in brandies may be indicative of added wine, either leached from the wood in the sherry casks commonly used for ageing brandy, or deliberately added. Sherry addition to brandy is limited to 2% or less under Australian legislation.

Ion chromatography enables rapid determination of anions in spirits. Most, except those from dissolved sulphur dioxide and organic acids, can be attributed to the purity of the water used in dilution. The measurement of total ion contents of spirits enable the use of Piper trilinear diagrams (Walton 1971) to determine the origin of groundwaters. Although this approach is unlikely to pinpoint the exact source of water in a liquor, it can give valuable information when deciding whether or not it has originated from a particular manufacturer.

5 Minor Organic Constituents of Illicit Spirits

The total chemical composition of alcoholic beverages is of interest mainly for flavour research. The list of aroma constituents detected in beer, wine and (unflavoured) spirits extends to some 1300 compounds (Nykanen and Suomalainen 1983). Since the majority of these are formed by the yeast during fermentation, most beverages (including illicit spirit) contain essentially the same compounds, although a few of them such as vitispirane in wine and brandy (Simpson et al. 1977), isovalevonitrile in beet molasses alcohol (Strating 1982) and 2-ethyl-3-methylbutanoic acid in rum (Lehtonen et al. 1977) appear to be specific to a single liquor type. As with the major volatile congeners, variations in the absolute levels and ratios of the minor constituents occur between beverages.

The comprehensive determination of the minor constituents is somewhat esoteric and too time-consuming for routine screening, and must be limited to special cases. Difficulties facing the analyst include (a) removal of minor constituents from ethanol and water (b) concentration without loss of volatiles or artefact formation (c) analysis (usually by high resolution gas chromatography) (d) identification, by mass-spectrometry or Fourier-transform infrared spectroscopy (e) quantification, either directly by external standards or using, say, the effective carbon number concept. It is not surprising therefore that extensive lists of quantitative data have not been published. Many of the above problems can be overcome to some extent by using the following approaches:

a) "Blind assay" technique. Saxberg et al. (1978) showed that they could detect the substitution of an expensive brand of Scotch whisky with spirit of lesser quality by pattern recognition, using multiple discriminant analysis of peaks in a gas chromatogram. Variables were described in terms of retention times and normalized areas, but their identity was unknown. Although the authors used a straight injection to produce 17 "features", this number could be greatly extended by a simple, but repeatable, concentrative step such as solvent extraction into iso-octane (2,2,4-trimethylpentane) using the method developed by Liddle and Bossard (1983): Spirit (100 ml at 37%) is mixed with 50 ml of saturated brine in a 200-ml volumetric flask. Iso-octane (2.0 ml) is added and the flask shaken vigorously for 10 min. Water is added to bring the level up into the neck of the flask. An aliquot of the supernatent organic phase for analysis by GC can be drawn off after allowing a few minutes for separation.

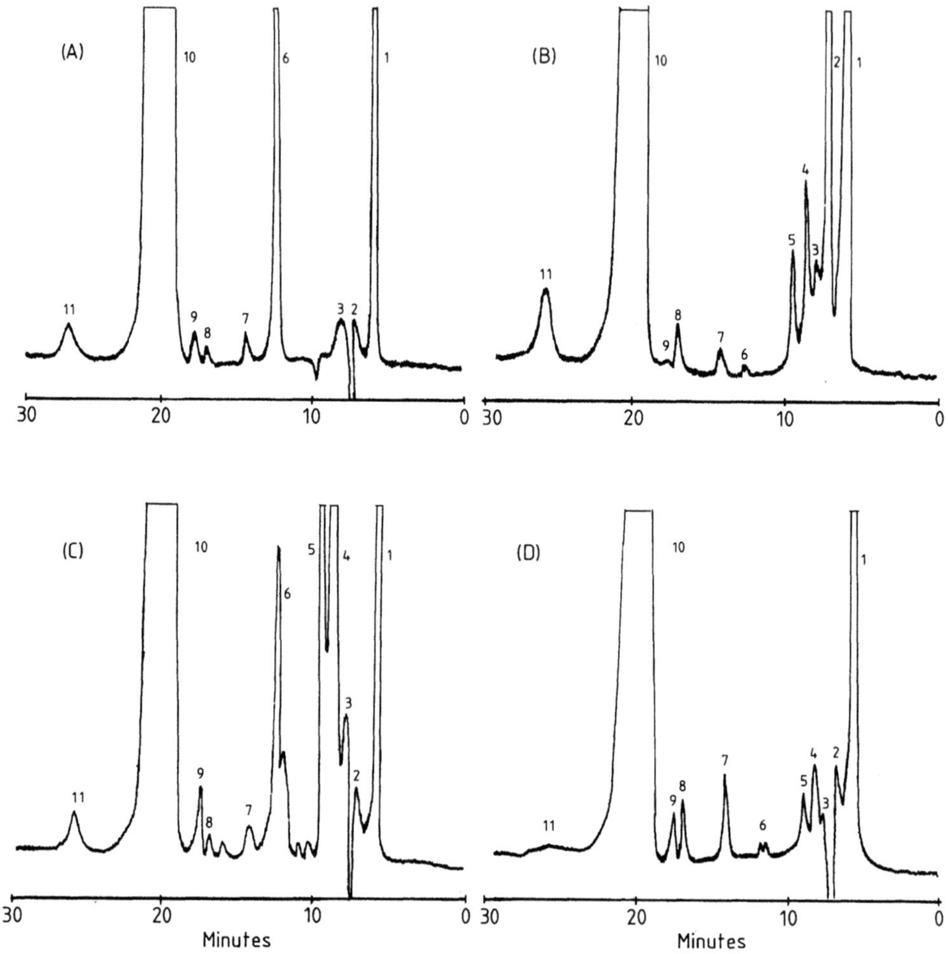

Fig. 8a–d. HPLC profile of brandies. The 11 major peaks shown here are sufficient to fingerprint most brandies. Column: Aminex HPX-87H ion exclusion, Eluant: 0.01 M H_3PO_4, Detector: RI, Tentative peak identities: *1* "polymeric material; *2, 3* unknown; *4* glucose; *5* fructose; *6* glycerol; *7* acetic acid; *8* acetaldehyde; *9* methanol; *10* ethanol; *11* 1-propanol. The fingerprint can be extended to include later-eluting major volatile congeners

The blind assay technique is particularly useful in cases where the objective is to compare a suspect spirit to one of known authenticity. We have successfully applied a quasi-blind assay using HPLC to monitor brandies (see Fig. 8). This technique, in conjunction with volatile congener levels, greatly extends the number of variables available for fingerprinting (Fig. 8). It has been useful in detecting the addition of sherry to brandy and whisky, in matching type (c) brandies with the original stock, and distinguishing between fruit wines and fruit-flavoured spirits for Customs classification purposes. Some caution must be exercised in drawing conclusions from peaks whose identity is unknown, since compositional changes may occur during the history of the liquor. It may become oxidized or sunstruck. Liddle and Bossard (1983) have reported on modifications of gin due to ageing.

b) Analysis of specific classes of compounds: Taylor (1968) distinguished home-made and marc brandy from legal matured brandy by the ratios of linoleic and oleic acids present. The method for determining unsaturated carbonyl compounds in grape marc (Williams and Strauss 1978) is also of value in identifying illicit grappas, since both use similar starting materials.

The possible misuse of synthetic flavours, which are readily available for culinary purposes from supermarkets and home-brewery outlets, is of concern to Customs Officers. Whisky, for example, is by far the most popular spirit drunk in Australia, and whilst the most common frauds continue to be watering down and substitution of an expensive variety with one that is cheaper, it is feasible that sophisticated concoctions based on nature-identical flavours remain undetected. Many methods which focus onto specific classes of compounds in whiskies have been published: Heterocyclic nitrogen compounds (Viro 1984); volatile sulphur compounds (Masuda and Nishimura 1981); phenolic compounds (Nishimura and Masuda 1971); and components derived from the barrel (Nishimura et al. 1983) to name but a few. It is likely that this specific approach will yield results more easily than attempting to analyze every peak in a chromatogram of the total extract of a suspect whisky.

6 Other Methods of Detection

HPLC has been used by MacGillivray and Hodgson (1985) to distinguish illicit from commercial liqueurs based on their sucrose : glucose : fructose profiles.

Many of the tests outlined in the preceding sections have involved modern instrumentation. Simple procedures will often suffice. Taste and odour, through they are subjective, are important criteria and are usually the initial basis of consumer complaints – only a fraction of which are justified (Cochrane 1986). Illicit spirits may give themselves away because of the amateurish way in which they are presented. Many type (a) illicit spirits have an uncommonly high ethanol content (Newton 1974). Some are hazy because of poor distillation and often the low price of illicit spirits must throw suspicion on their origin.

7 Conclusion

Most of the methods referred to in this report are well established and use instrumentation which is standard to the modern laboratory. Each can be used to produce a separate portion of the overall fingerprint of the spirit under analysis. The analyst concerned with the detection of illicit spirits should ideally build up his own library of data on authentic local products, although the published figures are useful and show that authentic liquor types are consistent within a range.

Anomalies in the fingerprint of a spirit indicate that it may be illicit. In order to detect these anomalies and then prove beyond reasonable doubt to the court that a siezed spirit is in fact not authentic may require the application of most or all of the methods we have listed, as well as considerable investigational work by the authorities responsible for the gathering of excise.

References

AOAC (1984) Official Methods of Analysis of the Association of Official Analytical Chemists. The Association of Official Analytical Chemists, Washington DC, 184–185

Bender MM, Rouhani I, Vines HM, Black CC (1973) $^{13}C/^{12}C$ ratio changes in crassulacean acid metabolism plants. Plant Physiol 52:427–430

Bober A, Haddaway LW (1963) Gas chromatographic identification of alcoholic beverages. J Gas Chromatogr 1(12):8–13

Bonte W, Decker J, Busse J (1978) Congener content of highproof alcoholic beverages. Blutalkohol 15:323–338

Bricout J, Menoret Y (1975) Teneur en isotopes stables du rhum et des principaux alcools de bouche. Ann Technol Agric 24(3–4):247–254

Calvin M, Bassham JA (1962) In: The photosynthesis of carbon compounds. Benjamin, New York

Casagrande S (1980) Dossagio dei distillati per gascromatografia. Riv Soc Ital Sci Aliment 9(912):109–120

Cavazza G (1975) La legislation du rhum et sa constitution chimiche. Ann Falsif Expert Chim 68:397–409

Cochrane GC (1986) Methods of analysis used for investigating whisky complaints. Anal Proc 23:357–359

Craig H (1961) Isotopic variations in meteoric waters. Science 13:1702–1703

Di Corcia A, Samperi R, Severini C (1980) Gas chromatographic column for the rapid determination of congeners in potable spirits. J Chromatogr 198:347–353

Filajdic M, Djukovic J (1973) Gas chromatographic determination of volatile constituents in Yugoslavian plum brandies. J Sci Food Agric 24:835–842

Gerhardt RD (1980) Trace element content of moonshine. Arch Environ Health 35(6):332–334

Hall PS (1976) The classification of whiskies by reference to their isopentanol/isobutanol ratio. J Assoc Publ Anal 14:41–45

Hatch MD, Slack CR, Johnson HS (1967) Further studies on a new pathway of photosynthetic carbon dioxide fixation in sugarcane and its occurrence in other plant species. Biochem J 102:417

Hildenbrand K (1982) Beitrag zur Beurteilung von Obstbranntweinen-Zwetschgenwasser aus dem Handel. Branntweinwirtschaft 122(1):2–8

Hoffman CM, Brunelle RL, Pro MJ, Martin GE (1968) Determination of trace component distribution in illicit spirits by neutron activation analysis, atomic absorption and gas liquid chromatography. J Assoc Off Anal Chem 51(3):580–586

Hogben R, Mular M (1976) Major congeners of Australian and imported brandies and other spirits as indicators of authenticity. J Sci Food Agric 27:1108–1114

Kaplan IR, Smith JW, Ruth E (1970) Carbon and sulphur concentration and isotopic composition in Apollo II lunar samples. Proc Apollo II Lunar Science Conference 2:1317–1329

Kreysa FJ, Buscemi PC (1977) Revenooers in Labcoats Chemtech 7:146–152

Lehtonen MJ, Gref BF, Puputti EV, Suomalainen H (1977) 2-Ethyl-3-methylbutyric acid: a new volatile fatty acid found in rum. J Agric Food Chem 25(4):953–955
Liddle PAP, Bossard A (1983) The modification of certain constituents of flavourings after addition to alcoholic beverages. In: Piggott JR (ed) Flavour of distilled beverages. Ellis, Horwood, Chichester
Lisle DB, Richards CP, Wardleworth DF (1978) The identification of distilled alcoholic beverages. J Inst Brew 84:93–96
MacGillivray BA, Hodgson BT (1985) A study of the characterisation of alcoholic beverages by sugar content using high performance liquid chromatography (HPLC). J Can Soc Forensic Sci 18:227–234
MacNamara K (1984) Rapid determination of major congeners in distilled beverages by direct analysis on bonded capillary columns. J High Res Chromatogr 7:641–643
Martin GJ, Martin ML, Mabon F, Michon J-J (1982) Identification of the origin of natural alcohols by natural abudance hydrogen-2 nuclear magnetic resonance. Anal Chem 54:2380–2382
Martin GJ, Zhang BL, Martin ML, Dupuy P (1983) Application of quantitative deuterium NMR to the study of isotope fractionation in the conversion of saccharides to ethanols. Biochem Biophys Res Commun 111:890
Masuda M, Nishimura K (1981) Changes in volatile sulphur compounds of whisky during aging. J Food Sci 47:101–105
Meranger JC, Somers E (1968) Determination of heavy metals in wines by atomic absorption Spectroscopy. J Assoc Off Anal Chem 51:922–925
Mesley RJ, Lisle DB, Richards CP, Wardleworth DF (1975) The analytical identification of rum. Ann Tech Agricole 24:361–370
Misselhorn M, Grafahrend W (1984) In: Nykanen I, Lehtonen P (eds) Flavour research of alcoholic beverages. Foundation for Biotechnical and Industrial Fermentation Research, pp 167–174
Narayanaswamy K, Golani HC (1977) Analysis of alcoholic liquors by gas liquid chromatography. ISI Bull 29:116–118, 121
Newton RP (1974) The analysis of illicit spirits. J Can Soc Forensic Sci 7:188–205
Ng TL, Woo SO (1980) Characterisation of Chinese alcoholic beverages by their congener contents. J Sci Food Agric 31:503–539
Nishimura K, Masuda M (1971) Minor constituents of whisky fusel oils. J Food Sci 36:819–822
Nishimura K, Ohnishi M, Masuda M, Koga K, Matsuyama M (1983) Reactions of wood components during maturation. In: Piggott JR (ed) Flavour of distilled beverages. Ellis Horwood, Chichester
Nosko S (1974) Evaluation of Williams-Christ brandy. Deutsche Lebensmittel-Rundschau 70:442–447
Nykanen L, Suomalainen H (1983) Aroma of Beer, wine and distilled alcoholic beverages. Akademie-Verlag, Berlin
Postel W, Adam L (1977) Gaschromatographische Charakterisierung von Whisky: Schottischer Whisky. Branntweinwirtschaft 117:229–234
Postel W, Adam L (1978) Gaschromatographische Charakterisierung von Whisky: Irischer Whisky. Branntweinwirtschaft 118:404–407
Reilly C (1973) Heavy metal contamination in home-produced beer and spirits. Ecol Food Nutr 2:43–47
Saxberg EH, Duewer DL, Booker JL, Kowalski BR (1978) Pattern recognition and blind assay techniques applied to forensic separation of whiskies. Anal Chim Acta 103:201–212
Schoeneman R, Dyer R (1968) Analytical profile of cistern room whiskies. J Assoc Off Anal Chem 51:973–987
Schoeneman R, Dyer R (1973) Analytical profile of Scotch whiskies. J Assoc Off Anal Chem 56:1–10
Schoeneman R, Dyer R, Earl E (1971) Analytical profile of straight Bourbon whiskies. J Assoc Off Anal Chem 54:1247–1261

Simpkins WA (1985) Congener profiles in the detection of illicit spirits. J Sci Food Agric 36:367–376
Simpkins WA, Rigby D (1982) Detection of the illicit extension of potable spirituous liquors using $^{13}C:^{12}C$ Ratios. J Sci Food Agric 33:898–903
Simpson RF, Strauss CR, Williams PJ (1977) Vitispirane: a C-13 spiro-ether in the aroma volatiles of grape juice, wines and distilled beverages. Chem Ind 15:663–664
Singer DD (1966a) The analysis and composition of potable spirits: determination of C_3, C_4 and C_5 alcohols in whisky and brandy by direct gas chromatography. Analyst 91:127–134
Singer DD (1966b) The proportion of 2-methylbutanol and 3-methylbutanol in some whiskies and brandies as determined by direct gas chromatography. Analyst 91:790–794
Smith BN, Epstein S (1971) Two categories of $^{13}C/^{12}C$ ratios for higher plants. Plant Physiol 47:380–383
Sofer Z (1980) Preparation of carbon dioxide for stable carbon isotope analysis of petroleum fractions. Anal Chem 52:1389–1391
Strating J, Westra WM (1982) Isovaleronitrile, a characteristic component of beet molasses alcohol. J Chromatogr 244:159–165
Suomalainen H (1971) Yeast and its effect on the flavour of alcoholic beverages. J Inst Brew 77:164–177
Taylor IS (1968) The objective characterization of illicit brandy. Food Technol Aust 20:376–377
Varja M (1972) Determination of iron, copper, calcium, magnesium and sodium in Hungarian alcoholic beverages by atomic absorption spectroscopy. Z Lebensm Unters Forsch 148:268–274
Viro M (1984) Heterocyclic nitrogen compounds in whisky and beer. Chromatographia 19:448–451
Walton WC (1971) Groundwater resource evaluation. McGraw-Hill, London, pp 453–455
Williams PJ, Strauss CR (1978) Spirits recovered from heap fermented grape marc: nature origin and removal of off-odour. J Sci Food Agric 29:527–533
Wong WW, Klein PD (1986) A review of techniques for the preparation of biological samples for mass spectrometric measurements of hydrogen-2/hydrogen-1 and oxygen-18/oxygen-16 isotope ratios. Mass Spectrom Rev 5:313–342

Determination of Sulfur Dioxide in Grapes and Wines

C. S. OUGH

1 Introduction

Very probably sulfur dioxide (SO_2) is analyzed in wines more often than any other component. For the best product the careful and judicious use of SO_2 is required. Most well-run wineries will check before and after the addition of SO_2 to first determine if it should be added, then to be sure the correct amount has been added. The usefulness of SO_2 in grape juice and wines has been reviewed many times. A few reviews are Kielhöfer 1963; Wucherpfennig 1975; Ough 1983; Schopfer and Aerny 1985. SO_2 performs several functions in the preservation of a grape juice or wine. It inhibits the polyphenol oxidase enzymes of the grape (White and Ough 1973; Ribéreau-Gayon et al. 1976b; Traverso-Rueda and Singleton 1973), prevents oxidation of the wine by binding with aldehydes and ketones (Lafon-Lafourcade 1985), preventing Maillard-type reactions, and reacts with phenols to prevent their further oxidation such as wine pinking (Simpson et al. 1983). The addition of SO_2 to a wine decreases the red color by adding to the 2- or 4-position on the anthocyanin.

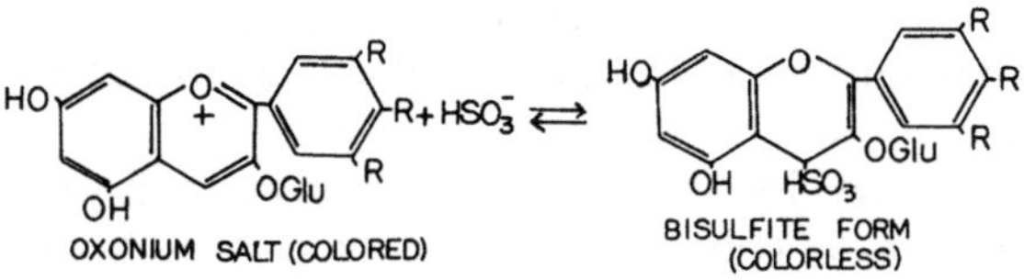

This addition prevents the resonance in the rings and the compound is colorless. Thus, a percentage of the color is lost. This addition compound is fairly unstable and the SO_2 in the free state is in equilibrium with the bound. The release of the bound is rapid if the amount of free is depleted. This reaction is mainly with monomeric pigments (Kampis and Asvany 1979). Somers et al. (1983) pointed out the obvious negative effects of free SO_2 on red wine color, but failed to be truly convincing as to the relationship to real wine quality. Cianferoni and Cianferoni (1977) proved again that addition of SO_2 to red musts caused increased color in the wine.

SO_2 also has useful but limited antimicrobial properties against yeast and bacteria (Beech and Thomas 1985; Lafon-Lafourcade 1985; Amerine et al. 1980; Ribéreau-Gayon et al. 1976a; Rehm and Whittmann 1962; Beech and Thomas 1985; King et al. 1981; Usseglio-Tomasset et al. 1981, 1982). The main active form of the SO_2 is the nonionized molecule. This is present in relatively small amounts at wine pH's (3.0 to 4.0). The amount of the SO_2 in that form can be approximately calculated from the use of the Henderson-Hasselbalch equation and on value of the free SO_2 (unbound) and using the approximation that there is none of the SO_2 in the SO_3^{2-} form (King et al. 1981; Ough 1985). The amount in the molecular form necessary to kill yeast and bacteria varies but for yeast generally 0.4 to 0.8 mg l^{-1} is sufficient for many yeasts; however, Sudraud and Chauvet (1985) suggested that 1.5 mg l^{-1} at the finish of fermentation and 1.2 mg l^{-1} during storage was required to prevent fermentation of residual sugar.

A certain amount of interest in SO_2 from a health point of view has always been present (Jaulmes 1970; Dalton-Bunnow 1985; Til et al. 1972; Feron and Wensvoort 1972). Recent episodes of reactions of highly sensitive individuals to SO_2 has caused the Food and Drug Administration to restudy the GRAS status of sulfites (Anonymous 1985). Members of the population that have threatening allergic-like reactions to SO_2 are cortisone-deficient asthma or enphysema sufferers. This has stirred activity in the production areas to improve methods and reduce the amounts of SO_2 used in wines. Legal limits, as well as actual levels of SO_2 in California wines, have lowered over the last 20 years by about one third (Ough 1985). Aerny et al. (1984) found similar trends in Swiss wines. Bioletti and Cruess (1912) made a real point of minimizing the use of SO_2. Shinohara and Watanabe (1974) used preclarification of the juice and pasteurization along with fining of the wines to lessen SO_2 use. Schopfer (1976) recognized that flash pasteurization and bentonite treatment of the must lowered the SO_2 requirements and suggested that no more than 100 mg l^{-1} total should be required, under proper handling conditions, in the bottled wine. Würdig (1976) and Paul (1976) also discussed ways and means to lower the SO_2 levels for wine use. Wucherpfennig et al. (1978) tried various treatments to spare SO_2 and concluded that SO_2 was essential to good wine-making. Haubs (1977) and Hofmann (1977) discussed ways to lessen the use of SO_2 on *Botrytis*-infected grapes. Valachovic (1985) had some unorthodox suggestions in handling wine. Margheri et al. (1978) considered that 45 mg l^{-1} free and 100 total at bottling could be obtained in most dry wines. Hydrogen sulfide [as $(NH_4)_2S$] was shown by Amati et al. (1978) to be as effective as equal amounts of SO_2 and superior to bentonite in preventing action of polyphenol oxidase. They also suggested use of pimaricin in place of SO_2. Hernandez (1985) said that 20 mg l^{-1} H_2S was superior to 100 mg l^{-1} of SO_2 in must treatment. Ubighi et al. (1982) used CaS and found it as an effective antioxidant as SO_2, but pointed out the possible problems which might result if no fugistatic additive was present. A study by Suzzi et al. (1985) suggested that yeast which produced SO_2 during fermentation stabilized white wines against browning; hence use of these yeasts would lessen the need for SO_2 additions (however, the total SO_2 needed may not change). Fining of the must plus treatment with 5-nitrofurylacrylic acid as a preservative lessened the need for SO_2 to 50 mg l^{-1} according to Valouyko et al. (1985). Asvany (1985), Gomes et al. (1985) and Usseglio-Tomasset (1985) discussed in general how to lower the SO_2 content. The conclusions from all these studies suggested clarification of juice, choice of proper yeast and substitution of H_2S and other fungicides for part of the SO_2.

The amount of SO_2 produced during fermentation by the yeast has been the subject of numerous investigations. Würdig (1985) in a review noted that the observation that yeasts produced SO_2 was in the literature as early as 1889. Würdig and Schlotter 1967; Dittrich and Staudenmayer 1968, 1970; Rankine and Pocock

1969 all reported on the fact that yeasts varied in their production of SO_2. Eschenbruch (1972a, b, c, 1974) started a systematic investigation of the factors affecting the biochemistry of SO_2 production. Minárk and Navara (1974) showed that elevated sulfate in the juice stimulated the production of SO_2. Minárik (1972) found that *Saccharomyces* produced less than 50 mg l^{-1} of SO_2. Dott et al. (1976, 1977) investigated the relationship between growth and fermentation and the uptake of sulfate between low and high SO_2 producing yeasts. Heinzel et al. (1976) showed differences in SO_2 production between yeasts from 6 mg l^{-1} to 295 mg l^{-1} and under aerobic conditions from 3.2 mg l^{-1} to 640 mg l^{-1}. Eschenbruch and Bonish (1976) found that pH influenced the SO_2 production of some strains and not others. Delfini et al. 1976; Delfini and Gaia 1977 found that the strain responses were very different when grown in Wickersham media compared to must. Poulard and Brelet (1978) looked at commercial yeast wine fermentations and found higher levels of SO_2 produced than with experimental yeasts. Feuillat and Bureau (1979) discussed the inhibition of SO_2 production by methionine and cysteine. Hartnell and Spedding (1979) found seven other sulfur-bearing compounds

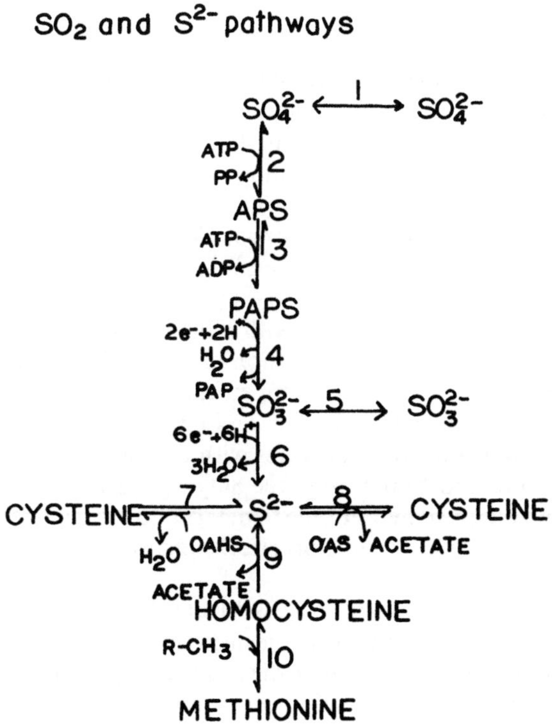

Fig. 1. Details of the sulfur pathways leading to formation of sulfite, sulfide, and sulfate in yeast. Permeases allow for transport into or out of the yeast. *1* Sulfate Permease; *2* ATP-Sulfurase; *3* APS-Kinase; *4* PAPS-Reductase; *5* Sulfite Permease; *6* Sulfite Reductase; *7* Serine Sulfhydrase; *8* OAS-Sulfhydrase; 9 OAHS-Sulfhydrase; *10* Methyl Transferase; *APS* Adenosine-5-Phosphosulfate; *PAPS* 3′-Phosphoadenosine-5′-Phosphosulfate; *OAS* 0-Acetylserine; *OAHS* 0-Acetylhomoserine

formed from sulfate by the yeast other than SO_2. A very complete review of sulfur metabolism by yeast is given by Bidan and Collon (1985).

Figure 1 gives the accepted pathway for the formation of SO_2 from sulfate.

There are a number of texts which describe the analytical procedure available for SO_2 determination in juice and wine. They include Amerine and Ough 1980; Horowitz 1980; Anonymous 1978; Ribéreau-Gayon 1976a. Wedzicha (1984) has a chapter on SO_2 analysis in foods, of which some applies to wine.

2 Free SO_2

Free SO_2 can range from 0 to over 100 mg l^{-1}, depending on the conditions and the amounts added. Most wines will be from 10 to 40 mg l^{-1} when bottled (measured after 24 h]. This is usually dependent on the condition of the wine, pH, and how long the wine is to be kept.

2.1 Iodiometric

The most commonly used method for measuring free SO_2 in wine is the Ripper procedure (Ripper 1892). It will probably remain the method for winery control analysis because of its simplicity. Wineries need a quick method to check on either additions to wine or on the amount of free SO_2 remaining in a wine at various points during the processing. This method serves this purpose well. Vahl and Converse (1980) collaboratively tested the method amongst California wineries and found large standard deviations in the test results and did not recommend its use. Brun (1978) had similar reservations.

Some of the errors are due to the reaction of iodine with other oxidizable substances such as phenols. Other errors are due to application of the method. Like any other nonstoichiometric determination time and temperature considerations must be regulated, as well as care in solution preparation and keeping.

Ascorbic acid is an interfering substance which titrates with the iodine. Burkhardt and Lay (1966) suggested a method to bind the free SO_2 with glyoxal and titrate in both a treated sample and an untreated sample to determine the true free SO_2 portion. Tanner and Sandoz (1972) made some practical suggestions for improvement.

Suggestions to use iodate instead of iodine and to back titrate the iodine produced with thiosulfate are advantageous in eliminating the need to continually restandardize the iodine solution. However, the restandardization is simple and the proposed improvement (Schneyder and Vlcek 1977) seems not to have many users.

Detection of the end point is difficult in that the starch end point slowly fades as the iodine reacts with other components. Suggestions to use an oxidation-reduction electrode were made by Vahl and Converse (1980) to detect the end point do not solve the problem, as the end point is dependent on the potential to give the blue starch end point, although it does allow a semi-standardization if time and temperature are controlled. Hagen (1977) reported on the temperature effects in this area, especially on errors due to reading obtained over the possible range in various laboratories. These differences account for 10 to 20% in free SO_2 analysis.

Ingram (1947), in studying the iodine-starch end point titrations of free SO_2, used a calomel electrode to obtain sharper end points. Villeton-Pachot et al. (1980) tested both potentiometric and coulometric electrochemical end point detection systems and found the coulometric to be the best. Both can be used on automatic titrators to an advantage. Pontallier et al. (1982) compared the automatic iodine titration with potentiometric end point determination to the aeration-oxidation method for red wines and found good agreement for older wines but not for new wines because of the problem of anthocyanin-bound SO_2. Siska (1977) used the amperometric end point with the iodate titration at 25 mV using a polarograph.

Sampling and sample care not only in this analysis but in all SO_2 analysis is extremely important. The SO_2 is a very volatile substance and easily lost at room temperature if the sample is exposed to air. The sample should be taken from a well-mixed source and kept in a full closed container until ready to analyze. Relatively few minutes' exposure of a sample pipetted into a flask can cause a measurable loss in SO_2. Below is a description for the standard Ripper procedure as suggested by Deibner 1953, 1959; Ribéreau-Gayon 1976a; Anonymous 1978; Amerine and Ough 1980.

Method

Prepare solutions: 0.1 $N\ I_3$ (add 12.7 g iodine plus 25 g potassium iodide in 1 l of H_2O, dilute with H_2O to desired strength for use, standardize with thiosulfate daily or when used, keep in a brown bottle tightly closed); 10% v/v H_2SO_4 (1:9 conc. sulfuric: H_2O); starch (dissolve 5 g of soluble starch in 1 l of H_2O containing 1 g of NaF).

Pertinent reactions

$$2Na_2S_2O_3 + I_2 \rightarrow 2NaI + Na_2S_4O_6 \text{ (standardization)}$$
$$SO_2 + 2H_2O + I_2 \rightarrow H_2SO_4 + 2HI \text{ (titration).}$$

Technique: Open sample container and pipet 50 ml of wine into a 250-ml Erlenmeyer flask. Add 10 ml of 10% H_2SO_4 and 5 ml of starch solution with swirling titrate the solution to a faint blue starch end point with the 0.02 $N\ I_3^-$. Wait 20 s and add more I_3^- if necessary. Record the value and calculate the results as SO_2 mg l^{-1}.

Calculation

$$SO_2\ \text{mg}\,l^{-1} = \frac{(\text{Vol } I_3^-)\ (N\ I_3^-)\ (32)\ (1000)}{\text{ml sample}}.$$

The iodate version of Schneyder and Vlcek (1977) is given below.

Method

Prepare solutions: Iodate solution (dissolve 0.11135 g potassium iodate with 200 ml of 2N H_2SO_4 and bring to 1 liter with H_2O); starch-iodide solution (dissolve 2.5 g soluble starch and 10 g KI in a 1-l flask and bring to volume with H_2O).

Pertinent reactions

$$8I^- + IO_3^- + 6H^+ \rightarrow 3I_3^- + 3H_2O$$
$$I_3^- + SO_2 + H_2O \rightarrow SO_3 + 3I^- + 2H^+.$$

Technique
Add 10 ml wine and 2.0 ml of iodide-starch solution to a 250 ml flask and titrate with the 0.0005203M KIO_3 solution to a blue starch end point.
Calculation
Free SO_2 mg l^{-1} = 10 × vol 0.0005203M KIO_3 or if not exactly that molality:

$$\text{Free } SO_2 \text{ mg l}^{-1} = \frac{(\text{Vol } IO_3^-)\,(N\, I_3^-)\,(32)\,(1000)}{\text{volume wine}}.$$

Ingram (1947) discussed the use of carbonyl group to bind the SO_2 to determine true free SO_2. The method, resuggested by Burkhardt and Lay (1966) to eliminate the effect of other oxidizable substances and changed slightly by Tanner and Sandoz (1972), is given below.
Method
Prepare solutions: 0.0156 *N* iodide-iodine solution (prepare 0.1 *N* I_3^- solution as described previously, standardize and dilute to required normality); glyoxal solution (40% in H_2O neutralized to pH 7.0 with NaOH); H_2SO_4 (1:4) and starch solution (2.5 g soluble starch and 1 g NaF in 1 l H_2O).
Pertinent reactions

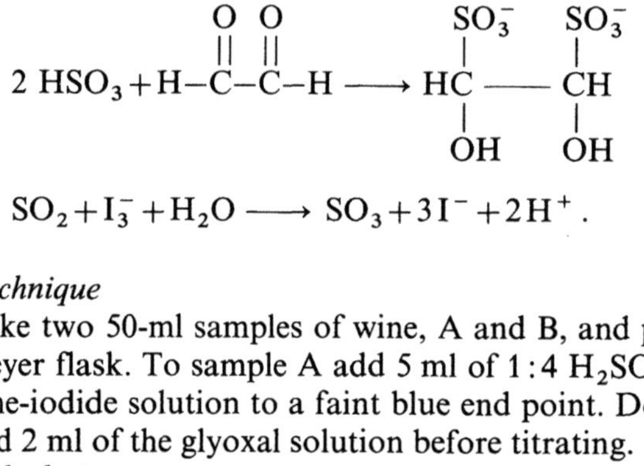

$$SO_2 + I_3^- + H_2O \longrightarrow SO_3 + 3I^- + 2H^+.$$

Technique
Take two 50-ml samples of wine, A and B, and place each into a 250-ml Erlenmeyer flask. To sample A add 5 ml of 1:4 H_2SO_4 and titrate with 0.0156 N iodine-iodide solution to a faint blue end point. Do the same for sample B except add 2 ml of the glyoxal solution before titrating.
Calculation
mg l^{-1} of free SO_2 = (A ml I_3^- – B ml I_3^-) (10). Reducing material other than SO_2 as SO_2 mg l^{-1} is (B ml I_3^-) (10). To calculate as ascorbic acid mg l^{-1} = (B ml I_3^-) (27.5).

2.2 Colorimetric

Red wines are difficult to titrate with iodine. The starch end point is masked by the pigment. In addition, the SO_2 reaction with the anthocyanins is shifted at the acid pH normally used for these titrations. Burroughs (1975) devised a clever colorimetric method to determine the true free SO_2. He determined the optical density at 520 nm on the untreated wine a_0, on 10 ml of wine plus 0.1 ml of 10% v/v acetaldehyde solution (after 30 min) a_1 and on 10 ml wine plus 0.1 ml of 3% sodium metabisulfite solution (after 10 min) a_2. Use a 2-mm or 5-mm cell. The free SO_2 is then equal to

$$K \text{ (equib contant) } \frac{(a_1 - a_0)}{(a_0 - a_2)} \text{ or as mg l of free } SO_2 = 3.8 \frac{(a_1 - a_0)}{(a_0 - a_2)}.$$

Paul (1975) was also working on a method related to this one. Bertrand (1976) studied the method and suggested it was not any better than the aeration-oxidation method; however, his tests on wine were at very low levels of free SO_2, where the chance of error was great and most of the pigment was the polymeric in form.

2.3 Aeration-Oxidation

This method consists of sparging an acidified juice or wine held in an ice bath with a stream of air. The entrained SO_2 passes through a condensor to remove the volatile acids and is then trapped in a hydrogen peroxide solution and the hydrogen ions produced by the oxidation of the SO_2 are titrated with sodium hydroxide. The method was proposed by Paul (1954, 1958); Kielhofer and Aumann (1957) confirmed it. Burroughs and Sparks (1963, 1964) used it for their work. The method has taken on several names. Jennings et al. (1978) called it the Tanner Method after Tanner (1963). Fujita et al. (1979) called it the Rankine Method after Rankine (1962), Amerine and Ough (1980) called it the Aeration-Oxidation Method which more appropriately describes the method. It is the usual OIV method (Anonymous 1978).

It has been tested by numerous workers including those mentioned above. Buechsenstein and Ough (1978), Wilson and Rankine (1977) made some follow-up studies to determine optimum flow rates of air. Kalus (1978) also noted the need for controlled sparging. The method has obvious drawbacks when applied to very young red wines. Otherwise, it is a very accurate and reproducible method. It is described below.

Method

Solution preparation: 0.01 *N* sodium hydroxide (prepare from stock solution and restandardize weekly with standard HC1); hydrogen peroxide, 0.3% (dilute 30% H_2O_2 with H_2O to 0.3%, make up daily and store in refrigerator); indicator (add 0.1 g methyl red and 0.05 g of methylene blue to a liter volumetric flask and make to volume with 50% v/v ethanol-water solution); phosphoric acid 25% (250 ml of 90% H_3PO_4 brought to 1 l with H_2O).

Pertinent reactions

$$SO_2 + H_2O_2 \rightarrow SO_3 + H_2O$$
$$H_2O + SO_3 \rightarrow 2H^+ + SO_4^{2-}$$
$$H^+ + OH^- \rightarrow H_2O.$$

Technique

Set up the glass apparatus as shown in Fig. 2. Pipet 10 ml of 0.3% H_2O_2 into the pear-shaped receiving flask. Add 3 drops of the indicator. Adjust the color to an

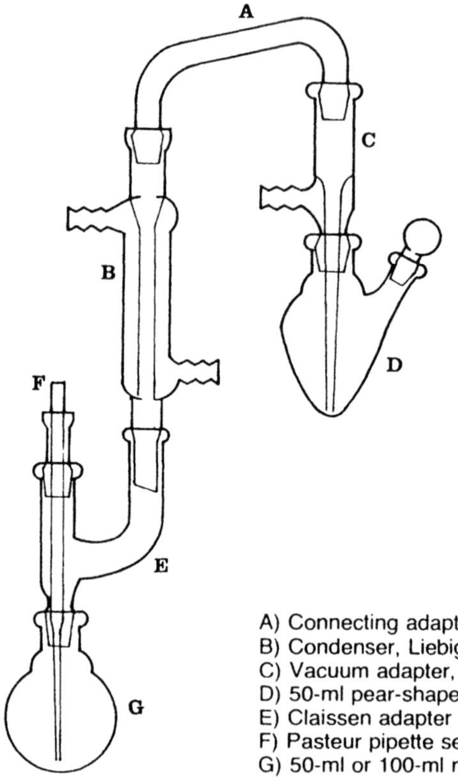

A) Connecting adapter, 14/20
B) Condenser, Liebig 110-mm 14/20 50-ml
C) Vacuum adapter, 14/20
D) 50-ml pear-shaped flask 14/20
E) Claissen adapter 14/20
F) Pasteur pipette sealed with 0 ring
G) 50-ml or 100-ml round-bottom flask 14/20

Fig. 2. Semi-micro apparatus for the determination of SO_2 by the aeration-oxidation method

olive green by the addition of a few drops of 0.01 *N* NaOH. Pipet 20 ml of wine and 10 ml of 25% phosphoric acid. Turn on cooling water in the condensor, start air flow of 1000 to 1500 cc min^{-1} sparging the sample. Run for 15 min and then remove the pear-shaped receiving flask, rinse to sparging tip off into it and titrate with 0.01 *N* NaOH to the olive green end point. A blank run should be made to determine if air is contaminated. A correction can be made.
Calculation:

$$\text{Free } SO_2\ (\text{mg l}^{-1}) = \frac{(N\ \text{NaOH})\ (\text{Vol. NaOH})\ (32)\ (1000)}{\text{mg of sample}}.$$

This method is good to a few mg l^{-1} of free SO_2.

2.4 Gas Sensing Electrodes

The gas sensing electrode consists of a membrane permeable to gas, inside solutions which allow the nonionized gas to ionize and a pH sensing electrode also in the inner solution.

The reactions are given below:

$$SO_2 + H_2O \rightleftarrows HSO_3^- + H^+$$
$$HSO_3^- + H_2O \rightleftarrows SO_3^{2-} + H^+ .$$

The inner solution is nearly saturated with bisulfite ion to maintain a uniform ionic strength. The SO_2 pentrating the membrane is ionized as in the equations shifting the H^+ concentration thus changing the pH. The Nernst equation described the electrochemical response.

$$E = E^0 + \frac{2.3\,RT}{F} \log_{10}[SO_2] .$$

At low concentrations SO_2 obeys Henry's Law. For free SO_2 the electrode is connected to a sensitive pH meter or equivalent potentiometer that can sense voltage differences to the third decimal and is dipped directly into wine or juice which has been acidified to pH 0.7 with acid. Some movement across the membrane must take place to assure the vapor pressure of the SO_2 at the membrane surface remain constant. Binder et al. (1975) tested the electrode and determined free SO_2 in various wines. They noted waiting times for equilibrium to be reached were 5 min for 1 mg l^{-1} and 10 s for 100 mg l^{-1} of SO_2. Also drift was reported due to evaporation.

Faulkner (1976) showed a flow-through system for continuous analysis of samples. Spedding and Stewart (1980) tested the electrode at juice and wine natural pHs and determined it could be used in wine. Experience in this author's laboratory (three different electrodes) and a similar experience of Caputi (1978 pers. communication) indicated that drift in the electrode was too severe for use in wines even though it worked well in model solutions.

2.5 Polarographic

This use of a polarograph to determine free SO_2 is not new, but past methods found the dropping mercury electrodes used were easily contaminated by contact with the wine. Bruno et al. (1979) suggested a method of extracting the wine in dimethylsulfoxide and making the measurements in the solvent. However, the method is time consuming taking about an hour/sample.

3 Total SO_2

Total SO_2 includes all the free SO_2 and any of the bound SO_2 which can be hydrolyzed in a basic solution. The compounds which form bound SO_2 are carbonyl molecules for the most part. The usual bisulfite complex is formed.

$$R{-}\overset{\overset{\displaystyle O}{\|}}{C}{-}R + HSO_3^- \rightleftarrows R{-}\underset{\underset{\displaystyle SO_3^-}{|}}{\overset{\overset{\displaystyle OH}{|}}{C}}{-}R_1 .$$

R_1 can be a carbon unit or a hydrogen.

Some of these compounds bind SO_2 very tightly, others less tightly. Burroughs and Sparks (1973 a) determined the equilibrium constants for a number of carbonyl bisulfite compounds. Among those determined were pyruvate, 2-ketoglutarate, L-xylosone, D-threo-2,5-hexo-diulose, 2,5-diketogluconate, 2-keto-gluconate and galacturonate in the pH range of wine. Burroughs and Spark (1973 b) further showed that moldy grapes had increased levels of these carbonyl compounds and bound more SO_2. This was further substantiated by Hidalgo (1978). Delfini et al. (1980) indicated that the addition of thiamine to the juice mediates the formation of α-ketoacids, thus lowering the need for as much total SO_2. Likewise, Farris et al. (1982) again pointed out that certain strains of yeast produce less α-ketoacids; hence, less total SO_2 is required. Farris et al. (1983) looked at 30 different *Saccharomyces cerevisiae* strains for α-ketoacid production. Piracci (1982) also pointed out the advantage of low amounts of α-ketoacids and acetaldehyde in having low total SO_2 values.

Most of the analytical methods for free SO_2 can be applied to total SO_2 by pre-hydrolysis of the carbonyl-bisulfite complexes. The equilibrium shifts rapidly at basic pH values so the complex is broken. In some cases the samples are heated under acid conditions and the rates of hydrolysis speed and the SO_2 is removed by entrainment or distillation.

3.1 Iodiometric-Ripper

Method

Prepare solutions: Sodium hydroxide 1*N* (40 g of NaOH to 1 l of H_2O). All others the same as for the free.

Pertinent reactions

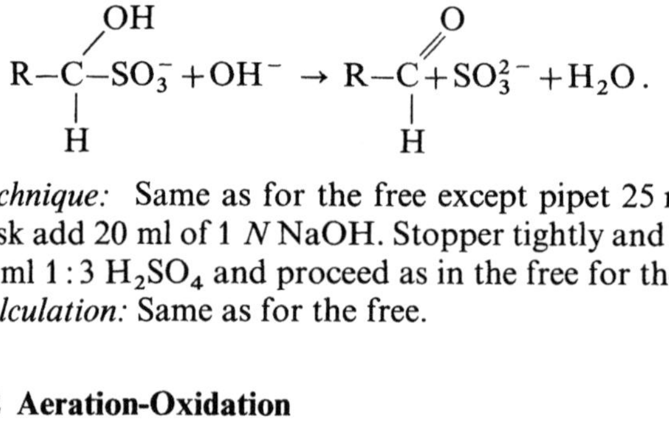

Technique: Same as for the free except pipet 25 ml wine or juice into a 250 ml flask add 20 ml of 1 *N* NaOH. Stopper tightly and let stand for 15 min. Then add 10 ml 1:3 H_2SO_4 and proceed as in the free for the Ripper.

Calculation: Same as for the free.

3.2 Aeration-Oxidation

Method

Preparation of solutions: Same as for free.

Pertinent reaction

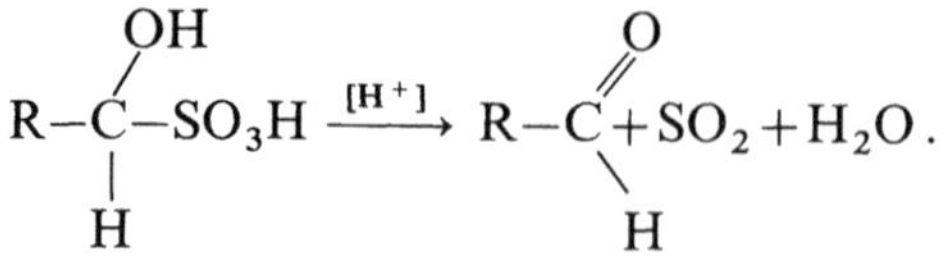

The heat speeds the rates allowing SO_2 to be removed by entainment in the air stream.
Technique: Set up as for free. Heat the flask with a microburner until a gentle reflux is obtained then start air flow and continue as for free.
Calculation: Same as for free.

3.3 Gas-Sensing Electrode and Polarographic

Both of these methods can be easily used for total SO_2 by simply hydrolyzing the bound SO_2 and then, with consideration of the dilution effect and after acidification, treat it as a free SO_2 determination. The Monier-Williams Method (Monier-Williams 1927a, b) as given by the AOAC (Horowitz 1980) was not applicable to wine. The aeration-oxidation method probably originated from this method, however.

Other methods which are variations include the Reith and Willems (1958), which used a complemetric EDTA titration of the remaining barium chloride after precipitation of the barium sulfate using eriochrome-T as the indicator. Good agreement with the direct titration of the H^+ formed. Tanner (1963) and Zonneveld and Meyer (1959) tested most of the possible variations and found the method described as aeration-oxidation to be most useful with base titration of the acid formed by the oxidation. Garcia et al. (1985) also compared methods and considered the aeration-oxidation method superior. Madison (1985) suggested using the method of Wedzicha and Johnson (1979) to measure the SO_2 by reaction with Ellman's Reagent.

3.4 Direct Distillation

Amerine and Ough (1980) gave a method for total SO_2 by directly distilling acidified wine and trapping the SO_2 in an iodine solution and then back titrating with thiosulfate to a starch endpoint. Deibner and Bénard (1953) discussed the distillation equipment and problems. Losses in iodine by evaporation make a blank determination necessary. It is rather time-consuming and not extremely reproducible. A variation method has been suggested again recently by Lissoni (1981).

3.5 Pararosaniline

The reaction of pararosaniline with SO_2 to give a colored compound is shown below as postulated by Dasgupta et al. (1980).

$$(NH_2Ph)_3C^+ + HCHO \rightarrow (NH_2Ph)_2{=}\overset{+}{C}{-}PhNHCH\ CH_2OH$$

$$\xrightarrow[-OH^-]{HSO_3^-} (NH_2Ph)_2{=}\overset{+}{C}{-}PhNH\ CH_2SO_3H, \quad \text{etc.}$$

The need for mercuric chloride is not required. The Shiffs base forms and allows the bisulfite addition. This work was preempted by Ogawa et al. (1979) when they determined that mercuric chloride was not required, trapped the entrained SO_2 from the aeration-oxidation method in alkaline solution and developed the color in a formaldehyde-pararosaniline solution. From pH 10–13 no oxidation of the SO_2 occurs in the trapping flask. However, it is required that N_2 gas be used in place of air for the entrainment to prevent oxidation. Interference from acetaldehyde and NO_2 can be eliminated by additions of dimedone and sodium azide, respectively.

Method

Prepare solutions: 0.1 *N* NaOH (from stock solution); 5% dimedone (5 g/100 ml ethanol); 1% sodium azide (0.1 g NaN_3/10 ml H_2O), 25% phosphoric acid (see oxidation-reduction for free SO_2); pararosaniline-formaldehyde solution (0.2 g pararosaniline hydrochloride in 100 ml H_2O. Hold overnight and then filter if necessary; then mix 20 ml plus 6 ml of HCl and make up to 100 ml with H_2O. Mix that with 100 ml of 0.2% formaldehyde solution); acetate buffer solution (1 M sodium acetate made up to pH 1.0 with 1 M HCl).

Chemical reaction: See previous discussion.

Technique: Pipet 8 ml of 0.1 *N* NaOH into receiving flask. Add 1 ml of 5% dimedone and 1 ml of 1% NaN_3, 2 ml ethanol, antifoam and 10 ml of 25% H_3PO_4 into distilling flask. Turn on N_2 at 0.5 to 0.6 l min^{-1} for 5 min. Add 20 ml of wine or juice. Turn on N_2 again and heat the flask with microburner after washing out receiving flask with a small amount of 0.2 *N* NaOH and adding 10 ml of 0.1 *N* NaOH. After 10 min remove the receiving flask; take 5 ml of solution add 4 ml of 1 M acetate buffer (pH 1.0) and 1 ml of pararosaniline solution, shake vigorously, let stand for 35 min at room temperature then read optical density at 560 nm. Prepare a curve from a standard sulfite solution with each point containing 0, 1, 2, 3 and 4 µg of SO_2/5 ml in 0.2 *N* NaOH.

The method which previously used $HgCl_2$ to trap the SO_2 and was described for wine by Hieke and Kreisel (1977) was compared to the above method and gave similar results. The method can also be used for free SO_2 by simply holding the distilling flask in an ice bath and proceeding as above.

Du Plessis and Tromp (1974) used a Technicon Autoanalyzer to automate the method. They did use $HgCl_2$ as the absorbent but the method could easily be modified to eliminate its use.

3.6 Continuous Flow

The above method could be considered a continuous flow system (DuPlessis and Tromp (1974). Ruzicka and Hansen (1981) detailed flow injection analysis (FIA) techniques. Grekas and Calokerinos (1985) reported a system in which acidified wine is heated and N_2 gas added in stream. The gas was removed continuously by a debubbler and the sulfite in the gas determined by molecular emission cavity analysis. It was sensitive to 4 µg ml^{-1}. A FIA where the SO_2 passes through a membrane into a pararosaniline solution and is colorimetrically measured was described. They found only 1% repeatability error at 10 mg l^{-1} concentration of free SO_2 (Möller and Winter 1985). They reported results close to those using the autoanalyzer method. List et al. (1986) found good results using the FIA and photometry with formaldehyde-pararosaniline color reagents for fruit juices and wine.

3.7 Enzymatic

The problems and the solution to the problems of an enzymatic analytical method for SO_2 were described by Beutler and Schütte (1983) and Beutler (1984). Up until then a good successful enzymatic method for SO_2 was not available. The method involves the following enzymatic steps:

$$SO_3^{-2} + O_2 + H_2O \xrightarrow[\text{oxidase}]{\text{sulfite}} SO_4^{-2} + H_2O_2$$

$$H_2O_2 + NADH + H^+ \xrightarrow{\text{NADH peroxidase}} 2H_2O + NAD^+ .$$

The problems are the reactions of SO_3^{-2} with H_2O_2

$$SO_3^{-2} + H_2O_2 \rightarrow SO_4^{-2} + H_2O$$

(very rapid reaction) and the inhibition of the NADH-peroxidase by SO_3^{-2}.

To avoid the chemical reaction of SO_3^{-2} and hydrogen peroxide the reaction is carried out in a basic media which slows the chemical reaction. The NADH-peroxidase has a high affinity towards the H_2O_2 ($K_m = 2.8 \times 10^{-5}$ mol l^{-1}, TRIS buffer 0.2 mol l^{-1}, pH 6.0). Sulfite oxidase has a $K_m = 2.4 \times 10^{-2}$ mol l^{-1} with TRIS buffer at pH 8.5

By proper buffer choices and sufficient enzyme concentrations, the reactions can proceed to completion in 20 min.

Schwedt and Bäurle (1985) compared this enzymatic method to the pararoseanaline method and the HPLC method. Good agreement for wine samples was obtained between the three methods. The method given is that of Beutler (1984).

Method

Prepare solutions: TEA buffer (0.6 mol l^{-1} = 5.57 g triethanolamine hydrochloride/50 ml – adjust pH to 8.0 with 0.1 *N* NaOH); NADH solution (7 mM = 25 mg $NADHNa_2$ and 50 mg $NaHCO_3$ dissolved in 50 ml H_2O); NADH-peroxidase suspension [4.5×10^4 U l^{-1} – dilute 0.1 ml of enzyme suspension with 0.2 ml $(NH_4)_2SO_4$ (3.0 M = 396 g l^{-1}, pH 7.0)] and sulfite oxidase suspension [5×10^3 U l^{-1} for application dilute 1.0 ml of enzyme suspension with $(NH_4)_2SO_4$ (3.0 M = 396 g l^{-1} at pH 6.0)]. Pipet into each of two 1-cm curvets (A = blank and B = sample) 1 ml of TEA buffer, 0.1 ml NADH solution, and 0.01 ml of NADH-peroxidase suspension; then into B add 0.1 ml sample (containing less than 150 µg of SO_2); into A add 2.0 ml H_2O and into B add 1.90 ml of H_2O. Mix and let stand for 5 min and read absorption at 365 nm and record. Add 0.05 ml of sulfite oxidase suspension and mix and read again after 20 to

30 min. Check after 2 min to be sure reaction is complete. If not read until it stops.

Calculate ΔE sulfite $= E_S^2 - E_S^1 - (E_B^2 - E_B^1)$

F = dilution factor (if required)

$$SO_2\ mg\,l^{-1} = \frac{202.4\ \Delta E\ \text{sulfite}}{0.34\ mM^{-1}\ mm^{-1}}\ F\,.$$

High concentration of ascorbic acid can interfere and it must be removed by pretreatment with ascorbic oxidase.

4 Gas Chromatographic

A highly polar and reactive substance such as SO_2 is difficult to determine by gas chromatographic analysis. However, several groups have reported success. Hamano et al. (1979) used a headspace technique to obtain a valid sample, then used a glass column with APS-1000 40/60 mesh to separate and detected the SO_2 with a Flame Photometric Detector (FPD) specific for sulfur. They tested the method against the aeration-oxidation method and obtained excellent agreement (Mitsuhashi et al. 1979).

Barnett and Davis (1983) and Davis et al. (1983) used Henry's Law to calculate the free SO_2 from a measure of the molecular SO_2 by headspace analysis. The method is an approximation in the sense that it ignores the contribution of the concentration of $SO_3^=$. Nevertheless it appears to be extremely accurate and reasonably rapid. It uses a Hall electrolytic detector in the sulfur mode. It is sensitive to concentrations of free SO_2 of 2 mg l^{-1} or more. The value of Henry's Law constant was found to be 27.5 and in agreement with the data of others.

The calculations involved from the approximation of the Henderson Hasselbalch equation gave a value

$$F = \frac{1}{[10^{(pH - pK_1)} + 1]}$$

where F is the fraction of free SO_2 in the molecular (SO_2) form.

$$\text{Mol}\ [SO_2] = \text{Free}\ [SO_2] \times F\,.$$

Using Henry's Law from the headspace analysis Mol $[SO_2]$ = gas $[SO_2] \times H$ where H is Henry's Law constant. Thus the free $[SO_2]$ in the liquid is

$$\text{Free}\ [SO_2] = \frac{\text{gas}\ [SO_2] \times H}{F}\,.$$

de Souza (1984) used a Teflon tube packed with Porapak QS to separate the SO_2 from other gases and then detected using a FPD. Good separation was obtained but no examples of quantification, other than of standards, was given.

The use of a Hall detector or a FPD unit limits the number of laboratories which might use these methods at the present time no matter how accurate or rapid they may be. For large wineries or regulatory units they seem worth investigation.

5 Other Methods

Undoubtedly new methods for SO_2 determination will be found. Lewis and Syty (1983) suggested the use of a UV absorption spectrophotometer to measure both free or total SO_2 by analyzing the vapor phase. They claim detection to 3 $\mu g\ ml^{-1}$. One of the more promising methods is ion chromatography (Sullivan and Smith 1985).

References

Aerny J, Regamey R, Crettenand J (1984) Nouvelle enquête sur la teneur en anhydride sulfureux total des vins suisses. Rev Suisse Vitic Arboric Hortic 16:217–221
Amati A, Guerzoni ME, Galassi S (1978) Richerche per limitare l'impiego della anidride solforosa in enologia. Vignevini 5(8):50–52
Amerine MA, Ough CS (1980) Methods for analysis of musts and wines. Wiley, New York, p 341
Amerine MA, Berg HW, Kunkee RE, Ough CS, Singleton VL, Webb AD (1980) Technology of winemaking. Avi Publishing, Westport Conn, p 794
Anonymous (1978) Recueil des methodes internatinales d'analyse des vins. Office International de la Vigne et du vin, Paris, p 348
Anonymous (1985) The reexamination of the GRAS status of sulfiting agents. Fed Am Soc Exp Biol (contract # FDA 223-83-2020) Dept. Health and Human Services. Wash DC, p 96
Asvany A (1985) Les technologies de vinification permettant de diminuer les doses de SO_2. Bull OIV 58:621–623
Barnett D, Davis EG (1983) A GC method for the determination of sulfur dioxide in food headspaces. J Chromatogr Sci 21:205–208
Beech FW, Thomas S (1985) Action antimicrobienne de l'anhydride sulfureux. Bull OIV 58:564–581
Bertrand GL (1976) Free sulfur dioxide in red wine: a comparison of analytical methods in relation to the thermodynamic activity. Am J Enol Vitic 27:106–110
Beutler HO (1984) A new enzymatic method for determination of sulfite in food. Food Chem 15:157–164
Beutler HO, Schütte I (1983) Eine enzymatische Methode zur Bestimmung von Sulfit in Lebensmitteln. Dtsch Lebensm Rundsch 79:323–330
Bidan P, Collon Y (1985) Metabolisime du soufre chez la levure. Bull OIV 58:544–563
Binder A, Ebel S, Kaal M, Thron T (1975) Quantitative Bestimmung von SO_2 in Wein durch direktpotentiometrische Messung. Dtsch Lebensm Rundsch 71:246–249
Bioletti FT, Cruess WV (1912) Enological investigation. II. Sulfurous acid in wine-making. Univ Calif Agric Exp Sta Bull 230:25–27
Brun S (1978) Étude collective sur l'analyse du vin. Résultats et interprétation. Ann Falsif Expert Chim 71:399–409
Bruno P, Caselli M, Di Fano A, Traini A (1979) Fast and simple polarographic method for determination of free and total sulphur dioxide in wines and other common beverages. Analyst 104:1083–1087

Buechsenstein JW, Ough CS (1978) SO_2 determination by aeration-oxidation: a comparison with Ripper. Am J Enol Vitic 29:161–164
Burkhardt R, Lay A (1966) Bestimmung der Ascorbinsäure mit Glykolaldehyd in Most und Weißweinen neben schwefliger Säure. Mitt Klosterneuburg 16:457–462
Burroughs LF (1975) Determining free sulfur dioxide in red wine. Am J Enol Vitic 26:25–29
Burroughs LF, Sparks AH (1963) The determination of total SO_2 content of ciders. Analyst 88:304–309
Burroughs LF, Sparks AH (1964) The determination of free SO_2 content of ciders. Analyst 89:55–60
Burroughs LF, Sparks AH (1973a) Sulphite-binding power of wines and ciders. I. Equilibrium constants for the dissociation of carbonyl bisulphite compounds. J Sci Food Agric 24:187–198
Burroughs LF, Sparks AH (1973b) Sulphite-binding power of wines and ciders. III. Determination of carbonyl compounds in a wine and calculation of its sulphite-binding power. J Sci Food Agric 24:207–217
Cianferoni R, Cianferoni L (1977) Trattamento d'urto con l'anidride solforosa ai fini di una migliore colorazione dei vini rossi. Vignevini 4(2):23–26
Dalton-Bunnow MF (1985) Review of sulfite sensitivity. Am J Hosp Pharm 42:2220–2226
Dasgupta PK, DeCesare K, Ullrey JC (1980) Determination of atmospheric sulfur dioxide with tetrachloromercurate (II) and the mechanism of the Schiff Reaction. Anal Chem 52:1912–1922
Davis EG, Barnett D, Moy PM (1983) Determination of molecular and free sulphur dioxide in foods by headspace gas chromatography. J Food Technol 18:233–240
Deibner L (1953) Sur les particularités du dosage iodometrique de petites quantités d'anhydride sulfureux libre et combiné a l'acetaldehyde en solution diluée et, en particulier dans les distillats des vins. Ann Technol Agric 2:207–242
Deibner L (1959) L'anhydride sulfureux et l'ion sulfurique dans les vins et les jus de raisin. Rev Ferment Ind Aliment 14:179–186, 227–250
Deibner L, Bénard P (1953) Recherches sur la separation qualitative à l'aide d'ion nouvel appareil distillatoire de l'anhydride sulfureux contenu dans les liquides organiques, et sur les conditions de sa stabilité dons les distillats. Ind Agric Aliment 70:11–15
Delfini C, Gaia P (1972) Indagine sulla produzione di anidride solforosa nel corso della fermentazione alcolica nei possiti Malvasia delle Lipari, Passito dé Caluso e Recioto della Valpolicella. Vini Ital 19:239–244
Delfini C, Gaia P, Bosia PD (1976) Formazione di anidride solforosa ed acido solfidrico da parte dei lieviti nel corso della fermetazione alcolica. Vini Ital 18:251–264
Delfini C, Castino M, Ciolfi G (1980) L'aggiunta di tiamina ai mosti per ridurre i chetoacidi ed accrescere l'efficacia della SO_2 nei vini. Riv Vitic Enol 33:572–589
de Souza TLC (1984) Supelpak-S: The GC separating column for sulfur gases. J Chromatogr Sci 22:470–472
Dittrich HH, Staudenmayer T (1968) SO_2-Bildung, Böckserbildung und Böckserbeseitigung. D Weinztg 104:707–709
Dittrich HH, Staudenmayer T (1970) Über die Zusammenhänge zwischen Sulfit-Bildung und der Schwefelwasserstoff-Bildung bei *S. cerevisiae*. Zentralbl Bakteriol Parasitenk Infektionsk Hyg 2 Naturwiss Abt 124:113–118
Dott W, Heinzel M, Trüper HG (1976) Sulfite formation by wine yeasts. I. Relationships between growth, fermentation and sulfite formation. Arch Microbiol 107:289–292
Dott W, Heinzel M, Trüper HG (1977) Sulfite formation by wine yeasts. IV. Active uptake of sulfate by "low" and "high" sulfite producing yeasts. Arch Microbiol 112:283–285
Du Plessis CS, Tromp A (1974) Automated determination of total sulphur dioxide in wine. Agrochemophysica 6:1–4
Eschenbruch R (1972a) Zur Substratabhängigkeit der H_2S- und SO_2-Bildung bei *Saccharomyces cerevisiae* Stämmen. Wein-Wiss 27:40–44
Eschenbruch R (1972b) Der Einfluß von Methionin und Cystein auf die SO_2-Bildung einiger Stämme von *Saccharomyces cerevisiae* bei der Vergärung von Traubenmost. Vitis 11:53–57

Eschenbruch R (1972c) Sulfate uptake and sulfite formation related to the methionine and/or cysteine content of grape must during fermentation by strains of *Saccharomyces cerevisiae*. Vitis 11:222–227

Eschenbruch R (1974) Sulfite and sulfite formation during wine making. A review. Am J Enol Vitic 25:157–161

Eschenbruch R, Bonish P (1976) The influence of pH on sulfite formation by yeasts. Arch Microbiol 107:229–231

Farris GA, Fatichenti F, Deiana P, Madau G, Cardu P, Serra M (1982) Selezione di stipiti di sacch cerevisiae bassé produttori di accettori di SO_2 prove di fermentazione in cantina. Riv Vitic Enol 35:376–384

Farris GA, Fatichenti F, Deiana P, Madau G (1983) Functional selection of low sulfur dioxide-acceptor produces among 30 *Saccharomyces cerevisiae* strains. J Ferment Technol 61:201–204

Faulkner SV (1976) A new approach to sulphur dioxide analyses. Process Biochem May:47–52

Feron VJ, Wensvoort P (1972) Gastric lesions in rats after the feeding of sulfite. Pathol Eur 7:103–111

Feuillat M, Bureau G (1979) Mécanisme de formation des sulfites dans le moût de raisin. Application à la caractérisation des souches de levures productrices d'anhydride sulfureux. CR Séances Acad Agric Fr 65:1359–1364

Fujita K, Ikuzawa M, Izumi T, Hamano T, Mitsuhashi Y, Matsuki Y, Adachi T, Nonogi H, Fuke T, Suzuki H, Toyoda M, Ito Y, Iwaida M (1979) Establishment of a modified Rankine method for the separate determination of free and combined sulphites in foods. III. Z Lebensm Unters Forsch 168:206–211

Garcia ASC, San Romão MV, Godenho MC (1985) O anidrido sulfuroso em mostos e vinhos. Estudo comparativo de métodos de análise. Cienc Tec Titivinic 4(1):5–19

Gomes JVM, Babo MFDS (1985) Les technologies de vinification permettant de diminuer les doses de SO_2. Bull OIV 58:624–636

Grekas N, Calokerinos AC (1985) Continuous flow molecular emission cavity analysis of sulfite and sulfur dioxide. Analyst 110:335–339

Hagen M (1977) Les erreurs de l'analyse ou les variations du SO_2 libre en fonction de la temperature. Rev Fr Oenol (66):55–57

Hamano T, Mitsuhashi Y, Matsuki Y, Ikuzawa M, Fujita K, Izumi T, Adachi T, Nonogi H, Fuke T, Suzuki H, Toyoda M, Ito Y, Iwaida M (1979) Application of gas chromatography for separate determination of free and combined sulfites in foods. I. Z Lebensm Unters Forsch 168:195–199

Hartnell PC, Spedding DJ (1979) Uptake and metabolism of ^{35}S-sulfate by wine yeast. Vitis 18:307–315

Haubs H (1977) Auswirkungen der Herabsetzung des SO_2-Gehaltes auf Lesetermin, Weinbereitung, Weintyp und Verkauf. Dtsch Weinbau 32:652–653

Heinzel M, Dott W, Trüper HG (1976) Störungen im Schwefelstoffwechsel als Ursache der SO_2-Bildung durch Weinhefen. Wein-Wiss 31:275–286

Hernandez R (1985) Les technologies de vinification permettant de diminuer les doses de SO_2. Bull OIV 58:617–620

Hidalgo L (1978) La pourriture des raisins moyens de protection et influence sur les caractères des vins. Ann Technol Agric 27:127–136

Hieki E, Kreisel A (1977) Automated analysis of beverages in particular wine. Part II. Automatable colorimetric determination of total sulfurous acid in grape must and wine, compared with the official method. Z Anal Chem 285:39–42

Hofmann A (1977) Verminderung der SO_2-Werte in der Kellereipraxis. Dtsch Weinbau 32:654–655

Horowitz W (1980) Official methods of analysis of the association of official analytical chemists. Association of Official Analytical Chemists. Washington DC, p 1018

Ingram M (1947) Investigation of errors arising in iodometric determination of free sulphurous acid by the acetone procedure. J Soc Chem Ind 66:105–115

Jaulmes P (1970) État actuel des techniques pour le remplacement de l'anhydride sulfureux. Bull OIV 43:1320–1333

Jennings N, Bunton NG, Crosby NT, Alliston TG (1978) A comparison of three methods of determination of sulfur dioxide in food and drink. J Assoc Public Anal 16:59–70
Kalus WH (1978) Fehlerquellen bei der Schwefeldioxidbestimmung nach der Destillationsmethode von Reith-Willems. Fresenius' Z Anal Chem 289:198–201
Kampis A, Asvany A (1979) A polimer színanyagok és a szabad kénessav ása a vörös borok színére. Borgazdasag 27:152–155
Kielhöfer E (1963) Etat et action de l'acide sulfureux dans les vins règles de son emploi. Ann Technol Agric 12 (Suppl No 1):77–91
Kielhöfer E, Aumann H (1957) Die Bestimmung der gesamten und freien schwefligen Säure im Wein, auch in Gegenwart von Ascorbinsäure. Mitt Klosterneuburg 7A:287–297
King AD Jr, Ponting JD, Sanshuck DW, Jackson R, Mihara K (1981) Factors affecting death of yeast by sulfur dioxide. J Food Prot 44:92–97
Lafon-Lafourcade S (1985) Rôle des microorganismes dans la formation de substances combinant le SO_2. Bull OIV 58:590–604
Lewis SF, Syty A (1983) Determination of sulfite in table wines by ultra violet absorption spectrophotometry in the gas phase. Atom Spectrosc 4:199–203
Lissoni M (1981) La rapida determinazione dell'anidride solforosa (SO_2) negli alimenti. Ind Aliment (Pinerolo) 20:284–285
List D, Ruwish I, Longhans P (1986) Einsatzmöglichkeiten der Fließinjektionsanalyse in der Fruchtsaftanalytik. Flüss Obst 53(1):10–14
Madison BL (1985) Colorimetric method for the determination of sulfites using Ellman's Reagent. Proceedings AOAC Task Force Meeting on Sulfite Residue Analysis Methods, Aug 15 (1985) Washington DC
Margheri G, Versini G, Sartori G (1978) Ricerche inerenti alla elaborazione dei vini binachi di qualita con basso tenore di SO_2. Vignevini 5(9):11–16
Minárik E (1972) SO_2-Bildung durch Sulfatreduktion bei verschiedenen Hefearten der Gattung *Saccharomyces*. Mitt Klosterneuburg 22:245–252
Minárk E, Navara A (1974) Effect of sulfate and sulfur amino acids levels on sulfite and sulfide formation by wine yeasts. Ann Microbiol Enzimol 24:21–36
Mitsuhashi Y, Hamano T, Hasegawa A, Tanaka K, Matsuki Y, Udachi T, Obara K, Nonogi H, Fuke T, Sudo M, Ikuzawa M, Fujita K, Izumi T, Ogawa S, Toyoda M, Ito Y, Iwaida M (1979) Comparative determination of free and combined sulphites in foods by the modified Rankine method and flame photometric detection gas chromatography. V. Z Lebensm Unters Forsch 168:299–304
Möller J, Winter B (1985) Application of flow injection techniques for the analysis of inorganic anions. Fresenius' Z Anal Chem 320:451–456
Monier-Williams GW (1927a) The determination of sulphur dioxide in foods. Rep Publ Health Med Subj No 43 Min Health, London
Monier-Williams GW (1927b) Determination of sulphur dioxide. Analyst 52:415–416
Ogawa S, Suzuki H, Toyoda M, Ito Y, Iwaida M, Nonogi H, Fuke T, Obara K, Adachi T, Fujita K, Ikuzawa M, Izumi T, Hamano T, Mitsuhashi Y, Matsuki Y (1979) Colorimetric microdetermination of sulfites in foods by use of the modified Rankine apparatus. IV. Z Lebensm Unters Forsch 168:293–298
Ough CS (1983) Sulfur dioxide and sulfites. In: Branen AL, Davidson PM (eds) Antimicrobials in Foods. Marcel Dekker, New York, pp 177–203
Ough CS (1985) Determination of sulfur dioxide in grapes and wines. J Assoc Off Anal Chem 69:5–7
Paul F (1954) Zuverlässige Bestimmungs-Methoden für Aldehyd und schwefelige Säure in Wein und Fruchtsäften unter Verwendung des Apparates von Lieb und Zacherl. Mitt Rebe Wein Klosterneuburg 4A:225–234
Paul F (1958) Die alkalimetrische Bestimmung der freien, gebundenen und gesamten schwefligen Säure mittels des Apparates von Lieb und Zacherl. Mitt Rebe Wein Klosterneuburg 8A:21–27
Paul F (1975) Formas de combinación del SO_2 durante la vinificación en blanco y en tinto. Sem Vitivinic 30(1.508–1.509):2845, 2847, 2849, 2851–2852
Paul F (1976) Évolution de l'anhydride sulfureux au cours de la fermentation alcoolique et possibilitiés techniques de diminution du vin en SO_2. Bull OIV 49:702–709

Piracci A (1982) L'importanza dell'acetaldeide e degli acidi chetonici nel ridurre l'efficiacia della SO_2 nei vini bianchi laziali. Vini Ital 24:98–104

Pontallier P, Callede JP, Ribéreau-Gayon P (1982) Determination of free SO_2 in red wines by automatic potentionmetric titration. Occurence of a specific behavior in new wines. Sci Aliment 2:329–339

Poulard A, Brelet M (1978) Les levures formatrices d'anhydride sulfureux. Vignes Vines (275):9–14

Rankine BC (1962) New method for determining sulphur dioxide in wine. Aust Wine Brew Spirit Rev 80(5):14, 16

Rankine BC, Pocock KF (1969) Influence of yeast strain on binding of sulfur dioxide in wines and on its formation during fermentation. J Sci Food Agric 20:104–109

Rehm HJ, Wittmann H (1962) Beitrag zur Kenntnis der antimikrobiellen Wirkung der schwefligen Säure. I. Mitt. Übersicht über einflußnehmende Faktoren auf die antimikrobielle Wirkung der schwefligen Säure. Z Lebensm Unters Forsch 118:413–429

Reith JF, Willems JJL (1958) Über die Bestimmung der schwefligen Säure in Lebensmitteln. Z Lebensm Unters Forsch 108:270–280

Ribéreau-Gayon J, Peynaud E, Sudraud P, Ribéreau-Gayon P (1976a) Traité d'oenologie sciences et techniques du vin. Tome I. Analyse et Contrôle des vins. Dunod, Paris, p 671

Ribéreau-Gayon J, Peynaud E, Ribéreau-Gayon P, Sudraud P (1976b) Traité d'oenologie sciences et techniques du vin. Tome 3. Vinifications Transformations du Vin. Dunod, Paris, p 719

Ripper M (1892) Schweflige Säure in Weinen und deren Bestimmung. J Prakt Chem 46:428–473

Ruzicka J, Hansen EH (1981) Flow injection analysis. Chemical Analysis, vol 62, Wiley, New York

Schneyder J, Vlcek G (1977) Maßanalytische Bestimmung der freien schwefligen Säure in Wein mit Jodsäure. Mitt Höheren Bundeslehr-Versuchsanst Wein-Obstbau (Klosterneuburg) 27:87–88

Schopfer JF (1976) Évolution de l'anhydride sulfureux au cours de la fermentation alcoolique et possibilities technique de diminution du vin en SO_2. Bull OIV 49:313–326

Schopfer JF, Aerny J (1985) Le rôle de l'anhydride sulfureux en vinification. Bull OIV 58:515–542

Schwedt G, Bäurle (1985) Methodenvergleiche (Photometrie, HPLC, Enzymatische Analyse) zur Bestimmung von Sulfit in Lebensmitteln. Fresenius' Z Anal Chem 322:350–353

Shinohara T, Watanabe M (1974) Experimental white wine making. Trials to reduce sulfur dioxide usage. J Soc Brew (Japan) 69:253–255

Simpson RF, Bennett SB, Miller GC (1983) Oxidative pinking of white wines: a note on the influence of sulphur dioxide and ascorbic acid. Food Technol (Aust) 35:34–36

Siska E (1977) Borászati termékek kénessavtartalmának meghatározása. Borgazdasag 25:107–110

Somers TC, Evans ME, Cellier KM (1983) Red wine quality and style: diversities of composition and adverse influences from free SO_2. Vitis 22:348–356

Spedding DJ, Stewart GM (1980) Use of sulphur dioxide gas sensing membrane probe in wines and juices at their natural pH. Analyst 105:1182–1187

Sudraud P, Chauvet S (1985) Activité antilevure de l'anhydride sulfureux moléculaire. Connaiss Vigne Vin 19:31–40

Sullivan DM, Smith RL (1985) Determination of sulfite in foods by ion chromatography. Food Technol (July) 45–48, 53

Suzzi G, Romano P, Zambonelli C (1985) *Saccharomyces* strain selection in minimizing SO_2 requirement during vinification. Am J Enol Vitic 36:199–202

Tanner H (1963) Die Bestimmung der gesamten schwefligen Säure in Getränken, Konzentraten und in Essigen. Mitt Geb Lebensmittelunters Hyg 54:158–174

Tanner H, Sandoz M (1972) Bestimmung der freien SO_2 neben Askorbinsäure und anderen Reduktionen unter Zuhilfenahme von Glyoxal. Schweiz Z Obst Weinbau 108:331–337

Til HP, Feron VJ, de Groot AP (1972) The toxicity of sulfite. I. Long-term feeding and multigeneration studies in rats. Food Cosmet Toxicol 10:291–310

Traverso-Rueda T, Singleton VL (1973) Catecholase activity in grape juice and its implications in winemaking. Am J Enol Vitic 24:103–109

Ubighi M, Castino M, Di Stefano R, Opessio MS (1982) Prove sull'impiego dell' H_2S in fase prefermentativa in sostituzione della SO_2. Riv Vitic Enol 35:485–510

Usseglio-Tomasset L (1985) Les technologies de vinification permettant de diminuer les doses de SO_2. Bull OIV 58:606–616

Usseglio-Tomasset L, Ciolfi G, Pagliara A (1981) Estimating sulfur dioxide resistance of yeasts. I. The delaying action on fermentation start. Vini Ital 23:78, 80–90

Usseglio-Tomasset L, Ciolfi G, di Stefano R (1982) The influence of the presence of anthocyanins on the antiseptic activity of sulfur dioxide towards yeasts. Vini Ital 24:86–94

Vahl JM, Converse JE (1980) Ripper procedure for determining sulfur dioxide in wines: Collaborative study. J Assoc Off Anal Chem 63:194–199

Valachovic M (1985) O potrebe a možnostiach zniženia obsahu SO_2 vo vine. Vinohrad 23:230–231

Valouyko GG, Pavlenko NM, Ogorodnik ST (1985) Les technologies de vinification permettant de diminuer les doses de SO_2. Bull OIV 58:637–644

Villeton-Pachot JP, Persin M, Gal JY (1980) Titrage coulométrique du dioxyde de soufre dans les vins avec détection électrochemique du point equivalent. Analysis 8:422–427

Wedzicha BL (1984) Chemistry of sulfur dioxide in foods. Elsevier Applied Science, London, p 381

Wedzicha BL, Johnson MK (1979) A variation of the Monier-Williams distillation technique for the determination of sulphur dioxide in ginger ale. Analyst 104:694–696

White BB, Ough CS (1973) Oxygen uptake studies on grape juice. Am J Enol Vitic 24:148–152

Wilson B, Rankine BC (1977) A note on measurement of free sulfur dioxide in white wines by aspiration. Aust Wine Brew Spirits Rev 96(3):14

Wucherpfennig K (1975) Bedeutung der schwefligen Säure für die Traubensaftherstellung und ihre lebensmittelrechtlichen Aspekte. Flüss Obst 42:451–456, 461–464

Wucherpfennig K, Schrobinger U, Keller U (1978) Versuche zur Einsparung von schwefliger Säure bei der Weinbereitung unter spezieller Berücksichtigung der Mostpasteurisation. Schweiz Z Obst Weinbau 114:24–33

Würdig G (1976) Évolution de l'anhydride sulfureux au cours de la fermentation alcoolique et possibilitiés techniques de diminution du vin en SO_2. Bull OIV 49:405–415

Würdig G (1985) Levures produisant du SO_2. Bull OIV 58:582–589

Würdig G, Schlotter HA (1967) SO_2-Bildung in gärenden Traubenmosten. Z Lebensm Unters Forsch 134:7–13

Zonneveld H, Meyer A (1959) Bestimmung der schwefligen Säure in Lebensmitteln, insbesondere in Trockengemüse. Z Lebensm Unters Forsch 109:198–205

Determination of Diethylene Glycol in Wine

L. S. Conte and A. Minguzzi

1 Introduction

1.1 Use

Diethylene glycol is used in antifreeze solution for sprinkler systems, water seals for gas tanks, etc. A solution of water with 40% diethylene glycol freezes at −18° C, with 50% at −28° C.

DEG is further used as lubrificating and finishing agent for wool, worsted, cotton, rayon, and silk. It is also used as solvent for vat dyes, in composition corks, glues, gelatin, casein, and pastes to prevent drying out.

In 1985, differing amounts of DEG were found in some wines, to which it had been added to increase the extract.

1.2 Empirical Formula

$C_4H_{10}O_3$ (mol. wt. 106.12)

Diethylene glycol

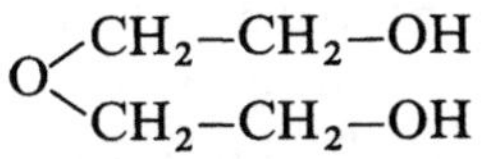

2-2′-oxybisethanol

1.3 Alternative Names

2-2′ oxybisethanol, B-B′-dioxyethylether, 2-hydroxyethylether

1.4 Preparation

Made by heating ethylene oxide and glycol

1.5 Physical Characteristics

Colorless, hygroscopic, pratically odorless liquid, slightly sweetish taste, d 20/20 = 1.118; solidifies at −10.45° C; melting point −6.5° C, boiling point 244–245° C, miscible with alcohol, ethyl ether, acetone, ethylene glycol; insoluble in benzene and carbon tetrachloride.

1.6 Toxicology

LD50 orally in rats is 8.54 g kg^{-1}, in guinea pigs is 6.61 g kg^{-1} (Smith et al. 1941). Lethal dose in humans is about 1.4 ml/kg (Rowe 1962; Browning 1965). Ingestion causes transient stimulation of central nervous system, followed by depression, vomiting, drowsiness, coma, respiratory failure, convulsions, renal damage, which may proceed to anuria, uremia and death.

2 Methods of Analysis

2.1 Abstracts

In this section, we report in abstract several methods for the analytical determination of diethylene glycol. The analytical techniques that are used are: gas chromatography (GLC), thin layer chromatography (TLC), high pressure liquid chromatography (HPLC) and nuclear magnetic resonance (NMR).

2.1.1 Gas Chromatography

Several hydroxylated compounds like carbohydrates, organic acids and polyalcohols are present in wine. All these compounds have physicochemical characteristics similar to those of diethylene glycol. The hydroxylated compounds are generally present in amounts of some grams per liter, while diethylene glycol is usually determined in very small amounts (ppm).

All the compounds cited give no problem of overlapping to the peak of DEG in gas chromatographic elution with capillary columns, but the great difference of concentration may present problems of column flooding with a reduction in the efficiency of separation.

For this reason, most of the authors have taken care to arrive at the GLC analysis with a fraction as free as possible from carbohydrates, organic acids, and polyalcohols.

This result was obtained by eliminating these compounds with chemical methods or with selective solvents.

The methods cited here may be divided into two main groups: methods which operate with polar capillary columns (Carbowax 20 M or similar) and methods which operate with nonpolar capillary columns (silicones or similar).

2.1.1.1 Gas Chromatography with Polar Capillary Columns

Bandion et al. (1985) proposed to determine ethanediol and diethylene glycol added to wine by GLC at 180° C with F.I.D. on a column (25 m × 0.30 mm) coated with Carbowax 20 M. Carrier gas was N (4 ml min^{-1}). Very sweet wines were diluted 1 : 1 with water.

Littmann (1985) mixed wine with Celite 542 and Na_2SO_4. Then the paste was packed into a column and DEG was eluted with ethyl ether. The column for GLC analysis was a fused silica (50 m × 0.32 mm) coated with FFAP and the oven temperature was programmed from 110 to 220° C at 10° C/min.

A Carbowax 20 M (25 m × 0.32 mm) was also used by Fostel (1985), with a temperature programming from 147 to 161° C at 2° C min^{-1}. The internal standard was butane 1,4 diol and the sensitiveness of the method was 1 ppm.

Kaiser and Rieder (1985) coupled a DB-WAX (Supeko) with a DB-1 in series, together with a pre-column of 15% OV 1 on Gas Chrom Q. The temperature programming was from 60 to 170° C at 5° C min^{-1}, then up to 270 at 20° C min^{-1}. The detection limit was 1 mg l^{-1}.

Other authors operate extractions and purifications before the injections into a polar column.

Wagner and Kreutzer (1985) used 10 ml of wine or a 1:1 aqueous dilution when sugars were more then 100 g l^{-1}. Then, 5 g of $Ba(OH)_2$ mixed with sea sand was added. Fifty ml of acetone was then added and the vessel was heated at 45° C for 5 min, then its contents was filtered through a fritted-glass filter, the filter was washed with acetone and the filtrate and washings were evaporated at 40° C under vacuum. A solution of the residue in 20% ethanol was analyzed by GLC on a column (2 m × 2 mm) packed with 10% Carbowax 20 M on Chromosorb W AW-DMCS, operating with temperature programming from 120 to 170° C at 10° C min^{-1}.

Bertrand (1985) proposed a method that allowed an extraction of more than 70% of DEG without the presence of carbohydrates. The internal standard was 1,3 propandiol. One ml of standard solution, 5 ml of acetone, 5 ml of ethyl ether and 2 g of Na_2CO_3 were added to 5 ml of wine. The injection of extracted diethylene glycol was made in two different ways: splitless (10 s) and split on a FFAP or CPWAX 57 CB (Chrompack) 10 m × 0.22 mm. The temperature was 80° C, programmed at 5° C min^{-1}.

This method does not eliminate organic acids and phenolic substances, while the method proposed by Hori et al. (1986) cleaned up the wine sample on a column of alumina with methanol as eluent. The gas chromatographic analysis was realized with a packed column (1.5 m × 3 mm) of 20% polyoxyethylene glycol 20 M on Chromosorb W AW-DMCS. The detection limit was 10 ppm. The method used by Hori was quick and took less than 10 min.

Wine was cleaned up by organic acids adjusting pH to 9 with KOH after saturation with NaCl in the method proposed by Werkhoff and Bretschneider (1986). An extraction with ethyl acetate followed and the extract, diluted to volume with ethyl acetate, was injected in a fused silica capillary column (60 m × 0.32 mm), coated with Durabond-WAX, fitted with a deactivated fused-silica retention gap pre-column (20 cm × 0.53 mm). The temperature was programmed from 80 to 220° C at 5° C min^{-1}. The calibration graph was rectilinear for up to 10 mg l^{-1} of diethylene glycol, recoveries of 65 mg l^{-1} from wine were quantitative and the coefficient of variation at this level was 3.9%.

The sample clean-up was obtained by Caccamo et al. (1986) by means of a Carbopack B cartridge passing 5 ml of water, then 1.5 ml of the sample to which 1 g l^{-1} of 1,4 butanediol had been added as internal standard. The gas chromatographic analysis of the eluate was performed on a glass column (1.0 m × 2 mm i.d.), packed with Carbopack C graphitized carbon black, modified with 0.8% of

THEED (Supelco). The oven temperature was 120° C and limit of sensitivity reported was 1 ppm.

2.1.1.2 Gas Chromatography with Nonpolar Capillary Column

The elimination of sugars at 120 g l^{-1} was obtained by Fuehrling and Wollenberg (1985) by means of a fermentation with yeasts at 40° C; then the sample was deionized with mixed ion-exchange resin AG 501-X8. The sample was then chromatographated on silica gel and eluted with acetone. The eluate, evaporated to dryness, was treated with N-methyl-N-trimethylsilylheptafluorobutyramine and analyzed by GLC on a fused silica capillary column (15 m × 0.25 mm), coated with OV 101.

The temperature programming was from 105 to 180° C and the recovery of DEG was estimated to be 100%.

2.1.2 Thin Layer Chromatography (TLC)

Lehmann and Ganz (1985) proposed a TLC method with a detection limit of 10 ppm. Diethylene glycol was separated from sugars and concentrated by application to an Extralut column with elution with CH_2Cl_2-propan-2-ol (17:3). The eluate was evaporated to dryness and the residue is dissolved in methanol for analysis by TLC on silicagel 60, with acetone-5N NH_3-$CHCl_3$ (8:1:1) as mobile phase and detected by spraying with a 5% solution of $K_2Cr_2O_7$ in H_2SO_4, the coloured spots were compared with those of standard solution of DEG (0.8 to 8 μg) separated under the same TLC conditions.

2.1.3 High Pressure Liquid Chromatography (HPLC)

The separation of DEG from glucose, fructose, glycerol, acetic acid, 2,3-buthylene glycol and ethanol was achieved by Van Rooyen and Van Wyk (1986) by means of an Aminex HPX-87 H 300 × 7.8 mm ion exclusion column (Bio-Rad) fitted with a Microguard ion-exclusion column 40 × 4.6 mm (Bio-Rad). The mobile phase was freshly distilled water with 0.013 N H_3PO_4. The column temperature was 50° C (constant), the flow rate 1.0 ml min^{-1} and the detection was made with a differential refractometer. The wine sample was passed through a Waters C18 Sep pack clean-up precolumn. The minimum detectable concentration of DEG was 0.1 g l^{-1}. A concentration as low as 0.01 g l^{-1} can be successfully detected after a fourfold concentration of the sample. The time of analysis was less than 20 min per sample, including sample preparation.

UV detector was also used by Pfeiffer and Radler (1985), with 6.5 mM H_2SO_4 as mobile phase (0.6 ml min^{-1}). The wine sample is passed through a membrane filter (0.45 μm) and ultrasonically mixed before analysis of a 5 μl portion at 65° C with the use of a Micro-Guard pre-column and an analytical column (30 cm × 7.8 mm) of Aminex HPX 87 H.

2.1.4 Nuclear Magnetic Resonance (NMR)

Rapp et al. (1986) describe a procedure for the structure-specific quantitative determination of diethylene glycol in wines with ^{13}C NMR spectroscopy. The analysis can be performed without pre-treatments of the sample with a detection limit of 10 mg l^{-1}. The time of analysis is about 30 min.

2.1.5 Mass Spectrometry

Several authors confirmed the GLC detection of diethylene glycol by means of mass spectrometry (Bandion et al. 1985; Fostel 1985; Holtzer 1985; Schubert 1985; Kuhlmann 1986).

They reported a mass spectrum of underivatized diethylene glycol according to the following fragmentation:

$$HO{-}CH_2{-}CH_2{-}O{-}CH_2 \wr \; CH_2{-}OH$$

M−18

M−31 ⟶ M−31−30

$$HO{-}CH_2{-}CH_2{-}O{-}CH_2^+ \longrightarrow HO{-}CH_2{-}CH_2^+ + HCHO\,.$$

The obtained mass-spectrum is reported in Fig. 1.

Watkins (1985) reported the fragmentation obtained by means of an Ion trap detector (ITD) for diethylene glycol as BSTFA. The identification was made by comparison with the 38752 spectra in NBS library as shown in Fig. 2.

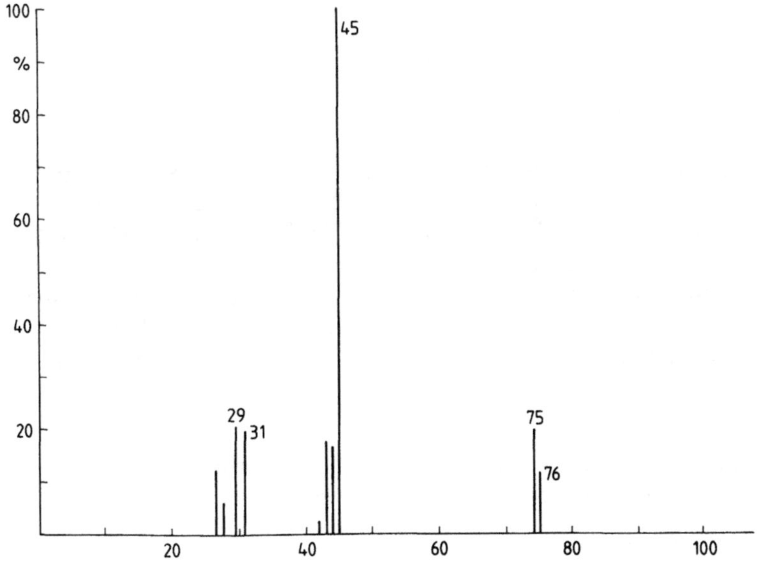

Fig. 1. Mass spectrum of diethylene glycol

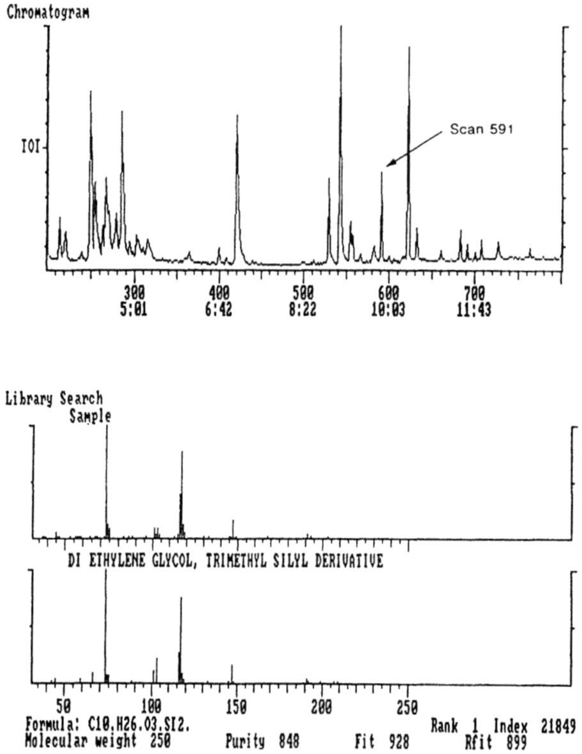

Fig. 2. Ion trap detector: Library search results for the identification of DEG as trimethylsilyl derivative. (Watkins 1985, Ion Trap Newsletter)

2.2 Extended Form Methods of Analysis

Some chromatographic methods are reported in great detail in this paragraph, so that it is possible to use the methods without consulting the original papers.

The proposed methods include four chromatographic methods: injection of wine directly in a capillary polar column, injection of an extract in a capillary polar column, injection of a purified extract in a capillary polar column and injection of a purified and derivatized extract in a capillary non-polar column.

Furthermore, an HPLC method that uses individual and coupled columns is described.

2.2.1 Gas Chromatographic Determination by Direct Injection of Wine into a Polar Capillary Column (Holtzer 1985)

2.2.1.1 Principle of the Method

The method provides the direct injection of untreated or diluted wine.

2.2.1.2 *Reagents*

Anhydrous ethanol
1,4-butylenglycole (I.5.)

2.2.1.3 *Apparatus*

Gas chromatograph with fused silica capillary column (25 m × 0.32 mm), coated with Carbowax 20 M, film thickness 0.3 μ.

2.2.1.4 *Determination*

One μl of wine is injected through a split-system (split ratio 1 : 20) with the following gas chromatographic parameters: carrier gas (He) = 2 ml min^{-1}, oven temperature programmed from 150 to 200° C at 8° C min^{-1}, injector temperature 200° C, detector (FID) temperature 250° C.

Sweet wines are diluted 1 : 1 with ethanol before injection.

2.2.1.5 *Notes*

The method has a sensitivity of 5 mg l^{-1} in the described operative conditions. The identification of DEG is confirmed by means of mass spectrometry.

For the length of the life of the capillary column, it must be remarked that the injection of untreated wine leads strong polar compounds into the column. For this reason, the purging of the columns at high temperature is necessary, this procedure involves a loss of the separation efficiency of the column. The loss of efficiency can be seen after 150–200 injections.

2.2.2 Gas Chromatographic Determination by Injection of a Wine Extract into a Polar Capillary Column (Kuhlmann 1986)

2.2.2.1 *Principle of the Method*

Diethylene glycol is extracted with chloroform and determination in carried out by means of capillary gas chromatography or GC-MS.

2.2.2.2 *Reagents*

Chloroform
Ethanol
Anhydrous sodium sulphate

2.2.2.3 *Apparatus*

Rotative evaporator

Capillary gas chromatograph with split-less injector system, equipped with a capillary column 25 m × 0.20 mm, coated with Carbowax 20 M, film thickness 0.2 μ.

2.2.2.4 Determination

Chloroform (0.5 ml) is added to 300 μl of wine in a 10 ml flask. The flask is vigorously shook and 1.7 g of anhydrous sodium sulfate is added. The mixture is shaken again for 1 min. After decantation, the liquid phase is transferred into a 25 ml flask containing 600 ml of ethanol. The solid phase is washed twice with 2 ml of chloroform each time. The volume is adjusted to 600 μl adding $CHCl_3$ after evaporation under vacuum.

GLC analysis follows, with a Carbowax 20 M capillary column in the following operative conditions: Pressure of carrier gas 28 PSI, temperature of injector 250° C (splitless for 0.3 min), temperature of detector (FID) 250° C. The temperature of the column was programmed from 60 to 200° C at 15° C min^{-1}.

2.2.2.5 Notes

The method is fast enough to be used in large screenings. With DEG concentration between 5 and 10 mg l^{-1} interference with other substances was observed. The minimum detectable quantity was 2 mg l^{-1} by means of GC and 0.2 mg l^{-1} by means of GC-MS.

In the case of identification with GC-MS, a reduction of sensitivity of about 1 mg l^{-1}, due to sodium sulfate was observed. The treatment at 600° C for 3 h of the salt solves this problem.

Sugars and organic acids did not appear in the chloroform extract, 1,4-butandiol was used as internal standard.

2.2.3 Gas Chromatographic Determination by Injection of a Purified Wine Extract into a Polar Capillary Column (Matta and Gaetano 1986)

2.2.3.1 Principle of the Method

Diethylene glycol is extracted from wine with ethanol and chloroform and determined by GLC.

2.2.3.2 Reagents

Activated charcoal
Sodium hydroxide 4N solution
Anhydrous ethanol
Choloroform
1,3-propandiol: acqueous solution 500 mg l^{-1} (I.S.)

2.2.3.3 Apparatus

Rotating evaporator

Gas chromatograph with fused silica capillary column (30 m × 0.25 mm), coated with Carbowax 20 M bonded phase, film thickness 0.25 μ.

2.2.3.4 Determination

Thirty ml of wine is bleached with 0.5 g activated charcoal for 5 min.; 2.5 ml of NaOH 4N solution, 1 ml of I.S. solution and 10 ml of ethanol are added to 10 ml of decolorated wine in a separatory funnel.

After shaking, 10 ml of chloroform is added and shaking is repeated for 30 s. Layers are allowed to separate and the organic phase is recovered. The extraction with chloroform is repeated twice and all the organic phases are joined together. The chloroform extract is evaporated to dryness with rotating evaporator under vacuum.

The dry residue is dissolved with 5 ml of chloroform, taking care to wet the whole inner surface of the flask to obtain a complete dissolution of DEG and internal standard.

One μl of this solution is injected in split-system (split ratio 1 : 10) with the following gas chromatographic parameters: flow of carrier gas (He) 1 ml min^{-1}, oven temperature: 80 to 155° C at 4° C min^{-1}, injector temperature was 200° C and detector (FID) temperature was 200° C.

2.2.3.5 Notes

The separation and the form of the peaks may be improved with the injection technique named "hot needle" and with the column at room temperature. The suggested temperature programming may be changed according to the instrumentation used.

2.2.4 Gas Chromatographic Determination by Injection of a Purified and Derivatized Extract into a Nonpolar Capillary Column (Conte et al. 1986)

2.2.4.1 Principle of the Method

Diethylene glycol is extracted with ethyl ether from deacidified wine and determined by GLC.

2.2.4.2 Reagents

Ethanolic solution (200 mg l^{-1}) of 1,4-butandiol
Mixed ionic exchange resin (Ionenaustauscher V Merck)
Anhydrous ethanol
Ethyl ether
Anhydrous pyridin
Hexamethyldisilazane
Trimethylchlorosilane
Benzene

2.2.4.3 Apparatus

Glass frittered filter
Rotating evaporator

Gaschromatograph with glass capillary column (25 m × 0.25 mm), coated with SE 52, film thickness 0,25 μ.

2.2.4.4 Determination

Twenty ml of wine are added with 1 ml of internal standard solution and 10 g of ion exchange resin. The mixture is left to react for 1 h, with occasional shaking. Then the mixture is filtered under vacuum and 10 ml of filtered solution, added with 100 ml anhydrous ethanol, are concentrated to dryness with rotating evaporator.

The dry residue is extracted with ethyl ether for 10 min and the liquid phase is transferred into a tube and the solvent is evaporated under nitrogen stream. The residue is redissolved in 250 μl of a mixture of pyridine, hexamethyldisilazane, and trimethylchlorosilane in the ratio of 10:2:1 (Sweeley et al. 1963).

The tube is sealed and after 15 min at room temperature, the pyridine is evaporated and 100 μl of benzene are added.

One μl is injected in split-system (split ratio 1:40) with the following gas chromatographic conditions: flow of carrier gas (He) 1.5 ml min^{-1}, oven temperature programmed from 100 to 290° C at 5° C min^{-1}, injector and detector (FID) temperature 290° C.

2.2.4.5 Notes

The method, tried on a white wine and on a red wine to which increasing amounts of diethylene glycol was added, showed good extraction levels and a good concordance between calculated concentration and those found.

The reproducibility was tried with 12 replications on a sample containing 50 ppm of DEG.

The minimum detectable quantity was 10 ppm, with a gas chromatographic trace as reported in Fig. 3.

A limit of the method is the difficulty to obtain quantitative extraction in the case of very sweet wine. The problem is due to the formation of a thick solid layer after the concentration to dryness of wine.

This layer is physically opposed to the extraction with ethyl ether. In order to overcome this difficulty, the authors of this method recently experimented the dissolution with pyridine of the dry residue, followed by sugar oximation and silanization, based on a method proposed by Versini et al. (1984) for the determination of polyalcohols in musts. The GLC trace obtained with this method is reported in Fig. 4.

2.2.5 High Pressure Liquid Chromatography Determination (Bonn 1985)

2.2.5.1 Principle of the Method

The method is based on the direct injection of wine in HPLC, using two coupled columns of ion-exchange resins with different counter ions.

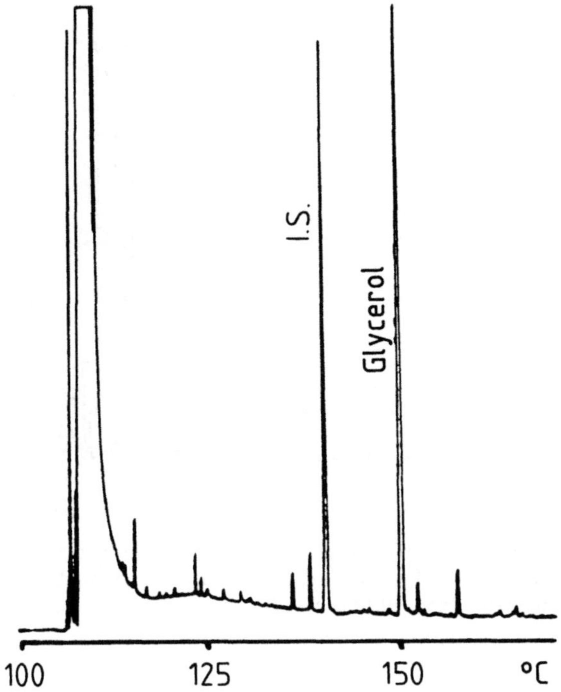

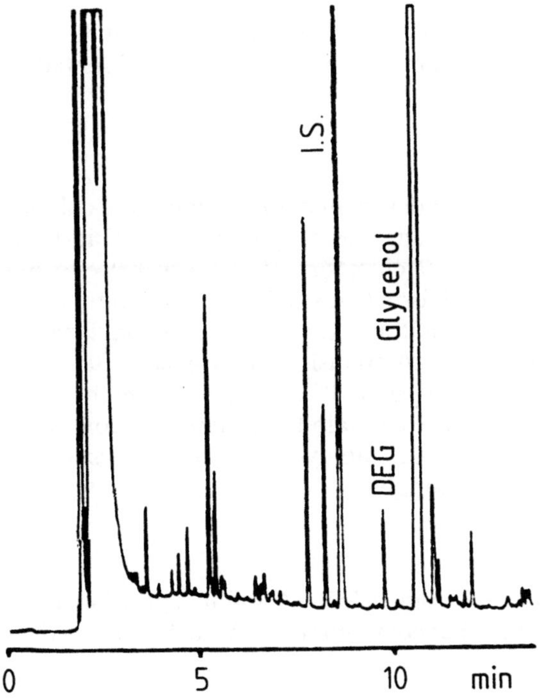

Fig. 3. Capillary GLC (SE 52. 25 m × 0.25 mm i.d.) of the extracted and sylanized DEG from a sample of wine: DEG = 10 ppm. (Conte et al. 1986a)

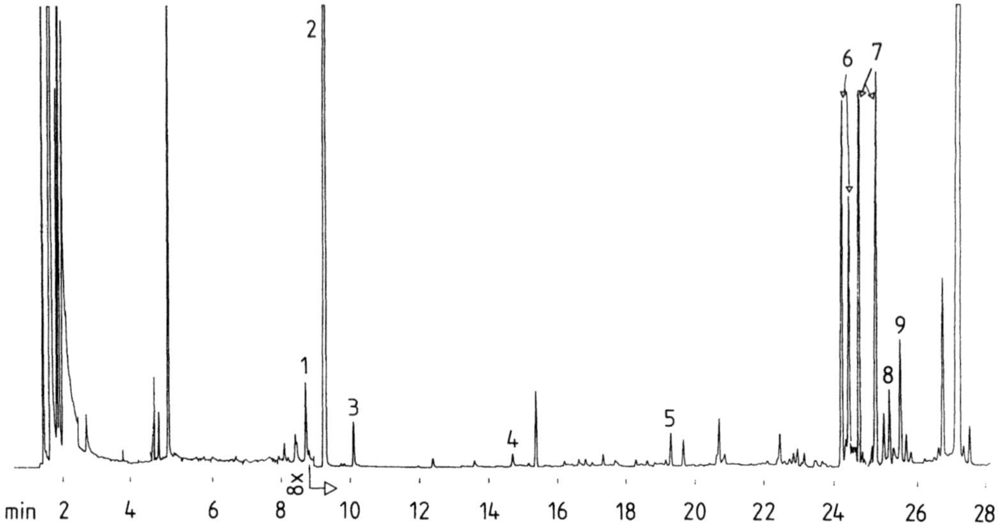

Fig. 4. Simultaneous determination of DEG (20 ppm), glycerol, sugars and organic acid in wine. Peaks: *1* DEG; *2* glycerol; *3* succinic acid; *4* tartaric acid; *5* malic acid; *6* fructose; *7* glucose; *8* mannitol; *9* sorbitol. (Conte et al. 1986, unpublished)

2.2.5.2 Reagents

Bi-distilled water

2.2.5.3 Apparatus

High performance liquid chromatography equipped with an integrated data system and column oven compartment, column switching valve and differential refractive index detector.

2.2.5.4 Determination

The sample was injected by a valve fitted with a 20 μl loop; the employed columns were: 300 × 7.8 mm i.d. HPX-87 H ion exchange resin, hydrogen form (Bio-Rad); a 100 × 7.8 mm i.d. cation exchange resin, calcium form (Bio-Rad); 300 × 7.8 mm i.d. HPX-87 P cation exchange resin, lead form (Bio-Rad); 300 × 7.8 mm i.d. calcium form Spherogel Carbohydrate N, sulfonated polystyrene-divinylbenzene resin (Beckmann). The pre-column was an ion-exclusion Micro-guard (Bio-Rad). The temperature of the column was optimized at 70° C to standardize the retention time of the considered compounds. To maintain the counter-ion concentration as constant as possible, a pre-column was employed. This also serves to keep back interfering ions possibly present in samples.

2.2.5.5 Notes

With the use of the sole cation exchange resin, calcium form, it is possible to obtain the separation of DEG from the other compounds in about 6 min. Further-

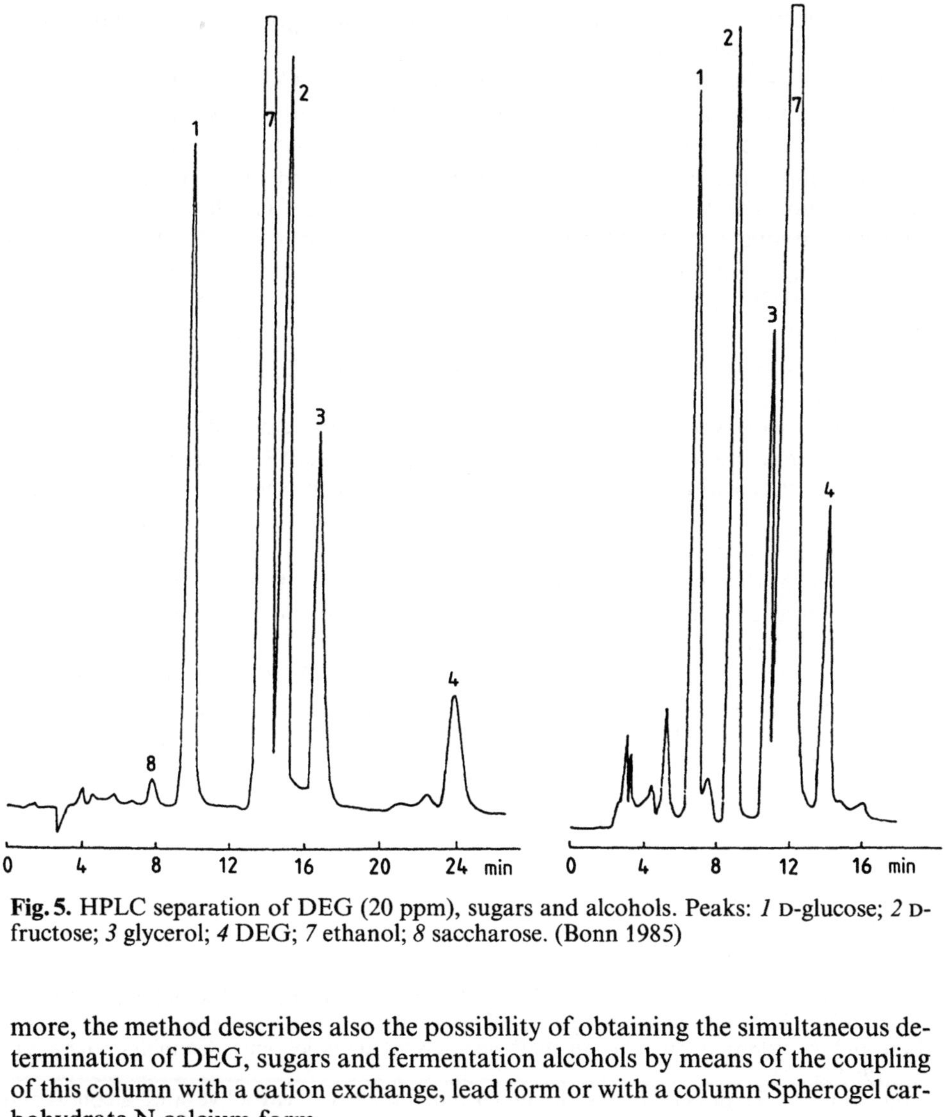

Fig. 5. HPLC separation of DEG (20 ppm), sugars and alcohols. Peaks: *1* D-glucose; *2* D-fructose; *3* glycerol; *4* DEG; *7* ethanol; *8* saccharose. (Bonn 1985)

more, the method describes also the possibility of obtaining the simultaneous determination of DEG, sugars and fermentation alcohols by means of the coupling of this column with a cation exchange, lead form or with a column Spherogel carbohydrate N calcium form.

In these two cases, the times of analysis are respectively 24 and 16 min (Fig. 5). The minimum detectable quantity is 20 ppm.

3 Concluding Considerations

On the whole, the experiences described in this chapter show that, by an analytical method (gas chromatography), the determination of diethylene glycol in wines is not an insurmountable problem: most of the phases and columns tested make good separation of DEG from the other components (glycerol, principally), and certainly other phases do the same.

The choice between polar stationary phases (type Carbowax) and nonpolar stationary phases (such as SE) is so to say so "a personal choice" and reflects two different philosophies: a laboratory that deals with enology only and that desires the maximum simplification of the work will advantageously use very polar phase (which is not always a synonym of fast analysis), thus avoiding derivatization. This will involve a reduced length of life of the column, added to the necessity of conditioning the column for a long time between one injection and the following. The latter philosophy will direct the choice to nonpolar columns, coupled with derivatization. Furthermore, this kind of column may advantageously be used for other determinations such as sterols, mono- and di-glycerides, sugars, and organic acids.

In other words, this choice is more suitable for the exigencies of control laboratories, engaged with a wide field of samples, where GLC is used in several fields, e.g., fats and oils. In these laboratories the preservation of the life of the column is an important problem and silylation represents an acquired routine technique of the laboratory.

The use of this method of derivatization, together with the choice of a nonpolar column allows the complete elution of all the injected compounds, without problems of pyrolysis or decomposition.

This aspect is of peculiar importance if the identification of DEG is realized by means of mass spectrometry, coupled with gas chromatography, as indicated by several authors. The use of silyl derivative saves the source of the mass spectrometer from contamination.

References

Bandion F, Valenta M, Kohlmann H (1985) Zum Nachweis Extrakterhöhender Zusätze zu Wein. Mitt Klosterneuburg 35:89–92

Bertrand A (1985) Recherche du diéthylène-glycol dans les vins. Connaiss Vigne Vin 19:191–195

Bonn G (1985) High-performance liquid chromatography of carbohydrates, alcohols and diethylene glycol on ion-exchange resins. J Chromatogr 350:381–387

Browning E (1965) Toxicity and metabolism of industrial solvents. Elsevier, New York

Caccamo F, Di Corcia A, Samperi R (1986) Determination of diethylene glycol in wine. Rapid purification with a Carbopack B cartridge and quantitation by gas chromatography. J Chromatogr 354:478–481

Conte LS, Frega N, Lercker G (1986a) Proposta di un metodo per la determinazione del dietilenglicole nei vini. Vignevini 13(1–2):23–25

Conte LS, Minguzzi A, Natali N (1986b) Determinazione del dietilenglicole nei vini: confronto fra diversi metodi di analisi. Vignevini 13(3):49–53

Fostel H (1985) Comments on the determination of diethylene glycol in wines (in German). Ernährung 9:783–786

Fuehrling D, Wollenberg H (1985) Zur Bestimmung kleiner Mengen von Diethylenglykol in Wein. Dtsch Lebensm Rundsch 81:325–328

Holtzer H (1985) Bestimmung von Diethylenglykol in Wein. Ernährung 9:568–569

Hori Y, Chonan T, Nishizawa M (1986) Rapid determination of diethylene glycol in wine by gas chromatography using alumina column clean-up (in Japanese) Shokuhin Eiseigaku Zasshi 27:187–189
Kaiser RE, Rieder RI (1985) Diethylenglykol in Wein. J High Resolut Chromatogr Chromatogr Commun 8:863–865
Kuhlmann F (1986) Schnellmethode zur Bestimmung von Diethylenglykol in Wein. Dtsch Lebensm Rundsch 82:84–85
Lehmann G, Ganz J (1985) Nachweis von Diethylenglykol in Wein. Z Lebensm Unters Forsch 181:362
Littmann S (1985) Zur Bestimmung kleiner Mengen von Diethylenglykol in Wein. Dtsch Lebensm Rundsch 81:328–329
Matta M, Gaetano G (1986) Determinazione del dietilenglicole nei vini. Riv Vitic Enol 39:27–34
Pfeiffer P, Radler F (1985) Bestimmung von Diethylenglykol mit der HPLC-Methode. Weinwirt Tech 8:234–235
Rapp A, Spraul M, Humpfer E (1986) ^{13}C NMR-spektroskopische Bestimmung von Diethylenglykol im Wein. Z Lebensm Unters Forsch 182:419–421
Rowe (1962) Industrial hygiene and toxicology, 2nd edn, vol II. Interscience, New York
Schubert R (1985) Diethylene glycol in wine: 1 ppm detection limit. Chem Rundsch 38:3
Smith R (1941) J Ind Hyg Toxicol, 23, 259-Literature cited in Merck Index (Ninth Edition), Ref 3735, p 499 (1976)
Sweeley CC, Bentley R, Makita M, Wells WW (1963) Gas-liquid chromatography of trimethylsilylderivates of sugars and related substances. J Am Chem Soc 85:2497–2507
Van Rooyen TJ, Van Wyk CJ (1986) A rapid quantitative HPLC method for determination of diethylene glycol. S Afr J Enol Vitic 7:36–38
Versini G, Dalla Serra A, Margheri G (1984) Polialcoli e zuccheri minori nei mosti concentrati rettificati. Possibili parametri di genuinità? Vignevini 11(3):41–47
Wagner K, Kreutzer P (1985) Gaschromatographische Bestimmung von Diethylenglykol in Wein. Weinwirt Tech 7:213
Watkins P (1985) Identification of "antifreeze" in wine. Ion Trap Newslett 1(2):5
Werkhoff P, Bretschneider WW (1986) Gaschromatographische Bestimmung von Diethylenglykol in Wein, Traubensaft und Traubensaftkonzentraten. Z Lebensm Unters Forsch 182:298–302

Subject Index

GPSR Compliance

The European Union's (EU) General Product Safety Regulation (GPSR) is a set of rules that requires consumer products to be safe and our obligations to ensure this.

If you have any concerns about our products, you can contact us on ProductSafety@springernature.com

In case Publisher is established outside the EU, the EU authorized representative is:

Springer Nature Customer Service Center GmbH
Europaplatz 3
69115 Heidelberg, Germany

Batch number: 10838062

Printed by Printforce, the Netherlands